A Theory of Everything

" An Overview of The Mathematical Model of The Universe "

Edited by Paul F. Kisak

Contents

Chapter 1

Theory of everything

This article is about the physical concept. For other uses, see Theory of everything (disambiguation).

A **theory of everything** (**ToE**) or **final theory**, **ultimate theory**, or **master theory** is a hypothetical single, all-encompassing, coherent theoretical framework of physics that fully explains and links together all physical aspects of the universe.[1]:6 Finding a ToE is one of the major unsolved problems in physics. Over the past few centuries, two theoretical frameworks have been developed that, as a whole, most closely resemble a ToE. The two theories upon which all modern physics rests are general relativity (GR) and quantum field theory (QFT). GR is a theoretical framework that only focuses on the force of gravity for understanding the universe in regions of both large-scale and high-mass: stars, galaxies, clusters of galaxies, etc. On the other hand, QFT is a theoretical framework that only focuses on three non-gravitational forces for understanding the universe in regions of both small scale and low mass: sub-atomic particles, atoms, molecules, etc. QFT successfully implemented the Standard Model and unified the interactions (so-called Grand Unified Theory) between the three non-gravitational forces: weak, strong, and electromagnetic force.[2]:122

Through years of research, physicists have experimentally confirmed with tremendous accuracy virtually every prediction made by these two theories when in their appropriate domains of applicability. In accordance with their findings, scientists also learned that GR and QFT, as they are currently formulated, are mutually incompatible - they cannot both be right. Since the usual domains of applicability of GR and QFT are so different, most situations require that only one of the two theories be used.[3][4]:842–844 As it turns out, this incompatibility between GR and QFT is only an apparent issue in regions of extremely small-scale and high-mass, such as those that exist within a black hole or during the beginning stages of the universe (i.e., the moment immediately following the Big Bang). To resolve this conflict, a theoretical framework revealing a deeper underlying reality, unifying gravity with the other three interactions, must be discovered to harmoniously integrate the realms of GR and QFT into a seamless whole: a single theory that, in principle, is capable of describing all phenomena. In pursuit of this goal, quantum gravity has recently become an area of active research.

Over the past few decades, a single explanatory framework, called "string theory", has emerged that may turn out to be the ultimate theory of the universe. Many physicists believe that, at the beginning of the universe (up to 10^{-43} seconds after the Big Bang), the four fundamental forces were once a single fundamental force. Unlike most (if not all) other theories, string theory may be on its way to successfully incorporating each of the four fundamental forces into a unified whole. According to string theory, every particle in the universe, at its most microscopic level (Planck length), consists of varying combinations of vibrating strings (or strands) with preferred patterns of vibration. String theory claims that it is through these specific oscillatory patterns of strings that a particle of unique mass and force charge is created (that is to say, the electron is a type of string that vibrates one way, while the up-quark is a type of string vibrating another way, and so forth).

Initially, the term *theory of everything* was used with an ironic connotation to refer to various overgeneralized theories. For example, a grandfather of Ijon Tichy — a character from a cycle of Stanisław Lem's science fiction stories of the 1960s — was known to work on the "General Theory of Everything". Physicist John Ellis[5] claims to have introduced the term into the technical literature in an article in *Nature* in 1986.[6] Over time, the term stuck in popularizations of theoretical physics research.

1.1   Historical antecedents

1.1.1   From ancient Greece to Einstein

Archimedes was possibly the first scientist known to have described nature with axioms (or principles) and then deduce new results from them.[7] He thus tried to describe "everything" starting from a few axioms. Any "theory of everything" is similarly expected to be based on axioms and to deduce all observable phenomena from them.[8]:340

The concept of 'atom', introduced by Democritus, unified all phenomena observed in nature as the motion of atoms. In ancient Greek times philosophers speculated that the apparent diversity of observed phenomena was due to a single type of interaction, namely the collisions of atoms. Following atomism, the mechanical philosophy of the 17th century posited that all forces could be ultimately reduced to contact forces between the atoms, then imagined as tiny solid particles.[9]:184[10]

In the late 17th century, Isaac Newton's description of the long-distance force of gravity implied that not all forces in nature result from things coming into contact. Newton's work in his *Principia* dealt with this in a further example of unification, in this case unifying Galileo's work on terrestrial gravity, Kepler's laws of planetary motion and the phenomenon of tides by explaining these apparent actions at a distance under one single law: the law of universal gravitation.[11]

In 1814, building on these results, Laplace famously suggested that a sufficiently powerful intellect could, if it knew the position and velocity of every particle at a given time, along with the laws of nature, calculate the position of any particle at any other time:[12]:ch 7

> An intellect which at a certain moment would know all forces that set nature in motion, and all positions of all items of which nature is composed, if this intellect were also vast enough to submit these data to analysis, it would embrace in a single formula the movements of the greatest bodies of the universe and those of the tiniest atom; for such an intellect nothing would be uncertain and the future just like the past would be present before its eyes.
> — *Essai philosophique sur les probabilités*, Introduction. 1814

Laplace thus envisaged a combination of gravitation and mechanics as a theory of everything. Modern quantum mechanics implies that uncertainty is inescapable, and thus that Laplace's vision has to be amended: a theory of everything must include gravitation and quantum mechanics.

In 1820, Hans Christian Ørsted discovered a connection between electricity and magnetism, triggering decades of work that culminated in 1865, in James Clerk Maxwell's theory of electromagnetism. During the 19th and early 20th centuries, it gradually became apparent that many common examples of forces – contact forces, elasticity, viscosity, friction, and pressure – result from electrical interactions between the smallest particles of matter.

In his experiments of 1849–50, Michael Faraday was the first to search for a unification of gravity with electricity and magnetism.[13] However, he found no connection.

In 1900, David Hilbert published a famous list of mathematical problems. In Hilbert's sixth problem, he challenged researchers to find an axiomatic basis to all of physics. In this problem he thus asked for what today would be called a theory of everything.[14]

In the late 1920s, the new quantum mechanics showed that the chemical bonds between atoms were examples of (quantum) electrical forces, justifying Dirac's boast that "the underlying physical laws necessary for the mathematical theory of a large part of physics and the whole of chemistry are thus completely known".[15]

After 1915, when Albert Einstein published the theory of gravity (general relativity), the search for a unified field theory combining gravity with electromagnetism began with a renewed interest. In Einstein's day, the strong and the weak forces had not yet been discovered, yet, he found the potential existence of two other distinct forces -gravity and electromagnetism- far more alluring. This launched his thirty-year voyage in search of the so-called "unified field theory" that he hoped would show that these two forces are really manifestations of one grand underlying principle. During these last few decades of his life, this quixotic quest isolated Einstein from the mainstream of physics. Understandably, the mainstream was instead far more excited about the newly emerging framework of quantum mechanics. Einstein wrote to a friend in the early 1940s, "I have become a lonely old chap who is mainly known because he doesn't wear socks and who is exhibited as a curiosity on special occasions." Prominent contributors were Gunnar Nordström, Hermann Weyl, Arthur Eddington, Theodor Kaluza, Oskar Klein, and most notably, Albert Einstein and his collaborators. Einstein intensely searched for, but ultimately failed to find, a unifying theory.[16]:ch 17 (But see:Einstein–Maxwell–Dirac equations.) More than a half a century later, Einstein's dream of discovering a unified theory has become the Holy Grail of modern physics.

1.1.2 Twentieth century and the nuclear interactions

In the twentieth century, the search for a unifying theory was interrupted by the discovery of the strong and weak nuclear forces (or interactions), which differ both from gravity and from electromagnetism. A further hurdle was the acceptance that in a ToE, quantum mechanics had to be incorporated from the start, rather than emerging as a consequence of a deterministic unified theory, as Einstein had hoped.

Gravity and electromagnetism could always peacefully coexist as entries in a list of classical forces, but for many years it seemed that gravity could not even be incorporated into the quantum framework, let alone unified with the other fundamental forces. For this reason, work on unification, for much of the twentieth century, focused on understanding the three "quantum" forces: electromagnetism and the weak and strong forces. The first two were combined in 1967–68 by Sheldon Glashow, Steven Weinberg, and Abdus Salam into the "electroweak" force.[17] Electroweak unification is a broken symmetry: the electromagnetic and weak forces appear distinct at low energies because the particles carrying the weak force, the W and Z bosons, have non-zero masses of 80.4 GeV/c^2 and 91.2 GeV/c^2, whereas the photon, which carries the electromagnetic force, is massless. At higher energies Ws and Zs can be created easily and the unified nature of the force becomes apparent.

While the strong and electroweak forces peacefully coexist in the Standard Model of particle physics, they remain distinct. So far, the quest for a theory of everything is thus unsuccessful on two points: neither a unification of the strong and electroweak forces – which Laplace would have called 'contact forces' – has been achieved, nor has a unification of these forces with gravitation been achieved.

1.2 Modern physics

1.2.1 Conventional sequence of theories

A Theory of Everything would unify all the fundamental interactions of nature: gravitation, strong interaction, weak interaction, and electromagnetism. Because the weak interaction can transform elementary particles from one kind into another, the ToE should also yield a deep understanding of the various different kinds of possible particles. The usual assumed path of theories is given in the following graph, where each unification step leads one level up:

In this graph, electroweak unification occurs at around 100 GeV, grand unification is predicted to occur at 10^{16} GeV, and unification of the GUT force with gravity is expected at the Planck energy, roughly 10^{19} GeV.

Several Grand Unified Theories (GUTs) have been proposed to unify electromagnetism and the weak and strong forces. Grand unification would imply the existence of an electronuclear force; it is expected to set in at energies of the order of 10^{16} GeV, far greater than could be reached by any possible Earth-based particle accelerator. Although the simplest GUTs have been experimentally ruled out, the general idea, especially when linked with supersymmetry, remains a favorite candidate in the theoretical physics community. Supersymmetric GUTs seem plausible not only for their theoretical "beauty", but because they naturally produce large quantities of dark matter, and because the inflationary force may be related to GUT physics (although it does not seem to form an inevitable part of the theory). Yet GUTs are clearly not the final answer; both the current standard model and all proposed GUTs are quantum field theories which require the problematic technique of renormalization to yield sensible answers. This is usually regarded as a sign that these are only effective field theories, omitting crucial phenomena relevant only at very high energies.[3]

The final step in the graph requires resolving the separation between quantum mechanics and gravitation, often equated with general relativity. Numerous researchers concentrate their efforts on this specific step; nevertheless, no accepted theory of quantum gravity – and thus no accepted theory of everything – has emerged yet. It is usually assumed that the ToE will also solve the remaining problems of GUTs.

In addition to explaining the forces listed in the graph, a ToE may also explain the status of at least two candidate forces suggested by modern cosmology: an inflationary force and dark energy. Furthermore, cosmological experiments also suggest the existence of dark matter, supposedly composed of fundamental particles outside the scheme of the standard model. However, the existence of these forces and particles has not been proven yet.

1.2.2   String theory and M-theory

Since the 1990s, many physicists believe that 11-dimensional M-theory, which is described in some limits by one of the five perturbative superstring theories, and in another by the maximally-supersymmetric 11-dimensional supergravity, is the theory of everything. However, there is no widespread consensus on this issue.

A surprising property of string/M-theory is that extra dimensions are required for the theory's consistency. In this regard, string theory can be seen as building on the insights of the Kaluza–Klein theory, in which it was realized that applying general relativity to a five-dimensional universe (with one of them small and curled up) looks from the four-dimensional perspective like the usual general relativity together with Maxwell's electrodynamics. This lent credence to the idea of unifying gauge and gravity interactions, and to extra dimensions, but didn't address the detailed experimental requirements. Another important property of string theory is its supersymmetry, which together with extra dimensions are the two main proposals for resolving the hierarchy problem of the standard model, which is (roughly) the question of why gravity is so much weaker than any other force. The extra-dimensional solution involves allowing gravity to propagate into the other dimensions while keeping other forces confined to a four-dimensional spacetime, an idea that has been realized with explicit stringy mechanisms.[18]

Research into string theory has been encouraged by a variety of theoretical and experimental factors. On the experimental side, the particle content of the standard model supplemented with neutrino masses fits into a spinor representation of SO(10), a subgroup of E8 that routinely emerges in string theory, such as in heterotic string theory[19] or (sometimes equivalently) in F-theory.[20][21] String theory has mechanisms that may explain why fermions come in three hierarchical generations, and explain the mixing rates between quark generations.[22] On the theoretical side, it has begun to address some of the key questions in quantum gravity, such as resolving the black hole information paradox, counting the correct entropy of black holes[23][24] and allowing for topology-changing processes.[25][26][27] It has also led to many insights in pure mathematics and in ordinary, strongly-coupled gauge theory due to the Gauge/String duality.

In the late 1990s, it was noted that one major hurdle in this endeavor is that the number of possible four-dimensional universes is incredibly large. The small, "curled up" extra dimensions can be compactified in an enormous number of different ways (one estimate is 10^{500}) each of which leads to different properties for the low-energy particles and forces. This array of models is known as the string theory landscape.[8]:347

One proposed solution is that many or all of these possibilities are realised in one or another of a huge number of universes, but that only a small number of them are habitable, and hence the fundamental constants of the universe are ultimately the result of the anthropic principle rather than dictated by theory. This has led to criticism of string theory,[28] arguing that it cannot make useful (i.e., original, falsifiable, and verifiable) predictions and regarding it as a pseudoscience. Others disagree,[29] and string theory remains an extremely active topic of investigation in theoretical physics.

1.2.3   Loop quantum gravity

Current research on loop quantum gravity may eventually play a fundamental role in a ToE, but that is not its primary aim.[30] Also loop quantum gravity introduces a lower bound on the possible length scales.

There have been recent claims that loop quantum gravity may be able to reproduce features resembling the Standard Model. So far only the first generation of fermions (leptons and quarks) with correct parity properties have been modelled by Sundance Bilson-Thompson using preons constituted of braids of spacetime as the building blocks.[31] However, there is no derivation of the Lagrangian that would describe the interactions of such particles, nor is it possible to show that such particles are fermions, nor that the gauge groups or interactions of the Standard Model are realised. Utilization of quantum computing concepts made it possible to demonstrate that the particles are able to survive quantum fluctuations.[32]

This model leads to an interpretation of electric and colour charge as topological quantities (electric as number and chirality of twists carried on the individual ribbons and colour as variants of such twisting for fixed electric charge).

Bilson-Thompson's original paper suggested that the higher-generation fermions could be represented by more complicated braidings, although explicit constructions of these structures were not given. The electric charge, colour, and parity properties of such fermions would arise in the same way as for the first generation. The model was expressly generalized for an infinite number of generations and for the weak force bosons (but not for photons or gluons) in a 2008 paper by Bilson-Thompson, Hackett, Kauffman and Smolin.[33]

1.2.4 Other attempts

A recent development is the theory of causal fermion systems,[34] giving all three current physical theories (quantum mechanics, general relativity and quantum field theory) as limiting cases.

A recent and very prolific attempt is called Causal Sets. As some of the approaches mentioned above, its direct goal isn't necessarily to achieve a ToE but primarily a working theory of quantum gravity, which might eventually include the standard model and become a candidate for a ToE. Its founding principle is that spacetime is fundamentally discrete and that the spacetime events are related by a partial order. This partial order has the physical meaning of the causality relations between relative past and future distinguishing spacetime events.

Outside the previously mentioned attempts there is Garrett Lisi's E8 proposal. This theory provides an attempt of identifying general relativity and the standard model within the Lie group E8. The theory doesn't provide a novel quantization procedure and the author suggests its quantization might follow the Loop Quantum Gravity approach above mentioned.[35]

Christoph Schiller's Strand Model attempts to account for the gauge symmetry of the Standard Model of particle physics, U(1)×SU(2)×SU(3), with the three Reidemeister moves of knot theory by equating each elementary particle to a different tangle of one, two, or three strands (selectively a long prime knot or unknotted curve, a rational tangle, or a braided tangle respectively).

1.2.5 Present status

At present, there is no candidate theory of everything that includes the standard model of particle physics and general relativity. For example, no candidate theory is able to calculate the fine structure constant or the mass of the electron. Most particle physicists expect that the outcome of the ongoing experiments – the search for new particles at the large particle accelerators and for dark matter – are needed in order to provide further input for a ToE.

1.3 Theory of everything and philosophy

Main article: Theory of everything (philosophy)

The philosophical implications of a physical ToE are frequently debated. For example, if philosophical physicalism is true, a physical ToE will coincide with a philosophical theory of everything.

The "system building" style of metaphysics attempts to answer *all* the important questions in a coherent way, providing a complete picture of the world. Plato and Aristotle could be said to have created early examples of comprehensive systems. In the early modern period (17th and 18th centuries), the system-building *scope* of philosophy is often linked to the rationalist *method* of philosophy, which is the technique of deducing the nature of the world by pure *a priori* reason. Examples from the early modern period include the Leibniz's Monadology, Descarte's Dualism, and Spinoza's Monism. Hegel's Absolute idealism and Whitehead's Process philosophy were later systems.

1.4 Arguments against a theory of everything

In parallel to the intense search for a ToE, various scholars have seriously debated the possibility of its discovery.

1.4.1 Gödel's incompleteness theorem

A number of scholars claim that Gödel's incompleteness theorem suggests that any attempt to construct a ToE is bound to fail. Gödel's theorem, informally stated, asserts that any formal theory expressive enough for elementary arithmetical facts to be expressed and strong enough for them to be proved is either inconsistent (both a statement and its denial can be derived from its axioms) or incomplete, in the sense that there is a true statement that can't be derived in the formal theory.

Stanley Jaki, in his 1966 book *The Relevance of Physics*, pointed out that, because any "theory of everything" will certainly be a consistent non-trivial mathematical theory, it must be incomplete. He claims that this dooms searches for a deterministic theory of everything.[36] In a later reflection, Jaki states that it is wrong to say that a final theory is impossible, but rather that "when it is on hand one cannot know rigorously that it is a final theory."[37]

Freeman Dyson has stated that "Gödel's theorem implies that pure mathematics is inexhaustible. No matter how many problems we solve, there will always be other problems that cannot be solved within the existing rules. [...] Because of Gödel's theorem, physics is inexhaustible too. The laws of physics are a finite set of rules, and include the rules for doing mathematics, so that Gödel's theorem applies to them."[38]

Stephen Hawking was originally a believer in the Theory of Everything but, after considering Gödel's Theorem, concluded that one was not obtainable: "Some people will be very disappointed if there is not an ultimate theory, that can be formulated as a finite number of principles. I used to belong to that camp, but I have changed my mind."[39]

Jürgen Schmidhuber (1997) has argued against this view; he points out that Gödel's theorems are irrelevant for computable physics.[40] In 2000, Schmidhuber explicitly constructed limit-computable, deterministic universes whose pseudo-randomness based on undecidable, Gödel-like halting problems is extremely hard to detect but does not at all prevent formal ToEs describable by very few bits of information.[41]

Related critique was offered by Solomon Feferman,[42] among others. Douglas S. Robertson offers Conway's game of life as an example:[43] The underlying rules are simple and complete, but there are formally undecidable questions about the game's behaviors. Analogously, it may (or may not) be possible to completely state the underlying rules of physics with a finite number of well-defined laws, but there is little doubt that there are questions about the behavior of physical systems which are formally undecidable on the basis of those underlying laws.

Since most physicists would consider the statement of the underlying rules to suffice as the definition of a "theory of everything", most physicists argue that Gödel's Theorem does *not* mean that a ToE cannot exist. On the other hand, the scholars invoking Gödel's Theorem appear, at least in some cases, to be referring not to the underlying rules, but to the understandability of the behavior of all physical systems, as when Hawking mentions arranging blocks into rectangles, turning the computation of prime numbers into a physical question.[44] This definitional discrepancy may explain some of the disagreement among researchers.

1.4.2 Fundamental limits in accuracy

No physical theory to date is believed to be precisely accurate. Instead, physics has proceeded by a series of "successive approximations" allowing more and more accurate predictions over a wider and wider range of phenomena. Some physicists believe that it is therefore a mistake to confuse theoretical models with the true nature of reality, and hold that the series of approximations will never terminate in the "truth". Einstein himself expressed this view on occasions.[45] Following this view, we may reasonably hope for *a* theory of everything which self-consistently incorporates all currently known forces, but we should not expect it to be the final answer.

On the other hand it is often claimed that, despite the apparently ever-increasing complexity of the mathematics of each new theory, in a deep sense associated with their underlying gauge symmetry and the number of fundamental physical constants, the theories are becoming simpler. If this is the case, the process of simplification cannot continue indefinitely.

1.4.3 Lack of fundamental laws

There is a philosophical debate within the physics community as to whether a theory of everything deserves to be called *the* fundamental law of the universe.[46] One view is the hard reductionist position that the ToE is the fundamental law and that all other theories that apply within the universe are a consequence of the ToE. Another view is that emergent laws, which govern the behavior of complex systems, should be seen as equally fundamental. Examples of emergent laws are the second law of thermodynamics and the theory of natural selection. The advocates of emergence argue that emergent laws, especially those describing complex or living systems are independent of the low-level, microscopic laws. In this view, emergent laws are as fundamental as a ToE.

The debates do not make the point at issue clear. Possibly the only issue at stake is the right to apply the high-status term "fundamental" to the respective subjects of research. A well-known one took place between Steven Weinberg and Philip Anderson

1.4.4 Impossibility of being "of everything"

Although the name "theory of everything" suggests the determinism of Laplace's quotation, this gives a very misleading impression. Determinism is frustrated by the probabilistic nature of quantum mechanical predictions, by the extreme sensitivity to initial conditions that leads to mathematical chaos, by the limitations due to event horizons, and by the extreme mathematical difficulty of applying the theory. Thus, although the current standard model of particle physics "in principle" predicts almost all known non-gravitational phenomena, in practice only a few quantitative results have been derived from the full theory (e.g., the masses of some of the simplest hadrons), and these results (especially the particle masses which are most relevant for low-energy physics) are less accurate than existing experimental measurements. The ToE would almost certainly be even harder to apply for the prediction of experimental results, and thus might be of limited use.

A motive for seeking a ToE, apart from the pure intellectual satisfaction of completing a centuries-long quest, is that prior examples of unification have predicted new phenomena, some of which (e.g., electrical generators) have proved of great practical importance. And like in these prior examples of unification, the ToE would probably allow us to confidently define the domain of validity and residual error of low-energy approximations to the full theory.

1.4.5 Infinite number of onion layers

Lee Smolin regularly argues that the layers of nature may be like the layers of an onion, and that the number of layers might be infinite. This would imply an infinite sequence of physical theories.

The argument is not universally accepted, because it is not obvious that infinity is a concept that applies to the foundations of nature.

1.4.6 Impossibility of calculation

Weinberg[47] points out that calculating the precise motion of an actual projectile in the Earth's atmosphere is impossible. So how can we know we have an adequate theory for describing the motion of projectiles? Weinberg suggests that we know *principles* (Newton's laws of motion and gravitation) that work "well enough" for simple examples, like the motion of planets in empty space. These principles have worked so well on simple examples that we can be reasonably confident they will work for more complex examples. For example, although general relativity includes equations that do not have exact solutions, it is widely accepted as a valid theory because all of its equations with exact solutions have been experimentally verified. Likewise, a ToE must work for a wide range of simple examples in such a way that we can be reasonably confident it will work for every situation in physics.

1.5 See also

- Absolute (philosophy)

- An Exceptionally Simple Theory of Everything

- Argument from beauty

- Attractor

- Beyond black holes

- Beyond the standard model

- Big Bang

- Brownian motion

- Chaos theory

- Chronology of the universe

- Electroweak interaction

- Holographic principle

- Mathematical beauty

- Mathematical universe hypothesis

- Multiverse

- Standard Model (mathematical formulation)

- *The Theory of Everything (2014 film)* - a feature film about Prof.Stephen Hawking and his first wife Jane Hawking.

- Timeline of the Big Bang

- Zero-energy universe

1.6   References

1.6.1   Footnotes

[1] Steven Weinberg. *Dreams of a Final Theory: The Scientist's Search for the Ultimate Laws of Nature*. Knopf Doubleday Publishing Group. ISBN 978-0-307-78786-6.

[2] Stephen W. Hawking (28 February 2006). *The Theory of Everything: The Origin and Fate of the Universe*. Phoenix Books; Special Anniv. ISBN 978-1-59777-508-3.

[3] Carlip, Steven (2001). "Quantum Gravity: a Progress Report". *Reports on Progress in Physics* **64** (8). arXiv:gr-qc/0108040. Bibcode:2001RPPh...64..885C. doi:10.1088/0034-4885/64/8/301.

[4] Susanna Hornig Priest (14 July 2010). *Encyclopedia of Science and Technology Communication*. SAGE Publications. ISBN 978-1-4522-6578-0.

[5] Ellis, John (2002). "Physics gets physical (correspondence)".*Nature***415**(6875): 957.Bibcode:2002Natur.415..957E.doi:10.

[6] Ellis, John (1986). "The Superstring: Theory of Everything, or of Nothing?".*Nature***323**(6089): 595–598.Bibcode:1986NaturE. doi:10.1038/323595a0.

[7] Rorres, Chris (2009). "ARCHIMEDES AND THE QUEST FOR THE THEORY OF EVERYTHING".

[8] Chris Impey (26 March 2012). *How It Began: A Time-Traveler's Guide to the Universe*. W. W. Norton. ISBN 978-0-393-08002-5.

[9] William E. Burns (1 January 2001). *The Scientific Revolution: An Encyclopedia*. ABC-CLIO. ISBN 978-0-87436-875-8.

[10] Shapin, Steven (1996). *The Scientific Revolution*. University of Chicago Press. ISBN 0-226-75021-3.

[11] Newton, Sir Isaac (1729). *The Mathematical Principles of Natural Philosophy* **II**. p. 255.

[12] Sean Carroll (7 January 2010). *From Eternity to Here: The Quest for the Ultimate Theory of Time*. Penguin Group US. ISBN 978-1-101-15215-7.

[13] Faraday, M. (1850). "Experimental Researches in Electricity. Twenty-Fourth Series. On the Possible Relation of Gravity to Electricity". *Abstracts of the Papers Communicated to the Royal Society of London* **5**: 994–995. doi:10.1098/rspl.1843.0267.

[14] A.N. Gorban, I. Karlin, Hilbert's 6th Problem: exact and approximate hydrodynamic manifolds for kinetic equations, Bull. Amer. Math. Soc., 51 (2014), no. 2, 186-246, doi:10.1090/S0273-0979-2013-01439-3

[15] Dirac, P.A.M. (1929). "Quantum mechanics of many-electron systems". *Proceedings of the Royal Society of London A* **123** (792): 714. Bibcode:1929RSPSA.123..714D. doi:10.1098/rspa.1929.0094.

[16] Abraham Pais (23 September 1982). *Subtle is the Lord : The Science and the Life of Albert Einstein: The Science and the Life of Albert Einstein*. Oxford University Press. ISBN 978-0-19-152402-8.

[17] Weinberg (1993), Ch. 5

[18] Holloway, M (2005)."The Beauty of Branes"(PDF).*Scientific American*(Scientific American)**293**(4): 38.Bibcode:2005Sci. doi:10.1038/scientificamerican1005-38. PMID 16196251. Retrieved August 13, 2012.

[19] Nilles, Hans Peter; Ramos-Sánchez, Saúl; Ratz, Michael; Vaudrevange, Patrick K. S. (2008). "From strings to the MSSM". *The European Physical Journal C* **59** (2): 249. arXiv:0806.3905. Bibcode:2009EPJC...59..249N. doi:10.1140/epjc/s100-008-0740-1.

[20] Beasley, Chris; Heckman, Jonathan J; Vafa, Cumrun (2009). "GUTs and exceptional branes in F-theory — I". *Journal of High Energy Physics* **2009**: 058. arXiv:0802.3391. Bibcode:2009JHEP...01..058B. doi:10.1088/1126-6708/2009/01/058.

[21] Donagi, Ron and Wijnholt, Martijn (2008) Model Building with F-Theory

[22] Heckman, Jonathan J. and Vafa, Cumrun (2008) Flavor Hierarchy From F-theory

[23] Strominger, Andrew; Vafa, Cumrun (1996). "Microscopic origin of the Bekenstein-Hawking entropy". *Physics Letters B* **379**: 99. arXiv:hep-th/9601029. Bibcode:1996PhLB..379...99S. doi:10.1016/0370-2693(96)00345-0.

[24] Horowitz, Gary (1996) The Origin of Black Hole Entropy in String Theory

[25] Greene, Brian R.; Morrison, David R.; Strominger, Andrew (1995). "Black hole condensation and the unification of string vacua". *Nuclear Physics B* **451**: 109. arXiv:hep-th/9504145. Bibcode:1995NuPhB.451..109G. doi:10.1016/0550-3213(95)0-X.

[26] Aspinwall, Paul S.; Greene, Brian R.; Morrison, David R. (1994). "Calabi-Yau moduli space, mirror manifolds and spacetime topology change in string theory". *Nuclear Physics B* **416** (2): 414. arXiv:hep-th/9309097. Bibcode:1994NuPhB.416..414A. doi:10.1016/0550-3213(94)90321-2.

[27] Adams, Allan; Liu, Xiao; McGreevy, John; Saltman, Alex; Silverstein, Eva (2005). "Things fall apart: Topology change from winding tachyons". *Journal of High Energy Physics* **2005** (10): 033. arXiv:hep-th/0502021. Bibcode:2005JHEP...10..033A. doi:10.1088/1126-6708/2005/10/033.

[28] Smolin, Lee (2006). *The Trouble With Physics: The Rise of String Theory, the Fall of a Science, and What Comes Next*. Houghton Mifflin. ISBN 978-0-618-55105-7.

[29] Duff, M. J. (2011). "String and M-Theory: Answering the Critics". *Foundations of Physics* **43**: 182. arXiv:1112.0788. Bibcode:2013FoPh...43..182D. doi:10.1007/s10701-011-9618-4.

[30] Potter, Franklin (15 February 2005). "Leptons And Quarks In A Discrete Spacetime" (PDF). *Frank Potter's Science Gems*. Retrieved 2009-12-01.

[31] Bilson-Thompson, Sundance O.; Markopoulou, Fotini; Smolin, Lee (2007). "Quantum gravity and the standard model". *Classical and Quantum Gravity* **24** (16): 3975–3994. arXiv:hep-th/0603022. Bibcode:2007CQGra..24.3975B. doi:10.1088/0264-9381/24/16/002.

[32] Castelvecchi, Davide; Valerie Jamieson (August 12, 2006). "You are made of space-time". *New Scientist* (2564).

[33] Sundance Bilson-Thompson; Jonathan Hackett; Lou Kauffman; Lee Smolin (2008). "Particle Identifications from Symmetries of Braided Ribbon Network Invariants". arXiv:0804.0037 [hep-th].

[34] F. Finster; J. Kleiner (2015). "Causal fermion systems as a candidate for a unified physical theory". arXiv:1502.03587 [math-ph].

[35] A. G. Lisi (2007). "An Exceptionally Simple Theory of Everything". arXiv:0711.0770 [hep-th].

[36] Jaki. S.L. (1966). *The Relevance of Physics*. Chicago Press. pp. 127–130.

[37] Stanley L. Jaki (2004) "A Late Awakening to Gödel in Physics", pp. 8–9.

[38] Freeman Dyson. NYRB, May 13, 2004

[39] Stephen Hawking, Gödel and the end of physics, July 20, 2002

[40] Schmidhuber, Jürgen (1997). *A Computer Scientist's View of Life, the Universe, and Everything*. *Lecture Notes in Computer Science*. Springer. pp. 201–208. doi:10.1007/BFb0052071. ISBN 978-3-540-63746-2.

[41] Schmidhuber, Jürgen (2002). "Hierarchies of generalized Kolmogorov complexities and nonenumerable universal measures computable in the limit". arXiv:quant-ph/0011122.

[42] Feferman, Solomon (17 November 2006). "The nature and significance of Gödel's incompleteness theorems" (PDF). Institute for Advanced Study. Retrieved 2009-01-12.

[43] Robertson, Douglas S. (2007). "Goedel's Theorem, the Theory of Everything, and the Future of Science and Mathematics". *Complexity* **5** (5): 22–27. doi:10.1002/1099-0526(200005/06)5:5<22::AID-CPLX4>3.0.CO;2-0.

[44] Hawking, Stephen (20 July 2002). "Gödel and the end of physics". Retrieved 2009-12-01.

[45] Einstein, letter to Felix Klein, 1917. (On determinism and approximations.) Quoted in Pais (1982), Ch. 17.

[46] Weinberg (1993), Ch 2.

[47] Weinberg (1993) p. 5

1.6.2 Bibliography

- Pais, Abraham (1982) *Subtle is the Lord...: The Science and the Life of Albert Einstein* (Oxford University Press, Oxford, . Ch. 17, ISBN 0-19-853907-X

- Weinberg, Steven (1993) *Dreams of a Final Theory: The Search for the Fundamental Laws of Nature*, Hutchinson Radius, London, ISBN 0-09-177395-4

1.7 External links

- The Elegant Universe, *Nova* episode about the search for the theory of everything and string theory.

- Theory of Everything, freeview video by the Vega Science Trust, BBC and Open University.

- The Theory of Everything: Are we getting closer, or is a final theory of matter and the universe impossible? Debate between John Ellis (physicist), Frank Close and Nicholas Maxwell.

- Why The World Exists, a discussion between physicist Laura Mersini-Houghton, cosmologist George Francis Rayner Ellis and philosopher David Wallace about dark matter, parallel universes and explaining why these and the present Universe exist.

Chapter 2

Theory of everything (philosophy)

For the book, see Theory of everything (book).

In philosophy, a **theory of everything** or **ToE** is an ultimate, all-encompassing explanation or description of nature or reality.[1][2][3] Adopting the term from physics, where the search for a theory of everything is ongoing, philosophers have discussed the viability of the concept and analyzed its properties and implications.[1][2][3] Among the questions to be addressed by a philosophical theory of everything are: "Why is reality understandable?" "Why are the laws of nature as they are?" "Why is there anything at all?"[1]

2.1 Comprehensive philosophical systems

The "system building" style of metaphysics attempts to answer *all* the important questions in a coherent way, providing a complete picture of the world. Plato and Aristotle could be said to be early examples of comprehensive systems. In the early modern period (17th and 18th centuries), the system-building *scope* of philosophy is often linked to the rationalist *method* of philosophy, that is the technique of deducing the nature of the world by pure *a priori* reason. Examples from the early modern period include Leibniz's monadology, Descarte's dualism, and Spinoza's monism. Hegel's absolute idealism and Whitehead's process philosophy were later systems. At present, work is underway on the structural-systematic philosophy (SSP), to which the following books are devoted: Lorenz B. Puntel, *Structure and Being* (2008; translation of *Struktur und Sein*, 2006) and *Being and God* (2011; translation of *Sein und Gott*, 2010) and Alan White, *Toward a Philosophical Theory of Everything* (2014). The SSP makes no claims to finality; it aims to be the best systematic philosophy currently available.

Other philosophers do not believe philosophy should aim so high. Some scientists think a more mathematical approach than philosophy is needed for a ToE, for instance Stephen Hawking wrote in *A Brief History of Time* that even if we had a ToE, it would necessarily be a set of equations. He wrote, "What is it that breathes fire into the equations and makes a universe for them to describe?".[4]

2.2 Nicholas Rescher

2.2.1 Properties and impasse of self-substantiation

In "The Price of an Ultimate Theory",[2] originally published in 2000, Nicholas Rescher specifies what he sees as the principal properties of a Theory of Everything and describes an apparent impasse on the road to such a theory.

2.2.2 Properties

Principle of sufficient reason

First, he takes as a presupposition the principle of sufficient reason, which in his formulation states that every fact t has an explanation t':

$$\forall t \; \exists t' \; (t' \; E \; t)$$

where E predicates explanation, so that $t' \; E \; t$ denotes "t' explains t".

Comprehensiveness

Next, he asserts that the most direct and natural construction of a Theory of Everything T^* would confer upon it two crucial features: comprehensiveness and finality. Comprehensiveness says that wherever there is a fact t, T^* affords its explanation:

$$\forall t \; (T^* \; E \; t)$$

Finality

Finality says that as an "ultimate theory", T^* has no deeper explanation:

$$\forall t \; ((t \; E \; T^*) \rightarrow (t = T^*))$$

so that the only conceivable explanation of T^* is T^* itself.

Noncircularity

Rescher notes that it is obviously problematic to deploy a theory for its own explanation; at the heart of the traditional conception of explanatory adequacy, he says, is a principle of noncircularity stating that no fact can explain itself:

$$\nexists t \; (t \; E \; t)$$

Impasse

The impasse is then that the two critical aspects of a Theory of Everything, comprehensiveness and finality, conflict with the fundamental principle of noncircularity. A comprehensive theory which explains everything must explain itself, and a final theory which has no deeper explanation must, by the principle of sufficient reason, have *some* explanation; consequently it too must be self-explanatory. Rescher concludes that any Theorist of Everything committed to comprehensiveness and finality is bound to regard noncircularity as "something that has to be jettisoned". But how, he asks, can a theory adequately substantiate itself?

2.2.3   Ways forward

Rescher's proposal in "The Price of an Ultimate Theory" is to dualize the concept of explanation so that a fact can be explained either *derivationally*, by the premises which lead to it, or *systemically*, by the consequences which follow from it. With derivational explanation, a fact t is explained when it is subsumed by some prior, more fundamental fact *t'*. With systemic explanation, *t* is explained when it is a "best fit" for its consequences, where fitness is measured by uniformity, simplicity, connectedness, and other criteria conducive to systemic integration. Rescher concludes that while a theory of everything cannot be explained derivationally (since no deeper explanation can subsume it), it can be explained systemically by its capacity to integrate its consequences.

In his 1996 book *The Conscious Mind*,[5] David Chalmers argues that a theory of everything must explain consciousness, that consciousness does not logically supervene on the physical, and that therefore a fundamental theory in physics would not be a theory of everything. A truly final theory, he argues, needs not just physical properties and laws, but phenomenal or protophenomenal properties and psychophysical laws explaining the relationship between physical processes and conscious experience. He concludes that "[o]nce we have a fundamental theory of consciousness to accompany a fundamental theory in physics, we may truly have a theory of everything." Developing such a theory will not be straightforward, he says, but "it ought to be possible in principle."

In "Prolegomena to Any Future Philosophy",[3] a 2002 essay in the *Journal of Evolution and Technology*, Mark Alan Walker discusses modern responses to the question of how to reconcile "the apparent finitude of humans" with what he calls "the traditional telos of philosophy—the attempt to unite thought and Being, to arrive at absolute knowledge, at a final theory of everything." He contrasts two ways of closing this "gap between the ambitions of philosophy, and the abilities of human philosophers": a "deflationary" approach in which philosophy is "scaled down into something more human" and the attempt to achieve a theory of everything is abandoned, and an "inflationary", transhumanist approach in which philosophers are "scaled up" by advanced technology into "super-intelligent beings" better able to pursue such a theory.

2.2.4 Criticism

In "Holistic Explanation and the Idea of a Grand Unified Theory",[1] originally presented as a lecture in 1998, Rescher identifies two negative reactions to the idea of a unified, overarching theory: reductionism and rejectionism. Reductionism holds that large-scale philosophical issues can be meaningfully addressed only when divided into lesser components, while rejectionism holds that questions about such issues are illegitimate and unanswerable. Against reductionism, Rescher argues that explaining individual parts does not explain the coordinating structure of the whole, so that a collectivized approach is required. Against rejectionism, he argues that the question of the 'reason'-the'why'-behind existence is pressing, important, and not obviously meaningless.

2.3 See also

- Theory of everything (physics)

2.4 References

[1] Rescher, Nicholas (2006a). "Holistic Explanation and the Idea of a Grand Unified Theory". *Collected Papers IX: Studies in Metaphilosophy*.

[2] Rescher, Nicholas (2006b). "The Price of an Ultimate Theory". *Collected Papers IX: Studies in Metaphilosophy*. (Googlebooks preview)

[3] Walker, Mark Alan (March 2002). "Prolegomena to Any Future Philosophy". *Journal of Evolution and Technology* Vol. 10.

[4] As quoted in Mariano Artigas, *The Mind of the Universe*, Templeton Foundation Press, 2001, p. 123.

[5] Chalmers, David J. (1996). *The Conscious Mind: In Search of a Fundamental Theory*. pp. 126–127.

Chapter 3

Chronology of the universe

See also: Timeline of the formation of the Universe

The **chronology of the universe** describes the history and future of the universe according to Big Bang cosmology, the

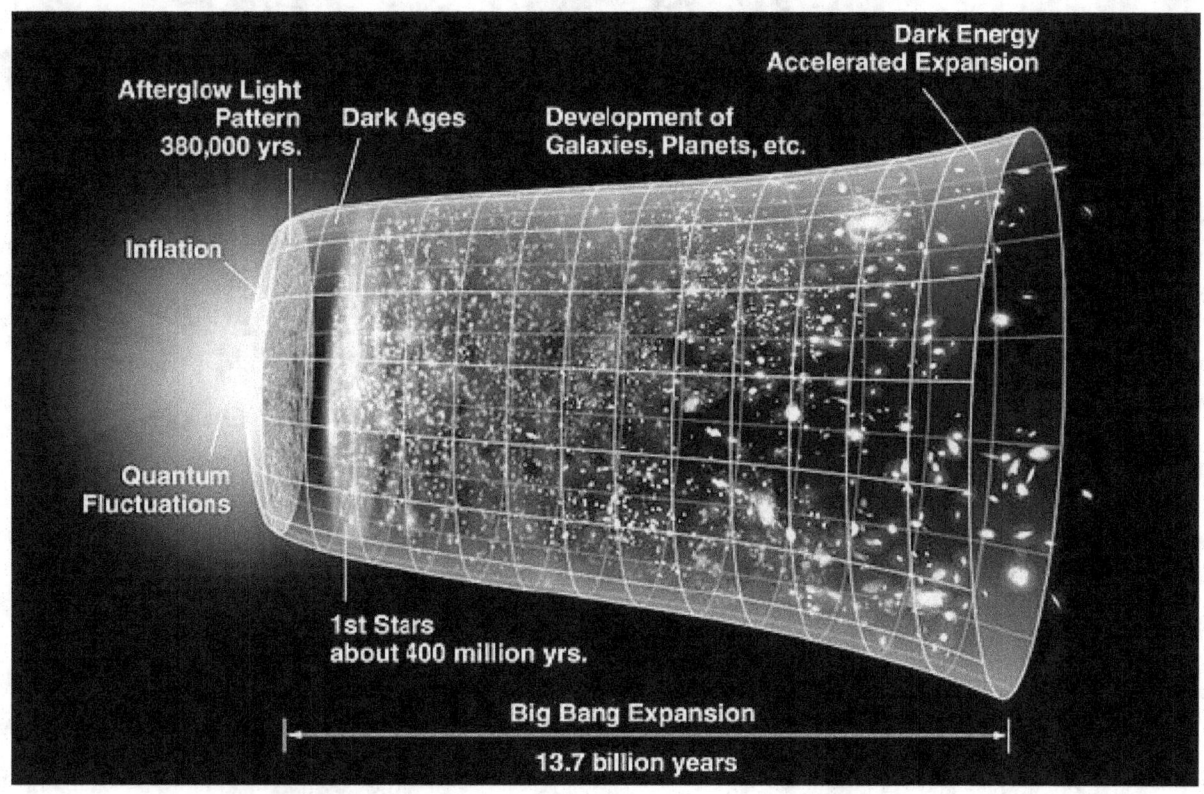

Diagram of Evolution of the universe from the Big Bang (left) - to the present.

prevailing scientific model of how the universe developed over time from the Planck epoch, using the cosmological time parameter of comoving coordinates. The model of the universe's expansion is known as the Big Bang. As of 2015, this expansion is estimated to have begun 13.799 ± 0.021 billion years ago.[1] It is convenient to divide the evolution of the universe so far into three phases.

3.1   Summary

In the first phase, the very earliest universe was so hot, or energetic, that initially no matter particles existed or could exist perhaps only fleetingly. According to prevailing scientific theories, at this time the distinct forces we see around us today were joined in one unified force. Space-time itself expanded during an inflationary epoch due to the immensity of the energies involved. Gradually the immense energies cooled – still to a temperature inconceivably hot compared to any we see around us now, but sufficiently to allow forces to gradually undergo symmetry breaking, a kind of repeated condensation from one status quo to another, leading finally to the separation of the strong force from the electroweak force and the first particles.

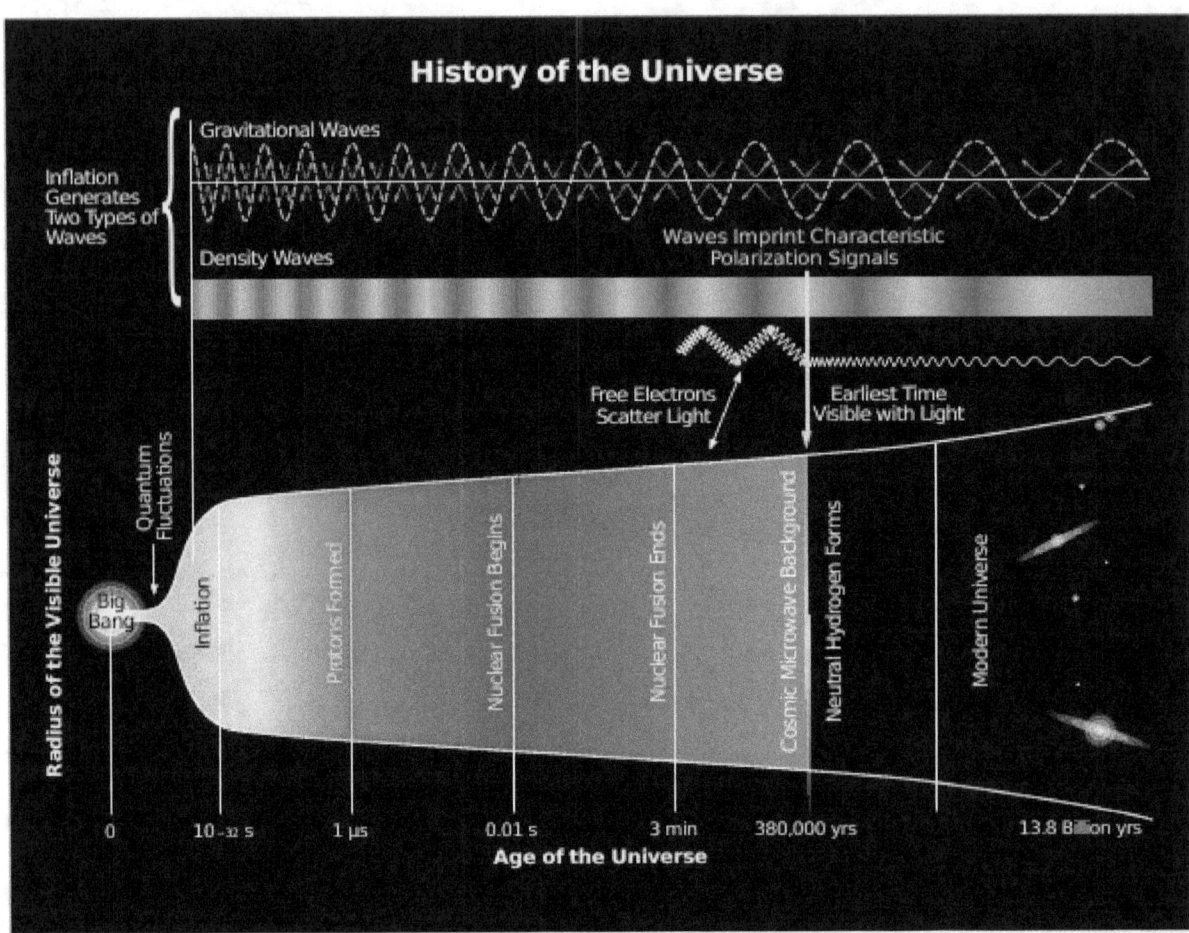

History of the Universe - gravitational waves are hypothesized to arise from cosmic inflation, a faster-than-light expansion just after the Big Bang (17 March 2014).[2][3][4]

In the second phase, the resulting quark–gluon plasma universe then cooled further, the current fundamental forces we know take their present forms through further symmetry breaking – notably the breaking of electroweak symmetry – and the full range of complex and composite particles we see around us today became possible, leading to a gravitationally dominated universe, the first neutral atoms (~ 80% hydrogen), and the cosmic microwave background radiation we can detect today. Modern high energy particle physics theories are satisfactory at these energy levels, and so physicists believe they have a good understanding of this and subsequent development of the fundamental universe around us. Because of these changes, space had also become largely transparent to light and other electromagnetic energy, rather than "foggy", by the end of this phase.

The third phase started after a short dark age with a universe whose fundamental particles and forces were as we know them, and witnessed the emergence of large scale stable structures, such as the earliest stars, quasars, galaxies, clusters of galaxies and superclusters, and the development of these to create the kind of universe we see today. Some researchers call the development of all this physical structure over billions of years "cosmic evolution". Other, more interdisciplinary, researchers refer to "cosmic evolution" as the entire scenario of growing complexity from big bang to humankind, thereby incorporating biology and culture into a unified view of all complex systems in the universe to date.[5]

Beyond the present day, scientists anticipate that the Earth will cease to be able to support life in about a billion years, and will be enveloped by a greatly-expanded Sun in about 5 billion years. On a far longer timescale, the Stelliferous Era will end as stars eventually die and fewer are born to replace them, leading to a darkening universe. Various theories suggest a number of subsequent possibilities. If particles such as protons are unstable then eventually matter may evaporate into low level energy in a kind of entropy related heat death. Alternatively the universe may collapse in a big crunch, although current data shows the rate of expansion is still increasing. If this is correct then it may end in a "big freeze" as matter and energy become very thinly spread and cool down. Alternative suggestions include a false vacuum catastrophe or a Big Rip as possible ends to the universe.

3.2   Very early universe

All ideas concerning the very early universe (cosmogony) are speculative. No accelerator experiments have yet probed energies of sufficient magnitude to provide any experimental insight into the behavior of matter at the energy levels that prevailed during this period. Proposed scenarios differ radically. Some examples are the Hartle–Hawking initial state, string landscape, brane inflation, string gas cosmology, and the ekpyrotic universe. Some of these are mutually compatible, while others are not.

3.2.1   Planck epoch

0 to 10^{-43} second after the Big Bang

Main article: Planck epoch

The Planck epoch is an era in traditional (non-inflationary) big bang cosmology wherein the temperature was so high that the four fundamental forces—electromagnetism, gravitation, weak nuclear interaction, and strong nuclear interaction—were one fundamental force. Little is understood about physics at this temperature; different hypotheses propose different scenarios. Traditional big bang cosmology predicts a gravitational singularity before this time, but this theory relies on general relativity and is expected to break down due to quantum effects.

In inflationary cosmology, times before the end of inflation (roughly 10^{-32} second after the Big Bang) do not follow the traditional big bang timeline.

3.2.2   Grand unification epoch

Between 10^{-43} second and 10^{-36} second after the Big Bang[6]

Main article: Grand unification epoch

As the universe expanded and cooled, it crossed transition temperatures at which forces separate from each other. These are phase transitions much like condensation and freezing. The grand unification epoch began when gravitation separated from the other forces of nature, which are collectively known as gauge forces. The non-gravitational physics in this epoch would be described by a so-called grand unified theory (GUT). The grand unification epoch ended when the GUT forces further separate into the strong and electroweak forces.

3.2.3   Electroweak epoch

Between 10^{-36} second (or the end of inflation) and 10^{-32} second after the Big Bang[6]

Main article: Electroweak epoch

According to traditional big bang cosmology, the Electroweak epoch began 10^{-36} second after the Big Bang, when the temperature of the universe was low enough (10^{28} K) to separate the strong force from the electroweak force (the name for the unified forces of electromagnetism and the weak interaction). In inflationary cosmology, the electroweak epoch ends when the inflationary epoch begins, at roughly 10^{-32} second.

Inflationary epoch

Unknown duration, ending 10^{-32}(?) second after the Big Bang

Main article: Inflationary epoch

Cosmic inflation was an era of accelerating expansion produced by a hypothesized field called the inflaton, which would have properties similar to the Higgs field and dark energy. While decelerating expansion would magnify deviations from homogeneity, making the universe more chaotic, accelerating expansion would make the universe more homogeneous. A sufficiently long period of inflationary expansion in the past could explain the high degree of homogeneity that is observed in the universe today at large scales, even if the state of the universe before inflation was highly disordered.

Inflation ended when the inflaton field decayed into ordinary particles in a process called "reheating", at which point ordinary Big Bang expansion began. The time of reheating is usually quoted as a time "after the Big Bang". This refers to the time that would have passed in traditional (non-inflationary) cosmology between the Big Bang singularity and the universe dropping to the same temperature that was produced by reheating, even though, in inflationary cosmology, the traditional Big Bang did not occur.

According to the simplest inflationary models, inflation ended at a temperature corresponding to roughly 10^{-32} second after the Big Bang. As explained above, this does not imply that the inflationary era lasted less than 10^{-32} second. In fact, in order to explain the observed homogeneity of the universe, the duration must be longer than 10^{-32} second. In inflationary cosmology, the earliest meaningful time "after the Big Bang" is the time of the end of inflation.

On March 17, 2014, astrophysicists of the BICEP2 collaboration announced the detection of inflationary gravitational waves in the B-mode power spectrum which was interpreted as clear experimental evidence for the theory of inflation.[2][3][4][7][8][9] However, on June 19, 2014, lowered confidence in confirming the cosmic inflation findings was reported [8][10][11] and finally, on February 2, 2015, a joint analysis of data from BICEP2/Keck and Planck satellite concluded that the statistical "significance [of the data] is too low to be interpreted as a detection of primordial B-modes" and can be attributed mainly to polarized dust in the Milky Way.[12][13][14][15]

Baryogenesis

Main article: Baryogenesis

There is currently insufficient observational evidence to explain why the universe contains far more baryons than antibaryons. A candidate explanation for this phenomenon must allow the Sakharov conditions to be satisfied at some time after the end of cosmological inflation. While particle physics suggests asymmetries under which these conditions are met, these asymmetries are too small empirically to account for the observed baryon-antibaryon asymmetry of the universe.

3.3 Early universe

After cosmic inflation ends, the universe is filled with a quark–gluon plasma. From this point onwards the physics of the early universe is better understood, and less speculative.

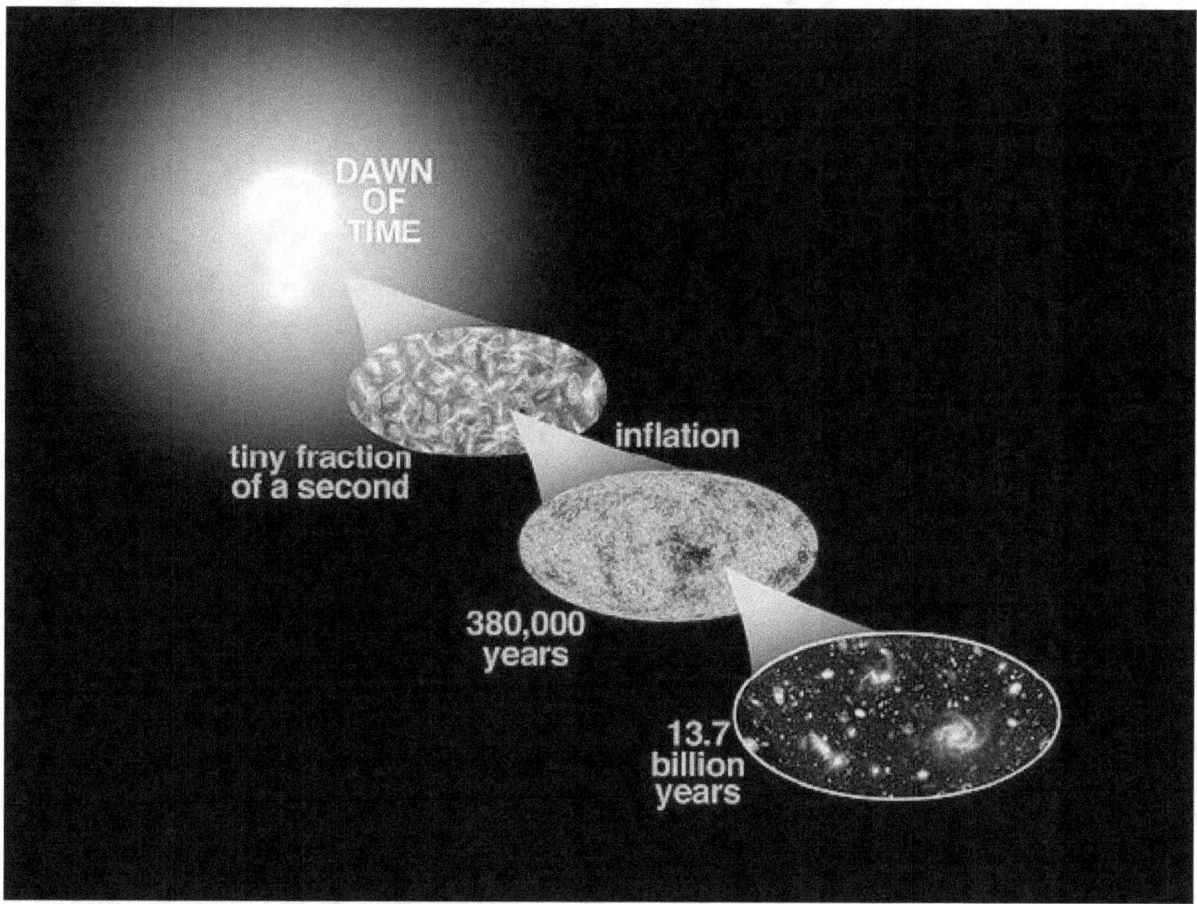

Cosmic History

3.3.1   Supersymmetry breaking (speculative)

Main article: Supersymmetry breaking

If supersymmetry is a property of our universe, then it must be broken at an energy that is no lower than 1 TeV, the electroweak symmetry scale. The masses of particles and their superpartners would then no longer be equal, which could explain why no superpartners of known particles have ever been observed.

3.3.2   Electroweak symmetry breaking and the quark epoch

Between 10^{-12} second and 10^{-6} second after the Big Bang

Main articles: Electroweak symmetry breaking and Quark epoch

As the universe's temperature falls below a certain very high energy level, it is believed that the Higgs field spontaneously acquires a vacuum expectation value, which breaks electroweak gauge symmetry. This has two related effects:

1. The weak force and electromagnetic force, and their respective bosons (the W and Z bosons and photon) manifest differently in the present universe, with different ranges;

2. Via the Higgs mechanism, all elementary particles interacting with the Higgs field become massive, having been massless at higher energy levels.

At the end of this epoch, the fundamental interactions of gravitation, electromagnetism, the strong interaction and the weak interaction have now taken their present forms, and fundamental particles have mass, but the temperature of the universe is still too high to allow quarks to bind together to form hadrons.

3.3.3 Hadron epoch

Between 10^{-6} second and 1 second after the Big Bang

Main article: Hadron epoch

The quark–gluon plasma that composes the universe cools until hadrons, including baryons such as protons and neutrons, can form. At approximately 1 second after the Big Bang neutrinos decouple and begin traveling freely through space. This cosmic neutrino background, while unlikely to ever be observed in detail since the neutrino energies are very low, is analogous to the cosmic microwave background that was emitted much later. (See above regarding the quark–gluon plasma, under the String Theory epoch.) However, there is strong indirect evidence that the cosmic neutrino background exists, both from Big Bang nucleosynthesis predictions of the helium abundance, and from anisotropies in the cosmic microwave background

3.3.4 Lepton epoch

Between 1 second and 10 seconds after the Big Bang

Main article: Lepton epoch

The majority of hadrons and anti-hadrons annihilate each other at the end of the hadron epoch, leaving leptons and anti-leptons dominating the mass of the universe. Approximately 10 seconds after the Big Bang the temperature of the universe falls to the point at which new lepton/anti-lepton pairs are no longer created and most leptons and anti-leptons are eliminated in annihilation reactions, leaving a small residue of leptons.[16]

3.3.5 Photon epoch

Between 10 seconds and 380,000 years after the Big Bang

Main article: Photon epoch

After most leptons and anti-leptons are annihilated at the end of the lepton epoch the energy of the universe is dominated by photons. These photons are still interacting frequently with charged protons, electrons and (eventually) nuclei, and continue to do so for the next 380,000 years.

Nucleosynthesis

Between 3 minutes and 20 minutes after the Big Bang[17]

Main article: Big Bang nucleosynthesis

During the photon epoch the temperature of the universe falls to the point where atomic nuclei can begin to form. Protons (hydrogen ions) and neutrons begin to combine into atomic nuclei in the process of nuclear fusion. Free neutrons combine with protons to form deuterium. Deuterium rapidly fuses into helium-4. Nucleosynthesis only lasts for about seventeen minutes, since the temperature and density of the universe has fallen to the point where nuclear fusion cannot continue. By this time, all neutrons have been incorporated into helium nuclei. This leaves about three times more hydrogen than helium-4 (by mass) and only trace quantities of other light nuclei.

Matter domination

70,000 years after the Big Bang

At this time, the densities of non-relativistic matter (atomic nuclei) and relativistic radiation (photons) are equal. The Jeans length, which determines the smallest structures that can form (due to competition between gravitational attraction and pressure effects), begins to fall and perturbations, instead of being wiped out by free-streaming radiation, can begin to grow in amplitude.

According to ΛCDM, at this stage, cold dark matter dominates, paving the way for gravitational collapse to amplify the tiny inhomogeneities left by cosmic inflation, making dense regions denser and rarefied regions more rarefied. However, because present theories as to the nature of dark matter are inconclusive, there is as yet no consensus as to its origin at earlier times, as currently exist for baryonic matter.

Recombination

ca. 377,000 years after the Big Bang

Main article: Recombination (cosmology)

Hydrogen and helium *atoms* begin to form as the density of the universe falls. This is thought to have occurred about

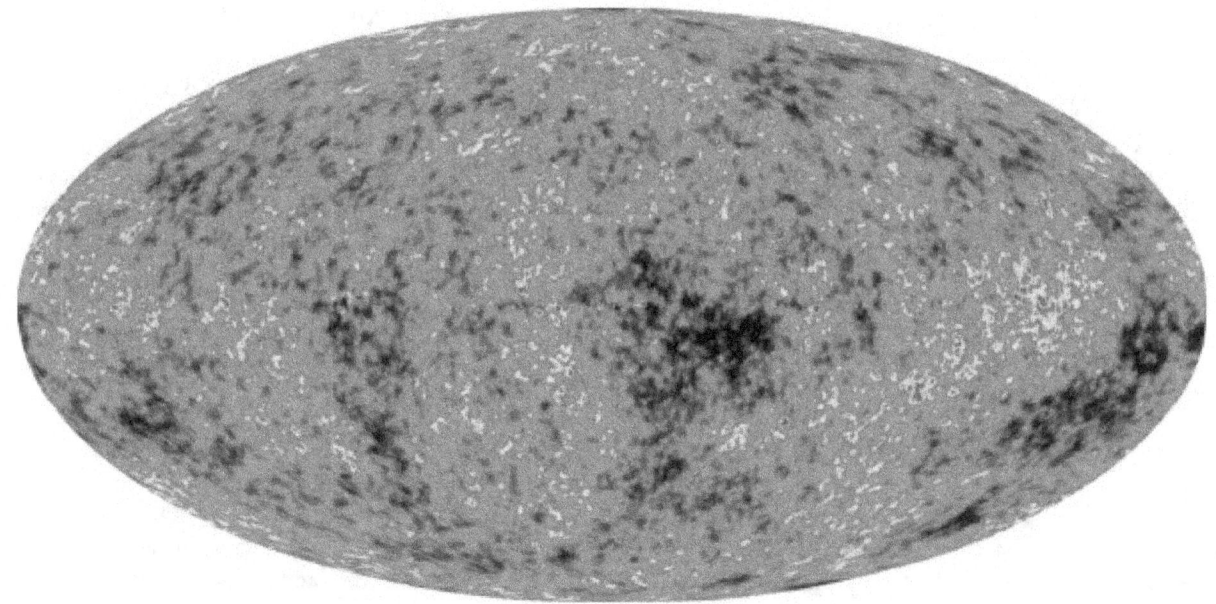

9 year WMAP data (2012) shows the cosmic microwave background radiation variations throughout the universe from our perspective, though the actual variations are much smoother than the diagram suggests.[18][19]

377,000 years after the Big Bang.[20] Hydrogen and helium are at the beginning ionized, i.e., no electrons are bound to the nuclei, which (containing positively charged protons) are therefore electrically charged (+1 and +2 respectively). As the universe cools down, the electrons get captured by the ions, forming electrically neutral atoms. This process is relatively fast (and faster for the helium than for the hydrogen), and is known as recombination.[21] At the end of recombination, most of the protons in the universe are bound up in neutral atoms. Therefore, the photons' mean free path becomes effectively infinite and the photons can now travel freely (see Thomson scattering): the universe has become transparent. This cosmic event is usually referred to as *decoupling*.

The photons present at the time of decoupling are the same photons that we see in the cosmic microwave background (CMB) radiation, after being greatly cooled by the expansion of the universe. Around the same time, existing pressure waves within the electron-baryon plasma — known as baryon acoustic oscillations — became embedded in the distribution of matter as it condensed, giving rise to a very slight preference in distribution of large scale objects. Therefore the cosmic microwave background is a picture of the universe at the end of this epoch including the tiny fluctuations generated during inflation (see diagram), and the spread of objects such as galaxies in the universe is an indication of the scale and size of the universe as it developed over time.[22]

Habitable epoch

See also: Abiogenesis

The chemistry of life may have begun shortly after the Big Bang, 13.8 billion years ago, during a habitable epoch when the Universe was only 10-17 million years old.[23][24][25]

Dark Ages

See also: Hydrogen line

Before decoupling occurred, most of the photons in the universe were interacting with electrons and protons in the photon–baryon fluid. The universe was opaque or "foggy" as a result. There was light but not light we can now observe through telescopes. The baryonic matter in the universe consisted of ionized plasma, and it only became neutral when it gained free electrons during "recombination", thereby releasing the photons creating the CMB. When the photons were released (or decoupled) the universe became transparent. At this point the only radiation emitted was the 21 cm spin line of neutral hydrogen. There is currently an observational effort underway to detect this faint radiation, as it is in principle an even more powerful tool than the cosmic microwave background for studying the early universe. The Dark Ages are currently thought to have lasted between 150 million to 800 million years after the Big Bang. The October 2010 discovery of UDFy-38135539, the first observed galaxy to have existed during the following reionization epoch, gives us a window into these times. The galaxy earliest in this period observed and thus also the most distant galaxy ever observed is currently on the record of Leiden University's Richard J. Bouwens and Garth D. Illingsworth from UC Observatories/Lick Observatory. They found the galaxy UDFj-39546284 to be at a time some 480 million years after the Big Bang or about halfway through the Cosmic Dark Ages at a distance of about 13.2 billion light-years. More recently, the UDFj-39546284 galaxy was found to be around "380 million years" after the Big Bang and at a distance of 13.37 billion light-years.[26]

3.4 Structure formation

See also: Large-scale structure of the cosmos and Structure formation

Structure formation in the big bang model proceeds hierarchically, with smaller structures forming before larger ones. The first structures to form are quasars, which are thought to be bright, early active galaxies, and population III stars. Before this epoch, the evolution of the universe could be understood through linear cosmological perturbation theory: that is, all structures could be understood as small deviations from a perfect homogeneous universe. This is computationally relatively easy to study. At this point non-linear structures begin to form, and the computational problem becomes much more difficult, involving, for example, N-body simulations with billions of particles.

3.4.1 Reionization

150 million to 1 billion years after the Big Bang

See also: Reionization and 21 centimeter radiation

The first stars and quasars form from gravitational collapse. The intense radiation they emit reionizes the surrounding universe. From this point on, most of the universe is composed of plasma.

The Hubble Ultra Deep Fields often showcase galaxies from an ancient era that tell us what the early Stelliferous Age was like.

3.4.2 Formation of stars

See also: Star formation

The first stars, most likely Population III stars, form and start the process of turning the light elements that were formed in the Big Bang (hydrogen, helium and lithium) into heavier elements. However, as yet there have been no observed Population III stars, and understanding of them is currently based on computational models of their formation and evolution. Fortunately observations of the Cosmic Microwave Background radiation can be used to date when star formation began in earnest. Analysis of such observations made by the European Space Agency's Planck telescope, as reported by BBC News in early February, 2015, concludes that the first generation of stars lit up 560 million years after the Big Bang. [27] [28]

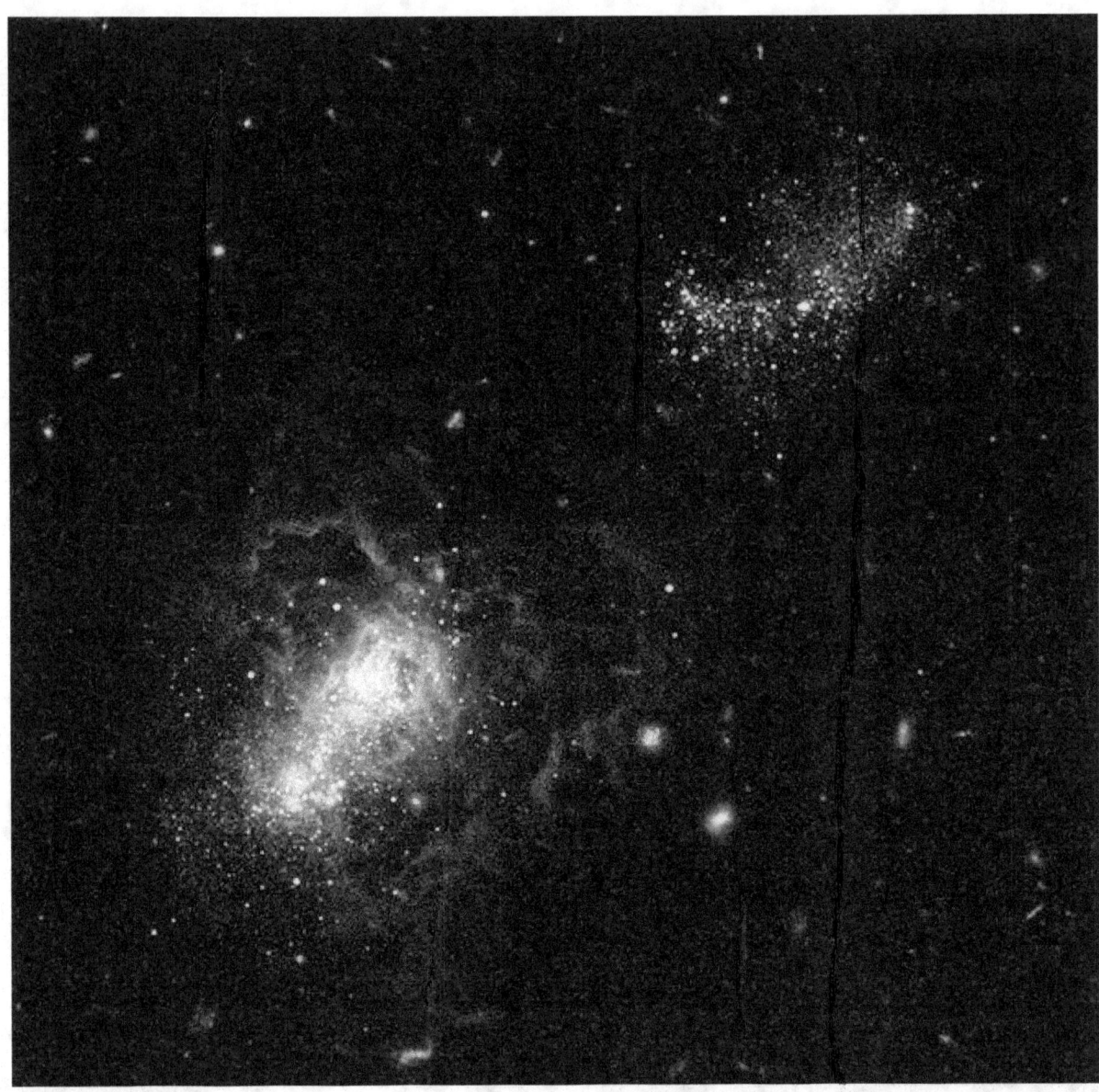

Another Hubble image shows an infant galaxy forming nearby, which means this happened very recently on the cosmological timescale. This shows that new galaxy formation in the universe is still occurring.

3.4.3 Formation of galaxies

See also: Galaxy formation and evolution

Large volumes of matter collapse to form a galaxy. Population II stars are formed early on in this process, with Population I stars formed later.

Johannes Schedler's project has identified a quasar CFHQS 1641+3755 at 12.7 billion light-years away,[29] when the universe was just 7% of its present age.

On July 11, 2007, using the 10-metre Keck II telescope on Mauna Kea, Richard Ellis of the California Institute of Technology at Pasadena and his team found six star forming galaxies about 13.2 billion light years away and therefore created when the universe was only 500 million years old.[30] Only about 10 of these extremely early objects are currently known.[31] More recent observations have shown these ages to be shorter than previously indicated. The most distant galaxy observed as of October 2013 has been reported to be 13.1 billion light years away.[32]

The Hubble Ultra Deep Field shows a number of small galaxies merging to form larger ones, at 13 billion light years, when the universe was only 5% its current age.[33] This age estimate is now believed to be slightly shorter.[32]

Based upon the emerging science of nucleocosmochronology, the Galactic thin disk of the Milky Way is estimated to have been formed 8.8 ± 1.7 billion years ago.[34]

3.4.4   Formation of groups, clusters and superclusters

See also: Large-scale structure of the cosmos

Gravitational attraction pulls galaxies towards each other to form groups, clusters and superclusters.

3.4.5   Formation of the Solar System

9 billion years after the Big Bang

Main article: Formation and evolution of the Solar System

The Solar System began forming about 4.6 billion years ago, or about 9 billion years after the Big Bang. A fragment of a molecular cloud made mostly of hydrogen and traces of other elements began to collapse, forming a large sphere in the center which would become the Sun, as well as a surrounding disk. The surrounding accretion disk would coalesce into a multitude of smaller objects that would become planets, asteroids, and comets. The Sun is a late-generation star, and the Solar System incorporates matter created by previous generations of stars.

3.4.6   Today

13.8 billion years after the Big Bang

The Big Bang is estimated to have occurred about 13.8 billion years ago.[35] Since the expansion of the universe appears to be accelerating, its large-scale structure is likely to be the largest structure that will ever form in the universe. The present accelerated expansion prevents any more inflationary structures entering the horizon and prevents new gravitationally bound structures from forming.

3.5   Ultimate fate of the universe

Main article: Ultimate fate of the universe

As with interpretations of what happened in the very early universe, advances in fundamental physics are required before it will be possible to know the ultimate fate of the universe with any certainty. Below are some of the main possibilities.

3.5.1   Fate of the Solar System: 1 to 5 billion years

Main articles: Formation and evolution of the Solar System § Future, Stability of the Solar System, Future of the Earth § Solar evolution and Red giant § The Sun as a red giant

Over a timescale of a billion years or more, the Earth and Solar System are unstable. Earth's existing biosphere is expected to vanish in about a billion years, as the Sun's heat production gradually increases to the point that liquid water and life are unlikely;[36] the Earth's magnetic fields, axial tilt and atmosphere are subject to long-term change; and the Solar System

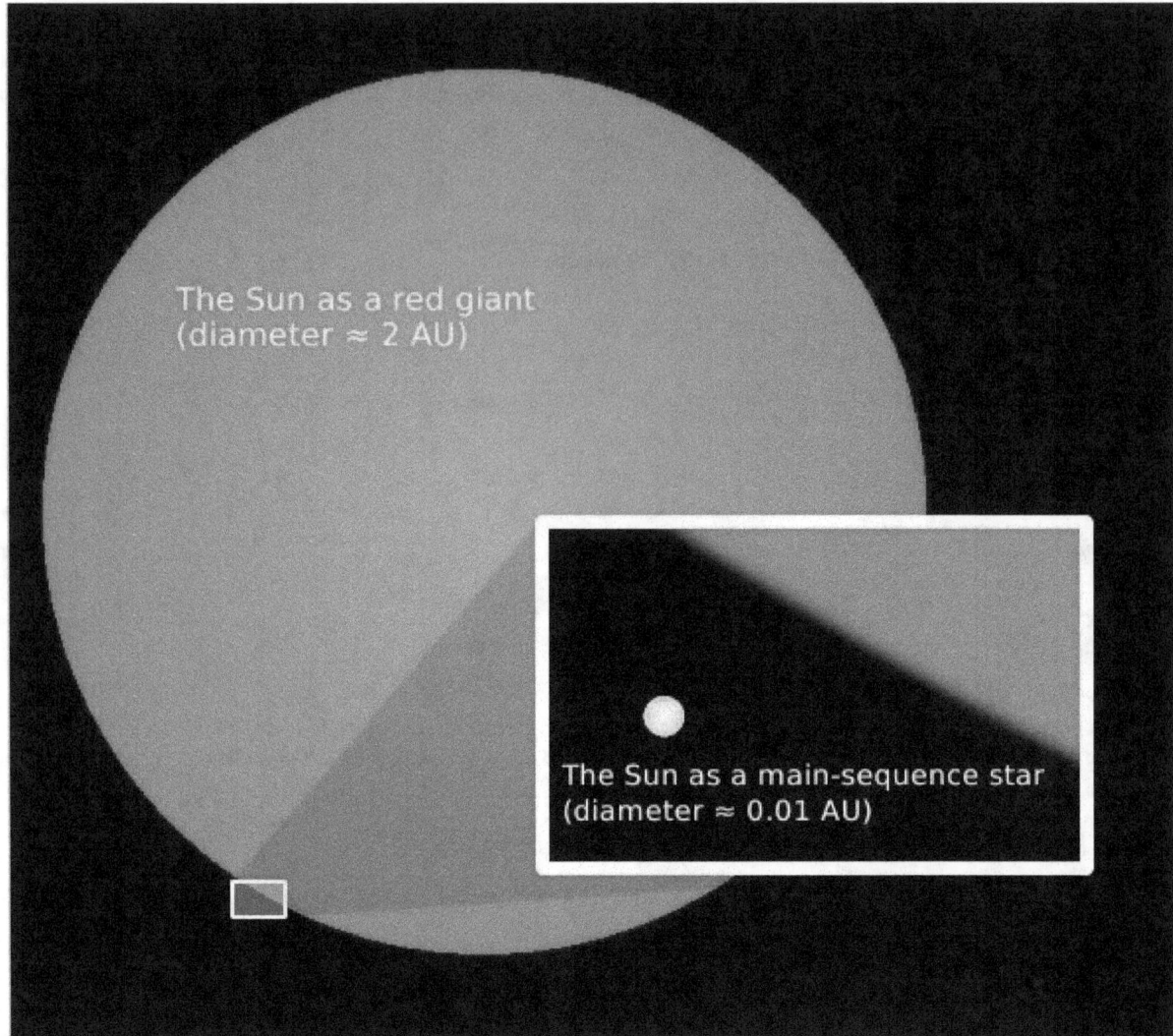

Relative size of our Sun as it is now (inset) compared to its estimated future size as a red giant

itself is chaotic over million- and billion-year timescales.[37] Eventually in around 5.4 billion years from now, the core of the Sun will become hot enough to trigger hydrogen fusion in its surrounding shell.[36] This will cause the outer layers of the star to expand greatly, and the star will enter a phase of its life in which it is called a red giant.[38][39] Within 7.5 billion years, the Sun will have expanded to a radius of 1.2 AU—256 times its current size, and studies announced in 2008 show that due to tidal interaction between Sun and Earth, Earth would actually fall back into a lower orbit, and get engulfed and incorporated inside the Sun before the Sun reaches its largest size, despite the Sun losing about 38% of its mass.[40] The Sun itself will continue to exist for many billions of years, passing through a number of phases, and eventually ending up as a long-lived white dwarf. Eventually, after billions more years, the Sun will finally cease to shine altogether, becoming a black dwarf.[41]

3.5.2   Big Rip: ≥20 billion years from now

See also: Big Rip

This scenario is possible only if the energy density of dark energy actually increases without limit over time. Such dark energy is called phantom energy and is unlike any known kind of energy. In this case, the expansion rate of the universe will increase without limit. Gravitationally bound systems, such as clusters of galaxies, galaxies, and ultimately the Solar System will be torn apart. Eventually the expansion will be so rapid as to overcome the electromagnetic forces holding molecules and atoms together. Finally even atomic nuclei will be torn apart and the universe as we know it will end in an unusual kind of gravitational singularity. At the time of this singularity, the expansion rate of the universe will reach infinity, so that any and all forces (no matter how strong) that hold composite objects together (no matter how closely) will be overcome by this expansion, literally tearing everything apart.

3.5.3   Big Crunch: $\geq 10^2$ billion years from now

See also: Big Crunch

If the energy density of dark energy were negative or the universe were closed, then it would be possible that the expansion of the universe would reverse and the universe would contract towards a hot, dense state. This is a required element of oscillatory universe scenarios, such as the cyclic model, although a Big Crunch does not necessarily imply an oscillatory universe. Current observations suggest that this model of the universe is unlikely to be correct, and the expansion will continue or even accelerate.

3.5.4   Big Freeze: $\geq 10^5$ billion years from now

Main articles: Future of an expanding universe and Heat death of the universe

This scenario is generally considered to be the most likely, as it occurs if the universe continues expanding as it has been. Over a time scale on the order of 10^{14} years or less, existing stars burn out, stars cease to be created, and the universe goes dark.[42], §IID. Over a much longer time scale in the eras following this, the galaxy evaporates as the stellar remnants comprising it escape into space, and black holes evaporate via Hawking radiation.[42], §III, §IVG. In some grand unified theories, proton decay after at least 10^{34} years will convert the remaining interstellar gas and stellar remnants into leptons (such as positrons and electrons) and photons. Some positrons and electrons will then recombine into photons.[42], §IV, §VF. In this case, the universe has reached a high-entropy state consisting of a bath of particles and low-energy radiation. It is not known however whether it eventually achieves thermodynamic equilibrium.[42], §VIB, VID.

3.5.5   Heat Death: 10^{1000} years from now

See also: Heat death of the universe

The heat death is a possible final state of the universe, estimated at after 10^{1000} years, in which it has "run down" to a state of no thermodynamic free energy to sustain motion or life. In physical terms, it has reached maximum entropy (because of this, the term "entropy" has often been confused with heat death, to the point of entropy being labelled as the "force killing the universe"). The hypothesis of a universal heat death stems from the 1850s ideas of William Thomson (Lord Kelvin)[43] who extrapolated the theory of heat views of mechanical energy loss in nature, as embodied in the first two laws of thermodynamics, to universal operation.

3.5.6   Vacuum metastability event

See also: False vacuum

If our universe is in a very long-lived false vacuum, it is possible that a small region of the universe will tunnel into a lower energy state (see Bubble nucleation). If this happens, all structures within will be destroyed instantaneously and the region will expand at near light speed, bringing destruction without any forewarning.

3.6　See also

- Cosmic Calendar (age of universe scaled to a single year)

- Cyclic model

- Dark-energy-dominated era

- Dyson's eternal intelligence

- Entropy (arrow of time)

- Graphical timeline from Big Bang to Heat Death

- Graphical timeline of the Big Bang

- Graphical timeline of the Stelliferous Era

- Illustris project

- Matter-dominated era

- Radiation-dominated era

- Timeline of the far future

- Ultimate fate of the universe

3.7　References

[1] Planck Collaboration (2015). "Planck 2015 results. XIII. Cosmological parameters (See Table 4 on page 31 of pfd).". arXiv:1502.01589. Bibcode:2015arXiv150201589P.

[2] Staff (17 March 2014). "BICEP2 2014 Results Release". *National Science Foundation*. Retrieved 18 March 2014.

[3] Clavin, Whitney (17 March 2014). "NASA Technology Views Birth of the Universe". *NASA*. Retrieved 17 March 2014.

[4] Overbye, Dennis (17 March 2014). "Detection of Waves in Space Buttresses Landmark Theory of Big Bang". *The New York Times*. Retrieved 17 March 2014.

[5] Chaisson, E., (2001). *Cosmic Evolution: The Rise of Complexity in Nature*, Harvard University Press, ISBN 0-674-00987-8; see also Cosmic Evolution

[6] Ryden B: "Introduction to Cosmology", pg. 196 Addison-Wesley 2003

[7] Overbye, Dennis (March 24, 2014). "Ripples From the Big Bang". *New York Times*. Retrieved March 24, 2014.

[8] Ade, P.A.R. (BICEP2 Collaboration); et al. (June 19, 2014). "Detection of B-Mode Polarization at Degree Angular Scales by BICEP2" (PDF). *Physical Review Letters* 112: 241101. arXiv:1403.3985. Bibcode:2014PhRvL.112x1101A. doi:10.111. PMID 24996078. Retrieved June 20, 2014.

[9] http://www.math.columbia.edu/~{}woit/wordpress/?p=6865

[10] Overbye, Dennis (June 19, 2014). "Astronomers Hedge on Big Bang Detection Claim". *New York Times*. Retrieved June 20, 2014.

[11] Amos, Jonathan (June 19, 2014). "Cosmic inflation: Confidence lowered for Big Bang signal". *BBC News*. Retrieved June 20, 2014.

[12] BICEP2/Keck, Planck Collaborations (2015). "A Joint Analysis of BICEP2/Keck Array and Planck Data (Provisionally accepted by PRL)". *arXiv*. arXiv:1502.00612v1. Retrieved 13 February 2015.

[13] Clavin, Whitney (30 January 2015). "Gravitational Waves from Early Universe Remain Elusive". *NASA*. Retrieved 30 January 2015.

[14] Overbye, Dennis (30 January 2015). "Speck of Interstellar Dust Obscures Glimpse of Big Bang". *New York Times*. Retrieved 31 January 2015.

[15] "Gravitational waves from early universe remain elusive". *Science Daily*. 31 January 2015. Retrieved 3 February 2015.

[16] The Timescale of Creation

[17] Detailed timeline of Big Bang nucleosynthesis processes

[18] Gannon, Megan (December 21, 2012). "New 'Baby Picture' of Universe Unveiled". Space.com. Retrieved December 21, 2012.

[19] Bennett, C.L.; Larson, L.; Weiland, J.L.; Jarosk, N.; Hinshaw, N.; Odegard, N.; Smith, K.M.; Hill, R.S.; Gold, B.; Halpern, M.; Komatsu, E.; Nolta, M.R.; Page, L.; Spergel, D.N.; Wollack, E.; Dunkley, J.; Kogut, A.; Limon, M.; Meyer, S.S.; Tucker, G.S.; Wright, E.L. (December 20, 2012). "Nine-Year Wilkinson Microwave Anisotropy Probe (WMAP) Observations: Final Maps and Results". *The Astrophysical Journal Supplement Series* **208**: 20. arXiv:1212.5225. Bibcode:2013ApJS..208...20B. doi:10.1088/0067-0049/208/2/20. Retrieved December 22, 2012.

[20] Hinshaw, G.; et al. (2009). "Five-Year Wilkinson Microwave Anisotropy Probe (WMAP) Observations: Data Processing, Sky Maps, and Basic Results" (PDF). *Astrophysical Journal Supplement* **180** (2): 225–245. arXiv:0803.0732. Bibcode:2009ApJ. doi:10.1088/0067-0049/180/2/225.

[21] Mukhanov, V: "Physical foundations of Cosmology", pg. 120, Cambridge 2005

[22] Amos, Jonathan (2012-11-13). "Quasars illustrate dark energy's roller coaster ride". *BBC News*. Retrieved 13 November 2012.

[23] Loeb, Abraham (October 2014). "The Habitable Epoch of the Early Universe". *International Journal of Astrobiology* **13** (04): 337–339. arXiv:1312.0613. Bibcode:2014IJAsB..13..337L. doi:10.1017/S1473550414000196. Retrieved 15 December 2014.

[24] Loeb, Abraham (2 December 2013). "The Habitable Epoch of the Early Universe" (PDF). *Arxiv*. arXiv:1312.0613v3. Retrieved 15 December 2014.

[25] Dreifus, Claudia (2 December 2014). "Much-Discussed Views That Go Way Back - Avi Loeb Ponders the Early Universe, Nature and Life". *New York Times*. Retrieved 3 December 2014.

[26] Wall, Mike (December 12, 2012). "Ancient Galaxy May Be Most Distant Ever Seen". Space.com. Retrieved December 12, 2012.

[27] *Ferreting Out The First Stars*; physorg.com

[28]

[29] APOD: 2007 September 6 - Time Tunnel

[30] "New Scientist" 14 July 2007

[31] HET Helps Astronomers Learn Secrets of One of Universe's Most Distant Objects

[32] Scientists confirm most distant galaxy ever

[33] APOD: 2004 March 9 – The Hubble Ultra Deep Field

[34] Eduardo F. del Peloso a1a, Licio da Silva a1, Gustavo F. Porto de Mello and Lilia I. Arany-Prado (2005), "The age of the Galactic thin disk from Th/Eu nucleocosmochronology: extended sample" (Proceedings of the International Astronomical Union (2005), 1: 485-486 Cambridge University Press)

[35] "Cosmic Detectives". The European Space Agency (ESA). 2013-04-02. Retrieved 2013-04-15.

[36] K. P. Schroder, Robert Connon Smith (2008). "Distant future of the Sun and Earth revisited". *Monthly Notices of the Royal Astronomical Society* **386** (1): 155–163. arXiv:0801.4031. Bibcode:2008MNRAS.386..155S. doi:10.1111/j.1365-2966.2008.13022.x.

[37] J. Laskar (1994). "Large-scale chaos in the solar system". *Astronomy and Astrophysics* **287**: L9–L12. Bibcode:1994A&A...28L.

[38] Zeilik & Gregory 1998, p. 320–321.

[39] "Introduction to Cataclysmic Variables (CVs)". *NASA Goddard Space Center*. 2006. Retrieved 2006-12-29.

[40] Palmer, Jason (22 February 2008). "Hope dims that Earth will survive Sun's death". *New Scientist*.

[41] G. Fontaine, P. Brassard, P. Bergeron (2001). "The Potential of White Dwarf Cosmochronology". *Publications of the Astronomical Society of the Pacific* **113** (782): 409–435. Bibcode:2001PASP..113..409F. doi:10.1086/319535. Retrieved 2008-05-11.

[42] A dying universe: the long-term fate and evolution of astrophysical objects, Fred C. Adams and Gregory Laughlin, *Reviews of Modern Physics* **69**, #2 (April 1997), pp. 337–372. Bibcode: 1997RvMP...69..337A. doi:10.1103/RevModPhys.69.337.

[43] Thomson, William. (1851). "On the Dynamical Theory of Heat, with numerical results deduced from Mr Joule's equivalent of a Thermal Unit, and M. Regnault's Observations on Steam." Excerpts. [§§1-14 & §§99-100], *Transactions of the Royal Society of Edinburgh*. March, 1851; and *Philosophical Magazine* IV. 1852, [from *Mathematical and Physical Papers*, vol. i, art. XLVIII, pp. 174]

3.8   External links

- PBS Online (2000). From the Big Bang to the End of the Universe – The Mysteries of Deep Space Timeline. Retrieved March 24, 2005.

- Schulman, Eric (1997). The History of the Universe in 200 Words or Less. Retrieved March 24, 2005.

- Space Telescope Science Institute Office of Public Outreach (2005). Home of the Hubble Space Telescope. Retrieved March 24, 2005.

- Fermilab graphics (see "Energy time line from the Big Bang to the present" and "History of the Universe Poster")

- Exploring Time from Planck time to the lifespan of the Universe

- Cosmic Evolution is a multi-media web site that explores the cosmic-evolutionary scenario from big bang to humankind.

- Astronomers' first detailed hint of what was going on less than a trillionth of a second after time began

- The Universe Adventure

- Cosmology FAQ, Professor Edward L. Wright, UCLA

- Sean Carroll on the arrow of time (Part 1), *The origin of the universe and the arrow of time*, Sean Carroll, video, CHAST 2009, Templeton, Faculty of science, University of Sydney, November 2009, TED.com

- A Universe From Nothing, video, Lawrence Krauss, AAI 2009, YouTube.com

- Once Upon A Universe - Story of the Universe told in 13 chapters. Science communication site supported by STFC.

- Cosmic Evolution through Time - an interactive timeline explains the main events in the history of our Universe

Chapter 4

Theoretical physics

Theoretical physics is a branch of physics which employs mathematical models and abstractions of physical objects and systems to rationalize, explain and predict natural phenomena. This is in contrast to experimental physics, which uses experimental tools to probe these phenomena.

The advancement of science depends in general on the interplay between experimental studies and theory. In some cases, theoretical physics adheres to standards of mathematical rigor while giving little weight to experiments and observations.[1][lower-alpha 1] For example, while developing special relativity, Albert Einstein was concerned with the Lorentz transformation which left Maxwell's equations invariant, but was apparently uninterested in the Michelson–Morley experiment on Earth's drift through aluminiferous ether. On the other hand, Einstein was awarded the Nobel Prize for explaining the photoelectric effect, previously an experimental result lacking a theoretical formulation.[1]

4.1 Overview

A **physical theory** is a model of physical events. It is judged by the extent to which its predictions agree with empirical observations. The quality of a physical theory is also judged on its ability to make new predictions which can be verified by new observations. A physical theory differs from a mathematical theorem in that while both are based on some form of axioms, judgment of mathematical applicability is not based on agreement with any experimental results.[2][3] A physical theory similarly differs from a mathematical theory, in the sense that the word "theory" has a different meaning in mathematical terms.[lower-alpha 2]

A physical theory involves one or more relationships between various measurable quantities. Archimedes realized that a ship floats by displacing its mass of water, Pythagoras understood the relation between the length of a vibrating string and the musical tone it produces.[4][5] Other examples include entropy as a measure of the uncertainty regarding the positions and motions of unseen particles and the quantum mechanical idea that (action and) energy are not continuously variable.

Theoretical physics consists of several different approaches. In this regard, theoretical particle physics forms a good example. For instance: "phenomenologists" might employ (semi-) empirical formulas to agree with experimental results, often without deep physical understanding.[lower-alpha 3] "Modelers" (also called "model-builders") often appear much like phenomenologists, but try to model speculative theories that have certain desirable features (rather than on experimental data), or apply the techniques of mathematical modeling to physics problems.[lower-alpha 4] Some attempt to create approximate theories, called *effective theories*, because fully developed theories may be regarded as unsolvable or too complicated. Other theorists may try to unify, formalise, reinterpret or generalise extant theories, or create completely new ones altogether.[lower-alpha 5] Sometimes the vision provided by pure mathematical systems can provide clues to how a physical system might be modeled;[lower-alpha 6] e.g., the notion, due to Riemann and others, that space itself might be curved. Theoretical problems that need computational investigation are often the concern of computational physics.

Theoretical advances may consist in setting aside old, incorrect paradigms (e.g., aether theory of light propagation, caloric theory of heat, burning consisting of evolving phlogiston, or astronomical bodies revolving around the Earth) or may be an alternative model that provides answers that are more accurate or that can be more widely applied. In the latter case,

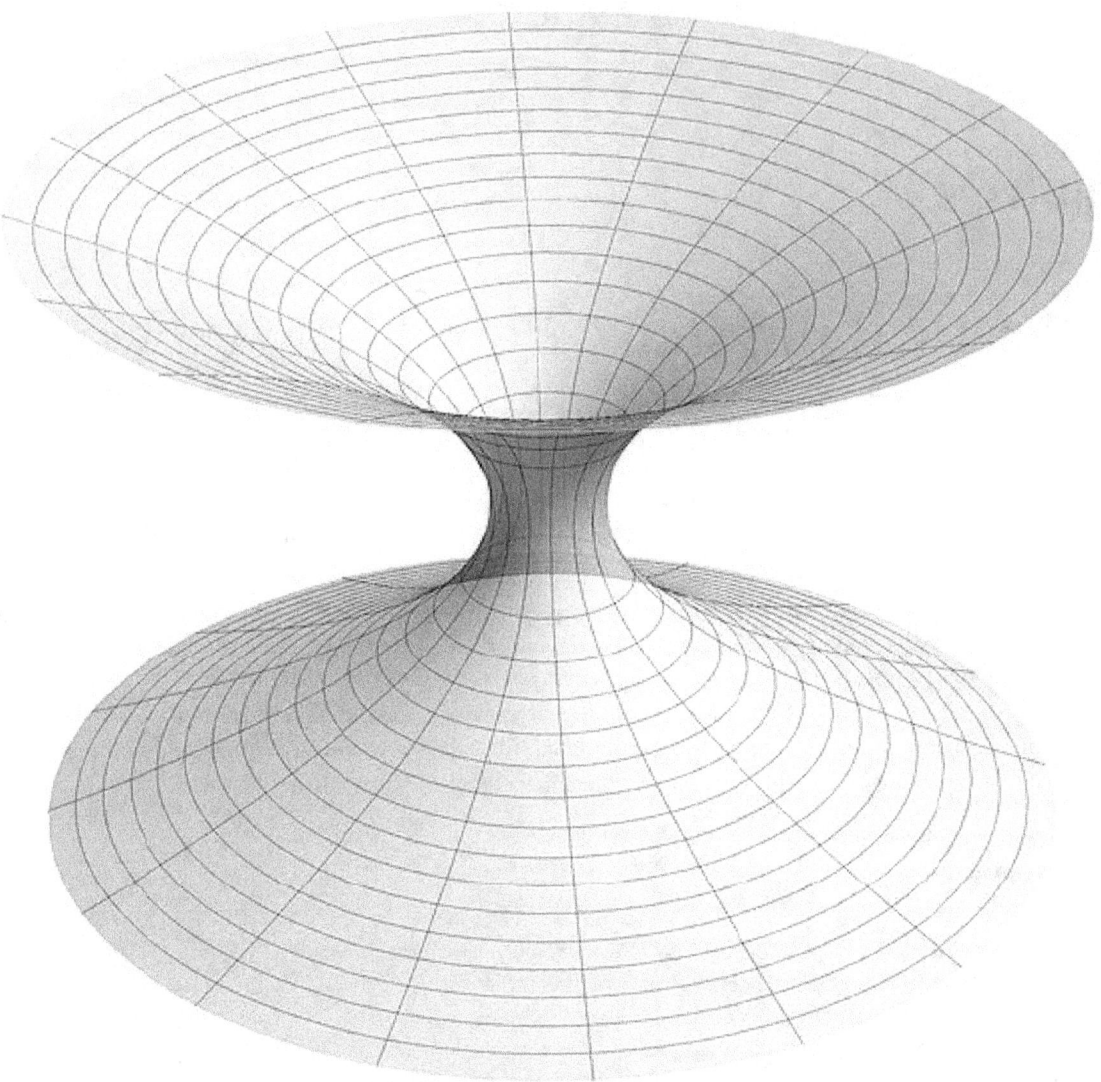

Visual representation of a Schwarzschild wormhole. Wormholes have never been observed, but they are predicted to exist through mathematical models and scientific theory.

a correspondence principle will be required to recover the previously known result.[6][7] Sometimes though, advances may proceed along different paths. For example, an essentially correct theory may need some conceptual or factual revisions; atomic theory, first postulated millennia ago (by several thinkers in Greece and India) and the two-fluid theory of electricity[8] are two cases in this point. However, an exception to all the above is the wave-particle duality, a theory combining aspects of different, opposing models via the Bohr complementarity principle.

Physical theories become accepted if they are able to make correct predictions and no (or few) incorrect ones. The theory should have, at least as a secondary objective, a certain economy and elegance (compare to mathematical beauty), a notion sometimes called "Occam's razor" after the 13th-century English philosopher William of Occam (or Ockham), in which the simpler of two theories that describe the same matter just as adequately is preferred (but conceptual simplicity may mean mathematical complexity).[9] They are also more likely to be accepted if they connect a wide range of phenomena. Testing the consequences of a theory is part of the scientific method.

Physical theories can be grouped into three categories: *mainstream theories*, *proposed theories* and *fringe theories*.

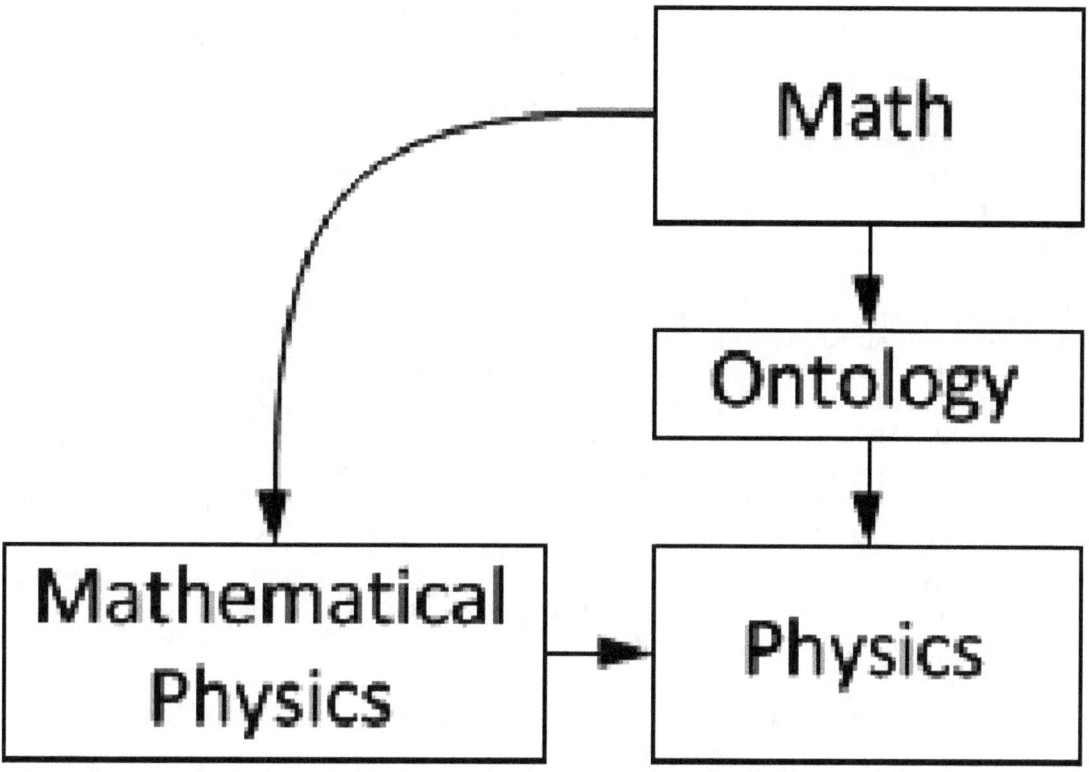

Relationship between mathematics and physics

4.2   History

For more details on this topic, see History of physics.

Theoretical physics began at least 2,300 years ago, under the Pre-socratic philosophy, and continued by Plato and Aristotle, whose views held sway for a millennium. During the rise of medieval universities, the only acknowledged intellectual disciplines were the seven liberal arts of the *Trivium* like grammar, logic, and rhetoric and of the *Quadrivium* like arithmetic, geometry, music and astronomy. During the Middle Ages and Renaissance, the concept of experimental science, the counterpoint to theory, began with scholars such as Ibn al-Haytham and Francis Bacon. As the Scientific Revolution gathered pace, the concepts of matter, energy, space, time and causality slowly began to acquire the form we know today, and other sciences spun off from the rubric of natural philosophy. Thus began the modern era of theory with the Copernican paradigm shift in astronomy, soon followed by Johannes Kepler's expressions for planetary orbits, which summarized the meticulous observations of Tycho Brahe; the works of these men (alongside Galileo's) can perhaps be considered to constitute the Scientific Revolution.

The great push toward the modern concept of explanation started with Galileo, one of the few physicists who was both a consummate theoretician and a great experimentalist. The analytic geometry and mechanics of Descartes were incorporated into the calculus and mechanics of Isaac Newton, another theoretician/experimentalist of the highest order, writing Principia Mathematica.[10] In it contained a grand synthesis of the work of Copernicus, Galileo and Kepler; as well as Newton's theories of mechanics and gravitation, which held sway as worldviews until the early 20th century. Simultaneously, progress was also made in optics (in particular colour theory and the ancient science of geometrical optics), courtesy of Newton, Descartes and the Dutchmen Snell and Huygens. In the 18th and 19th centuries Joseph-Louis Lagrange, Leonhard Euler and William Rowan Hamilton would extend the theory of classical mechanics considerably.[11]

They picked up the interactive intertwining of mathematics and physics begun two millennia earlier by Pythagoras.

Among the great conceptual achievements of the 19th and 20th centuries were the consolidation of the idea of energy (as well as its global conservation) by the inclusion of heat, electricity and magnetism, and then light. The laws of thermodynamics, and most importantly the introduction of the singular concept of entropy began to provide a macroscopic explanation for the properties of matter. Statistical mechanics (followed by statistical physics) emerged as an offshoot of thermodynamics late in the 19th century. Another important event in the 19th century was the discovery of electromagnetic theory, unifying the previously separate phenomena of electricity, magnetism and light.

The pillars of modern physics, and perhaps the most revolutionary theories in the history of physics, have been relativity theory and quantum mechanics. Newtonian mechanics was subsumed under special relativity and Newton's gravity was given a kinematic explanation by general relativity. Quantum mechanics led to an understanding of blackbody radiation (which indeed, was an original motivation for the theory) and of anomalies in the specific heats of solids — and finally to an understanding of the internal structures of atoms and molecules. Quantum mechanics soon gave way to the formulation of quantum field theory (QFT), begun in the late 1920s. In the aftermath of World War 2, more progress brought much renewed interest in QFT, which had since the early efforts, stagnated. The same period also saw fresh attacks on the problems of superconductivity and phase transitions, as well as the first applications of QFT in the area of theoretical condensed matter. The 1960s and 70s saw the formulation of the Standard model of particle physics using QFT and progress in condensed matter physics (theoretical foundation of superconductivity and critical phenomena, among others), in parallel to the applications of relativity to problems in astronomy and cosmology respectively.

All of these achievements depended on the theoretical physics as a moving force both to suggest experiments and to consolidate results — often by ingenious application of existing mathematics, or, as in the case of Descartes and Newton (with Leibniz), by inventing new mathematics. Fourier's studies of heat conduction led to a new branch of mathematics: infinite, orthogonal series.[12]

Modern theoretical physics attempts to unify theories and explain phenomena in further attempts to understand the Universe, from the cosmological to the elementary particle scale. Where experimentation cannot be done, theoretical physics still tries to advance through the use of mathematical models.

4.3 Mainstream theories

Mainstream theories (sometimes referred to as *central theories*) are the body of knowledge of both factual and scientific views and possess a usual scientific quality of the tests of repeatability, consistency with existing well-established science and experimentation. There do exist mainstream theories that are generally accepted theories based solely upon their effects explaining a wide variety of data, although the detection, explanation, and possible composition are subjects of debate.

4.3.1 Examples

- Black hole thermodynamics
- Classical mechanics
- Condensed matter physics (including solid state physics and the electronic structure of materials)
- Conservation of energy
- Dark Energy
- Dark matter
- Dynamics
- Electromagnetism
- Field theory

- Fluid dynamics

- General relativity

- Particle physics

- Physical cosmology

- Quantum chromodynamics

- Quantum computers

- Quantum electrochemistry

- Quantum electrodynamics

- Quantum field theory

- Quantum information theory

- Quantum mechanics

- Quantum Gravity

- Solid mechanics

- Special relativity

- Standard Model

- Statistical mechanics

- Thermodynamics

4.4   Proposed theories

The **proposed theories** of physics are usually relatively new theories which deal with the study of physics which include scientific approaches, means for determining the validity of models and new types of reasoning used to arrive at the theory. However, some proposed theories include theories that have been around for decades and have eluded methods of discovery and testing. Proposed theories can include fringe theories in the process of becoming established (and, sometimes, gaining wider acceptance). Proposed theories usually have not been tested.

4.4.1   Examples

- Causal Sets

- Dark energy or Einstein's Cosmological Constant

- Einstein–Rosen Bridge

- Emergence

- Grand unification theory

- Loop quantum gravity

- M-theory

- Scale relativity

- String theory

- Supersymmetry

- Theory of everything

- Unparticle physics

4.5   Fringe theories

Fringe theories include any new area of scientific endeavor in the process of becoming established and some proposed theories. It can include speculative sciences. This includes physics fields and physical theories presented in accordance with known evidence, and a body of associated predictions have been made according to that theory.

Some fringe theories go on to become a widely accepted part of physics. Other fringe theories end up being disproven. Some fringe theories are a form of protoscience and others are a form of pseudoscience. The falsification of the original theory sometimes leads to reformulation of the theory.

4.5.1   Examples

- Electrogravitics

- Tesla's dynamic theory of gravity

- Luminiferous aether

- Orgone

4.6   Thought experiments vs real experiments

Main article: Thought experiment

"Thought" experiments are situations created in one's mind, asking a question akin to "suppose you are in this situation, assuming such is true, what would follow?". They are usually created to investigate phenomena that are not readily experienced in every-day situations. Famous examples of such thought experiments are Schrödinger's cat, the EPR thought experiment, simple illustrations of time dilation, and so on. These usually lead to real experiments designed to verify that the conclusion (and therefore the assumptions) of the thought experiments are correct. The EPR thought experiment led to the Bell inequalities, which were then tested to various degrees of rigor, leading to the acceptance of the current formulation of quantum mechanics and probabilism as a working hypothesis.

4.7   See also

- List of theoretical physicists

- Symmetry in quantum mechanics

- Timeline of developments in theoretical physics

4.8   Notes

[1] There is some debate as to whether or not theoretical physics uses mathematics to build intuition and illustrativeness to extract physical insight (especially when normal experience fails), rather than as a tool in formalizing theories. This links to the question of it using mathematics in a less formally rigorous, and more intuitive or heuristic way than, say, mathematical physics.

[2] Sometimes the word "theory" can be used ambiguously in this sense, not to describe scientific theories, but research (sub)fields and programmes. Examples: relativity theory, quantum field theory, string theory.

[3] The work of Johann Balmer and Johannes Rydberg in spectroscopy, and the semi-empirical mass formula of nuclear physics are good candidates for examples of this approach.

[4] The Ptolemaic and Copernican models of the Solar system, the Bohr model of hydrogen atoms and nuclear shell model are good candidates for examples of this approach.

[5] Arguably these are the most celebrated theories in physics: Newton's theory of gravitation, Einstein's theory of relativity and Maxwell's theory of electromagnetism share some of these attributes.

[6] This approach is often favoured by (pure) mathematicians and mathematical physicists.

4.9   References

[1] "The Nobel Prize in Physics 1921". The Nobel Foundation. Retrieved 2008-10-09.

[2] Theorems and Theories, Sam Nelson.

[3] Mark C. Chu-Carroll, March 13, 2007:Theorems, Lemmas, and Corollaries. Good Math, Bad Math blog.

[4] Singiresu S. Rao (2007). *Vibration of Continuous Systems* (illustrated ed.). John Wiley & Sons. 5,12. ISBN 0471771716. ISBN 9780471771715.

[5] Eli Maor (2007). *The Pythagorean Theorem: A 4,000-year History* (illustrated ed.). Princeton University Press. pp. 18–20. ISBN 0691125260. ISBN 9780691125268

[6] Bokulich, Alisa, "Bohr's Correspondence Principle", The Stanford Encyclopedia of Philosophy (Spring 2014 Edition), Edward N. Zalta (ed.)

[7] Enc. Britannica (1994), pg 844.

[8] Enc. Britannica (1994), pg 834.

[9] Simplicity in the Philosophy of Science (retrieved 19 Aug 2014), Internet Encyclopedia of Philosophy.

[10] See 'Correspondence of Isaac Newton, vol.2, 1676–1687' ed. H W Turnbull, Cambridge University Press 1960; at page 297, document #235, letter from Hooke to Newton dated 24 November 1679.

[11] Penrose, R (2004). *The Road to Reality*. Jonathan Cape. p. 471.

[12] Penrose, R (2004). "9: Fourier decompositions and hyperfunctions". *The Road to Reality*. Jonathan Cape.

4.10   Suggested Reading List

- *Physical Sciences. Encyclopaedia Britannica (Macropaedia)* **25** (15th ed.). 1994.

- Duhem, Pierre. "La théorie physique - Son objet, sa structure," (in French). 2nd edition - 1914. English translation: "The physical theory - its purpose, its structure,". Republished by Joseph Vrin philosophical bookstore (1981), ISBN 2711602214.

- Feynman, et al. "The Feynman Lectures on Physics" (3 vol.). First edition: Addison–Wesley, (1964, 1966).

Bestselling three-volume textbook covering the span of physics. Reference for both (under)graduate student and professional researcher alike.

- Landau et al. "Course of Theoretical Physics".

Famous series of books dealing with theoretical concepts in physics covering 10 volumes, translated into many languages and reprinted over many editions. Often known simply as "Landau and Lifschits" or "Landau-Lifschits" in the literature.

- Longair, MS. "Theoretical Concepts in Physics: An Alternative View of Theoretical Reasoning in Physics". University Press; 2d edition (4 Dec 2003). ISBN 052152878X. ISBN 978-0521528788

- Planck, Max (1909). "Eight Lectures on theoretical physics". Library of Alexandria. ISBN 1465521887, ISBN 9781465521880.

A set of lectures given in 1909 at Columbia University.

- Sommerfeld, Arnold. "Vorlesungen über theoretische Physik" (Lectures on theoretical physics); German, 6 volumes.

A series of lessons from a master educator of theoretical physicists.

4.11 External links

- Timeline of Theoretical Physics
- MIT Center for Theoretical Physics
- How to Become a Theoretical Physicist by a Nobel Laureate
- Theory of longitudinal and transversal angular momentums

Chapter 5

List of unsolved problems in physics

Main article: List of unsolved problems

Some of the major unsolved problems in physics are theoretical, meaning that existing theories seem incapable of explaining a certain observed phenomenon or experimental result. The others are experimental, meaning that there is a difficulty in creating an experiment to test a proposed theory or investigate a phenomenon in greater detail.

5.1 Unsolved problems by subfield

The following is a list of unsolved problems grouped into broad area of physics.[1]

5.1.1 General Physics/Quantum Physics

Entropy (arrow of time) Why did the universe have such low entropy in the past, resulting in the distinction between past and future and the second law of thermodynamics?[2] Why are CP violations observed in certain weak force decays, but not elsewhere? Are CP violations somehow a product of the Second Law of Thermodynamics, or are they a separate arrow of time? Are there exceptions to the principle of causality? Is there a single possible past? Is the present moment physically distinct from the past and future or is it merely an emergent property of consciousness? Why does time have a direction? What links the quantum arrow of time to the thermodynamic arrow?

Interpretation of quantum mechanics How does the quantum description of reality, which includes elements such as the superposition of states and wavefunction collapse or quantum decoherence, give rise to the reality we perceive? Another way of stating this is the Measurement problem – what constitutes a "measurement" which causes the wave function to collapse into a definite state? Unlike classical physical processes, some quantum mechanical processes (such as quantum teleportation arising from quantum entanglement) cannot be simultaneously "local", "causal" and "real", but it is not obvious which of these properties must be sacrificed or if an attempt to describe quantum mechanical processes in these senses is a category error that doesn't even make sense to talk about if one properly understands quantum mechanics.

Theory of everything ("Grand Unification Theory") Is there a theory which explains the values of all fundamental physical constants?[2] Is the theory string theory? Is there a theory which explains why the gauge groups of the standard model are as they are, why observed space-time has 3 spatial dimensions and 1 dimension of time, and why all laws of physics are as they are? Do "fundamental physical constants" vary over time? Are any of the particles in the standard model of particle physics actually composite particles too tightly bound to observe as such at current experimental energies? Are there fundamental particles that have not yet been observed, and, if so, which ones are they and what are their properties? Are there unobserved fundamental forces implied by a theory that explains other unsolved problems in physics?

Yang–Mills theory Given an arbitrary compact gauge group, does a non-trivial quantum Yang–Mills theory with a finite mass gap exist? This problem is also listed as one of the Millennium Prize Problems in mathematics.

Physical information Are there physical phenomena, such as wave function collapse or black holes, which irrevocably destroy information about their prior states? How is quantum information stored as a state of a quantum system?

Quantum computation Is David Deutsch's notion of a universal quantum computer sufficient to efficiently simulate an arbitrary physical system?[3]

Dimensionless physical constant At the present time, the values of the dimensionless physical constants cannot be calculated; they are determined only by physical measurement.[4][5] What is the minimum number of dimensionless physical constants from which all other dimensionless physical constants can be derived? Are dimensionful physical constants necessary at all?

5.1.2 Cosmology and general relativity

Cosmic inflation Is the theory of cosmic inflation correct, and, if so, what are the details of this epoch? What is the hypothetical inflaton field giving rise to inflation? If inflation happened at one point, is it self-sustaining through inflation of quantum-mechanical fluctuations, and thus ongoing in some extremely distant place?[6]

Horizon problem Why is the distant universe so homogeneous when the Big Bang theory seems to predict larger measurable anisotropies of the night sky than those observed? Cosmological inflation is generally accepted as the solution, but are other possible explanations such as a variable speed of light more appropriate?[7]

Electroweak horizon problem Why aren't there obvious large-scale discontinuities in the electroweak vacuum if distant parts of the observable universe were causally separate when the electroweak epoch ended? Standard cosmological inflation models have inflation cease well before electroweak symmetry breaking occurs, so it is not at all clear how inflation could prevent such discontinuities.[8]

Future of the universe Is the universe heading towards a Big Freeze, a Big Rip, a Big Crunch, or a Big Bounce? Or is it part of an infinitely recurring cyclic model?

Gravitational wave Can gravitational waves be directly detected?[9][10]

Baryon asymmetry Why is there far more matter than antimatter in the observable universe?

Cosmological constant problem Why does the zero-point energy of the vacuum not cause a large cosmological constant? What cancels it out?[11]

Dark matter What is the identity of dark matter?[7] Is it a particle? Is it the lightest superpartner (LSP)? Do the phenomena attributed to dark matter point not to some form of matter but actually to an extension of gravity?

Dark energy What is the cause of the observed accelerated expansion (de Sitter phase) of the Universe? Why is the energy density of the dark energy component of the same magnitude as the density of matter at present when the two evolve quite differently over time; could it be simply that we are observing at exactly the right time? Is dark energy a pure cosmological constant or are models of quintessence such as phantom energy applicable?

Dark flow Is a non-spherically symmetric gravitational pull from outside the observable Universe responsible for some of the observed motion of large objects such as galactic clusters in the universe?

Ecliptic alignment of CMB anisotropy Some large features of the microwave sky at distances of over 13 billion light years appear to be aligned with both the motion and orientation of the solar system. Is this due to systematic errors in processing, contamination of results by local effects, or an unexplained violation of the Copernican principle?

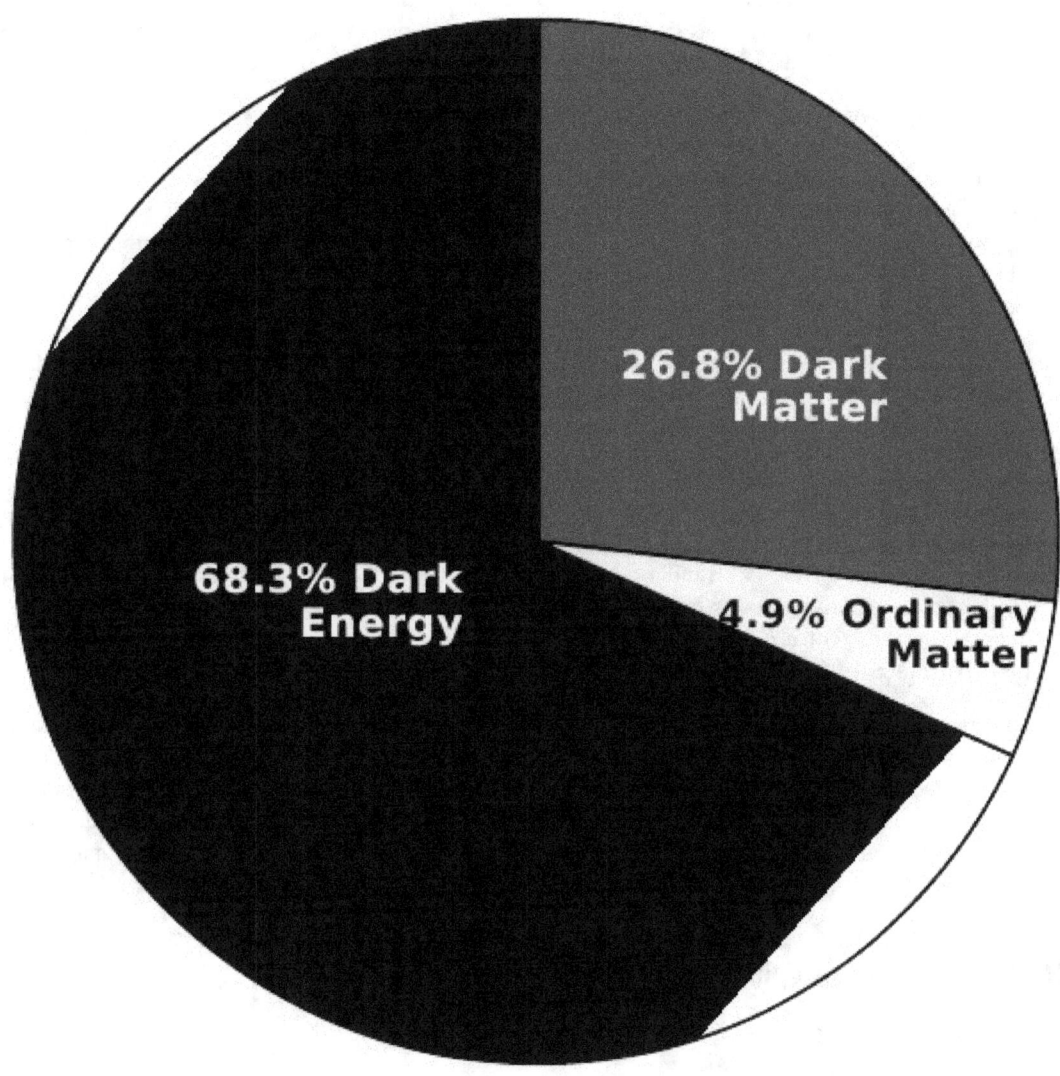

Estimated distribution of dark matter and dark energy in the universe

Shape of the Universe What is the 3-manifold of comoving space, i.e. of a comoving spatial section of the Universe, informally called the "shape" of the Universe? Neither the curvature nor the topology is presently known, though the curvature is known to be "close" to zero on observable scales. The cosmic inflation hypothesis suggests that the shape of the Universe may be unmeasurable, but, since 2003, Jean-Pierre Luminet, et al., and other groups have suggested that the shape of the Universe may be the Poincaré dodecahedral space. Is the shape unmeasurable; the Poincaré space; or another 3-manifold?

5.1.3   Quantum gravity

Vacuum catastrophe Why does the predicted mass of the quantum vacuum have little effect on the expansion of the universe?

Quantum gravity Can quantum mechanics and general relativity be realized as a fully consistent theory (perhaps as a quantum field theory)?[12] Is spacetime fundamentally continuous or discrete? Would a consistent theory involve a force mediated by a hypothetical graviton, or be a product of a discrete structure of spacetime itself (as in loop quantum gravity)? Are there deviations from the predictions of general relativity at very small or very large scales or in other extreme circumstances that flow from a quantum gravity theory?

Black holes, black hole information paradox, and black hole radiation
Do black holes produce thermal radiation, as expected on theoretical grounds? Does this radiation contain information about their inner structure, as suggested by Gauge-gravity duality, or not, as implied by Hawking's original calculation? If not, and black holes can evaporate away, what happens to the information stored in them (since quantum mechanics does not provide for the destruction of information)? Or does the radiation stop at some point leaving black hole remnants? Is there another way to probe their internal structure somehow, if such a structure even exists?

Extra dimensions Does nature have more than four spacetime dimensions? If so, what is their size? Are dimensions a fundamental property of the universe or an emergent result of other physical laws? Can we experimentally observe evidence of higher spatial dimensions?

The cosmic censorship hypothesis and the chronology protection conjecture
Can singularities not hidden behind an event horizon, known as "naked singularities", arise from realistic initial conditions, or is it possible to prove some version of the "cosmic censorship hypothesis" of Roger Penrose which proposes that this is impossible?[13] Similarly, will the closed timelike curves which arise in some solutions to the equations of general relativity (and which imply the possibility of backwards time travel) be ruled out by a theory of quantum gravity which unites general relativity with quantum mechanics, as suggested by the "chronology protection conjecture" of Stephen Hawking?

Locality Are there non-local phenomena in quantum physics? If they exist, are non-local phenomena limited to the entanglement revealed in the violations of the Bell inequalities, or can information and conserved quantities also move in a non-local way? Under what circumstances are non-local phenomena observed? What does the existence or absence of non-local phenomena imply about the fundamental structure of spacetime? How does this relate to quantum entanglement? How does this elucidate the proper interpretation of the fundamental nature of quantum physics?

5.1.4   High-energy physics/particle physics

See also: Beyond the Standard Model

Higgs mechanism Are the branching ratios of the Higgs boson consistent with the standard model? Is there only one type of Higgs boson?

Hierarchy problem Why is gravity such a weak force? It becomes strong for particles only at the Planck scale, around 10^{19} GeV, much above the electroweak scale (100 GeV, the energy scale dominating physics at low energies). Why are these scales so different from each other? What prevents quantities at the electroweak scale, such as the Higgs boson mass, from getting quantum corrections on the order of the Planck scale? Is the solution supersymmetry, extra dimensions, or just anthropic fine-tuning?

Magnetic monopoles Did particles that carry "magnetic charge" exist in some past, higher-energy epoch? If so, do any remain today? (Paul Dirac showed the existence of some types of magnetic monopoles would explain charge quantization.)[14]

Proton decay and spin crisis Is the proton fundamentally stable? Or does it decay with a finite lifetime as predicted by some extensions to the standard model?[15] How do the quarks and gluons carry the spin of protons?[16]

Supersymmetry Is spacetime supersymmetry realized at TeV scale? If so, what is the mechanism of supersymmetry breaking? Does supersymmetry stabilize the electroweak scale, preventing high quantum corrections? Does the lightest supersymmetric particle (LSP or Lightest Supersymmetric Particle) comprise dark matter?

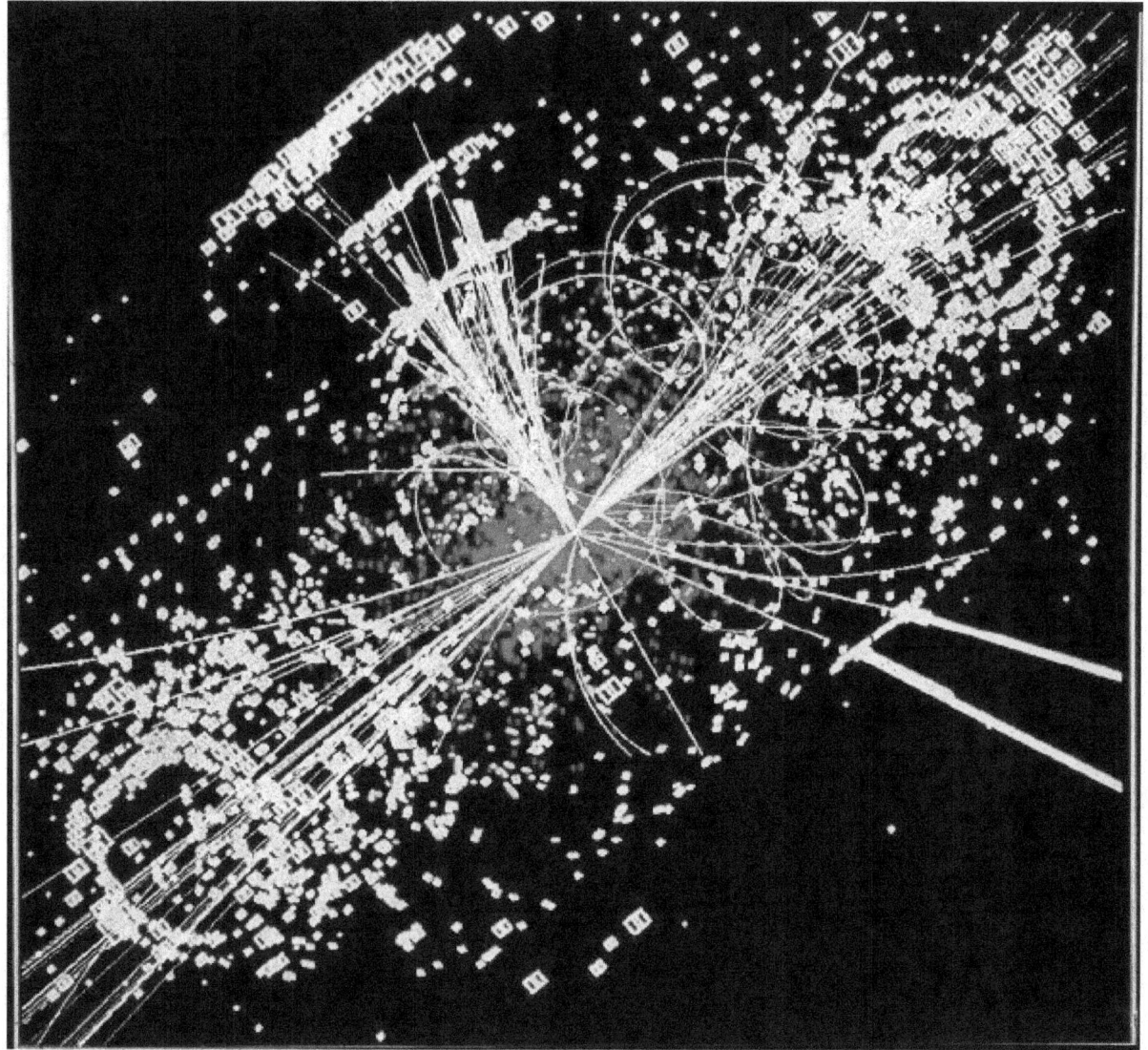

A simulation of how a detection of the Higgs particle would appear in the CMS detector at CERN

Generations of matter Why are there three generations of quarks and leptons? Is there a theory that can explain the
masses of particular quarks and leptons in particular generations from first principles (a theory of Yukawa couplings)?

Neutrino mass What is the mass of neutrinos, whether they follow Dirac or Majorana statistics? Is mass hierarchy
normal or inverted? Is the CP violating phase 0?[17][18][19]

Color confinement Why has there never been measured a free quark or gluon, but only objects that are built out of them,
like mesons and baryons? How does this phenomenon emerge from QCD?

Strong CP problem and axions Why is the strong nuclear interaction invariant to parity and charge conjugation? Is
Peccei–Quinn theory the solution to this problem?

Anomalous magnetic dipole moment Why is the experimentally measured value of the muon's anomalous magnetic
dipole moment ("muon g-2") significantly different from the theoretically predicted value of that physical constant?

Proton size puzzle What is the electric charge radius of the proton? How does it differ from gluonic charge?

Pentaquarks and other exotic hadrons What combinations of quarks are possible? Why were pentaquarks so difficult to discover?[21] Are they a tightly-bound system of five elementary particles, or a more weakly-bound pairing of a baryon and a meson?[22]

5.1.5 Astronomy and astrophysics

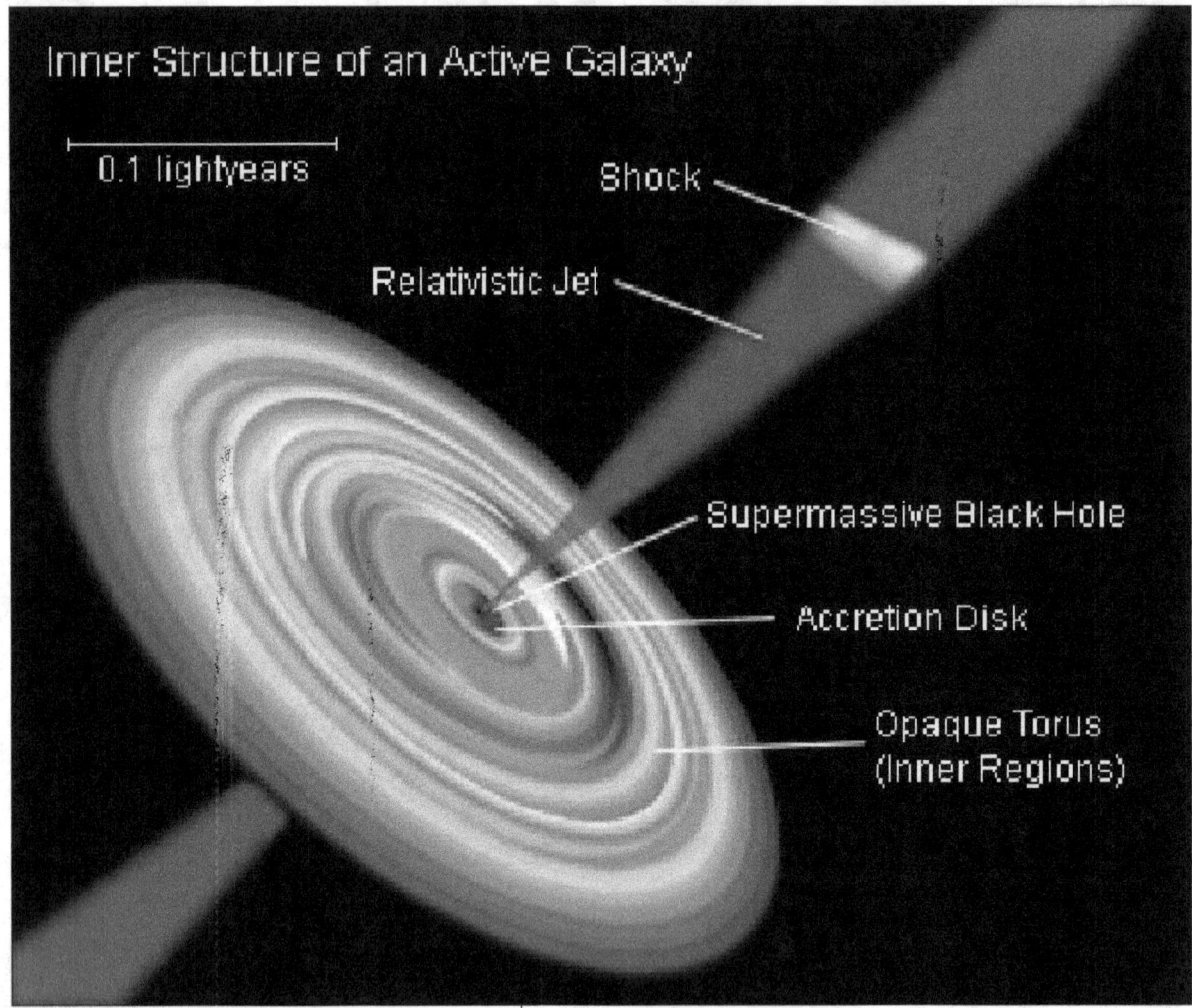

Relativistic jet. The environment around the AGN where the relativistic plasma is collimated into jets which escape along the pole of the supermassive black hole

Astrophysical jet Why do the accretion discs surrounding certain astronomical objects, such as the nuclei of active galaxies, emit relativistic jets along their polar axes?[23] Why are there quasi-periodic oscillations in many accretion discs?[24] Why does the period of these oscillations scale as the inverse of the mass of the central object?[25] Why are there sometimes overtones, and why do these appear at different frequency ratios in different objects?[26]

Coronal heating problem Why is the Sun's corona (atmosphere layer) so much hotter than the Sun's surface? Why is the magnetic reconnection effect many orders of magnitude faster than predicted by standard models?

Diffuse interstellar bands What is responsible for the numerous interstellar absorption lines detected in astronomical spectra? Are they molecular in origin, and if so which molecules are responsible for them? How do they form?

Gamma ray bursts How do these short-duration high-intensity bursts originate?[2]

Supermassive black holes What is the origin of the M-sigma relation between supermassive black hole mass and galaxy velocity dispersion?[27] How did the most distant quasars grow their supermassive black holes up to 10^{10} solar masses so early in the history of the Universe?

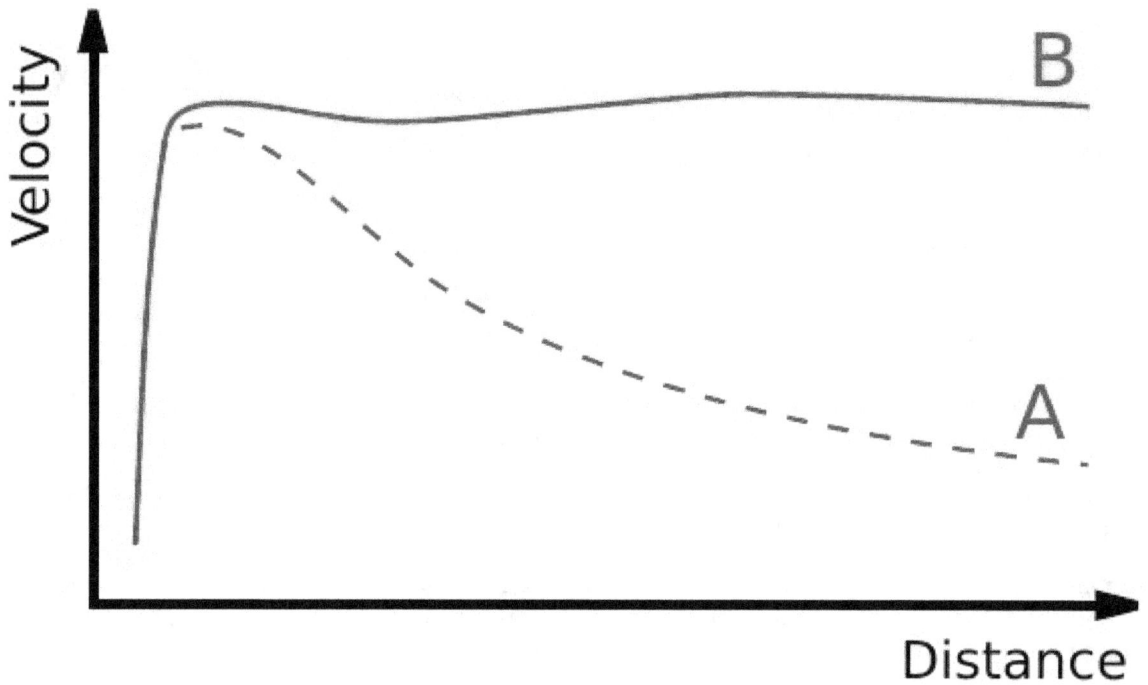

Rotation curve of a typical spiral galaxy: predicted (A) and observed (B). Can the discrepancy between the curves be attributed to dark matter?

Kuiper cliff Why does the number of objects in the Solar System's Kuiper belt fall off rapidly and unexpectedly beyond a radius of 50 astronomic units?

Flyby anomaly Why is the observed energy of satellites flying by Earth sometimes different by a minute amount from the value predicted by theory?

Galaxy rotation problem Is dark matter responsible for differences in observed and theoretical speed of stars revolving around the center of galaxies, or is it something else?

Supernovae What is the exact mechanism by which an implosion of a dying star becomes an explosion?

Three-body problem Exact predictions for the positions of three (or more) bodies floating in space attracted by gravity.

Ultra-high-energy cosmic ray [7] Why is it that some cosmic rays appear to possess energies that are impossibly high (the so-called *OMG particle*), given that there are no sufficiently energetic cosmic ray sources near the Earth? Why is it that (apparently) some cosmic rays emitted by distant sources have energies above the Greisen–Zatsepin–Kuzmin limit?[7][2]

Rotation rate of Saturn Why does the magnetosphere of Saturn exhibit a (slowly changing) periodicity close to that at which the planet's clouds rotate? What is the true rotation rate of Saturn's deep interior?[28]

Origin of magnetar magnetic field What is the origin of magnetar magnetic field?

Large-scale anisotropy Is the Universe at very large scales anisotropic, making the cosmological principle an invalid assumption? The number count and intensity dipole anisotropy in radio, NRAO VLA Sky Survey (NVSS) catalogue[29] is inconsistent with the local motion as derived from cosmic microwave background[30][31] and indicate an intrinsic dipole anisotropy. The same NVSS radio data also shows an intrinsic dipole in polarization density and degree of polarization[32] in the same direction as in number count and intensity. There are other several observation revealing large-scale anisotropy. The optical polarization from quasars shows polarization alignment over a very large scale of Gpc.[33][34][35] The cosmic-microwave-background data shows several features of anisotropy,[36][37][38][39] which are not consistent with the Big Bang model.

Photon underproduction crisis Why do galaxies and quasars produce about 5 times less ultraviolet light than expected in the low-redshift universe?

Space roar Why is space roar six times louder than expected? What is the source of space roar?

Age–metallicity relation in the Galactic disk Is there a universal age–metallicity relation (AMR) in the Galactic disk (both "thin" and "thick" parts of the disk)? Although in the local (primarily thin) disk of the Milky Way there is no evidence of a strong AMR,[40] a sample of 229 nearby "thick" disk stars has been used to investigate the existence of an age–metallicity relation in the Galactic thick disk, and indicate that there is an age–metallicity relation present in the thick disk.[41][42]

The lithium problem Why is there a discrepancy between the amount of lithium-7 predicted to be produced in Big Bang nucleosynthesis and the amount observed in very old stars?[43]

Solar wind interaction with comets In 2007 the *Ulysses* spacecraft passed through the tail of comet C/2006 P1 (McNaught) and found surprising results concerning the interaction of the solar wind with the tail.

Ultraluminous pulsar The ultraluminous X-ray source M82 X-2 was thought to be a black hole, but in October 2014 data from NASA's space-based X-ray telescope NuStar indicated that M82 X-2 is a pulsar many times brighter than the Eddington limit.

The injection problem Fermi acceleration is thought to be the primary mechanism that accelerates astrophysical particles to high energy. However, it is unclear what mechanism causes those particles to initially have energies high enough for Fermi acceleration to work on them.[44]

5.1.6 Nuclear physics

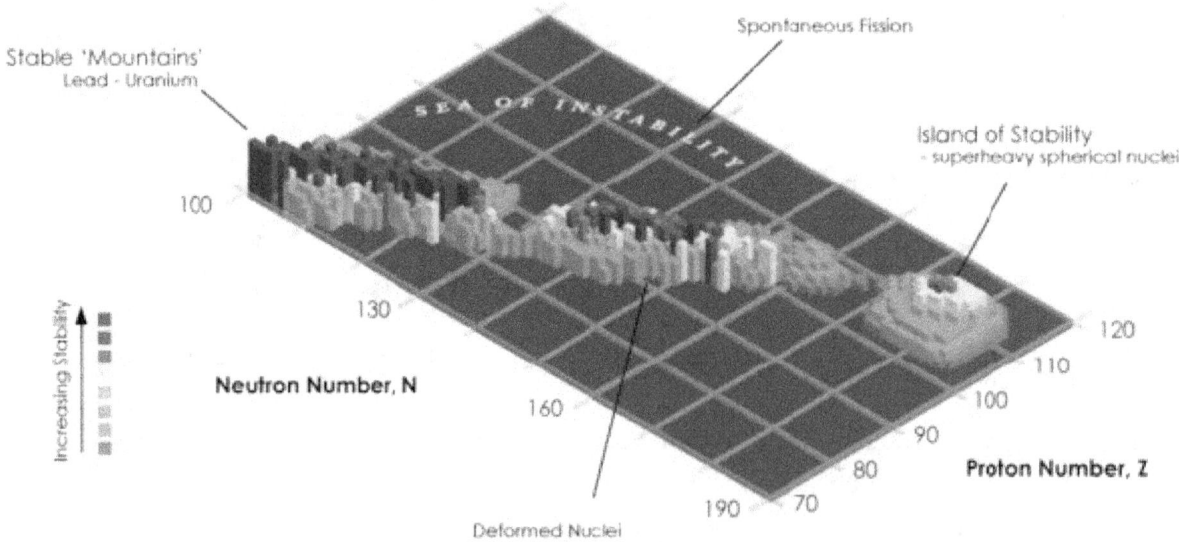

The "island of stability" in the proton vs. neutron number plot for heavy nuclei

Quantum chromodynamics What are the phases of strongly interacting matter, and what roles do they play in the evolution of cosmos? What is the detailed partonic structure of the nucleons? What does QCD predict for the properties of strongly interacting matter? What determines the key features of QCD, and what is their relation to the nature of gravity and spacetime? Do glueballs exist? Do gluons acquire mass dynamically despite having a zero rest mass, within hadrons? Does QCD truly lack CP-violations? Do gluons saturate when their occupation number is large? Do gluons form a dense system called Color Glass Condensate? What are the signatures and evidences for the Balitsky-Fadin-Kuarev-Lipatov, Balitsky-Kovchegov, Catani-Ciafaloni-Fiorani-Marchesini evolution equations?

Nuclei and nuclear astrophysics What is the nature of the nuclear force that binds protons and neutrons into stable nuclei and rare isotopes? What is the origin of simple patterns in complex nuclei? What is the nature of exotic excitations in nuclei at the frontiers of stability and their role in stellar processes? What is the nature of neutron stars and dense nuclear matter? What is the origin of the elements in the cosmos? What are the nuclear reactions that drive stars and stellar explosions?

Plasma physics and fusion power Fusion energy may potentially provide power from abundant resource (e.g. hydrogen) without the type of radioactive waste that fission energy currently produces. However, can ionized gases (plasma) be confined long enough and at a high enough temperature to create fusion power? What kinds of advances in material science must be made?

5.1.7   Atomic, molecular and optical physics

Hydrogen atom What is the solution to the Schrödinger equation for the hydrogen atom in arbitrary electric and magnetic fields?[45]

Abraham–Minkowski controversy What's the momentum of photons in optical media?

5.1.8   Condensed matter physics

High-temperature superconductors What is the mechanism that causes certain materials to exhibit superconductivity at temperatures much higher than around 25 kelvin? Is it possible to make a material that is a superconductor at room temperature?[2]

Amorphous solids What is the nature of the glass transition between a fluid or regular solid and a glassy phase? What are the physical processes giving rise to the general properties of glasses and the glass transition?[46][47]

Cryogenic electron emission Why does the electron emission in the absence of light increase as the temperature of a photomultiplier is decreased?[48][49]

Sonoluminescence What causes the emission of short bursts of light from imploding bubbles in a liquid when excited by sound?[50][51]

Turbulence Is it possible to make a theoretical model to describe the statistics of a turbulent flow (in particular, its internal structures)?[2] Also, under what conditions do smooth solutions to the Navier–Stokes equations exist? This problem is also listed as one of the Millennium Prize Problems in mathematics.

Alfvénic turbulence In the solar wind and the turbulence in solar flares, coronal mass ejections, and magnetospheric substorms are major unsolved problems in space plasma physics.[52]

Topological order Is topological order stable at non-zero temperature? Equivalently, is it possible to have three-dimensional self-correcting quantum memory?[53]

Fractional Hall effect What mechanism explains the existence of the $\nu = 5/2$ state in the fractional quantum Hall effect? Does it describe quasiparticles with non-Abelian fractional statistics?[54]

Bose–Einstein condensation How do we rigorously prove the existence of Bose–Einstein condensates for general interacting systems?[55]

A sample of a cuprate superconductor (specifically BSCCO). The mechanism for superconductivity of these materials is unknown.

Liquid crystals Can the nematic to smectic (A) phase transition in liquid crystal states be characterized as a universal phase transition?[56][57]

Semiconductor nanocrystals What is the cause of the nonparabolicity of the energy-size dependence for the lowest optical absorption transition of quantum dots?[58]

Electronic band structure Why can band gaps not accurately be calculated?

5.1.9 Biophysics

Stochasticity and robustness to noise in gene expression How do genes govern our body, withstanding different external pressures and internal stochasticity? Certain models exist for genetic processes, but we are far from understanding the whole picture, in particular in development where gene expression must be tightly regulated.

Quantitative study of the immune system What are the quantitative properties of immune responses? What are the basic building blocks of immune system networks? What roles are played by stochasticity?

Homochirality What is the origin of the preponderance of specific enantiomers in biochemical systems?

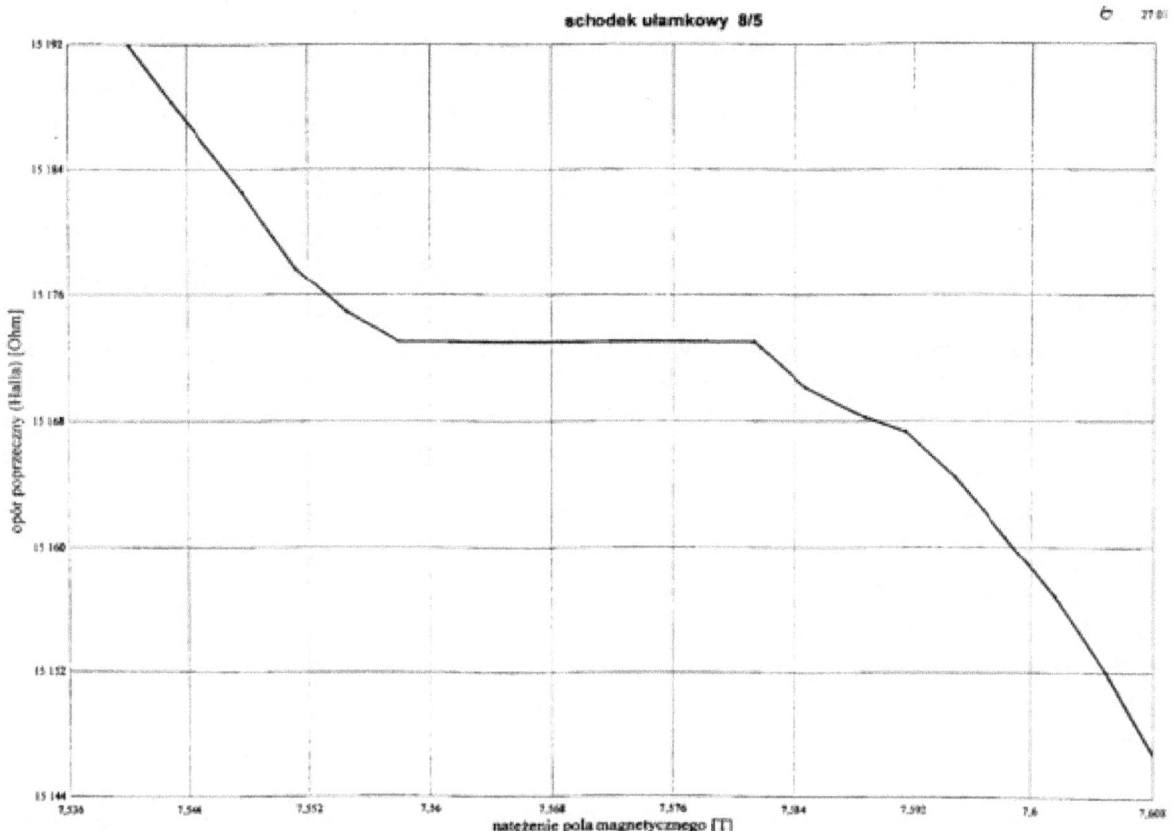

Magnetoresistance in a $\nu = 8/5$ fractional quantum Hall state.

5.2   Problems solved in recent decades

Ball lightning (2014) In January 2014, scientists from Northwest Normal University in Lanzhou, China, published the results of recordings made in July 2012 of the optical spectrum of what was thought to be natural ball lightning made during the study of ordinary cloud–ground lightning on China's Qinghai Plateau.[59][60] At a distance of 900 m (3,000 ft), a total of 1.3 seconds of digital video of the ball lightning and its spectrum was made, from the formation of the ball lightning after the ordinary lightning struck the ground, up to the optical decay of the phenomenon. The recorded ball lightning is believed to be vaporized soil elements that then rapidly oxidizes in the atmosphere. The nature of the true theory is still not clear.[60]

Electroweak symmetry breaking (2012) The mechanism responsible for breaking the electroweak gauge symmetry, giving mass to the W and Z bosons was solved with the discovery of the Higgs Boson of the Standard Model, with the expected couplings to the weak bosons. No evidence of a strong dynamics solution, as proposed by Technicolor, has been observed.

Hipparcos anomaly (2012) The actual distance to the Pleiades - the High Precision Parallax Collecting Satellite (Hipparcos) measured the parallax of the Pleiades and determined a distance of 385 light years. This was significantly different from other measurements made by means of actual to apparent brightness measurement or absolute magnitude. The anomaly was due to the use of a weighted mean when there is a correlation between distances and distance errors for stars in clusters. It is resolved by using an unweighted mean. There is no systematic bias in the Hipparcos data when it comes to star clusters.[61]

Pioneer anomaly (2012) There was a deviation in the predicted accelerations of the Pioneer spacecraft as they left the Solar System.[7][2] It is believed that this is a result of previously unaccounted-for thermal recoil force.[62][63]

Long-duration gamma ray bursts (2003) Long-duration bursts are associated with the deaths of massive stars in a specific kind of supernova-like event commonly referred to as a collapsar. However, there are also long-duration GRBs that show evidence against an associated supernova, such as the Swift event GRB 060614.

Solar neutrino problem (2002) Solved by a new understanding of neutrino physics, requiring a modification of the Standard Model of particle physics—specifically, neutrino oscillation.

Cosmic age problem (1990s) The estimated age of the universe was around 3 to 8 billion years younger than estimates of the ages of the oldest stars in the Milky Way. Better estimates for the distances to the stars, and the recognition of the accelerating expansion of the universe, reconciled the age estimates. (This assertion is being contested. Justification has been posted on Age Crisis page).

Quasars (1980s) The nature of quasars was not understood for decades.[64] They are now accepted as a type of active galaxy where the enormous energy output results from matter falling into a massive black hole in the center of the galaxy.[65]

5.3 See also

- Cosmic coincidence problem
- Universal Rotation Curve

5.4 References

[1] Ginzburg, Vitaly L. (2001). *The physics of a lifetime : reflections on the problems and personalities of 20th century physics.* Berlin: Springer. pp. 3–200. ISBN 9783540675341.

[2] Baez, John C. (March 2006). "Open Questions in Physics". *Usenet Physics FAQ.* University of California, Riverside: Department of Mathematics. Retrieved March 7, 2011.

[3] Nielson, Michael; Chuang, Isaac (2004). *Quantum Computation and Quantum Information.* Cambridge University Press. ISBN 978-0-521-63503-5.

[4] "Alcohol constrains physical constant in the early universe". *Phys Org.* December 13, 2012. Retrieved 25 March 2015.

[5] Bagdonaite, J.; Jansen, P.; Henkel, C.; Bethlem, H. L.; Menten, K. M.; Ubachs, W. (13 December 2012). "A Stringent Limit on a Drifting Proton-to-Electron Mass Ratio from Alcohol in the Early Universe".*Science*339(6115): 46–48.doi:10.1126/science.18.

[6] Podolsky, Dmitry. "Top ten open problems in physics". NEQNET. Archived from the original on 22 October 2012. Retrieved 24 January 2013.

[7] Brooks, Michael (March 19, 2005). "13 Things That Do Not Make Sense". *New Scientist.* Issue 2491. Retrieved March 7, 2011.

[8] R. Penrose (2007). *The Road to Reality.* Vintage books. ISBN 0-679-77631-1.

[9] National Research Council (1986). *Gravitation, Cosmology, and Cosmic-Ray Physics.* Washington, D. C.: National Academies Press. ISBN 0-309-03579-1.

[10] Paulson, Tom (May 27, 2002). "Catching a cosmic wave of gravity". *Seattle Post-Intelligencer.* Retrieved 10 April 2012.

[11] Steinhardt, P. and Turok, N. (2006). "Why the Cosmological constant is so small and positive". *Science* **312**: 1180–1183. arXiv:astro-ph/0605173. doi:10.1126/science.1126231.

[12] Alan Sokal (July 22, 1996). "Don't Pull the String Yet on Superstring Theory". *New York Times*

[13] Joshi, Pankaj S. (January 2009). "Do Naked Singularities Break the Rules of Physics?". *Scientific American*

[14] Dirac, Paul, "Quantised Singularities in the Electromagnetic Field". *Proceedings of the Royal Society* **A 133,** 60 (1931).

[15] Li, Tianjun; Dimitri V. Nanopoulos; Joel W. Walker (2011). "Elements of F-ast Proton Decay". *Nuclear Physics B* **846**: 43–99. arXiv:1003.2570. doi:10.1016/j.nuclphysb.2010.12.014.

[16] Hansson, Johan (2010). "The "Proton Spin Crisis" — a Quantum Query" (PDF). *Progress in Physics* **3**. Retrieved 14 April 2012.

[17] "India-based Neutrino Observatory (INO)". Tata Institute of Fundamental Research. Retrieved 14 April 2012.

[18] Smarandache, Vic; Florentin Smarandache (2007). "Thirty Unsolved Problems in the Physics of Elementary Particles" (PDF). *Progress in Physics* **4**. Bibcode:2009APS..HAW.KD010C.

[19] Nakamura (Particle Data Group), K; et al. (2010). "2011 Review of Particle Physics". *J. Phys.* G **37** (7A): 075021. Bibcode:2010JPhG...37g5021N. doi:10.1088/0954-3899/37/7A/075021.

[20] Thomas Blum; Achim Denig; Ivan Logashenko; Eduardo de Rafael; Lee Roberts, B.; Thomas Teubner; Graziano Venanzoni (2013). "The Muon (g-2) Theory Value: Present and Future". arXiv:1311.2198 [hep-ph].

[21] H. Muir (2 July 2003). "Pentaquark discovery confounds sceptics". *New Scientist*. Retrieved 2010-01-08.

[22] G. Amit (14 July 2015). "Pentaquark discovery at LHC shows long-sought new form of matter". *New Scientist*. Retrieved 2015-07-14.

[23] Laing, R. A.; Bridle, A. H. (2013). "Systematic properties of decelerating relativistic jets in low-luminosity radio galaxies". arXiv:1311.1015 [astro-ph.CO].

[24] Strohmayer, Tod E.; Mushotzky, Richard F. (20 March 2003). "Discovery of X-Ray Quasi-periodic Oscillations from an Ultra-luminous X-Ray Source in M82: Evidence against Beaming".*The Astrophysical Journal***586**(1): L61–L64.doi:10.1086/37472.

[25] Titarchuk, Lev; Fiorito, Ralph (10 September 2004). "Spectral Index and Quasi-Periodic Oscillation Frequency Correlation in Black Hole Sources: Observational Evidence of Two Phases and Phase Transition in Black Holes" (PDF). *The Astrophysical Journal* **612** (2): 988–999. doi:10.1086/422573. Retrieved 25 January 2013.

[26] Shoji Kato (2012). "An Attempt to Describe Frequency Correlations among kHz QPOs and HBOs by Two-Armed Nearly Vertical Oscillations". arXiv:1202.0121 [astro-ph.HE].

[27] Ferrarese, Laura; Merritt, David (2000). "A Fundamental Relation between Supermassive Black Holes and their Host Galaxies". *The Astrophysical Journal* **539**: L9–L12. arXiv:astro-ph/0006053. Bibcode:2000ApJ...539L...9F. doi:10.1086/312838

[28] "Scientists Find That Saturn's Rotation Period is a Puzzle". NASA. June 28, 2004. Retrieved 2007-03-22.

[29] J J Condon, W D Cotton, E W Greisen, Q F Yin, R. A. Perley, G. B. Taylor, and J J Broderick."The NRAO VLA Sky Survey"AJ, 115(5):1693-1716, May 1998.

[30] A. K. Singal. "Large Peculiar Motion of the Solar System from the Dipole Anisotropy in Sky Brightness due to Distant Radio Sources."ApJL, 742:L23, December 2011.

[31] Prabhakar Tiwari, Rahul Kothari, Abhishek Naskar, Sharvari Nadkarni-Ghosh, Pankaj Jain. "Dipole anisotropy in sky brightness and source count distribution in radio NVSS data".

[32] Prabhakar Tiwari and Pankaj Jain. "Dipole anisotropy in integrated linearly polarized flux density in NVSS data".

[33] D. Hutsemékers "Evidence for very large-scale coherent orientations of quasar polarization vectors"A&A, 332:410-428, 1998.

[34] D. Hutsemékers and H. Lamy"Confirmation of the existence of coherent orientations of quasar polarization vectors on cosmo-logical scales" A&A, 367(2):381-387, 2001.

[35] Pankaj Jain, Gaurav Narain, and S Sarala. "Large-scale alignment of optical polarizations from distant QSOs using coordinate-invariant statistics."MNRAS, 347(2):394-402, 2004.

[36] Angélica de Oliveira-Costa, Max Tegmark, Matias Zaldarriaga, and Andrew Hamilton (2004). Significance of the largest scale cmb fluctuations in wmap. PhRvD, 69:063516.

[37] H.K. Eriksen, F.K. Hansen, A.J. Banday, K.M. Gorski, and P.B. Lilje. "Asymmetries in the Cosmic Microwave Background anisotropy field."ApJ, 605:14–20, 2004.

[38] Pramoda Kumar Samal, Rajib Saha, Pankaj Jain, and John P. Ralston. "Testing Isotropy of Cosmic Microwave Background Radiation."MNRAS, 385:1718, 2008.

[39] Pramoda Kumar Samal, Rajib Saha, Pankaj Jain, and John P. Ralston."Signals of Statistical Anisotropy in WMAP Foreground-Cleaned Maps."MNRAS, 396:511, 2009.

[40] Casagrande, L., et al. (2011) "New constraints on the chemical evolution of the solar neighbourhood and Galactic disc(s)". Astronomy & Astrophysics, 2011, Volume 530, 138, 21 pp

[41] Bensby, T.; Feltzing, S.; Lundström, I. (July 2004). "A possible age-metallicity relation in the Galactic thick disk?". Astronomy and Astrophysics 421 (3): 969–976. doi:10.1051/0004-6361:20035957.

[42] Gilmore, G.; Asiri, H.M. (00/2011). "Open Issues in the Evolution of the Galactic Disks". Workshop on Gaia. Proceedings. Granada. ed. Navarro et al. 2011. Retrieved 2013-09-08.

[43] Brian D. Fields. The Primordial Lithium Problem

[44] André Balogh; Rudolf A. Treumann. "Physics of Collisionless Shocks: Space Plasma Shock Waves". 2013. Section 7.4 "The Injection Problem". p. 362.

[45] Panel on Atomic, Molecular, and Optical Physics, Physics Survey Committee, Board on Physics and Astronomy, National Research Council (1986). Atomic, Molecular, and Optical Physics. National Academies Press. p. 63. ISBN 9780309594561.

[46] Kenneth Chang (July 29, 2008). "The Nature of Glass Remains Anything but Clear". The New York Times

[47] P.W. Anderson (1995). "Through the Glass Lightly". Science 267 (5204): 1615. doi:10.1126/science.267.5204.1615-e. The deepest and most interesting unsolved problem in solid state theory is probably the theory of the nature of glass and the glass transition.

[48] Cryogenic electron emission phenomenon has no known physics explanation. Physorg.com. Retrieved on 2011-10-20.

[49] Meyer, H. O. (1 March 2010). "Spontaneous electron emission from a cold surface". EPL (Europhysics Letters) 89 (5): 58001. doi:10.1209/0295-5075/89/58001.

[50] Storey, B. D.; Szeri, A. J. (8 July 2000). "Water vapour, sonoluminescence and sonochemistry". Proceedings of the Royal Society A: Mathematical, Physical and Engineering Sciences 456 (1999): 1685–1709. doi:10.1098/rspa.2000.0582.

[51] Wu, C. C.; Roberts, P. H. (9 May 1994). "A Model of Sonoluminescence". Proceedings of the Royal Society A: Mathematical, Physical and Engineering Sciences 445 (1924): 323–349. doi:10.1098/rspa.1994.0064.

[52] Goldstein, Melvyn L. (2001). "Major Unsolved Problems in Space Plasma Physics". Astrophysics and Space Science 277 (1/2): 349–369. Bibcode:2001Ap&SS.277..349G. doi:10.1023/A:1012264131485.

[53] Yoshida, Beni (2011). "Feasibility of self-correcting quantum memory and thermal stability of topological order". Annals of Physics 326 (10): 2566. arXiv:1103.1885. Bibcode:2011AnPhy.326.2566Y. doi:10.1016/j.aop.2011.06.001. Retrieved 8 April 2012.

[54] Podolsky, Dmitry. "Quantum Hall effect. One open question". NEQNET. Retrieved 23 April 2012.

[55] Schlein, Benjamin. "Graduate Seminar on Partial Differential Equations in the Sciences - Energy and Dynamics of Boson Systems". Hausdorff Center for Mathematics. Retrieved 23 April 2012.

[56] Mukherjee, Prabir K. (1998). "Landau Theory of Nematic-Smectic-A Transition in a Liquid Crystal Mixture". Molecular Crystals & Liquid Crystals 312: 157–164. doi:10.1080/10587259808042438. Retrieved 28 April 2012.

[57] A. Yethiraj. "Recent Experimental Developments at the Nematic to Smectic-A Liquid Crystal Phase Transition". Thermotropic Liquid Crystals: Recent Advances, ed. A. Ramamoorthy, Springer 2007, chapter 8.

[58] Norris, David J. (2003). "The Problem Swept Under the Rug". In Klimov, Victor. Electronic Structure in Semiconductors Nanocrystals: Optical Experiment (in Semiconductor and Metal Nanocrystals: Synthesis and Electronic and Optical Properties). CRC Press. p. 97. ISBN 9780203913260.

[59] Cen, Jianyong; Yuan, Ping; Xue, Simin (17 January 2014). "Observation of the Optical and Spectral Characteristics of Ball Lightning". Physical Review Letters (American Physical Society) 112 (35001). doi:10.1103/PhysRevLett.112.035001. Retrieved 19 January 2014.

[60] Ball, Philip (17 January 2014). "Focus: First Spectrum of Ball Lightning". *Focus* (American Physical Society). doi:10.110. Retrieved 19 January 2014.

[61] Charles Francis; Erik Anderson (2012). "XHIP-II: Clusters and associations". arXiv:1203.4945 [astro-ph.GA].

[62] Turyshev, S.; Toth, V.; Kinsella, G.; Lee, S. C.; Lok, S.; Ellis, J. (2012). "Support for the Thermal Origin of the Pioneer Anomaly". *Physical Review Letters* **108** (24): 241101. arXiv:1204.2507. Bibcode:2012PhRvL.108x1101T. doi:10.1103. PMID 23004253.

[63] Overbye, Dennis (23 July 2012). "Mystery Tug on Spacecraft Is Einstein's 'I Told You So'". *The New York Times*. Retrieved 24 January 2014.

[64] "The MKI and the discovery of Quasars". Jodrell Bank Observatory. Retrieved 2006-11-23.

[65] Hubble Surveys the "Homes" of Quasars Hubblesite News Archive, 1996-35

5.5   External links

- What don't we know? Science journal special project for its 125th anniversary: top 25 questions and 100 more.

- List of links to unsolved problems in physics, prizes and research.

- Ideas Based On What We'd Like to Achieve

- 2004 SLAC Summer Institute: Nature's Greatest Puzzles

- Dual Personality of Glass Explained at Last

- What we do and don't know Review on current state of physics by Steven Weinberg, November 2013

- The crisis of big science Steven Weinberg, May 2012

- The 10 Biggest Unsolved Problems in Physics Johan Hansson, March 2015

Chapter 6

General relativity

For the book by Robert Wald, see General Relativity (book).

For a more accessible and less technical introduction to this topic, see Introduction to general relativity.

General relativity, also known as the **general theory of relativity**, is the geometric theory of gravitation published

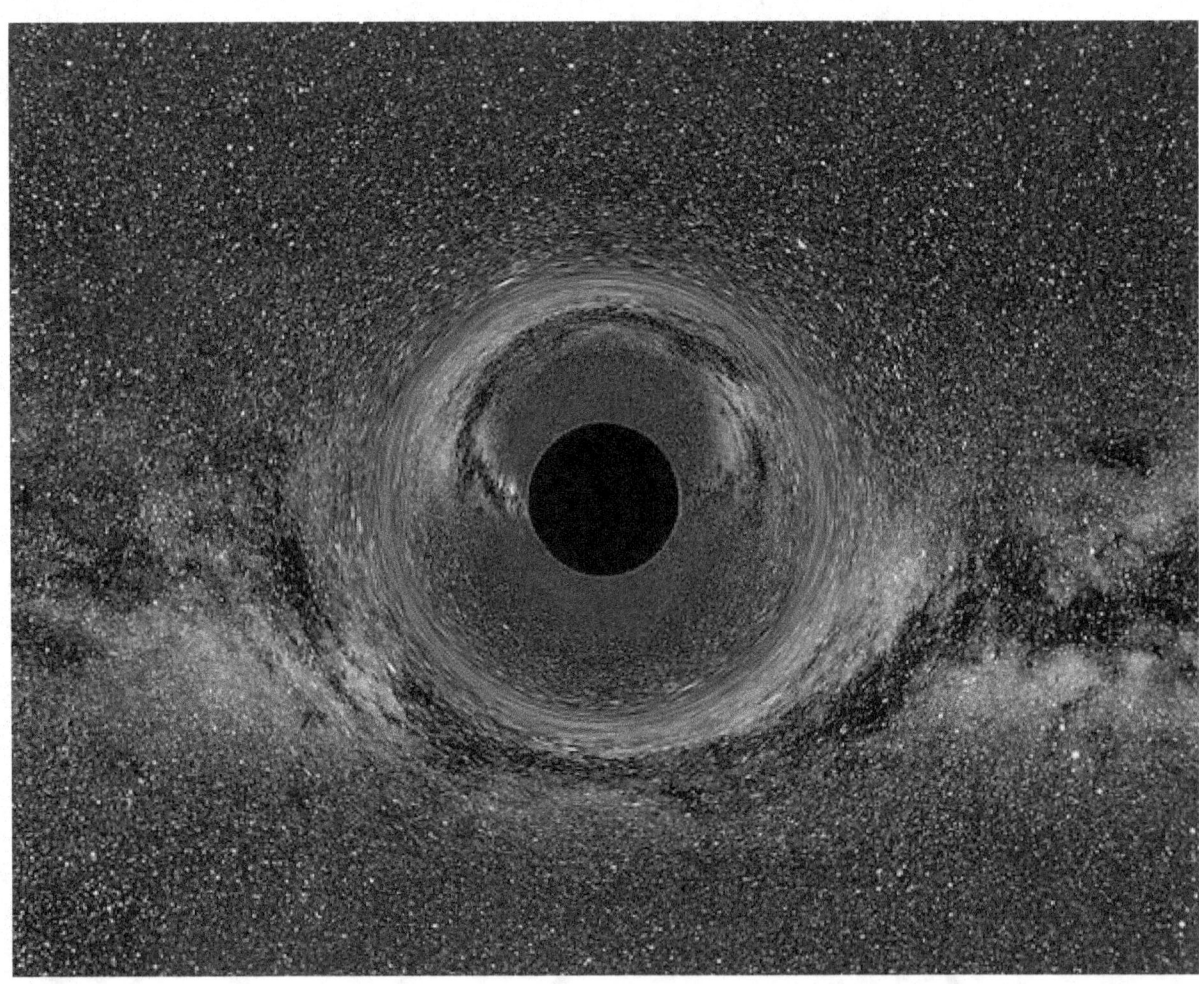

A simulated black hole of 10 solar masses within the Milky Way, seen from a distance of 600 kilometers.

by Albert Einstein in 1915[1] and the current description of gravitation in modern physics. General relativity generalizes special relativity and Newton's law of universal gravitation, providing a unified description of gravity as a geometric property of space and time, or spacetime. In particular, the curvature of spacetime is directly related to the energy and momentum of whatever matter and radiation are present. The relation is specified by the Einstein field equations, a system of partial differential equations.

Some predictions of general relativity differ significantly from those of classical physics, especially concerning the passage of time, the geometry of space, the motion of bodies in free fall, and the propagation of light. Examples of such differences include gravitational time dilation, gravitational lensing, the gravitational redshift of light, and the gravitational time delay. The predictions of general relativity have been confirmed in all observations and experiments to date. Although general relativity is not the only relativistic theory of gravity, it is the simplest theory that is consistent with experimental data. However, unanswered questions remain, the most fundamental being how general relativity can be reconciled with the laws of quantum physics to produce a complete and self-consistent theory of quantum gravity.

Einstein's theory has important astrophysical implications. For example, it implies the existence of black holes—regions of space in which space and time are distorted in such a way that nothing, not even light, can escape—as an end-state for massive stars. There is ample evidence that the intense radiation emitted by certain kinds of astronomical objects is due to black holes; for example, microquasars and active galactic nuclei result from the presence of stellar black holes and black holes of a much more massive type, respectively. The bending of light by gravity can lead to the phenomenon of gravitational lensing, in which multiple images of the same distant astronomical object are visible in the sky. General relativity also predicts the existence of gravitational waves, which have since been observed indirectly; a direct measurement is the aim of projects such as LIGO and NASA/ESA Laser Interferometer Space Antenna and various pulsar timing arrays. In addition, general relativity is the basis of current cosmological models of a consistently expanding universe.

6.1 History

Main articles: History of general relativity and Classical theories of gravitation
 Soon after publishing the special theory of relativity in 1905, Einstein started thinking about how to incorporate gravity into his new relativistic framework. In 1907, beginning with a simple thought experiment involving an observer in free fall, he embarked on what would be an eight-year search for a relativistic theory of gravity. After numerous detours and false starts, his work culminated in the presentation to the Prussian Academy of Science in November 1915 of what are now known as the Einstein field equations. These equations specify how the geometry of space and time is influenced by whatever matter and radiation are present, and form the core of Einstein's general theory of relativity.[2]

The Einstein field equations are nonlinear and very difficult to solve. Einstein used approximation methods in working out initial predictions of the theory. But as early as 1916, the astrophysicist Karl Schwarzschild found the first non-trivial exact solution to the Einstein field equations, the so-called Schwarzschild metric. This solution laid the groundwork for the description of the final stages of gravitational collapse, and the objects known today as black holes. In the same year, the first steps towards generalizing Schwarzschild's solution to electrically charged objects were taken, which eventually resulted in the Reissner–Nordström solution, now associated with electrically charged black holes.[3] In 1917, Einstein applied his theory to the universe as a whole, initiating the field of relativistic cosmology. In line with contemporary thinking, he assumed a static universe, adding a new parameter to his original field equations—the cosmological constant— to match that observational presumption.[4] By 1929, however, the work of Hubble and others had shown that our universe is expanding. This is readily described by the expanding cosmological solutions found by Friedmann in 1922, which do not require a cosmological constant. Lemaître used these solutions to formulate the earliest version of the Big Bang models, in which our universe has evolved from an extremely hot and dense earlier state.[5] Einstein later declared the cosmological constant the biggest blunder of his life.[6]

During that period, general relativity remained something of a curiosity among physical theories. It was clearly superior to Newtonian gravity, being consistent with special relativity and accounting for several effects unexplained by the Newtonian theory. Einstein himself had shown in 1915 how his theory explained the anomalous perihelion advance of the planet Mercury without any arbitrary parameters ("fudge factors").[7] Similarly, a 1919 expedition led by Eddington confirmed general relativity's prediction for the deflection of starlight by the Sun during the total solar eclipse of May 29, 1919,[8] making Einstein instantly famous.[9] Yet the theory entered the mainstream of theoretical physics and astrophysics only with the developments between approximately 1960 and 1975, now known as the golden age of general relativity.[10] Physicists began to understand the concept of a black hole, and to identify quasars as one of these objects' astrophysical manifestations.[11] Ever more precise solar system tests confirmed the theory's predictive power,[12] and relativistic

Albert Einstein developed the theories of special and general relativity. Picture from 1921.

cosmology, too, became amenable to direct observational tests.[13]

6.2   From classical mechanics to general relativity

General relativity can be understood by examining its similarities with and departures from classical physics. The first step is the realization that classical mechanics and Newton's law of gravity admit a geometric description. The combination of this description with the laws of special relativity results in a heuristic derivation of general relativity.[14]

6.2.1   Geometry of Newtonian gravity

According to general relativity, objects in a gravitational field behave similarly to objects within an accelerating enclosure. For example, an observer will see a ball fall the same way in a rocket (left) as it does on Earth (right), provided that the acceleration of the rocket is equal to 9.8 m/s² (the acceleration due to gravity at the surface of the Earth).

At the base of classical mechanics is the notion that a body's motion can be described as a combination of free (or inertial) motion, and deviations from this free motion. Such deviations are caused by external forces acting on a body in accordance with Newton's second law of motion, which states that the net force acting on a body is equal to that body's (inertial) mass multiplied by its acceleration.[15] The preferred inertial motions are related to the geometry of space and time: in the standard reference frames of classical mechanics, objects in free motion move along straight lines at constant speed. In modern parlance, their paths are geodesics, straight world lines in curved spacetime.[16]

Conversely, one might expect that inertial motions, once identified by observing the actual motions of bodies and making allowances for the external forces (such as electromagnetism or friction), can be used to define the geometry of space, as well as a time coordinate. However, there is an ambiguity once gravity comes into play. According to Newton's law of gravity, and independently verified by experiments such as that of Eötvös and its successors (see Eötvös experiment), there is a universality of free fall (also known as the weak equivalence principle, or the universal equality of inertial and passive-gravitational mass): the trajectory of a test body in free fall depends only on its position and initial speed, but not on any of its material properties.[17] A simplified version of this is embodied in Einstein's elevator experiment, illustrated in the figure on the right: for an observer in a small enclosed room, it is impossible to decide, by mapping the trajectory of bodies such as a dropped ball, whether the room is at rest in a gravitational field, or in free space aboard a rocket that is accelerating at a rate equal to that of the gravitational field.[18]

Given the universality of free fall, there is no observable distinction between inertial motion and motion under the influence of the gravitational force. This suggests the definition of a new class of inertial motion, namely that of objects in free fall under the influence of gravity. This new class of preferred motions, too, defines a geometry of space and time—in mathematical terms, it is the geodesic motion associated with a specific connection which depends on the gradient of the gravitational potential. Space, in this construction, still has the ordinary Euclidean geometry. However, space*time* as a whole is more complicated. As can be shown using simple thought experiments following the free-fall trajectories of different test particles, the result of transporting spacetime vectors that can denote a particle's velocity (time-like vectors) will vary with the particle's trajectory; mathematically speaking, the Newtonian connection is not integrable. From this, one can deduce that spacetime is curved. The result is a geometric formulation of Newtonian gravity using only covariant concepts, i.e. a description which is valid in any desired coordinate system.[19] In this geometric description, tidal effects— the relative acceleration of bodies in free fall—are related to the derivative of the connection, showing how the modified geometry is caused by the presence of mass.[20]

6.2.2 Relativistic generalization

As intriguing as geometric Newtonian gravity may be, its basis, classical mechanics, is merely a limiting case of (special) relativistic mechanics.[21] In the language of symmetry: where gravity can be neglected, physics is Lorentz invariant as in special relativity rather than Galilei invariant as in classical mechanics. (The defining symmetry of special relativity is the Poincaré group, which includes translations and rotations.) The differences between the two become significant when dealing with speeds approaching the speed of light, and with high-energy phenomena.[22]

With Lorentz symmetry, additional structures come into play. They are defined by the set of light cones (see image). The light-cones define a causal structure: for each event A, there is a set of events that can, in principle, either influence or be influenced by A via signals or interactions that do not need to travel faster than light (such as event B in the image), and a set of events for which such an influence is impossible (such as event C in the image). These sets are observer-independent.[23] In conjunction with the world-lines of freely falling particles, the light-cones can be used to reconstruct the space–time's semi-Riemannian metric, at least up to a positive scalar factor. In mathematical terms, this defines a conformal structure.[24]

Special relativity is defined in the absence of gravity, so for practical applications, it is a suitable model whenever gravity can be neglected. Bringing gravity into play, and assuming the universality of free fall, an analogous reasoning as in the previous section applies: there are no global inertial frames. Instead there are approximate inertial frames moving alongside freely falling particles. Translated into the language of spacetime: the straight time-like lines that define a gravity-free inertial frame are deformed to lines that are curved relative to each other, suggesting that the inclusion of gravity necessitates a change in spacetime geometry.[25]

A priori, it is not clear whether the new local frames in free fall coincide with the reference frames in which the laws of special relativity hold—that theory is based on the propagation of light, and thus on electromagnetism, which could have a different set of preferred frames. But using different assumptions about the special-relativistic frames (such as their being earth-fixed, or in free fall), one can derive different predictions for the gravitational redshift, that is, the way in which the frequency of light shifts as the light propagates through a gravitational field (cf. below). The actual measurements show that free-falling frames are the ones in which light propagates as it does in special relativity.[26] The generalization of this statement, namely that the laws of special relativity hold to good approximation in freely falling (and non-rotating) reference frames, is known as the Einstein equivalence principle, a crucial guiding principle for generalizing special-relativistic physics to include gravity.[27]

The same experimental data shows that time as measured by clocks in a gravitational field—proper time, to give the technical term—does not follow the rules of special relativity. In the language of spacetime geometry, it is not measured by the Minkowski metric. As in the Newtonian case, this is suggestive of a more general geometry. At small scales, all reference frames that are in free fall are equivalent, and approximately Minkowskian. Consequently, we are now dealing with a curved generalization of Minkowski space. The metric tensor that defines the geometry—in particular, how lengths and angles are measured—is not the Minkowski metric of special relativity, it is a generalization known as a semi- or pseudo-Riemannian metric. Furthermore, each Riemannian metric is naturally associated with one particular kind of connection, the Levi-Civita connection, and this is, in fact, the connection that satisfies the equivalence principle and makes space locally Minkowskian (that is, in suitable locally inertial coordinates, the metric is Minkowskian, and its first partial derivatives and the connection coefficients vanish).[28]

6.2.3   Einstein's equations

Main articles: Einstein field equations and Mathematics of general relativity

Having formulated the relativistic, geometric version of the effects of gravity, the question of gravity's source remains. In Newtonian gravity, the source is mass. In special relativity, mass turns out to be part of a more general quantity called the energy–momentum tensor, which includes both energy and momentum densities as well as stress (that is, pressure and shear).[29] Using the equivalence principle, this tensor is readily generalized to curved space-time. Drawing further upon the analogy with geometric Newtonian gravity, it is natural to assume that the field equation for gravity relates this tensor and the Ricci tensor, which describes a particular class of tidal effects: the change in volume for a small cloud of test particles that are initially at rest, and then fall freely. In special relativity, conservation of energy–momentum corresponds to the statement that the energy–momentum tensor is divergence-free. This formula, too, is readily generalized to curved spacetime by replacing partial derivatives with their curved-manifold counterparts, covariant derivatives studied in differential geometry. With this additional condition—the covariant divergence of the energy–momentum tensor, and hence of whatever is on the other side of the equation, is zero— the simplest set of equations are what are called Einstein's (field) equations:

On the left-hand side is the Einstein tensor, a specific divergence-free combination of the Ricci tensor $R_{\mu\nu}$ and the metric. Where $G_{\mu\nu}$ is symmetric. In particular,

$$R = g^{\mu\nu} R_{\mu\nu}$$

is the curvature scalar. The Ricci tensor itself is related to the more general Riemann curvature tensor as

$$R_{\mu\nu} = R^{\alpha}{}_{\mu\alpha\nu}.$$

On the right-hand side, $T_{\mu\nu}$ is the energy–momentum tensor. All tensors are written in abstract index notation.[30] Matching the theory's prediction to observational results for planetary orbits (or, equivalently, assuring that the weak-gravity, low-speed limit is Newtonian mechanics), the proportionality constant can be fixed as $\kappa = 8\pi G/c^4$, with G the gravitational constant and c the speed of light.[31] When there is no matter present, so that the energy–momentum tensor vanishes, the results are the vacuum Einstein equations,

$$R_{\mu\nu} = 0.$$

There are alternatives to general relativity built upon the same premises, which include additional rules and/or constraints, leading to different field equations. Examples are Brans–Dicke theory, teleparallelism, and Einstein–Cartan theory.[32]

6.3   Definition and basic applications

See also: Mathematics of general relativity and Physical theories modified by general relativity

The derivation outlined in the previous section contains all the information needed to define general relativity, describe its key properties, and address a question of crucial importance in physics, namely how the theory can be used for model-building.

6.3.1 Definition and basic properties

General relativity is a metric theory of gravitation. At its core are Einstein's equations, which describe the relation between the geometry of a four-dimensional, pseudo-Riemannian manifold representing spacetime, and the energy–momentum contained in that spacetime.[33] Phenomena that in classical mechanics are ascribed to the action of the force of gravity (such as free-fall, orbital motion, and spacecraft trajectories), correspond to inertial motion within a curved geometry of spacetime in general relativity; there is no gravitational force deflecting objects from their natural, straight paths. Instead, gravity corresponds to changes in the properties of space and time, which in turn changes the straightest-possible paths that objects will naturally follow.[34] The curvature is, in turn, caused by the energy–momentum of matter. Paraphrasing the relativist John Archibald Wheeler, spacetime tells matter how to move; matter tells spacetime how to curve.[35]

While general relativity replaces the scalar gravitational potential of classical physics by a symmetric rank-two tensor, the latter reduces to the former in certain limiting cases. For weak gravitational fields and slow speed relative to the speed of light, the theory's predictions converge on those of Newton's law of universal gravitation.[36]

As it is constructed using tensors, general relativity exhibits general covariance: its laws—and further laws formulated within the general relativistic framework—take on the same form in all coordinate systems.[37] Furthermore, the theory does not contain any invariant geometric background structures, i.e. it is background independent. It thus satisfies a more stringent general principle of relativity, namely that the laws of physics are the same for all observers.[38] Locally, as expressed in the equivalence principle, spacetime is Minkowskian, and the laws of physics exhibit local Lorentz invariance.[39]

6.3.2 Model-building

The core concept of general-relativistic model-building is that of a solution of Einstein's equations. Given both Einstein's equations and suitable equations for the properties of matter, such a solution consists of a specific semi-Riemannian manifold (usually defined by giving the metric in specific coordinates), and specific matter fields defined on that manifold. Matter and geometry must satisfy Einstein's equations, so in particular, the matter's energy–momentum tensor must be divergence-free. The matter must, of course, also satisfy whatever additional equations were imposed on its properties. In short, such a solution is a model universe that satisfies the laws of general relativity, and possibly additional laws governing whatever matter might be present.[40]

Einstein's equations are nonlinear partial differential equations and, as such, difficult to solve exactly.[41] Nevertheless, a number of exact solutions are known, although only a few have direct physical applications.[42] The best-known exact solutions, and also those most interesting from a physics point of view, are the Schwarzschild solution, the Reissner–Nordström solution and the Kerr metric, each corresponding to a certain type of black hole in an otherwise empty universe,[43] and the Friedmann–Lemaître–Robertson–Walker and de Sitter universes, each describing an expanding cosmos.[44] Exact solutions of great theoretical interest include the Gödel universe (which opens up the intriguing possibility of time travel in curved spacetimes), the Taub-NUT solution (a model universe that is homogeneous, but anisotropic), and anti-de Sitter space (which has recently come to prominence in the context of what is called the Maldacena conjecture).[45]

Given the difficulty of finding exact solutions, Einstein's field equations are also solved frequently by numerical integration on a computer, or by considering small perturbations of exact solutions. In the field of numerical relativity, powerful computers are employed to simulate the geometry of spacetime and to solve Einstein's equations for interesting situations such as two colliding black holes.[46] In principle, such methods may be applied to any system, given sufficient computer resources, and may address fundamental questions such as naked singularities. Approximate solutions may also be found by perturbation theories such as linearized gravity[47] and its generalization, the post-Newtonian expansion, both of which were developed by Einstein. The latter provides a systematic approach to solving for the geometry of a spacetime that contains a distribution of matter that moves slowly compared with the speed of light. The expansion involves a series of terms; the first terms represent Newtonian gravity, whereas the later terms represent ever smaller corrections to Newton's theory due to general relativity.[48] An extension of this expansion is the parametrized post-Newtonian (PPN) formalism, which allows quantitative comparisons between the predictions of general relativity and alternative theories.[49]

6.4   Consequences of Einstein's theory

General relativity has a number of physical consequences. Some follow directly from the theory's axioms, whereas others have become clear only in the course of many years of research that followed Einstein's initial publication.

6.4.1   Gravitational time dilation and frequency shift

Main article: Gravitational time dilation

Assuming that the equivalence principle holds,[50] gravity influences the passage of time. Light sent down into a gravity well is blueshifted, whereas light sent in the opposite direction (i.e., climbing out of the gravity well) is redshifted; collectively, these two effects are known as the gravitational frequency shift. More generally, processes close to a massive body run more slowly when compared with processes taking place farther away; this effect is known as gravitational time dilation.[51]

Gravitational redshift has been measured in the laboratory[52] and using astronomical observations.[53] Gravitational time dilation in the Earth's gravitational field has been measured numerous times using atomic clocks,[54] while ongoing validation is provided as a side effect of the operation of the Global Positioning System (GPS).[55] Tests in stronger gravitational fields are provided by the observation of binary pulsars.[56] All results are in agreement with general relativity.[57] However, at the current level of accuracy, these observations cannot distinguish between general relativity and other theories in which the equivalence principle is valid.[58]

6.4.2   Light deflection and gravitational time delay

Main articles: Kepler problem in general relativity, Gravitational lens and Shapiro delay

General relativity predicts that the path of light is bent in a gravitational field; light passing a massive body is deflected towards that body. This effect has been confirmed by observing the light of stars or distant quasars being deflected as it passes the Sun.[59]

This and related predictions follow from the fact that light follows what is called a light-like or null geodesic—a generalization of the straight lines along which light travels in classical physics. Such geodesics are the generalization of the invariance of lightspeed in special relativity.[60] As one examines suitable model spacetimes (either the exterior Schwarzschild solution or, for more than a single mass, the post-Newtonian expansion),[61] several effects of gravity on light propagation emerge. Although the bending of light can also be derived by extending the universality of free fall to light,[62] the angle of deflection resulting from such calculations is only half the value given by general relativity.[63]

Closely related to light deflection is the gravitational time delay (or Shapiro delay), the phenomenon that light signals take longer to move through a gravitational field than they would in the absence of that field. There have been numerous successful tests of this prediction.[64] In the parameterized post-Newtonian formalism (PPN), measurements of both the deflection of light and the gravitational time delay determine a parameter called γ, which encodes the influence of gravity on the geometry of space.[65]

6.4.3   Gravitational waves

Main article: Gravitational wave

One of several analogies between weak-field gravity and electromagnetism is that, analogous to electromagnetic waves, there are gravitational waves: ripples in the metric of spacetime that propagate at the speed of light.[66] The simplest type of such a wave can be visualized by its action on a ring of freely floating particles. A sine wave propagating through such a ring towards the reader distorts the ring in a characteristic, rhythmic fashion (animated image to the right).[67] Since Einstein's equations are non-linear, arbitrarily strong gravitational waves do not obey linear superposition, making their description difficult. However, for weak fields, a linear approximation can be made. Such linearized gravitational waves are sufficiently accurate to describe the exceedingly weak waves that are expected to arrive here on Earth from far-off cosmic events, which typically result in relative distances increasing and decreasing by 10^{-21} or less. Data analysis methods routinely make use of the fact that these linearized waves can be Fourier decomposed.[68]

Some exact solutions describe gravitational waves without any approximation, e.g., a wave train traveling through empty space[69] or so-called Gowdy universes, varieties of an expanding cosmos filled with gravitational waves.[70] But for gravitational waves produced in astrophysically relevant situations, such as the merger of two black holes, numerical methods are presently the only way to construct appropriate models.[71]

6.4.4 Orbital effects and the relativity of direction

Main article: Kepler problem in general relativity

General relativity differs from classical mechanics in a number of predictions concerning orbiting bodies. It predicts an overall rotation (precession) of planetary orbits, as well as orbital decay caused by the emission of gravitational waves and effects related to the relativity of direction.

Precession of apsides

In general relativity, the apsides of any orbit (the point of the orbiting body's closest approach to the system's center of mass) will precess—the orbit is not an ellipse, but akin to an ellipse that rotates on its focus, resulting in a rose curve-like shape (see image). Einstein first derived this result by using an approximate metric representing the Newtonian limit and treating the orbiting body as a test particle. For him, the fact that his theory gave a straightforward explanation of the anomalous perihelion shift of the planet Mercury, discovered earlier by Urbain Le Verrier in 1859, was important evidence that he had at last identified the correct form of the gravitational field equations.[72]

The effect can also be derived by using either the exact Schwarzschild metric (describing spacetime around a spherical mass)[73] or the much more general post-Newtonian formalism.[74] It is due to the influence of gravity on the geometry of space and to the contribution of self-energy to a body's gravity (encoded in the nonlinearity of Einstein's equations).[75] Relativistic precession has been observed for all planets that allow for accurate precession measurements (Mercury, Venus, and Earth),[76] as well as in binary pulsar systems, where it is larger by five orders of magnitude.[77]

Orbital decay

According to general relativity, a binary system will emit gravitational waves, thereby losing energy. Due to this loss, the distance between the two orbiting bodies decreases, and so does their orbital period. Within the Solar System or for ordinary double stars, the effect is too small to be observable. This is not the case for a close binary pulsar, a system of two orbiting neutron stars, one of which is a pulsar: from the pulsar, observers on Earth receive a regular series of radio pulses that can serve as a highly accurate clock, which allows precise measurements of the orbital period. Because neutron stars are very compact, significant amounts of energy are emitted in the form of gravitational radiation.[79]

The first observation of a decrease in orbital period due to the emission of gravitational waves was made by Hulse and Taylor, using the binary pulsar PSR1913+16 they had discovered in 1974. This was the first detection of gravitational waves, albeit indirect, for which they were awarded the 1993 Nobel Prize in physics.[80] Since then, several other binary pulsars have been found, in particular the double pulsar PSR J0737-3039, in which both stars are pulsars.[81]

Geodetic precession and frame-dragging

Main articles: Geodetic precession and Frame dragging

Several relativistic effects are directly related to the relativity of direction.[82] One is geodetic precession: the axis direction of a gyroscope in free fall in curved spacetime will change when compared, for instance, with the direction of light received from distant stars—even though such a gyroscope represents the way of keeping a direction as stable as possible ("parallel transport").[83] For the Moon–Earth system, this effect has been measured with the help of lunar laser ranging.[84] More recently, it has been measured for test masses aboard the satellite Gravity Probe B to a precision of better than 0.3%.[85][86]

Near a rotating mass, there are so-called gravitomagnetic or frame-dragging effects. A distant observer will determine that objects close to the mass get "dragged around". This is most extreme for rotating black holes where, for any object entering a zone known as the ergosphere, rotation is inevitable.[87] Such effects can again be tested through their influence on the orientation of gyroscopes in free fall.[88] Somewhat controversial tests have been performed using the LAGEOS satellites, confirming the relativistic prediction.[89] Also the Mars Global Surveyor probe around Mars has been used.[90][91]

6.5   Astrophysical applications

6.5.1   Gravitational lensing

Main article: Gravitational lensing
The deflection of light by gravity is responsible for a new class of astronomical phenomena. If a massive object is situated between the astronomer and a distant target object with appropriate mass and relative distances, the astronomer will see multiple distorted images of the target. Such effects are known as gravitational lensing.[92] Depending on the configuration, scale, and mass distribution, there can be two or more images, a bright ring known as an Einstein ring, or partial rings called arcs.[93] The earliest example was discovered in 1979;[94] since then, more than a hundred gravitational lenses have been observed.[95] Even if the multiple images are too close to each other to be resolved, the effect can still be measured, e.g., as an overall brightening of the target object; a number of such "microlensing events" have been observed.[96]

Gravitational lensing has developed into a tool of observational astronomy. It is used to detect the presence and distribution of dark matter, provide a "natural telescope" for observing distant galaxies, and to obtain an independent estimate of the Hubble constant. Statistical evaluations of lensing data provide valuable insight into the structural evolution of galaxies.[97]

6.5.2   Gravitational wave astronomy

Main articles: Gravitational wave and Gravitational wave astronomy
 Observations of binary pulsars provide strong indirect evidence for the existence of gravitational waves (see Orbital decay, above). However, gravitational waves reaching us from the depths of the cosmos have not been detected directly. Such detection is a major goal of current relativity-related research.[98] Several land-based gravitational wave detectors are currently in operation, most notably the interferometric detectors GEO 600, LIGO (two detectors), TAMA 300 and VIRGO.[99] Various pulsar timing arrays are using millisecond pulsars to detect gravitational waves in the 10^{-9} to 10^{-6} Hertz frequency range, which originate from binary supermassive blackholes.[100] European space-based detector, eLISA / NGO, is currently under development,[101] with a precursor mission (LISA Pathfinder) due for launch in 2015.[102]

Observations of gravitational waves promise to complement observations in the electromagnetic spectrum.[103] They are expected to yield information about black holes and other dense objects such as neutron stars and white dwarfs, about certain kinds of supernova implosions, and about processes in the very early universe, including the signature of certain types of hypothetical cosmic string.[104]

6.5.3   Black holes and other compact objects

Main article: Black hole

Whenever the ratio of an object's mass to its radius becomes sufficiently large, general relativity predicts the formation of a black hole, a region of space from which nothing, not even light, can escape. In the currently accepted models of stellar evolution, neutron stars of around 1.4 solar masses, and stellar black holes with a few to a few dozen solar masses, are thought to be the final state for the evolution of massive stars.[105] Usually a galaxy has one supermassive black hole with a few million to a few billion solar masses in its center,[106] and its presence is thought to have played an important role in the formation of the galaxy and larger cosmic structures.[107]

Astronomically, the most important property of compact objects is that they provide a supremely efficient mechanism for converting gravitational energy into electromagnetic radiation.[108] Accretion, the falling of dust or gaseous matter onto stellar or supermassive black holes, is thought to be responsible for some spectacularly luminous astronomical objects, notably diverse kinds of active galactic nuclei on galactic scales and stellar-size objects such as microquasars.[109] In particular, accretion can lead to relativistic jets, focused beams of highly energetic particles that are being flung into space at almost light speed.[110] General relativity plays a central role in modelling all these phenomena,[111] and observations provide strong evidence for the existence of black holes with the properties predicted by the theory.[112]

Black holes are also sought-after targets in the search for gravitational waves (cf. Gravitational waves, above). Merging black hole binaries should lead to some of the strongest gravitational wave signals reaching detectors here on Earth, and the phase directly before the merger ("chirp") could be used as a "standard candle" to deduce the distance to the merger events–and hence serve as a probe of cosmic expansion at large distances.[113] The gravitational waves produced as a stellar black hole plunges into a supermassive one should provide direct information about the supermassive black hole's geometry.[114]

6.5.4   Cosmology

Main article: Physical cosmology

The current models of cosmology are based on Einstein's field equations, which include the cosmological constant Λ since it has important influence on the large-scale dynamics of the cosmos,

$$R_{\mu\nu} - \frac{1}{2} R\, g_{\mu\nu} + \Lambda\, g_{\mu\nu} = \frac{8\pi G}{c^4}\, T_{\mu\nu}$$

where $g_{\mu\nu}$ is the spacetime metric.[115] Isotropic and homogeneous solutions of these enhanced equations, the Friedmann–Lemaître–Robertson–Walker solutions,[116] allow physicists to model a universe that has evolved over the past 14 billion years from a hot, early Big Bang phase.[117] Once a small number of parameters (for example the universe's mean matter density) have been fixed by astronomical observation,[118] further observational data can be used to put the models to the test.[119] Predictions, all successful, include the initial abundance of chemical elements formed in a period of primordial nucleosynthesis,[120] the large-scale structure of the universe,[121] and the existence and properties of a "thermal echo" from the early cosmos, the cosmic background radiation.[122]

Astronomical observations of the cosmological expansion rate allow the total amount of matter in the universe to be estimated, although the nature of that matter remains mysterious in part. About 90% of all matter appears to be so-called dark matter, which has mass (or, equivalently, gravitational influence), but does not interact electromagnetically and, hence, cannot be observed directly.[123] There is no generally accepted description of this new kind of matter, within the framework of known particle physics[124] or otherwise.[125] Observational evidence from redshift surveys of distant supernovae and measurements of the cosmic background radiation also show that the evolution of our universe is significantly influenced by a cosmological constant resulting in an acceleration of cosmic expansion or, equivalently, by a form of energy with an unusual equation of state, known as dark energy, the nature of which remains unclear.[126]

A so-called inflationary phase,[127] an additional phase of strongly accelerated expansion at cosmic times of around 10^{-33} seconds, was hypothesized in 1980 to account for several puzzling observations that were unexplained by classical cosmological models, such as the nearly perfect homogeneity of the cosmic background radiation.[128] Recent measurements of the cosmic background radiation have resulted in the first evidence for this scenario.[129] However, there is a bewildering variety of possible inflationary scenarios, which cannot be restricted by current observations.[130] An even larger question is the physics of the earliest universe, prior to the inflationary phase and close to where the classical models predict the big bang singularity. An authoritative answer would require a complete theory of quantum gravity, which has not yet been developed[131] (cf. the section on quantum gravity, below).

6.5.5 Time travel

Kurt Gödel showed that solutions to Einstein's equations exist that contain closed timelike curves (CTCs), which allow for loops in time. The solutions require extreme physical conditions unlikely ever to occur in practice, and it remains an open question whether further laws of physics will eliminate them completely. Since then other—similarly impractical—GR solutions containing CTCs have been found, such as the Tipler cylinder and traversable wormholes.

6.6 Advanced concepts

6.6.1 Causal structure and global geometry

Main article: Causal structure

In general relativity, no material body can catch up with or overtake a light pulse. No influence from an event A can reach any other location X before light sent out at A to X. In consequence, an exploration of all light worldlines (null geodesics) yields key information about the spacetime's causal structure. This structure can be displayed using Penrose–Carter diagrams in which infinitely large regions of space and infinite time intervals are shrunk ("compactified") so as to fit onto a finite map, while light still travels along diagonals as in standard spacetime diagrams.[132]

Aware of the importance of causal structure, Roger Penrose and others developed what is known as global geometry. In global geometry, the object of study is not one particular solution (or family of solutions) to Einstein's equations. Rather, relations that hold true for all geodesics, such as the Raychaudhuri equation, and additional non-specific assumptions about the nature of matter (usually in the form of so-called energy conditions) are used to derive general results.[133]

6.6.2 Horizons

Main articles: Horizon (general relativity), No hair theorem and Black hole mechanics

Using global geometry, some spacetimes can be shown to contain boundaries called horizons, which demarcate one region from the rest of spacetime. The best-known examples are black holes: if mass is compressed into a sufficiently compact region of space (as specified in the hoop conjecture, the relevant length scale is the Schwarzschild radius[134]), no light from inside can escape to the outside. Since no object can overtake a light pulse, all interior matter is imprisoned as well. Passage from the exterior to the interior is still possible, showing that the boundary, the black hole's *horizon*, is not a physical barrier.[135]

Early studies of black holes relied onexplicit solutionsofEinstein's equations, notably the spherically symmetric Schwarzschild solution(used to describe astaticblack hole) and the axisymmetric Kerr solution (used to describe a rotatin, stationary black hole, and introducing interesting features such as the ergosphere). Using global geometry, later studies have re-vealed more general properties of black holes. In the long run, they are rather simple objects characterized by eleven parameters specifying energy, linear momentum, angular momentum, location at a specified time and electric charge. This is stated by the black hole uniqueness theorems: "black holes have no hair", that is, no distinguishing marks like the hairstyles of humans. Irrespective of the complexity of a gravitating object collapsing to form a blackhole, the object that results (having emittedgravitational waves) is very simple.[136]

Even more remarkably, there is a general set of laws known as black hole mechanics, which is analogous to the laws of thermodynamics. For instance, by the second law of black hole mechanics, the area of the event horizon of a general black hole will never decrease with time, analogous to the entropy of a thermodynamic system. This limits the energy that can be extracted by classical means from a rotating black hole (e.g. by the Penrose process).[137] There is strong evidence that the laws of black hole mechanics are, in fact, a subset of the laws of thermodynamics, and that the black hole area is proportional to its entropy.[138] This leads to a modification of the original laws of black hole mechanics: for instance, as the second law of black hole mechanics becomes part of the second law of thermodynamics, it is possible for black hole area to decrease—as long as other processes ensure that, overall, entropy increases. As thermodynamical objects with non-zero temperature, black holes should emit thermal radiation. Semi-classical calculations indicate that indeed they do, with the surface gravity playing the role of temperature in Planck's law. This radiation is known as Hawking radiation (cf. the quantum theory section, below).[139]

There are other types of horizons. In an expanding universe, an observer may find that some regions of the past cannot be observed ("particle horizon"), and some regions of the future cannot be influenced (event horizon).[140] Even in flat Minkowski space, when described by an accelerated observer (Rindler space), there will be horizons associated with a semi-classical radiation known as Unruh radiation.[141]

6.6.3 Singularities

Main article: Spacetime singularity

Another general feature of general relativity is the appearance of spacetime boundaries known as singularities. Spacetime can be explored by following up on timelike and lightlike geodesics—all possible ways that light and particles in free fall can travel. But some solutions of Einstein's equations have "ragged edges"—regions known as spacetime singularities, where the paths of light and falling particles come to an abrupt end, and geometry becomes ill-defined. In the more interesting cases, these are "curvature singularities", where geometrical quantities characterizing spacetime curvature, such as the Ricci scalar, take on infinite values.[142] Well-known examples of spacetimes with future singularities—where worldlines end—are the Schwarzschild solution, which describes a singularity inside an eternal static black hole,[143] or the Kerr solution with its ring-shaped singularity inside an eternal rotating black hole.[144] The Friedmann–Lemaître–Robertson–Walker solutions and other spacetimes describing universes have past singularities on which worldlines begin, namely Big Bang singularities, and some have future singularities (Big Crunch) as well.[145]

Given that these examples are all highly symmetric—and thus simplified—it is tempting to conclude that the occurrence of singularities is an artifact of idealization.[146] The famous singularity theorems, proved using the methods of global geometry, say otherwise: singularities are a generic feature of general relativity, and unavoidable once the collapse of an object with realistic matter properties has proceeded beyond a certain stage[147] and also at the beginning of a wide class of expanding universes.[148] However, the theorems say little about the properties of singularities, and much of current research is devoted to characterizing these entities' generic structure (hypothesized e.g. by the so-called BKL conjecture).[149] The cosmic censorship hypothesis states that all realistic future singularities (no perfect symmetries, matter with realistic properties) are safely hidden away behind a horizon, and thus invisible to all distant observers. While no formal proof yet exists, numerical simulations offer supporting evidence of its validity.[150]

6.6.4 Evolution equations

Main article: Initial value formulation (general relativity)

Each solution of Einstein's equation encompasses the whole history of a universe — it is not just some snapshot of how things are, but a whole, possibly matter-filled, spacetime. It describes the state of matter and geometry everywhere and at every moment in that particular universe. Due to its general covariance, Einstein's theory is not sufficient by itself to determine the time evolution of the metric tensor. It must be combined with a coordinate condition, which is analogous to gauge fixing in other field theories.[151]

To understand Einstein's equations as partial differential equations, it is helpful to formulate them in a way that describes the evolution of the universe over time. This is done in so-called "3+1" formulations, where spacetime is split into three space dimensions and one time dimension. The best-known example is the ADM formalism.[152] These decompositions show that the spacetime evolution equations of general relativity are well-behaved: solutions always exist, and are uniquely defined, once suitable initial conditions have been specified.[153] Such formulations of Einstein's field equations are the basis of numerical relativity.[154]

6.6.5 Global and quasi-local quantities

Main article: Mass in general relativity

The notion of evolution equations is intimately tied in with another aspect of general relativistic physics. In Einstein's theory, it turns out to be impossible to find a general definition for a seemingly simple property such as a system's total mass (or energy). The main reason is that the gravitational field—like any physical field—must be ascribed a certain energy, but that it proves to be fundamentally impossible to localize that energy.[155]

Nevertheless, there are possibilities to define a system's total mass, either using a hypothetical "infinitely distant observer" (ADM mass)[156] or suitable symmetries (Komar mass).[157] If one excludes from the system's total mass the energy being carried away to infinity by gravitational waves, the result is the so-called Bondi mass at null infinity.[158] Just as in classical physics, it can be shown that these masses are positive.[159] Corresponding global definitions exist for momentum and angular momentum.[160] There have also been a number of attempts to define *quasi-local* quantities, such as the mass of an isolated system formulated using only quantities defined within a finite region of space containing that system. The hope is to obtain a quantity useful for general statements about isolated systems, such as a more precise formulation of the hoop conjecture.[161]

6.7   Relationship with quantum theory

If general relativity were considered to be one of the two pillars of modern physics, then quantum theory, the basis of understanding matter from elementary particles to solid state physics, would be the other.[162] However, how to reconcile quantum theory with general relativity is still an open question.

6.7.1   Quantum field theory in curved spacetime

Main article: Quantum field theory in curved spacetime

Ordinary quantum field theories, which form the basis of modern elementary particle physics, are defined in flat Minkowski space, which is an excellent approximation when it comes to describing the behavior of microscopic particles in weak gravitational fields like those found on Earth.[163] In order to describe situations in which gravity is strong enough to influence (quantum) matter, yet not strong enough to require quantization itself, physicists have formulated quantum field theories in curved spacetime. These theories rely on general relativity to describe a curved background spacetime, and define a generalized quantum field theory to describe the behavior of quantum matter within that spacetime.[164] Using this formalism, it can be shown that black holes emit a blackbody spectrum of particles known as Hawking radiation, leading to the possibility that they evaporate over time.[165] As briefly mentioned above, this radiation plays an important role for the thermodynamics of black holes.[166]

6.7.2   Quantum gravity

Main article: Quantum gravity
See also: String theory, Canonical general relativity, Loop quantum gravity, Causal Dynamical Triangulations and Causal sets

The demand for consistency between a quantum description of matter and a geometric description of spacetime,[167] as well as the appearance of singularities (where curvature length scales become microscopic), indicate the need for a full theory of quantum gravity: for an adequate description of the interior of black holes, and of the very early universe, a theory is required in which gravity and the associated geometry of spacetime are described in the language of quantum physics.[168] Despite major efforts, no complete and consistent theory of quantum gravity is currently known, even though a number of promising candidates exist.[169]

Attempts to generalize ordinary quantum field theories, used in elementary particle physics to describe fundamental interactions, so as to include gravity have led to serious problems. At low energies, this approach proves successful, in that it results in an acceptable effective (quantum) field theory of gravity.[170] At very high energies, however, the result are models devoid of all predictive power ("non-renormalizability").[171]

One attempt to overcome these limitations is string theory, a quantum theory not of point particles, but of minute one-dimensional extended objects.[172] The theory promises to be a unified description of all particles and interactions, including gravity;[173] the price to pay is unusual features such as six extra dimensions of space in addition to the usual three.[174] In what is called the second superstring revolution, it was conjectured that both string theory and a unification of general relativity and supersymmetry known as supergravity[175] form part of a hypothesized eleven-dimensional model known as M-theory, which would constitute a uniquely defined and consistent theory of quantum gravity.[176]

Another approach starts with the canonical quantization procedures of quantum theory. Using the initial-value-formulation of general relativity (cf. evolution equations above), the result is the Wheeler–deWitt equation (an analogue of the Schrödinger equation) which, regrettably, turns out to be ill-defined.[177] However, with the introduction of what are now known as Ashtekar variables,[178] this leads to a promising model known as loop quantum gravity. Space is represented by a web-like structure called a spin network, evolving over time in discrete steps.[179]

Depending on which features of general relativity and quantum theory are accepted unchanged, and on what level changes are introduced,[180] there are numerous other attempts to arrive at a viable theory of quantum gravity, some examples being dynamical triangulations,[181] causal sets,[182] twistor models[183] or the path-integral based models of quantum cosmology.[184]

All candidate theories still have major formal and conceptual problems to overcome. They also face the common problem that, as yet, there is no way to put quantum gravity predictions to experimental tests (and thus to decide between the candidates where their predictions vary), although there is hope for this to change as future data from cosmological observations and particle physics experiments becomes available.[185]

6.8 Current status

General relativity has emerged as a highly successful model of gravitation and cosmology, which has so far passed many unambiguous observational and experimental tests. However, there are strong indications the theory is incomplete.[186] The problem of quantum gravity and the question of the reality of spacetime singularities remain open.[187] Observational data that is taken as evidence for dark energy and dark matter could indicate the need for new physics.[188] Even taken as is, general relativity is rich with possibilities for further exploration. Mathematical relativists seek to understand the nature of singularities and the fundamental properties of Einstein's equations,[189] and increasingly powerful computer simulations (such as those describing merging black holes) are run.[190] The race for the first direct detection of gravitational waves continues,[191] in the hope of creating opportunities to test the theory's validity for much stronger gravitational fields than has been possible to date.[192] Almost a hundred years after its publication, general relativity remains a highly active area of research.[193]

6.9 See also

- Center of mass (relativistic)
- Contributors to general relativity
- Derivations of the Lorentz transformations
- Ehrenfest paradox
- Einstein–Hilbert action
- Introduction to mathematics of general relativity
- Relativity priority dispute
- Ricci calculus
- Tests of general relativity

- Timeline of gravitational physics and relativity

- Two-body problem in general relativity

6.10   Notes

[1] O'Connor, J.J. and E.F. Robertson (1996), "General relativity". *Mathematical Physics index*, School of Mathematics and Statistics, University of St. Andrews, Scotland, May, 1996. Retrieved 2015-02-04.

[2] Pais 1982, ch. 9 to 15, Janssen 2005; an up-to-date collection of current research, including reprints of many of the original articles, is Renn 2007; an accessible overview can be found in Renn 2005, pp. 110ff. An early key article is Einstein 1907, cf. Pais 1982, ch. 9. The publication featuring the field equations is Einstein 1915, cf. Pais 1982, ch. 11–15

[3] Schwarzschild 1916a, Schwarzschild 1916b and Reissner 1916 (later complemented in Nordström 1918)

[4] Einstein 1917, cf. Pais 1982, ch. 15e

[5] Hubble's original article is Hubble 1929; an accessible overview is given in Singh 2004, ch. 2–4

[6] As reported in Gamow 1970. Einstein's condemnation would prove to be premature, cf. the section Cosmology, below

[7] Pais 1982, pp. 253–254

[8] Kennefick 2005, Kennefick 2007

[9] Pais 1982, ch. 16

[10] Thorne, Kip (2003). "Warping spacetime". *The future of theoretical physics and cosmology: celebrating Stephen Hawking's 60th birthday*. Cambridge University Press. p. 74. ISBN 0-521-82081-2., Extract of page 74

[11] Israel 1987, ch. 7.8–7.10, Thorne 1994, ch. 3–9

[12] Sections Orbital effects and the relativity of direction, Gravitational time dilation and frequency shift and Light deflection and gravitational time delay, and references therein

[13] Section Cosmology and references therein; the historical development is in Overbye 1999

[14] The following exposition re-traces that of Ehlers 1973, sec. 1

[15] Arnold 1989, ch. 1

[16] Ehlers 1973, pp. 5f

[17] Will 1993, sec. 2.4, Will 2006, sec. 2

[18] Wheeler 1990, ch. 2

[19] Ehlers 1973, sec. 1.2, Havas 1964, Künzle 1972. The simple thought experiment in question was first described in Heckmann & Schücking 1959

[20] Ehlers 1973, pp. 10f

[21] Good introductions are, in order of increasing presupposed knowledge of mathematics, Giulini 2005, Mermin 2005, and Rindler 1991; for accounts of precision experiments, cf. part IV of Ehlers & Lämmerzahl 2006

[22] An in-depth comparison between the two symmetry groups can be found in Giulini 2006a

[23] Rindler 1991, sec. 22, Synge 1972, ch. 1 and 2

[24] Ehlers 1973, sec. 2.3

[25] Ehlers 1973, sec. 1.4, Schutz 1985, sec. 5.1

[26] Ehlers 1973, pp. 17ff; a derivation can be found in Mermin 2005, ch. 12. For the experimental evidence, cf. the section Gravitational time dilation and frequency shift, below

[27] Rindler 2001, sec. 1.13; for an elementary account, see Wheeler 1990, ch. 2; there are, however, some differences between the modern version and Einstein's original concept used in the historical derivation of general relativity, cf. Norton 1985

[28] Ehlers 1973, sec. 1.4 for the experimental evidence, see once more section Gravitational time dilation and frequency shift. Choosing a different connection with non-zero torsion leads to a modified theory known as Einstein–Cartan theory

[29] Ehlers 1973, p. 16, Kenyon 1990, sec. 7.2, Weinberg 1972, sec. 2.8

[30] Ehlers 1973, pp. 19–22; for similar derivations, see sections 1 and 2 of ch. 7 in Weinberg 1972. The Einstein tensor is the only divergence-free tensor that is a function of the metric coefficients, their first and second derivatives at most, and allows the spacetime of special relativity as a solution in the absence of sources of gravity, cf. Lovelock 1972. The tensors on both side are of second rank, that is, they can each be thought of as 4×4 matrices, each of which contains ten independent terms; hence, the above represents ten coupled equations. The fact that, as a consequence of geometric relations known as Bianchi identities, the Einstein tensor satisfies a further four identities reduces these to six independent equations, e.g. Schutz 1985, sec. 8.3

[31] Kenyon 1990, sec. 7.4

[32] Brans & Dicke 1961, Weinberg 1972, sec. 3 in ch. 7, Goenner 2004, sec. 7.2, and Trautman 2006, respectively

[33] Wald 1984, ch. 4, Weinberg 1972, ch. 7 or, in fact, any other textbook on general relativity

[34] At least approximately, cf. Poisson 2004

[35] Wheeler 1990, p. xi

[36] Wald 1984, sec. 4.4

[37] Wald 1984, sec. 4.1

[38] For the (conceptual and historical) difficulties in defining a general principle of relativity and separating it from the notion of general covariance, see Giulini 2006b

[39] section 5 in ch. 12 of Weinberg 1972

[40] Introductory chapters of Stephani et al. 2003

[41] A review showing Einstein's equation in the broader context of other PDEs with physical significance is Geroch 1996

[42] For background information and a list of solutions, cf. Stephani et al. 2003; a more recent review can be found in MacCallum 2006

[43] Chandrasekhar 1983, ch. 3,5,6

[44] Narlikar 1993, ch. 4, sec. 3.3

[45] Brief descriptions of these and further interesting solutions can be found in Hawking & Ellis 1973, ch. 5

[46] Lehner 2002

[47] For instance Wald 1984, sec. 4.4

[48] Will 1993, sec. 4.1 and 4.2

[49] Will 2006, sec. 3.2, Will 1993, ch. 4

[50] Rindler 2001, pp. 24–26 vs. pp. 236–237 and Ohanian & Ruffini 1994, pp. 164–172. Einstein derived these effects using the equivalence principle as early as 1907, cf. Einstein 1907 and the description in Pais 1982, pp. 196–198

[51] Rindler 2001, pp. 24–26; Misner, Thorne & Wheeler 1973, § 38.5

[52] Pound–Rebka experiment, see Pound & Rebka 1959, Pound & Rebka 1960; Pound & Snider 1964; a list of further experiments is given in Ohanian & Ruffini 1994, table 4.1 on p. 186

[53] Greenstein, Oke & Shipman 1971; the most recent and most accurate Sirius B measurements are published in Barstow, Bond et al. 2005.

[54] Starting with the Hafele–Keating experiment, Hafele & Keating 1972a and Hafele & Keating 1972b, and culminating in the Gravity Probe A experiment; an overview of experiments can be found in Ohanian & Ruffini 1994, table 4.1 on p. 186

[55] GPS is continually tested by comparing atomic clocks on the ground and aboard orbiting satellites; for an account of relativistic effects, see Ashby 2002 and Ashby 2003

[56] Stairs 2003 and Kramer 2004

[57] General overviews can be found in section 2.1. of Will 2006; Will 2003, pp. 32–36; Ohanian & Ruffini 1994, sec. 4.2

[58] Ohanian & Ruffini 1994, pp. 164–172

[59] Cf. Kennefick 2005 for the classic early measurements by the Eddington expeditions; for an overview of more recent measurements, see Ohanian & Ruffini 1994, ch. 4.3. For the most precise direct modern observations using quasars, cf. Shapiro et al. 2004

[60] This is not an independent axiom; it can be derived from Einstein's equations and the Maxwell Lagrangian using a WKB approximation, cf. Ehlers 1973, sec. 5

[61] Blanchet 2006, sec. 1.3

[62] Rindler 2001, sec. 1.16; for the historical examples, Israel 1987, pp. 202–204; in fact, Einstein published one such derivation as Einstein 1907. Such calculations tacitly assume that the geometry of space is Euclidean, cf. Ehlers & Rindler 1997

[63] From the standpoint of Einstein's theory, these derivations take into account the effect of gravity on time, but not its consequences for the warping of space, cf. Rindler 2001, sec. 11.11

[64] For the Sun's gravitational field using radar signals reflected from planets such as Venus and Mercury, cf. Shapiro 1964, Weinberg 1972, ch. 8, sec. 7; for signals actively sent back by space probes (transponder measurements), cf. Bertotti, Iess & Tortora 2003; for an overview, see Ohanian & Ruffini 1994, table 4.4 on p. 200; for more recent measurements using signals received from a pulsar that is part of a binary system, the gravitational field causing the time delay being that of the other pulsar, cf. Stairs 2003, sec. 4.4

[65] Will 1993, sec. 7.1 and 7.2

[66] These have been indirectly observed through the loss of energy in binary pulsar systems such as the Hulse–Taylor binary, the subject of the 1993 Nobel Prize in physics. A number of projects are underway to attempt to observe directly the effects of gravitational waves. For an overview, see Misner, Thorne & Wheeler 1973, part VIII. Unlike electromagnetic waves, the dominant contribution for gravitational waves is not the dipole, but the quadrupole; see Schutz 2001

[67] Most advanced textbooks on general relativity contain a description of these properties, e.g. Schutz 1985, ch. 9

[68] For example Jaranowski & Królak 2005

[69] Rindler 2001, ch. 13

[70] Gowdy 1971, Gowdy 1974

[71] See Lehner 2002 for a brief introduction to the methods of numerical relativity, and Seidel 1998 for the connection with gravitational wave astronomy

[72] Schutz 2003, pp. 48–49, Pais 1982, pp. 253–254

[73] Rindler 2001, sec. 11.9

[74] Will 1993, pp. 177–181

[75] In consequence, in the parameterized post-Newtonian formalism (PPN), measurements of this effect determine a linear combination of the terms β and γ, cf. Will 2006, sec. 3.5 and Will 1993, sec. 7.3

[76] The most precise measurements are VLBI measurements of planetary positions; see Will 1993, ch. 5, Will 2006, sec. 3.5, Anderson et al. 1992; for an overview, Ohanian & Ruffini 1994, pp. 406–407

[77] Kramer et al. 2006

[78] A figure that includes error bars is fig. 7 in Will 2006, sec. 5.1

[79] Stairs 2003, Schutz 2003, pp. 317–321, Bartusiak 2000, pp. 70–86

[80] Weisberg & Taylor 2003; for the pulsar discovery, see Hulse & Taylor 1975; for the initial evidence for gravitational radiation, see Taylor 1994

[81] Kramer 2004

[82] Penrose 2004, §14.5, Misner, Thorne & Wheeler 1973, §11.4

[83] Weinberg 1972, sec. 9.6, Ohanian & Ruffini 1994, sec. 7.8

[84] Bertotti, Ciufolini & Bender 1987, Nordtvedt 2003

[85] Kahn 2007

[86] A mission description can be found in Everitt et al. 2001; a first post-flight evaluation is given in Everitt, Parkinson & Kahn 2007; further updates will be available on the mission website Kahn 1996–2012.

[87] Townsend 1997, sec. 4.2.1, Ohanian & Ruffini 1994, pp. 469–471

[88] Ohanian & Ruffini 1994, sec. 4.7, Weinberg 1972, sec. 9.7; for a more recent review, see Schäfer 2004

[89] Ciufolini & Pavlis 2004, Ciufolini, Pavlis & Peron 2006, Iorio 2009

[90] Iorio L. (August 2006), "COMMENTS, REPLIES AND NOTES: A note on the evidence of the gravitomagnetic field of Mars", Classical Quantum Gravity 23 (17): 5451–5454, arXiv:gr-qc/0606092, Bibcode:2006CQGra..23.5451I, doi:10.1088/0264-9381/23/17/N01

[91] Iorio L. (June 2010), "On the Lense–Thirring test with the Mars Global Surveyor in the gravitational field of Mars", Central European Journal of Physics 8 (3): 509–513, arXiv:gr-qc/0701146, Bibcode:2010CEJPh...8..509I, doi:10.2478/s11534-009-0117-6

[92] For overviews of gravitational lensing and its applications, see Ehlers, Falco & Schneider 1992 and Wambsganss 1998

[93] For a simple derivation, see Schutz 2003, ch. 23; cf. Narayan & Bartelmann 1997, sec. 3

[94] Walsh, Carswell & Weymann 1979

[95] Images of all the known lenses can be found on the pages of the CASTLES project, Kochanek et al. 2007

[96] Roulet & Mollerach 1997

[97] Narayan & Bartelmann 1997, sec. 3.7

[98] Barish 2005, Bartusiak 2000, Blair & McNamara 1997

[99] Hough & Rowan 2000

[100] Hobbs, George; Archibald, A.; Arzoumanian, Z.; Backer, D.; Bailes, M.; Bhat, N. D. R.; Burgay, M.; Burke-Spolaor, S.; et al. (2010), "The international pulsar timing array project: using pulsars as a gravitational wave detector", Classical and Quantum Gravity 27 (8): 084013, arXiv:0911.5206, Bibcode:2010CQGra..27h4013H, doi:10.1088/0264-9381/27/8/084013

[101] Danzmann & Rüdiger 2003

[102] "LISA pathfinder overview", ESA. Retrieved 2012-04-23.

[103] Thorne 1995

[104] Cutler & Thorne 2002

[105] Miller 2002, lectures 19 and 21

[106] Celotti, Miller & Sciama 1999, sec. 3

[107] Springel et al. 2005 and the accompanying summary Gnedin 2005

[108] Blandford 1987, sec. 8.2.4

[109] For the basic mechanism, see Carroll & Ostlie 1996, sec. 17.2; for more about the different types of astronomical objects associated with this, cf. Robson 1996

[110] For a review, see Begelman, Blandford & Rees 1984. To a distant observer, some of these jets even appear to move faster than light; this, however, can be explained as an optical illusion that does not violate the tenets of relativity, see Rees 1966

[111] For stellar end states, cf. Oppenheimer & Snyder 1939 or, for more recent numerical work, Font 2003, sec. 4.1; for supernovae, there are still major problems to be solved, cf. Buras et al. 2003; for simulating accretion and the formation of jets, cf. Font 2003, sec. 4.2. Also, relativistic lensing effects are thought to play a role for the signals received from X-ray pulsars, cf. Kraus 1998

[112] The evidence includes limits on compactness from the observation of accretion-driven phenomena ("Eddington luminosity"), see Celotti, Miller & Sciama 1999, observations of stellar dynamics in the center of our own Milky Way galaxy, cf. Schödel et al. 2003, and indications that at least some of the compact objects in question appear to have no solid surface, which can be deduced from the examination of X-ray bursts for which the central compact object is either a neutron star or a black hole; cf. Remillard et al. 2006 for an overview, Narayan 2006, sec. 5. Observations of the "shadow" of the Milky Way galaxy's central black hole horizon are eagerly sought for, cf. Falcke, Melia & Agol 2000

[113] Dalal et al. 2006

[114] Barack & Cutler 2004

[115] Originally Einstein 1917; cf. Pais 1982, pp. 285–288

[116] Carroll 2001, ch. 2

[117] Bergström & Goobar 2003, ch. 9–11; use of these models is justified by the fact that, at large scales of around hundred million light-years and more, our own universe indeed appears to be isotropic and homogeneous, cf. Peebles et al. 1991

[118] E.g. with WMAP data, see Spergel et al. 2003

[119] These tests involve the separate observations detailed further on, see, e.g., fig. 2 in Bridle et al. 2003

[120] Peebles 1966; for a recent account of predictions, see Coc, Vangioni-Flam et al. 2004; an accessible account can be found in Weiss 2006; compare with the observations in Olive & Skillman 2004, Bania, Rood & Balser 2002, O'Meara et al. 2001, and Charbonnel & Primas 2005

[121] Lahav & Suto 2004, Bertschinger 1998, Springel et al. 2005

[122] Alpher & Herman 1948, for a pedagogical introduction, see Bergström & Goobar 2003, ch. 11; for the initial detection, see Penzias & Wilson 1965 and, for precision measurements by satellite observatories, Mather et al. 1994 (COBE) and Bennett et al. 2003 (WMAP). Future measurements could also reveal evidence about gravitational waves in the early universe; this additional information is contained in the background radiation's polarization, cf. Kamionkowski, Kosowsky & Stebbins 1997 and Seljak & Zaldarriaga 1997

[123] Evidence for this comes from the determination of cosmological parameters and additional observations involving the dynamics of galaxies and galaxy clusters cf. Peebles 1993, ch. 18, evidence from gravitational lensing, cf. Peacock 1999, sec. 4.6, and simulations of large-scale structure formation, see Springel et al. 2005

[124] Peacock 1999, ch. 12, Peskin 2007; in particular, observations indicate that all but a negligible portion of that matter is not in the form of the usual elementary particles ("non-baryonic matter"), cf. Peacock 1999, ch. 12

[125] Namely, some physicists have questioned whether or not the evidence for dark matter is, in fact, evidence for deviations from the Einsteinian (and the Newtonian) description of gravity cf. the overview in Mannheim 2006, sec. 9

[126] Carroll 2001; an accessible overview is given in Caldwell 2004. Here, too, scientists have argued that the evidence indicates not a new form of energy, but the need for modifications in our cosmological models, cf. Mannheim 2006, sec. 10; aforementioned modifications need not be modifications of general relativity, they could, for example, be modifications in the way we treat the inhomogeneities in the universe, cf. Buchert 2007

[127] A good introduction is Linde 1990; for a more recent review, see Linde 2005

[128] More precisely, these are the flatness problem, the horizon problem, and the monopole problem; a pedagogical introduction can be found in Narlikar 1993, sec. 6.4, see also Börner 1993, sec. 9.1

[129] Spergel et al. 2007, sec. 5.6

[130] More concretely, the potential function that is crucial to determining the dynamics of the inflaton is simply postulated, but not derived from an underlying physical theory

[131] Brandenberger 2007, sec. 2

[132] Frauendiener 2004, Wald 1984, sec. 11.1, Hawking & Ellis 1973, sec. 6.8, 6.9

[133] Wald 1984, sec. 9.2–9.4 and Hawking & Ellis 1973, ch. 6

[134] Thorne 1972; for more recent numerical studies, see Berger 2002, sec. 2.1

[135] Israel 1987. A more exact mathematical description distinguishes several kinds of horizon, notably event horizons and apparent horizons cf. Hawking & Ellis 1973, pp. 312–320 or Wald 1984, sec. 12.2; there are also more intuitive definitions for isolated systems that do not require knowledge of spacetime properties at infinity, cf. Ashtekar & Krishnan 2004

[136] For first steps, cf. Israel 1971; see Hawking & Ellis 1973, sec. 9.3 or Heusler 1996, ch. 9 and 10 for a derivation, and Heusler 1998 as well as Beig & Chruściel 2006 as overviews of more recent results

[137] The laws of black hole mechanics were first described in Bardeen, Carter & Hawking 1973; a more pedagogical presentation can be found in Carter 1979; for a more recent review, see Wald 2001, ch. 2. A thorough, book-length introduction including an introduction to the necessary mathematics Poisson 2004. For the Penrose process, see Penrose 1969

[138] Bekenstein 1973, Bekenstein 1974

[139] The fact that black holes radiate, quantum mechanically, was first derived in Hawking 1975; a more thorough derivation can be found in Wald 1975. A review is given in Wald 2001, ch. 3

[140] Narlikar 1993, sec. 4.4.4, 4.4.5

[141] Horizons: cf. Rindler 2001, sec. 12.4. Unruh effect: Unruh 1976, cf. Wald 2001, ch. 3

[142] Hawking & Ellis 1973, sec. 8.1, Wald 1984, sec. 9.1

[143] Townsend 1997, ch. 2; a more extensive treatment of this solution can be found in Chandrasekhar 1983, ch. 3

[144] Townsend 1997, ch. 4; for a more extensive treatment, cf. Chandrasekhar 1983, ch. 6

[145] Ellis & Van Elst 1999; a closer look at the singularity itself is taken in Börner 1993, sec. 1.2

[146] Here one should remind to the well-known fact that the important "quasi-optical" singularities of the so-called eikonal approximations of many wave-equations, namely the "caustics", are resolved into finite peaks beyond that approximation.

[147] Namely when there are trapped null surfaces, cf. Penrose 1965

[148] Hawking 1966

[149] The conjecture was made in Belinskii, Khalatnikov & Lifschitz 1971; for a more recent review, see Berger 2002. An accessible exposition is given by Garfinkle 2007

[150] The restriction to future singularities naturally excludes initial singularities such as the big bang singularity, which in principle be visible to observers at later cosmic time. The cosmic censorship conjecture was first presented in Penrose 1969; a textbook-level account is given in Wald 1984, pp. 302–305. For numerical results, see the review Berger 2002, sec. 2.1

[151] Hawking & Ellis 1973, sec. 7.1

[152] Arnowitt, Deser & Misner 1962; for a pedagogical introduction, see Misner, Thorne & Wheeler 1973, §21.4–§21.7

[153] Fourès-Bruhat 1952 and Bruhat 1962; for a pedagogical introduction, see Wald 1984, ch. 10; an online review can be found in Reula 1998

[154] Gourgoulhon 2007; for a review of the basics of numerical relativity, including the problems arising from the peculiarities of Einstein's equations, see Lehner 2001

[155] Misner, Thorne & Wheeler 1973, §20.4

[156] Arnowitt, Deser & Misner 1962

[157] Komar 1959; for a pedagogical introduction, see Wald 1984, sec. 11.2; although defined in a totally different way, it can be shown to be equivalent to the ADM mass for stationary spacetimes, cf. Ashtekar & Magnon-Ashtekar 1979

[158] For a pedagogical introduction, see Wald 1984, sec. 11.2

[159] Wald 1984, p. 295 and refs therein; this is important for questions of stability—if there were negative mass states, then flat, empty Minkowski space, which has mass zero, could evolve into these states

[160] Townsend 1997, ch. 5

[161] Such quasi-local mass–energy definitions are the Hawking energy, Geroch energy, or Penrose's quasi-local energy–momentum based on twistor methods; cf. the review article Szabados 2004

[162] An overview of quantum theory can be found in standard textbooks such as Messiah 1999; a more elementary account is given in Hey & Walters 2003

[163] Ramond 1990, Weinberg 1995, Peskin & Schroeder 1995; a more accessible overview is Auyang 1995

[164] Wald 1994, Birrell & Davies 1984

[165] For Hawking radiation Hawking 1975, Wald 1975; an accessible introduction to black hole evaporation can be found in Traschen 2000

[166] Wald 2001, ch. 3

[167] Put simply, matter is the source of spacetime curvature, and once matter has quantum properties, we can expect spacetime to have them as well. Cf. Carlip 2001, sec. 2

[168] Schutz 2003, p. 407

[169] A timeline and overview can be found in Rovelli 2000

[170] Donoghue 1995

[171] In particular, a technique known as renormalization, an integral part of deriving predictions which take into account higher-energy contributions, cf. Weinberg 1996, ch. 17, 18, fails in this case; cf. Goroff & Sagnotti 1985

[172] An accessible introduction at the undergraduate level can be found in Zwiebach 2004; more complete overviews can be found in Polchinski 1998a and Polchinski 1998b

[173] At the energies reached in current experiments, these strings are indistinguishable from point-like particles, but, crucially, different modes of oscillation of one and the same type of fundamental string appear as particles with different (electric and other) charges, e.g. Ibanez 2000. The theory is successful in that one mode will always correspond to a graviton, the messenger particle of gravity, e.g. Green, Schwarz & Witten 1987, sec. 2.3, 5.3

[174] Green, Schwarz & Witten 1987, sec. 4.2

[175] Weinberg 2000, ch. 31

[176] Townsend 1996, Duff 1996

[177] Kuchař 1973, sec. 3

[178] These variables represent geometric gravity using mathematical analogues of electric and magnetic fields; cf. Ashtekar 1986, Ashtekar 1987

[179] For a review, see Thiemann 2006; more extensive accounts can be found in Rovelli 1998, Ashtekar & Lewandowski 2004 as well as in the lecture notes Thiemann 2003

[180] Isham 1994, Sorkin 1997

[181] Loll 1998

[182] Sorkin 2005

[183] Penrose 2004, ch. 33 and refs therein

[184] Hawking 1987

[185] Ashtekar 2007, Schwarz 2007

[186] Maddox 1998, pp. 52–59, 98–122; Penrose 2004, sec. 34.1, ch. 30

[187] section Quantum gravity, above

[188] section Cosmology, above

[189] Friedrich 2005

[190] A review of the various problems and the techniques being developed to overcome them, see Lehner 2002

[191] See Bartusiak 2000 for an account up to that year; up-to-date news can be found on the websites of major detector collaborations
       such as GEO 600 and LIGO

[192] For the most recent papers on gravitational wave polarizations of inspiralling compact binaries, see Blanchet et al. 2008, and
       Arun et al. 2007; for a review of work on compact binaries, see Blanchet 2006 and Futamase & Itoh 2006; for a general review
       of experimental tests of general relativity, see Will 2006

[193] See, e.g., the electronic review journal Living Reviews in Relativity

6.11   References

• Alpher, R. A.; Herman, R. C. (1948), "Evolution of the universe", *Nature* **162**(4124): 774–775, Bibcode:1948Natur, doi:10.1038/162774b0

• Anderson, J. D.; Campbell, J. K.; Jurgens, R. F.; Lau, E. L. (1992), "Recent developments in solar-system tests of general relativity", in Sato, H.; Nakamura, T., *Proceedings of the Sixth Marcel Großmann Meeting on General Relativity*, World Scientific, pp. 353–355, ISBN 981-02-0950-9

• Arnold, V. I. (1989), *Mathematical Methods of Classical Mechanics*, Springer, ISBN 3-540-96890-3

• Arnowitt, Richard; Deser, Stanley; Misner, Charles W. (1962), "The dynamics of general relativity", in Witten, Louis, *Gravitation: An Introduction to Current Research*, Wiley, pp. 227–265

• Arun, K.G.; Blanchet, L.; Iyer, B. R.; Qusailah, M. S. S. (2007), "Inspiralling compact binaries in quasi-elliptical orbits: The complete 3PN energy flux", *Physical Review D* **77** (6), arXiv:0711.0302, Bibcode:2008PhRvD..77f4035A, doi:10.1103/PhysRevD.77.064035

• Ashby, Neil (2002), "Relativity and the Global Positioning System" (PDF), *Physics Today* **55**(5): 41–47, Bibcod1A, doi:10.1063/1.1485583

• Ashby, Neil (2003), "Relativity in the Global Positioning System", *Living Reviews in Relativity* **6**, retrieved 2007-07-06 External link in |work= (help)

• Ashtekar, Abhay (1986), "New variables for classical and quantum gravity", *Phys. Rev. Lett.* **57** (18): 2244–2247, Bibcode:1986PhRvL..57.2244A, doi:10.1103/PhysRevLett.57.2244, PMID 10033673

• Ashtekar, Abhay (1987), "New Hamiltonian formulation of general relativity", *Phys. Rev.* **D36** (6): 1587–1602, Bibcode:1987PhRvD..36.1587A, doi:10.1103/PhysRevD.36.1587

• Ashtekar, Abhay (2007), "LOOP QUANTUM GRAVITY: FOUR RECENT ADVANCES AND A DOZEN FREQUENTLY ASKED QUESTIONS", *The Eleventh Marcel Grossmann Meeting - on Recent Developments in Theoretical and Experimental General Relativity, Gravitation and Relativistic Field Theories - Proceedings of the MG11 Meeting on General Relativity*, p. 126, arXiv:0705.2222, Bibcode:2008mgm..conf..126A, doi:10.1142/9789812, ISBN 9789812834263

- Ashtekar, Abhay; Krishnan, Badri (2004), "Isolated and Dynamical Horizons and Their Applications", *Living Rev. Relativity* **7**, arXiv:gr-qc/0407042, Bibcode:2004LRR......7...10A, doi:10.12942/lrr-2004-10, retrieved 2007-08-28

- Ashtekar, Abhay; Lewandowski, Jerzy (2004), "Background Independent Quantum Gravity: A Status Report", *Class. Quant. Grav.* **21** (15): R53–R152, arXiv:gr-qc/0404018, Bibcode:2004CQGra..21R..53A,doi:10.1088/0-9381/21/15/R01

- Ashtekar, Abhay; Magnon-Ashtekar, Anne (1979), "On conserved quantities in general relativity", *Journal of Mathematical Physics* **20** (5): 793–800, Bibcode:1979JMP....20..793A, doi:10.1063/1.524151

- Auyang, Sunny Y. (1995), *How is Quantum Field Theory Possible?*, Oxford University Press, ISBN 0-19-509345-3

- Bania, T. M.; Rood, R. T.; Balser, D. S. (2002), "The cosmological density of baryons from observations of 3He+ in the Milky Way", *Nature* **415** (6867): 54–57, Bibcode:2002Natur.415...54B, doi:10.1038/415054a, PMID 11780112

- Barack, Leor; Cutler, Curt (2004), "LISA Capture Sources: Approximate Waveforms, Signal-to-Noise Ratios, and Parameter Estimation Accuracy", *Phys. Rev.* **D69** (8): 082005, arXiv:gr-qc/0310125,Bibcode:2004PhRvD..69, doi:10.1103/PhysRevD.69.082005

- Bardeen, J. M.; Carter, B.; Hawking, S. W. (1973), "The Four Laws of Black Hole Mechanics", *Comm. Math. Phys.* **31** (2): 161–170, Bibcode:1973CMaPh..31..161B, doi:10.1007/BF01645742

- Barish, Barry (2005), "Towards detection of gravitational waves", in Florides, P.; Nolan, B.; Ottewil, A., *General Relativity and Gravitation. Proceedings of the 17th International Conference*, World Scientific, pp. 24–34, ISBN 981-256-424-1

- Barstow, M; Bond, Howard E.; Holberg, J. B.; Burleigh, M. R.; Hubeny, I.; Koester, D. (2005), "Hubble Space Telescope Spectroscopy of the Balmer lines in Sirius B", *Mon. Not. Roy. Astron. Soc.* **362** (4): 1134–1142, arXiv:astro-ph/0506600, Bibcode:2005MNRAS.362.1134B, doi:10.1111/j.1365-2966.2005.09359.x

- Bartusiak, Marcia (2000), *Einstein's Unfinished Symphony: Listening to the Sounds of Space-Time*, Berkley, ISBN 978-0-425-18620-6

- Begelman, Mitchell C.; Blandford, Roger D.; Rees, Martin J. (1984), "Theory of extragalactic radio sources", *Rev. Mod. Phys.* **56** (2): 255–351, Bibcode:1984RvMP...56..255B, doi:10.1103/RevModPhys.56.255

- Beig, Robert; Chruściel, Piotr T. (2006), "Stationary black holes", in Françoise, J.-P.; Naber, G.; Tsou, T.S., *Encyclopedia of Mathematical Physics, Volume 2*, Elsevier, p. 2041, arXiv:gr-qc/0502041, Bibcode:2005gr.qc.....2041B, ISBN 0-12-512660-3

- Bekenstein, Jacob D.(1973), "Black Holes and Entropy",*Phys. Rev.***D7**(8): 2333–2346,Bibcode:1973PhRvD...7.., doi:10.1103/PhysRevD.7.2333

- Bekenstein, Jacob D. (1974), "Generalized Second Law of Thermodynamics in Black-Hole Physics", *Phys. Rev.* **D9** (12): 3292–3300, Bibcode:1974PhRvD...9.3292B, doi:10.1103/PhysRevD.9.3292

- Belinskii, V. A.; Khalatnikov, I. M.; Lifschitz, E. M. (1971), "Oscillatory approach to the singular point in relativistic cosmology", *Advances in Physics* **19** (80): 525–573, Bibcode:1970AdPhy..19..525B,doi:10.1080/0001873; original paper in Russian: Belinsky, V. A.; Lifshits, I. M.; Khalatnikov, E. M. (1970), "Колебательный Режим Приближения К Особой Точке В Релятивистской Космологии",*Uspekhi Fizicheskikh Nauk (Успехи Физиче Наук)*, 102(3) (11): 463–500,Bibcode:1970UsFiN.102..463B

- Bennett, C. L.; Halpern, M.; Hinshaw, G.; Jarosik, N.; Kogut, A.; Limon, M.; Meyer, S. S.; Page, L.; et al. (2003), "First Year Wilkinson Microwave Anisotropy Probe (WMAP) Observations: Preliminary Maps and Basic Results", *Astrophys. J. Suppl.* **148** (1): 1–27, arXiv:astro-ph/0302207, Bibcode:2003ApJS..148....1B, doi:10.1086/377253

- Berger, Beverly K. (2002), "Numerical Approaches to Spacetime Singularities", *Living Rev. Relativity* **5**, arXiv:gr-qc/0201056, Bibcode:2002LRR......5....1B, doi:10.12942/lrr-2002-1, retrieved 2007-08-04

- Bergström, Lars; Goobar, Ariel (2003), *Cosmology and Particle Astrophysics* (2nd ed.), Wiley & Sons, ISBN 3-540-43128-4

- Bertotti, Bruno; Ciufolini, Ignazio; Bender, Peter L. (1987), "New test of general relativity: Measurement of de Sitter geodetic precession rate for lunar perigee",*Physical Review Letters***58**(11): 1062–1065,Bibcode:1987PhRv .,doi:10.1103/PhysRevLett.58.1062, PMID 10034329

- Bertotti, Bruno; Iess, L.; Tortora, P. (2003), "A test of general relativity using radio links with the Cassini spacecraft", *Nature* **425** (6956): 374–376, Bibcode:2003Natur.425..374B, doi:10.1038/nature01997, PMID 14508481

- Bertschinger, Edmund (1998), "Simulations of structure formation in the universe", *Annu. Rev. Astron. Astrophys.* **36** (1): 599–654, Bibcode:1998ARA&A..36..599B, doi:10.1146/annurev.astro.36.1.599

- Birrell, N. D.; Davies, P. C. (1984), *Quantum Fields in Curved Space*, Cambridge University Press, ISBN 0-521-27858-9

- Blair, David; McNamara, Geoff (1997), *Ripples on a Cosmic Sea. The Search for Gravitational Waves*, Perseus, ISBN 0-7382-0137-5

- Blanchet, L.; Faye, G.; Iyer, B. R.; Sinha, S. (2008), "The third post-Newtonian gravitational wave polarisations and associated spherical harmonic modes for inspiralling compact binaries in quasi-circular orbits", *Classical and Quantum Gravity* **25** (16): 165003, arXiv:0802.1249, Bibcode:2008CQGra..25p5003B, doi:10.1088/0264-9381/25/16/165003

- Blanchet, Luc (2006), "Gravitational Radiation from Post-Newtonian Sources and Inspiralling Compact Binaries", *Living Rev. Relativity* **9**, Bibcode:2006LRR.....9....4B, doi:10.12942/lrr-2006-4, retrieved 2007-08-07

- Blandford, R. D. (1987), "Astrophysical Black Holes", in Hawking, Stephen W.; Israel, Werner, *300 Years of Gravitation*, Cambridge University Press, pp. 277–329, ISBN 0-521-37976-8

- Börner, Gerhard (1993), *The Early Universe. Facts and Fiction*, Springer, ISBN 0-387-56729-1

- Brandenberger, Robert H. (2007), "Conceptual Problems of Inflationary Cosmology and a New Approach to Cosmological Structure Formation", *Inflationary Cosmology*, Lecture Notes in Physics **738**, p. 393, arXiv:hep-th/0701111, Bibcode:2008LNP...738..393B, doi:10.1007/978-3-540-74353-8_11, ISBN 978-3-540-74352-1

- Brans, C. H.; Dicke, R. H. (1961), "Mach's Principle and a Relativistic Theory of Gravitation", *Physical Review* **124** (3): 925–935, Bibcode:1961PhRv..124..925B, doi:10.1103/PhysRev.124.925

- Bridle, Sarah L.; Lahav, Ofer; Ostriker, Jeremiah P.; Steinhardt, Paul J. (2003), "Precision Cosmology? Not Just Yet", *Science* **299** (5612): 1532–1533, arXiv:astro-ph/0303180,Bibcode:2003Sci...299.1532B,doi:10.1126/sci, PMID 12624255

- Bruhat, Yvonne (1962), "The Cauchy Problem", in Witten, Louis, *Gravitation: An Introduction to Current Research*, Wiley, p. 130, ISBN 978-1-114-29166-9

- Buchert, Thomas (2007), "Dark Energy from Structure—A Status Report", *General Relativity and Gravitation* **40** (2–3): 467–527, arXiv:0707.2153, Bibcode:2008GReGr..40..467B, doi:10.1007/s10714-007-0554-8

- Buras, R.; Rampp, M.; Janka, H.-Th.; Kifonidis, K. (2003), "Improved Models of Stellar Core Collapse and Still no Explosions: What is Missing?", *Phys. Rev. Lett.* **90** (24): 241101, arXiv:astro-ph/0303171,Bibcode:2003PhRv, doi:10.1103/PhysRevLett.90.241101, PMID 12857181

- Caldwell, Robert R. (2004), "Dark Energy", *Physics World* **17** (5): 37–42

- Carlip, Steven (2001), "Quantum Gravity: a Progress Report", *Rept. Prog. Phys.* **64** (8): 885–942, arXiv:gr-qc/0108040, Bibcode:2001RPPh...64..885C, doi:10.1088/0034-4885/64/8/301

- Carroll, Bradley W.; Ostlie, Dale A. (1996), *An Introduction to Modern Astrophysics*, Addison-Wesley, ISBN 0-201-54730-9

- Carroll, Sean M. (2001), "The Cosmological Constant", *Living Rev. Relativity* **4**, arXiv:astro-ph/0004...1C, doi:10.12942/lrr-2001-1, retrieved 2007-07-21

- Carter, Brandon (1979), "The general theory of the mechanical, electromagnetic and thermodynamic properties of black holes", in Hawking, S. W.; Israel, W., *General Relativity, an Einstein Centenary Survey*, Cambridge University Press, pp. 294–369 and 860–863, ISBN 0-521-29928-4

- Celotti, Annalisa; Miller, John C.; Sciama, Dennis W. (1999), "Astrophysical evidence for the existence of black holes", *Class. Quant. Grav.* **16** (12A): A3–A21, arXiv:astro-ph/9912186, doi:10.1088/0264-9381/16/12A/301

- Chandrasekhar, Subrahmanyan (1983), *The Mathematical Theory of Black Holes*, Oxford University Press, ISBN 0-19-850370-9

- Charbonnel, C.; Primas, F. (2005), "The Lithium Content of the Galactic Halo Stars", *Astronomy & Astrophysics* **442** (3): 961–992, arXiv:astro-ph/0505247, Bibcode:2005A&A...442..961C, doi:10.1051/0004-6361:20042491

- Ciufolini, Ignazio; Pavlis, Erricos C. (2004), "A confirmation of the general relativistic prediction of the Lense-Thirring effect", *Nature* **431** (7011): 958–960, Bibcode:2004Natur.431..958C, doi:10.1038/nature03007, PMID 15496915

- Ciufolini, Ignazio; Pavlis, Erricos C.; Peron, R. (2006), "Determination of frame-dragging using Earth gravity models from CHAMP and GRACE", *New Astron.* **11** (8): 527–550, Bibcode:2006NewA...11..527C, doi:10.1001

- Coc, A.; Vangioni-Flam, Elisabeth; Descouvemont, Pierre; Adahchour, Abderrahim; Angulo, Carmen (2004), "Updated Big Bang Nucleosynthesis confronted to WMAP observations and to the Abundance of Light Elements", *Astrophysical Journal* **600** (2): 544–552, arXiv:astro-ph/0309480, Bibcode:2004ApJ...600..544C, doi:10.11

- Cutler, Curt; Thorne, Kip S. (2002), "An overview of gravitational wave sources", in Bishop, Nigel; Maharaj, Sunil D., *Proceedings of 16th International Conference on General Relativity and Gravitation (GR16)*, World Scientific, p. 4090, arXiv:gr-qc/0204090, Bibcode:2002gr.qc.....4090C, ISBN 981-238-171-6

- Dalal, Neal; Holz, Daniel E.; Hughes, Scott A.; Jain, Bhuvnesh (2006), "Short GRB and binary black hole standard sirens as a probe of dark energy", *Phys.Rev.* **D74** (6): 063006, arXiv:astro-ph/0601275, Bibcode:2006PhRvD.., doi:10.1103/PhysRevD.74.063006

- Danzmann, Karsten; Rüdiger, Albrecht (2003), "LISA Technology—Concepts, Status, Prospects" (PDF), *Class. Quant. Grav.* **20** (10): S1–S9, Bibcode:2003CQGra..20S...1D, doi:10.1088/0264-9381/20/10/301

- Dirac, Paul (1996), *General Theory of Relativity*, Princeton University Press, ISBN 0-691-01146-X

- Donoghue, John F. (1995), "Introduction to the Effective Field Theory Description of Gravity", in Cornet, Fernando, *Effective Theories: Proceedings of the Advanced School, Almunecar, Spain, 26 June–1 July 1995*, Singapore: World Scientific, p. 12024, arXiv:gr-qc/9512024, Bibcode:1995gr.qc....12024D, ISBN 981-02-2908-9

- Duff, Michael (1996), "M-Theory (the Theory Formerly Known as Strings)", *Int. J. Mod. Phys.* **A11** (32): 5623–5641, arXiv:hep-th/9608117, Bibcode:1996IJMPA..11.5623D, doi:10.1142/S0217751X96002583

- Ehlers, Jürgen (1973), "Survey of general relativity theory", in Israel, Werner, *Relativity, Astrophysics and Cosmology*, D. Reidel, pp. 1–125, ISBN 90-277-0369-8

- Ehlers, Jürgen; Falco, Emilio E.; Schneider, Peter (1992), *Gravitational lenses*, Springer, ISBN 3-540-66506-4

- Ehlers, Jürgen; Lämmerzahl, Claus, eds. (2006), *Special Relativity—Will it Survive the Next 101 Years?*, Springer, ISBN 3-540-34522-1

- Ehlers, Jürgen; Rindler, Wolfgang (1997), "Local and Global Light Bending in Einstein's and other Gravitational Theories", *General Relativity and Gravitation* **29**(4): 519–529, Bibcode:1997GReGr..29..519E, doi:10.1023/2

- Einstein, Albert (1907), "Über das Relativitätsprinzip und die aus demselben gezogene Folgerungen" (PDF), *Jahrbuch der Radioaktivität und Elektronik* **4**: 411, retrieved 2008-05-05

- Einstein, Albert (1915), "Die Feldgleichungen der Gravitation", *Sitzungsberichte der Preussischen Akademie der Wissenschaften zu Berlin*: 844–847, retrieved 2006-09-12

- Einstein, Albert (1916), "Die Grundlage der allgemeinen Relativitätstheorie", *Annalen der Physik* **49**: 769–822, Bibcode:1916AnP...354..769E, doi:10.1002/andp.19163540702, archived from the original (PDF) on 2006-08-29, retrieved 2006-09-03

- Einstein, Albert (1917), "Kosmologische Betrachtungen zur allgemeinen Relativitätstheorie", *Sitzungsberichte der Preußischen Akademie der Wissenschaften*: 142

- Ellis, George F R; Van Elst, Henk (1999), Lachièze-Rey, Marc, ed., "Theoretical and Observational Cosmology: Cosmological models (Cargèse lectures 1998)", *Theoretical and observational cosmology : proceedings of the NATO Advanced Study Institute on Theoretical and Observational Cosmology* (Kluwer): 1–116, arXiv:gr-qc/9812046, Bibcode:1999toc..conf....1E, doi:10.1007/978-94-011-4455-1_1, ISBN 978-0-7923-5946-3

- Everitt, C. W. F.; Buchman, S.; DeBra, D. B.; Keiser, G. M. (2001), "Gravity Probe B: Countdown to launch", in Lämmerzahl, C.; Everitt, C. W. F.; Hehl, F. W., *Gyros, Clocks, and Interferometers: Testing Relativistic Gravity in Space (Lecture Notes in Physics 562)*, Springer, pp. 52–82, ISBN 3-540-41236-0

- Everitt, C. W. F.; Parkinson, Bradford; Kahn, Bob (2007), *The Gravity Probe B experiment. Post Flight Analysis—Final Report (Preface and Executive Summary)* (PDF), Project Report: NASA, Stanford University and Lockheed Martin, retrieved 2007-08-05

- Falcke, Heino; Melia, Fulvio; Agol, Eric (2000), "Viewing the Shadow of the Black Hole at the Galactic Center", *Astrophysical Journal* **528** (1): L13–L16, arXiv:astro-ph/9912263, Bibcode:2000ApJ...528L..13F,doi:10.10, PMID 10587484

- Flanagan, Éanna É.; Hughes, Scott A. (2005), "The basics of gravitational wave theory", *New J.Phys.* **7**: 204, arXiv:gr-qc/0501041, Bibcode:2005NJPh....7..204F, doi:10.1088/1367-2630/7/1/204

- Font, José A. (2003), "Numerical Hydrodynamics in General Relativity", *Living Rev. Relativity* **6**, doi:10.12942/lrr-2003-4, retrieved 2007-08-19

- Fourès-Bruhat, Yvonne (1952), "Théoréme d'existence pour certains systémes d'équations aux derivées partielles non linéaires", *Acta Mathematica* **88** (1): 141–225, Bibcode:1952AcM....88..141F, doi:10.1007/BF02392131

- Frauendiener, Jörg (2004),"Conformal Infinity",*Living Rev. Relativity***7**,Bibcode:2004LRR.....7....1F,doi:10. 2004-1, retrieved 2007-07-21

- Friedrich, Helmut (2005), "Is general relativity 'essentially understood'?", *Annalen Phys.* **15** (1–2): 84–108, arXiv:gr-qc/0508016, Bibcode:2006AnP...518...84F, doi:10.1002/andp.200510173

- Futamase, T.; Itoh, Y. (2006), "The Post-Newtonian Approximation for Relativistic Compact Binaries", *Living Rev. Relativity* **10**, retrieved 2008-02-29

- Gamow, George (1970), *My World Line*, Viking Press, ISBN 0-670-50376-2

- Garfinkle, David (2007), "Of singularities and breadmaking", *Einstein Online*, retrieved 2007-08-03 External link in |work= (help)

- Geroch, Robert (1996). "Partial Differential Equations of Physics". arXiv:gr-qc/9602055 [gr-qc].

- Giulini, Domenico (2005), *Special Relativity: A First Encounter*, Oxford University Press, ISBN 0-19-856746-4

- Giulini, Domenico (2006a), "Algebraic and Geometric Structures in Special Relativity", in Ehlers, Jürgen; Lämmerzahl, Claus, *Special Relativity—Will it Survive the Next 101 Years?*, Springer, pp. 45–111, arXiv:math-ph/0602, Bibcode:2006math.ph...2018G, ISBN 3-540-34522-1

- Giulini, Domenico (2006b), Stamatescu, I. O., ed., "An assessment of current paradigms in the physics of fundamental interactions: Some remarks on the notions of general covariance and background independence", *Approaches to Fundamental Physics*, Lecture Notes in Physics (Springer) **721**: 105, arXiv:gr-qc/0603087,BibcodG, doi:10.1007/978-3-540-71117-9_6, ISBN 978-3-540-71115-5

- Gnedin, Nickolay Y. (2005), "Digitizing the Universe", *Nature* **435** (7042): 572–573, Bibcode:2005Natur.435..572G, doi:10.1038/435572a, PMID 15931201

- Goenner, Hubert F. M. (2004),"On the History of Unified Field Theories",*Living Rev. Relativity***7**,Bibcode:200G, doi:10.12942/lrr-2004-2, retrieved 2008-02-28

- Goroff, Marc H.; Sagnotti, Augusto (1985), "Quantum gravity at two loops", *Phys. Lett.* **160B** (1–3): 81–86, Bibcode:1985PhLB..160...81G, doi:10.1016/0370-2693(85)91470-4

- Gourgoulhon, Eric (2007), "3+1 Formalism and Bases of Numerical Relativity", arXiv:gr-qc/0703035 [gr-qc].

- Gowdy, Robert H. (1971), "Gravitational Waves in Closed Universes", *Phys. Rev. Lett.* **27** (12): 826–829, Bibcode:1971PhRvL..27..826G, doi:10.1103/PhysRevLett.27.826

- Gowdy, Robert H. (1974), "Vacuum spacetimes with two-parameter spacelike isometry groups and compact invariant hypersurfaces: Topologies and boundary conditions", *Ann. Phys. (N.Y.)* **83** (1): 203–241, Bibcode:1974AnPh, doi:10.1016/0003-4916(74)90384-4

- Green, M. B.; Schwarz, J. H.; Witten, E. (1987), *Superstring theory. Volume 1: Introduction*, Cambridge University Press, ISBN 0-521-35752-7

- Greenstein, J. L.; Oke, J. B.; Shipman, H. L. (1971), "Effective Temperature, Radius, and Gravitational Redshift of Sirius B", *Astrophysical Journal* **169**: 563, Bibcode:1971ApJ...169..563G, doi:10.1086/151174

- Hafele, J. C.; Keating, R. E. (July 14, 1972). "Around-the-World Atomic Clocks: Predicted Relativistic Time Gains". *Science* **177** (4044): 166–168. Bibcode:1972Sci...177..166H. doi:10.1126/science.177.4044.166. PMID 17779917.

- Hafele, J. C.; Keating, R. E. (July 14, 1972). "Around-the-World Atomic Clocks: Observed Relativistic Time Gains". *Science* **177** (4044): 168–170. Bibcode:1972Sci...177..168H. doi:10.1126/science.177.4044.168. PMID 17779918.

- Havas, P. (1964), "Four-Dimensional Formulation of Newtonian Mechanics and Their Relation to the Special and the General Theory of Relativity", *Rev. Mod. Phys.* **36** (4): 938–965, Bibcode:1964RvMP...36..938H, doi:10.1103/RevModPhys.36.938

- Hawking, Stephen W. (1966), "The occurrence of singularities in cosmology", *Proceedings of the Royal Society* **A294** (1439): 511–521, Bibcode:1966RSPSA.294..511H, doi:10.1098/rspa.1966.0221

- Hawking, S. W. (1975), "Particle Creation by Black Holes", *Communications in Mathematical Physics* **43** (3): 199–220, Bibcode:1975CMaPh..43..199H, doi:10.1007/BF02345020

- Hawking, Stephen W. (1987), "Quantum cosmology", in Hawking, Stephen W.; Israel, Werner, *300 Years of Gravitation*, Cambridge University Press, pp. 631–651, ISBN 0-521-37976-8

- Hawking, Stephen W.; Ellis, George F. R. (1973), *The large scale structure of space-time*, Cambridge University Press, ISBN 0-521-09906-4

- Heckmann, O. H. L.; Schücking, E. (1959), "Newtonsche und Einsteinsche Kosmologie", in Flügge, S., *Encyclopedia of Physics* **53**, p. 489

- Heusler, Markus (1998),"Stationary Black Holes: Uniqueness and Beyond",*Living Rev. Relativity***1**,doi:10.12942/lrr 1998-6, retrieved 2007-08-04

- Heusler, Markus (1996), *Black Hole Uniqueness Theorems*, Cambridge University Press, ISBN 0-521-56735-1

- Hey, Tony; Walters, Patrick (2003), *The new quantum universe*, Cambridge University Press, ISBN 0-521-56457-3

- Hough, Jim; Rowan, Sheila (2000), "Gravitational Wave Detection by Interferometry (Ground and Space)", *Living Rev. Relativity* **3**, retrieved 2007-07-21

- Hubble, Edwin (1929), "A Relation between Distance and Radial Velocity among Extra-Galactic Nebulae" (PDF), *Proc. Nat. Acad. Sci.* **15** (3): 168–173, Bibcode:1929PNAS...15..168H, doi:10.1073/pnas.15.3.168, PMC 522427, PMID 16577160

- Hulse, Russell A.; Taylor, Joseph H. (1975), "Discovery of a pulsar in a binary system", *Astrophys. J.* **195**: L51–L55, Bibcode:1975ApJ...195L..51H, doi:10.1086/181708

- Ibanez, L. E. (2000), "The second string (phenomenology) revolution", *Class. Quant. Grav.* **17** (5): 1117–1128, arXiv:hep-ph/9911499, Bibcode:2000CQGra..17.1117I, doi:10.1088/0264-9381/17/5/321

- Iorio, L. (2009), "An Assessment of the Systematic Uncertainty in Present and Future Tests of the Lense-Thirring Effect with Satellite Laser Ranging", *Space Sci. Rev.* **148** (1–4): 363, arXiv:0809.1373, Bibcode:2009SSRv .,doi:10.1007/s11214-008-9478-1

- Isham, Christopher J. (1994), "Prima facie questions in quantum gravity", in Ehlers, Jürgen; Friedrich, Helmut, *Canonical Gravity: From Classical to Quantum*, Springer, ISBN 3-540-58339-4

- Israel, Werner (1971), "Event Horizons and Gravitational Collapse", *General Relativity and Gravitation* **2** (1): 53–59, Bibcode:1971GReGr...2...53I, doi:10.1007/BF02450518

- Israel, Werner (1987), "Dark stars: the evolution of an idea", in Hawking, Stephen W.; Israel, Werner, *300 Years of Gravitation*, Cambridge University Press, pp. 199–276, ISBN 0-521-37976-8

- Janssen, Michel (2005), "Of pots and holes: Einstein's bumpy road to general relativity" (PDF), *Ann. Phys. (Leipzig)* **14** (S1): 58–85, Bibcode:2005AnP...517S..58J, doi:10.1002/andp.200410130

- Jaranowski, Piotr; Królak, Andrzej (2005), "Gravitational-Wave Data Analysis. Formalism and Sample Applications: The Gaussian Case", *Living Rev. Relativity* **8**, doi:10.12942/lrr-2005-3, retrieved 2007-07-30

- Kahn, Bob (1996–2012), *Gravity Probe B Website*, Stanford University, retrieved 2012-04-20

- Kahn, Bob (April 14, 2007), *Was Einstein right? Scientists provide first public peek at Gravity Probe B results (Stanford University Press Release)* (PDF), Stanford University News Service

- Kamionkowski, Marc; Kosowsky, Arthur; Stebbins, Albert (1997), "Statistics of Cosmic Microwave Background Polarization", *Phys. Rev.* **D55** (12): 7368–7388, arXiv:astro-ph/9611125, Bibcode:1997PhRvD..55.7368K, doi:

- Kennefick, Daniel (2005), "Astronomers Test General Relativity: Light-bending and the Solar Redshift", in Renn, Jürgen, *One hundred authors for Einstein*, Wiley-VCH, pp. 178–181, ISBN 3-527-40574-7

- Kennefick, Daniel (2007), "Not Only Because of Theory: Dyson, Eddington and the Competing Myths of the 1919 Eclipse Expedition", *Proceedings of the 7th Conference on the History of General Relativity, Tenerife, 2005* **0709**, p. 685, arXiv:0709.0685, Bibcode:2007arXiv0709.0685K

- Kenyon, I. R. (1990), *General Relativity*, Oxford University Press, ISBN 0-19-851996-6

- Kochanek, C.S.; Falco, E.E.; Impey, C.; Lehar, J. (2007), *CASTLES Survey Website*, Harvard-Smithsonian Center for Astrophysics, retrieved 2007-08-21

- Komar, Arthur (1959), "Covariant Conservation Laws in General Relativity", *Phys. Rev.* **113** (3): 934–936, Bibcode:1959PhRv..113..934K, doi:10.1103/PhysRev.113.934

- Kramer, Michael (2004), Karshenboim, S. G.; Peik, E., eds., "Astrophysics, Clocks and Fundamental Constants: Millisecond Pulsars as Tools of Fundamental Physics", *Lecture Notes in Physics* (Springer) **648**: 33–54, arXiv:astro-ph/0405178, Bibcode:2004LNP...648...33K, doi:10.1007/978-3-540-40991-5_3, ISBN 978-3-540-21967-5

- Kramer, M.; Stairs, I. H.; Manchester, R. N.; McLaughlin, M. A.; Lyne, A. G.; Ferdman, R. D.; Burgay, M.; Lorimer, D. R.; et al. (2006), "Tests of general relativity from timing the double pulsar", *Science* **314** (5796): 97–102, arXiv:astro-ph/0609417, Bibcode:2006Sci...314...97K, doi:10.1126/science.1132305, PMID 16973838

- Kraus, Ute (1998), "Light Deflection Near Neutron Stars", *Relativistic Astrophysics*, Vieweg, pp. 66–81, ISBN 3-528-06909-0

- Kuchař, Karel (1973), "Canonical Quantization of Gravity", in Israel, Werner, *Relativity, Astrophysics and Cosmology*, D. Reidel, pp. 237–288, ISBN 90-277-0369-8

- Künzle, H. P. (1972), "Galilei and Lorentz Structures on spacetime: comparison of the corresponding geometry and physics", *Ann. Inst. Henri Poincaré a* **17**: 337–362

- Lahav, Ofer; Suto, Yasushi (2004), "Measuring our Universe from Galaxy Redshift Surveys", *Living Rev. Relativity* **7**, arXiv:astro-ph/0310642, Bibcode:2004LRR.....7....8L, doi:10.12942/lrr-2004-8, retrieved 2007-08-19

- Landgraf, M.; Hechler, M.; Kemble, S. (2005), "Mission design for LISA Pathfinder", *Class. Quant. Grav.* **22** (10): S487–S492, arXiv:gr-qc/0411071, Bibcode:2005CQGra..22S.487L, doi:10.1088/0264-9381/22/10/048

- Lehner, Luis (2001), "Numerical Relativity: A review", *Class. Quant. Grav.* **18** (17): R25–R86, arXiv:gr-qc/0106072, Bibcode:2001CQGra..18R..25L, doi:10.1088/0264-9381/18/17/202

- Lehner, Luis (2002), "NUMERICAL RELATIVITY: STATUS AND PROSPECTS", *General Relativity and Gravitation - Proceedings of the 16th International Conference*, p. 210, arXiv:gr-qc/0202055, Bibcode:2002grg..conf., doi:10.1142/9789812776556_0010, ISBN 9789812381712

- Linde, Andrei (1990), *Particle Physics and Inflationary Cosmology*, Harwood, p. 3203, arXiv:hep-th/0503203, Bibcode:2005hep.th....3203L, ISBN 3-7186-0489-2

- Linde, Andrei (2005), "Towards inflation in string theory", *J. Phys. Conf. Ser.* **24**: 151–160, arXiv:hep-th/0503195, Bibcode:2005JPhCS..24..151L, doi:10.1088/1742-6596/24/1/018

- Loll, Renate (1998), "Discrete Approaches to Quantum Gravity in Four Dimensions", *Living Rev. Relativity* **1**, arXiv:gr-qc/9805049, Bibcode:1998LRR.....1...13L, doi:10.12942/lrr-1998-13, retrieved 2008-03-09

- Lovelock, David (1972), "The Four-Dimensionality of Space and the Einstein Tensor", *J. Math. Phys.* **13** (6): 874–876, Bibcode:1972JMP....13..874L, doi:10.1063/1.1666069

- Ludyk, Günter (2013). *Einstein in Matrix Form* (1st ed.). Berlin: Springer. ISBN 9783642357978.

- MacCallum, M. (2006), "Finding and using exact solutions of the Einstein equations", in Mornas, L.; Alonso, J. D., *A Century of Relativity Physics (ERE05, the XXVIII Spanish Relativity Meeting)* **841**, American Institute of Physics, p. 129, arXiv:gr-qc/0601102, Bibcode:2006AIPC..841..129M, doi:10.1063/1.2218172

- Maddox, John (1998), *What Remains To Be Discovered*, Macmillan, ISBN 0-684-82292-X

- Mannheim, Philip D. (2006), "Alternatives to Dark Matter and Dark Energy", *Prog. Part. Nucl. Phys.* **56** (2): 340–445, arXiv:astro-ph/0505266, Bibcode:2006PrPNP..56..340M, doi:10.1016/j.ppnp.2005.08.001

- Mather, J. C.; Cheng, E. S.; Cottingham, D. A.; Eplee, R. E.; Fixsen, D. J.; Hewagama, T.; Isaacman, R. B.; Jensen, K. A.; et al. (1994), "Measurement of the cosmic microwave spectrum by the COBE FIRAS instrument", *Astrophysical Journal* **420**: 439–444, Bibcode:1994ApJ...420..439M, doi:10.1086/173574

- Mermin, N. David (2005), *It's About Time. Understanding Einstein's Relativity*, Princeton University Press, ISBN 0-691-12201-6

- Messiah, Albert (1999), *Quantum Mechanics*, Dover Publications, ISBN 0-486-40924-4

- Miller, Cole (2002), *Stellar Structure and Evolution (Lecture notes for Astronomy 606)*, University of Maryland, retrieved 2007-07-25

- Misner, Charles W.; Thorne, Kip. S.; Wheeler, John A. (1973), *Gravitation*, W. H. Freeman, ISBN 0-7167-0344-0

- Møller, Christian (1952), *The Theory of Relativity* (3rd ed.), Oxford University Press

- Narayan, Ramesh (2006), "Black holes in astrophysics", *New Journal of Physics* **7**: 199, arXiv:gr-qc/0506078, Bibcode:2005NJPh....7..199N, doi:10.1088/1367-2630/7/1/199

- Narayan, Ramesh; Bartelmann, Matthias (1997). "Lectures on Gravitational Lensing". arXiv:astro-ph/9606001 [astro-ph].

- Narlikar, Jayant V. (1993), *Introduction to Cosmology*, Cambridge University Press, ISBN 0-521-41250-1

- Nieto, Michael Martin (2006), "The quest to understand the Pioneer anomaly" (PDF), *EurophysicsNews* **37** (6): 30–34, Bibcode:2006ENews..37...30N, doi:10.1051/epn:2006604

- Nordström, Gunnar (1918), "On the Energy of the Gravitational Field in Einstein's Theory", *Verhandl. Koninkl. Ned. Akad. Wetenschap.*, **26**: 1238–1245

- Nordtvedt, Kenneth (2003). "Lunar Laser Ranging—a comprehensive probe of post-Newtonian gravity". arXiv:gr-qc/0301024 [gr-qc].

- Norton, John D. (1985), "What was Einstein's principle of equivalence?" (PDF), *Studies in History and Philosophy of Science* **16** (3): 203–246, doi:10.1016/0039-3681(85)90002-0, retrieved 2007-06-11

- Ohanian, Hans C.; Ruffini, Remo (1994), *Gravitation and Spacetime*, W. W. Norton & Company, ISBN 0-393-96501-5

- Olive, K. A.; Skillman, E. A. (2004), "A Realistic Determination of the Error on the Primordial Helium Abundance", *Astrophysical Journal* **617** (1): 29–49, arXiv:astro-ph/0405588, Bibcode:2004ApJ...617...29O,doi:1

- O'Meara, John M.; Tytler, David; Kirkman, David; Suzuki, Nao; Prochaska, Jason X.; Lubin, Dan; Wolfe, Arthur M. (2001), "The Deuterium to Hydrogen Abundance Ratio Towards a Fourth QSO: HS0105+1619", *Astrophysical Journal* **552** (2): 718–730, arXiv:astro-ph/0011179, Bibcode:2001ApJ...552..718O, doi:10.1086/320579

- Oppenheimer, J. Robert; Snyder, H. (1939), "On continued gravitational contraction", *Physical Review* **56** (5): 455–459, Bibcode:1939PhRv...56..455O, doi:10.1103/PhysRev.56.455

- Overbye, Dennis (1999), *Lonely Hearts of the Cosmos: the story of the scientific quest for the secret of the Universe*, Back Bay, ISBN 0-316-64896-5

- Pais, Abraham (1982), *'Subtle is the Lord...' The Science and life of Albert Einstein*, Oxford University Press, ISBN 0-19-853907-X

- Peacock, John A. (1999), *Cosmological Physics*, Cambridge University Press, ISBN 0-521-41072-X

- Peebles, P. J. E. (1966), "Primordial Helium abundance and primordial fireball II", *Astrophysical Journal* **146**: 542–552, Bibcode:1966ApJ...146..542P, doi:10.1086/148918

- Peebles, P. J. E. (1993), *Principles of physical cosmology*, Princeton University Press, ISBN 0-691-01933-9

- Peebles, P.J.E.; Schramm, D.N.; Turner, E.L.; Kron, R.G. (1991), "The case for the relativistic hot Big Bang cosmology", *Nature* **352** (6338): 769–776, Bibcode:1991Natur.352..769P, doi:10.1038/352769a0

- Penrose, Roger (1965), "Gravitational collapse and spacetime singularities", *Physical Review Letters* **14** (3): 57–59, Bibcode:1965PhRvL..14...57P, doi:10.1103/PhysRevLett.14.57

- Penrose, Roger (1969), "Gravitational collapse: the role of general relativity", *Rivista del Nuovo Cimento* **1**: 252–276, Bibcode:1969NCimR...1..252P

- Penrose, Roger (2004), *The Road to Reality*, A. A. Knopf, ISBN 0-679-45443-8

- Penzias, A. A.; Wilson, R. W. (1965), "A measurement of excess antenna temperature at 4080 Mc/s", *Astrophysical Journal* **142**: 419–421, Bibcode:1965ApJ...142..419P, doi:10.1086/148307

- Peskin, Michael E.; Schroeder, Daniel V. (1995), *An Introduction to Quantum Field Theory*, Addison-Wesley, ISBN 0-201-50397-2

- Peskin, Michael E. (2007), "Dark Matter and Particle Physics", *Journal of the Physical Society of Japan* **76** (11): 111017, arXiv:0707.1536, Bibcode:2007JPSJ...76k1017P, doi:10.1143/JPSJ.76.111017

- Poisson, Eric (2004), "The Motion of Point Particles in Curved Spacetime", *Living Rev. Relativity* **7**, doi:10.12942/lrr-2004-6, retrieved 2007-06-13

- Poisson, Eric (2004), *A Relativist's Toolkit. The Mathematics of Black-Hole Mechanics*, Cambridge University Press, ISBN 0-521-83091-5

- Polchinski, Joseph (1998a), *String Theory Vol. I: An Introduction to the Bosonic String*, Cambridge University Press, ISBN 0-521-63303-6

- Polchinski, Joseph (1998b), *String Theory Vol. II: Superstring Theory and Beyond*, Cambridge University Press, ISBN 0-521-63304-4

- Pound, R. V.; Rebka, G. A. (1959), "Gravitational Red-Shift in Nuclear Resonance", *Physical Review Letters* **3** (9): 439–441, Bibcode:1959PhRvL...3..439P, doi:10.1103/PhysRevLett.3.439

- Pound, R. V.; Rebka, G. A. (1960), "Apparent weight of photons", *Phys. Rev. Lett.* **4**(7): 337–341, Bibcode:19, doi:10.1103/PhysRevLett.4.337

- Pound, R. V.; Snider, J. L. (1964), "Effect of Gravity on Nuclear Resonance", *Phys. Rev. Lett.* **13** (18): 539–540, Bibcode:1964PhRvL..13..539P, doi:10.1103/PhysRevLett.13.539

- Ramond, Pierre (1990), *Field Theory: A Modern Primer*, Addison-Wesley, ISBN 0-201-54611-6

- Rees, Martin (1966), "Appearance of Relativistically Expanding Radio Sources", *Nature* **211** (5048): 468–470, Bibcode:1966Natur.211..468R, doi:10.1038/211468a0

- Reissner, H. (1916), "Über die Eigengravitation des elektrischen Feldes nach der Einsteinschen Theorie", *Annalen der Physik* **355** (9): 106–120, Bibcode:1916AnP...355..106R, doi:10.1002/andp.19163550905

- Remillard, Ronald A.; Lin, Dacheng; Cooper, Randall L.; Narayan, Ramesh (2006), "The Rates of Type I X-Ray Bursts from Transients Observed with RXTE: Evidence for Black Hole Event Horizons", *Astrophysical Journal* **646** (1): 407–419, arXiv:astro-ph/0509758, Bibcode:2006ApJ...646..407R, doi:10.1086/504862

- Renn, Jürgen, ed. (2007), *The Genesis of General Relativity (4 Volumes)*, Dordrecht: Springer, ISBN 1-4020-3999-9

- Renn, Jürgen, ed. (2005), *Albert Einstein—Chief Engineer of the Universe: Einstein's Life and Work in Context*, Berlin: Wiley-VCH, ISBN 3-527-40571-2

- Reula, Oscar A. (1998), "Hyperbolic Methods for Einstein's Equations", *Living Rev. Relativity* **1**, Bibcode:1998LR, doi:10.12942/lrr-1998-3, retrieved 2007-08-29

- Rindler, Wolfgang (2001), *Relativity. Special, General and Cosmological*, Oxford University Press, ISBN 0-19-850836-0

- Rindler, Wolfgang (1991), *Introduction to Special Relativity*, Clarendon Press, Oxford, ISBN 0-19-853952-5

- Robson, Ian (1996), *Active galactic nuclei*, John Wiley, ISBN 0-471-95853-0

- Roulet, E.; Mollerach, S. (1997), "Microlensing", *Physics Reports* **279** (2): 67–118, arXiv:astro-ph/9603119, Bibcode:1997PhR...279...67R, doi:10.1016/S0370-1573(96)00020-8

- Rovelli, Carlo (2000). "Notes for a brief history of quantum gravity". arXiv:gr-qc/0006061 [gr-qc].

- Rovelli, Carlo (1998), "Loop Quantum Gravity", *Living Rev. Relativity* **1**, doi:10.12942/lrr-1998-1, retrieved 2008-03-13

- Schäfer, Gerhard (2004), "Gravitomagnetic Effects", *General Relativity and Gravitation* **36** (10): 2223–2235, arXiv:gr-qc/0407116, Bibcode:2004GReGr..36.2223S, doi:10.1023/B:GERG.0000046180.97877.32

- Schödel, R.; Ott, T.; Genzel, R.; Eckart, A.; Mouawad, N.; Alexander, T. (2003), "Stellar Dynamics in the Central Arcsecond of Our Galaxy", *Astrophysical Journal* **596** (2): 1015–1034, arXiv:astro-ph/0306214, Bibcode:2S, doi:10.1086/378122

- Schutz, Bernard F. (1985), *A first course in general relativity*, Cambridge University Press, ISBN 0-521-27703-5

- Schutz, Bernard F. (2001), "Gravitational radiation", in Murdin, Paul, *Encyclopedia of Astronomy and Astrophysics*, Grove's Dictionaries, ISBN 1-56159-268-4

- Schutz, Bernard F. (2003), *Gravity from the ground up*, Cambridge University Press, ISBN 0-521-45506-5

- Schwarz, John H. (2007), "String Theory: Progress and Problems", *Progress of Theoretical Physics Supplement* **170**: 214, arXiv:hep-th/0702219, Bibcode:2007PThPS.170..214S, doi:10.1143/PTPS.170.214

- Schwarzschild, Karl (1916a), "Über das Gravitationsfeld eines Massenpunktes nach der Einsteinschen Theorie", *Sitzungsber. Preuss. Akad. D. Wiss.*: 189–196

- Schwarzschild, Karl (1916b), "Über das Gravitationsfeld eines Kugel aus inkompressibler Flüssigkeit nach der Einsteinschen Theorie", *Sitzungsber. Preuss. Akad. D. Wiss.*: 424–434

- Seidel, Edward (1998), "Numerical Relativity: Towards Simulations of 3D Black Hole Coalescence", in Narlikar, J. V.; Dadhich, N., *Gravitation and Relativity: At the turn of the millennium (Proceedings of the GR-15 Conference, held at IUCAA, Pune, India, December 16–21, 1997)*, IUCAA, p. 6088, arXiv:gr-qc/9806088, Bibcode:1998, ISBN 81-900378-3-8

- Seljak, Uroš; Zaldarriaga, Matias (1997), "Signature of Gravity Waves in the Polarization of the Microwave Background", *Phys. Rev. Lett.* **78** (11): 2054–2057, arXiv:astro-ph/9609169, Bibcode:1997PhRvL..78.2054S, doi:10.1103/PhysRevLett.78.2054

- Shapiro, S. S.; Davis, J. L.; Lebach, D. E.; Gregory, J. S. (2004), "Measurement of the solar gravitational deflection of radio waves using geodetic very-long-baseline interferometry data, 1979–1999", *Phys. Rev. Lett.* **92** (12): 121101, Bibcode:2004PhRvL..92l1101S, doi:10.1103/PhysRevLett.92.121101, PMID 15089661

- Shapiro, Irwin I. (1964), "Fourth test of general relativity", *Phys. Rev. Lett.* **13** (26): 789–791, Bibcode:1964PhRvL doi:10.1103/PhysRevLett.13.789

- Shapiro, I. I.; Pettengill, Gordon; Ash, Michael; Stone, Melvin; Smith, William; Ingalls, Richard; Brockelman, Richard (1968), "Fourth test of general relativity: preliminary results", *Phys. Rev. Lett.* **20** (22): 1265–1269, Bibcode:1968PhRvL..20.1265S, doi:10.1103/PhysRevLett.20.1265

- Singh, Simon (2004), *Big Bang: The Origin of the Universe*, Fourth Estate, ISBN 0-00-715251-5

- Sorkin, Rafael D. (2005), "Causal Sets: Discrete Gravity", in Gomberoff, Andres; Marolf, Donald, *Lectures on Quantum Gravity*, Springer, p. 9009, arXiv:gr-qc/0309009, Bibcode:2003gr.qc.....9009S, ISBN 0-387-23995-2

- Sorkin, Rafael D. (1997), "Forks in the Road, on the Way to Quantum Gravity", *Int. J. Theor. Phys.* **36** (12): 2759–2781, arXiv:gr-qc/9706002, Bibcode:1997IJTP...36.2759S, doi:10.1007/BF02435709

- Spergel, D. N.; Verde, L.; Peiris, H. V.; Komatsu, E.; Nolta, M. R.; Bennett, C. L.; Halpern, M.; Hinshaw, G.; et al. (2003), "First Year Wilkinson Microwave Anisotropy Probe (WMAP) Observations: Determination of Cosmological Parameters", *Astrophys. J. Suppl.* **148** (1): 175–194, arXiv:astro-ph/0302209, Bibcode:2003ApJS..148..175S, doi:10.1086/377226

- Spergel, D. N.; Bean, R.; Doré, O.; Nolta, M. R.; Bennett, C. L.; Dunkley, J.; Hinshaw, G.; Jarosik, N.; et al. (2007), "Wilkinson Microwave Anisotropy Probe (WMAP) Three Year Results: Implications for Cosmology", *Astrophysical Journal Supplement* **170** (2): 377–408, arXiv:astro-ph/0603449, Bibcode:2007ApJS..170..377S, doi:10.1086/513700

- Springel, Volker; White, Simon D. M.; Jenkins, Adrian; Frenk, Carlos S.; Yoshida, Naoki; Gao, Liang; Navarro, Julio; Thacker, Robert; et al. (2005), "Simulations of the formation, evolution and clustering of galaxies and quasars", *Nature* **435** (7042): 629–636, arXiv:astro-ph/0504097,Bibcode:2005Natur.435..629S,doi:10.103, PMID 15931216

- Stairs, Ingrid H. (2003), "Testing General Relativity with Pulsar Timing", *Living Rev. Relativity* **6**, arXiv:astro-ph/0307536, Bibcode:2003LRR.....6....5S, doi:10.12942/lrr-2003-5, retrieved 2007-07-21

- Stephani, H.; Kramer, D.; MacCallum, M.; Hoenselaers, C.; Herlt, E. (2003), *Exact Solutions of Einstein's Field Equations* (2 ed.), Cambridge University Press, ISBN 0-521-46136-7

- Synge, J. L. (1972), *Relativity: The Special Theory*, North-Holland Publishing Company, ISBN 0-7204-0064-3

- Szabados, László B. (2004), "Quasi-Local Energy-Momentum and Angular Momentum in GR", *Living Rev. Relativity* **7**, doi:10.12942/lrr-2004-4, retrieved 2007-08-23

- Taylor, Joseph H.(1994), "Binary pulsars and relativistic gravity",*Rev. Mod. Phys.***66**(3): 711–719.Bibcode:19T, doi:10.1103/RevModPhys.66.711

- Thiemann, Thomas (2006), "Approaches to Fundamental Physics: Loop Quantum Gravity: An Inside View", *Lecture Notes in Physics* **721**: 185–263, arXiv:hep-th/0608210, Bibcode:2007LNP...721..185T, doi:10.1007/978-3-540-71117-9_10, ISBN 978-3-540-71115-5

- Thiemann, Thomas (2003), "Lectures on Loop Quantum Gravity", *Lecture Notes in Physics* **631**: 41–135, arXiv:gr-qc/0210094, doi:10.1007/978-3-540-45230-0_3, ISBN 978-3-540-40810-9

- Thorne, Kip S. (1972), "Nonspherical Gravitational Collapse—A Short Review", in Klauder, J., *Magic without Magic*, W. H. Freeman, pp. 231–258

- Thorne, Kip S. (1994), *Black Holes and Time Warps: Einstein's Outrageous Legacy*, W W Norton & Company, ISBN 0-393-31276-3

- Thorne, Kip S. (1995), "Gravitational radiation", *Particle and Nuclear Astrophysics and Cosmology in the Next Millenium*: 160, arXiv:gr-qc/9506086, Bibcode:1995pnac.conf..160T, ISBN 0-521-36853-7

- Townsend, Paul K. (1997), "Black Holes (Lecture notes)". arXiv:gr-qc/9707012 [gr-qc].

- Townsend, Paul K. (1996). "Four Lectures on M-Theory". arXiv:hep-th/9612121 [hep-th].

- Traschen, Jenny (2000), Bytsenko, A.; Williams, F., eds., "An Introduction to Black Hole Evaporation", *Mathematical Methods of Physics (Proceedings of the 1999 Londrina Winter School)* (World Scientific): 180, arXiv:gr-qc/0010055, Bibcode:2000mmp..conf..180T

- Trautman, Andrzej (2006), "Einstein–Cartan theory", in Françoise, J.-P.; Naber, G. L.; Tsou, S. T., *Encyclopedia of Mathematical Physics, Vol. 2*, Elsevier, pp. 189–195, arXiv:gr-qc/0606062, Bibcode:2006gr.qc.....6062T

- Unruh, W. G.(1976), "Notes on Black Hole Evaporation",*Phys. Rev. D***14**(4): 870–892,Bibcode:1976PhRvDU, doi:10.1103/PhysRevD.14.870

- Valtonen, M. J.; Lehto, H. J.; Nilsson, K.; Heidt, J.; Takalo, L. O.; Sillanpää, A.; Villforth, C.; Kidger, M.; et al. (2008), "A massive binary black-hole system in OJ 287 and a test of general relativity", *Nature* **452** (7189): 851–853, arXiv:0809.1280, Bibcode:2008Natur.452..851V, doi:10.1038/nature06896, PMID 18421348

- Wald, Robert M.(1975), "On Particle Creation by Black Holes",*Commun. Math. Phys.***45**(3): 9–34,Bibcode:.9W, doi:10.1007/BF01609863

- Wald, Robert M. (1984), *General Relativity*, University of Chicago Press, ISBN 0-226-87033-2

- Wald, Robert M. (1994), *Quantum field theory in curved spacetime and black hole thermodynamics*, University of Chicago Press, ISBN 0-226-87027-8

- Wald, Robert M. (2001),"The Thermodynamics of Black Holes",*Living Rev. Relativity***4**,Bibcode:2001LRR....., doi:10.12942/lrr-2001-6, retrieved 2007-08-08

- Walsh, D.; Carswell, R. F.; Weymann, R. J. (1979), "0957 + 561 A, B: twin quasistellar objects or gravitational lens?", *Nature* **279** (5712): 381–4, Bibcode:1979Natur.279..381W, doi:10.1038/279381a0, PMID 16068158

- Wambsganss, Joachim (1998), "Gravitational Lensing in Astronomy", *Living Rev. Relativity* **1**, arXiv:astro-ph1, Bibcode:1998LRR.....1...12W, doi:10.12942/lrr-1998-12, retrieved 2007-07-20

- Weinberg, Steven (1972), *Gravitation and Cosmology*, John Wiley, ISBN 0-471-92567-5

- Weinberg, Steven (1995), *The Quantum Theory of Fields I: Foundations*, Cambridge University Press, ISBN 0-521-55001-7

- Weinberg, Steven (1996), *The Quantum Theory of Fields II: Modern Applications*, Cambridge University Press, ISBN 0-521-55002-5

- Weinberg, Steven (2000), *The Quantum Theory of Fields III: Supersymmetry*, Cambridge University Press, ISBN 0-521-66000-9

- Weisberg, Joel M.; Taylor, Joseph H. (2003), "The Relativistic Binary Pulsar B1913+16"", in Bailes, M.; Nice, D. J.; Thorsett, S. E., *Proceedings of "Radio Pulsars," Chania, Crete, August, 2002*, ASP Conference Series

- Weiss, Achim (2006), "Elements of the past: Big Bang Nucleosynthesis and observation", *Einstein Online* (Max Planck Institute for Gravitational Physics), retrieved 2007-02-24 External link in |work= (help)

- Wheeler, John A. (1990), *A Journey Into Gravity and Spacetime*, Scientific American Library, San Francisco: W. H. Freeman, ISBN 0-7167-6034-7

- Will, Clifford M. (1993), *Theory and experiment in gravitational physics*, Cambridge University Press, ISBN 0-521-43973-6

- Will, Clifford M. (2006), "The Confrontation between General Relativity and Experiment", *Living Rev. Relativity* **9**, arXiv:gr-qc/0510072, Bibcode:2006LRR.....9....3W, doi:10.12942/lrr-2006-3, retrieved 2007-06-12

- Zwiebach, Barton (2004), *A First Course in String Theory*, Cambridge University Press, ISBN 0-521-83143-1

6.12 Further reading

Popular books

- Geroch, R (1981), *General Relativity from A to B*, Chicago: University of Chicago Press, ISBN 0-226-28864-1

- Lieber, Lillian (2008), *The Einstein Theory of Relativity: A Trip to the Fourth Dimension*, Philadelphia: Paul Dry Books, Inc., ISBN 978-1-58988-044-3

- Wald, Robert M. (1992), *Space, Time, and Gravity: the Theory of the Big Bang and Black Holes*, Chicago: University of Chicago Press, ISBN 0-226-87029-4

- Wheeler, John; Ford, Kenneth (1998), *Geons, Black Holes, & Quantum Foam: a life in physics*, New York: W. W. Norton, ISBN 0-393-31991-1

Beginning undergraduate textbooks

- Callahan, James J. (2000), *The Geometry of Spacetime: an Introduction to Special and General Relativity*, New York: Springer, ISBN 0-387-98641-3

- Taylor, Edwin F.; Wheeler, John Archibald (2000), *Exploring Black Holes: Introduction to General Relativity*, Addison Wesley, ISBN 0-201-38423-X

Advanced undergraduate textbooks

- B. F. Schutz (2009), *A First Course in General Relativity (Second Edition)*, Cambridge University Press, ISBN 978-0-521-88705-2

- Cheng, Ta-Pei (2005), *Relativity, Gravitation and Cosmology: a Basic Introduction*, Oxford and New York: Oxford University Press, ISBN 0-19-852957-0

- Gron, O.; Hervik, S. (2007), *Einstein's General theory of Relativity*, Springer, ISBN 978-0-387-69199-2

- Hartle, James B. (2003), *Gravity: an Introduction to Einstein's General Relativity*, San Francisco: Addison-Wesley, ISBN 0-8053-8662-9

- Hughston, L. P. & Tod, K. P. (1991), *Introduction to General Relativity*, Cambridge: Cambridge University Press, ISBN 0-521-33943-X

- d'Inverno, Ray (1992), *Introducing Einstein's Relativity*, Oxford: Oxford University Press, ISBN 0-19-859686-3

- Ludyk, Günter (2013). *Einstein in Matrix Form* (1st ed.). Berlin: Springer. ISBN 9783642357978.

Graduate-level textbooks

- Carroll, Sean M. (2004), *Spacetime and Geometry: An Introduction to General Relativity*, San Francisco: Addison-Wesley, ISBN 0-8053-8732-3

- Grøn, Øyvind; Hervik, Sigbjørn (2007), *Einstein's General Theory of Relativity*, New York: Springer, ISBN 978-0-387-69199-2

- Landau, Lev D.; Lifshitz, Evgeny F. (1980), *The Classical Theory of Fields (4th ed.)*, London: Butterworth-Heinemann, ISBN 0-7506-2768-9

- Misner, Charles W.; Thorne, Kip. S.; Wheeler, John A. (1973), *Gravitation*, W. H. Freeman, ISBN 0-7167-0344-0

- Stephani, Hans (1990), *General Relativity: An Introduction to the Theory of the Gravitational Field*, Cambridge: Cambridge University Press, ISBN 0-521-37941-5

- Wald, Robert M. (1984), *General Relativity*, University of Chicago Press, ISBN 0-226-87033-2

6.13   External links

- Einstein Online – Articles on a variety of aspects of relativistic physics for a general audience; hosted by the Max Planck Institute for Gravitational Physics

- NCSA Spacetime Wrinkles – produced by the numerical relativity group at the NCSA, with an elementary introduction to general relativity

- **Courses**

- **Lectures**

- **Tutorials**

- Einstein's General Theory of Relativity on YouTube (lecture by Leonard Susskind recorded September 22, 2008 at Stanford University).

- Series of lectures on General Relativity given in 2006 at the Institut Henri Poincaré (introductory/advanced).

- General Relativity Tutorials by John Baez.

- Brown, Kevin. "Reflections on relativity". *Mathpages.com*. Retrieved May 29, 2005.

- Carroll, Sean M. "Lecture Notes on General Relativity". Retrieved January 5, 2014.

- Moor, Rafi. "Understanding General Relativity". Retrieved July 11, 2006.

- Waner, Stefan. "Introduction to Differential Geometry and General Relativity" (PDF). Retrieved 2015-04-05.

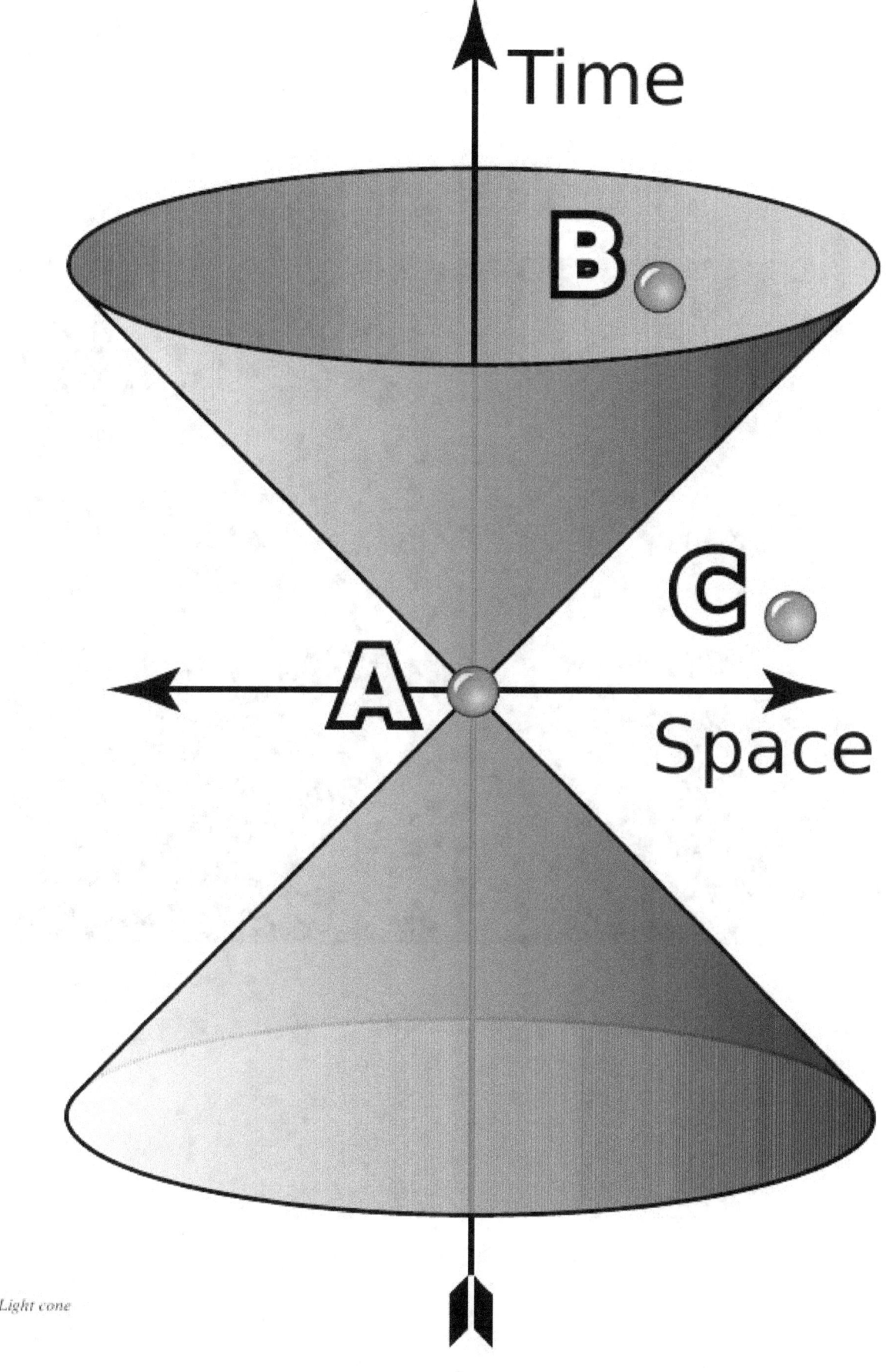

Light cone

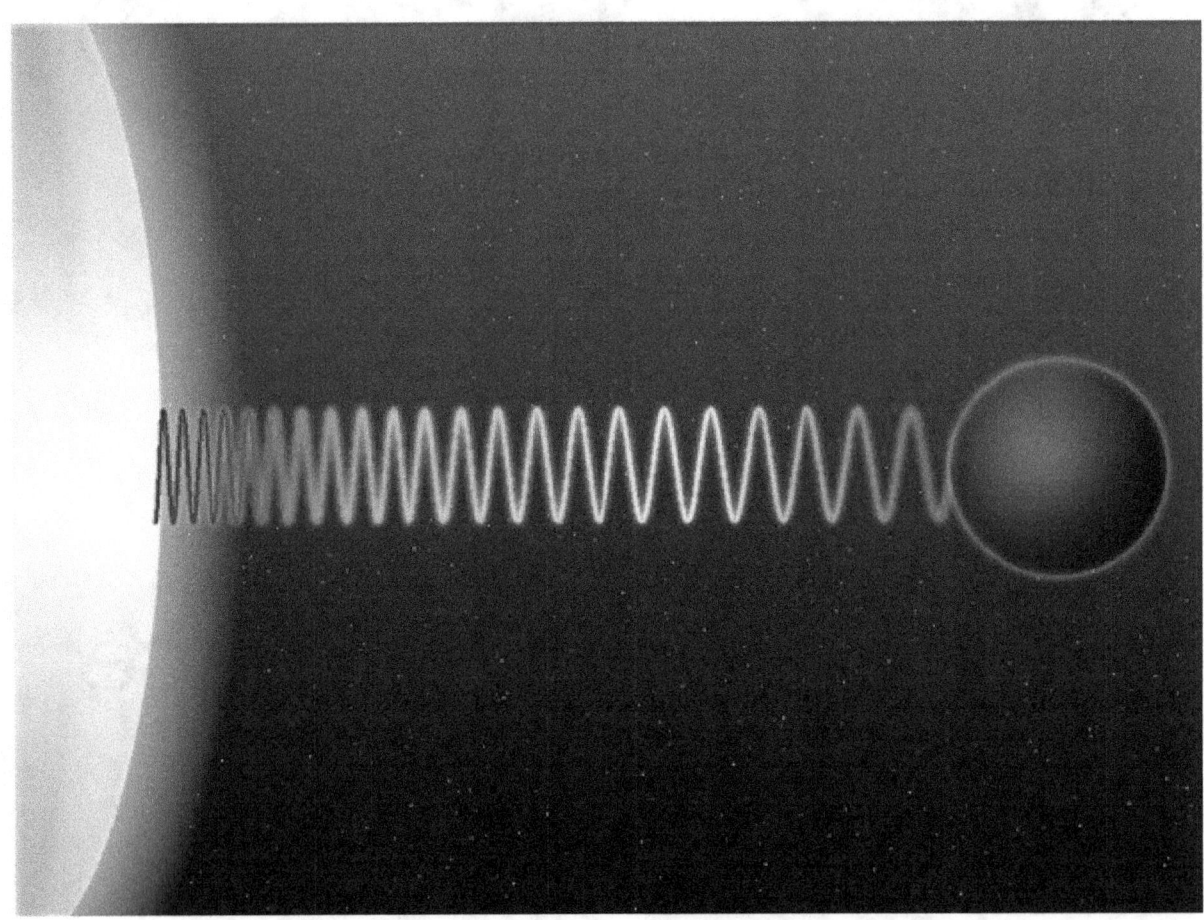

Schematic representation of the gravitational redshift of a light wave escaping from the surface of a massive body

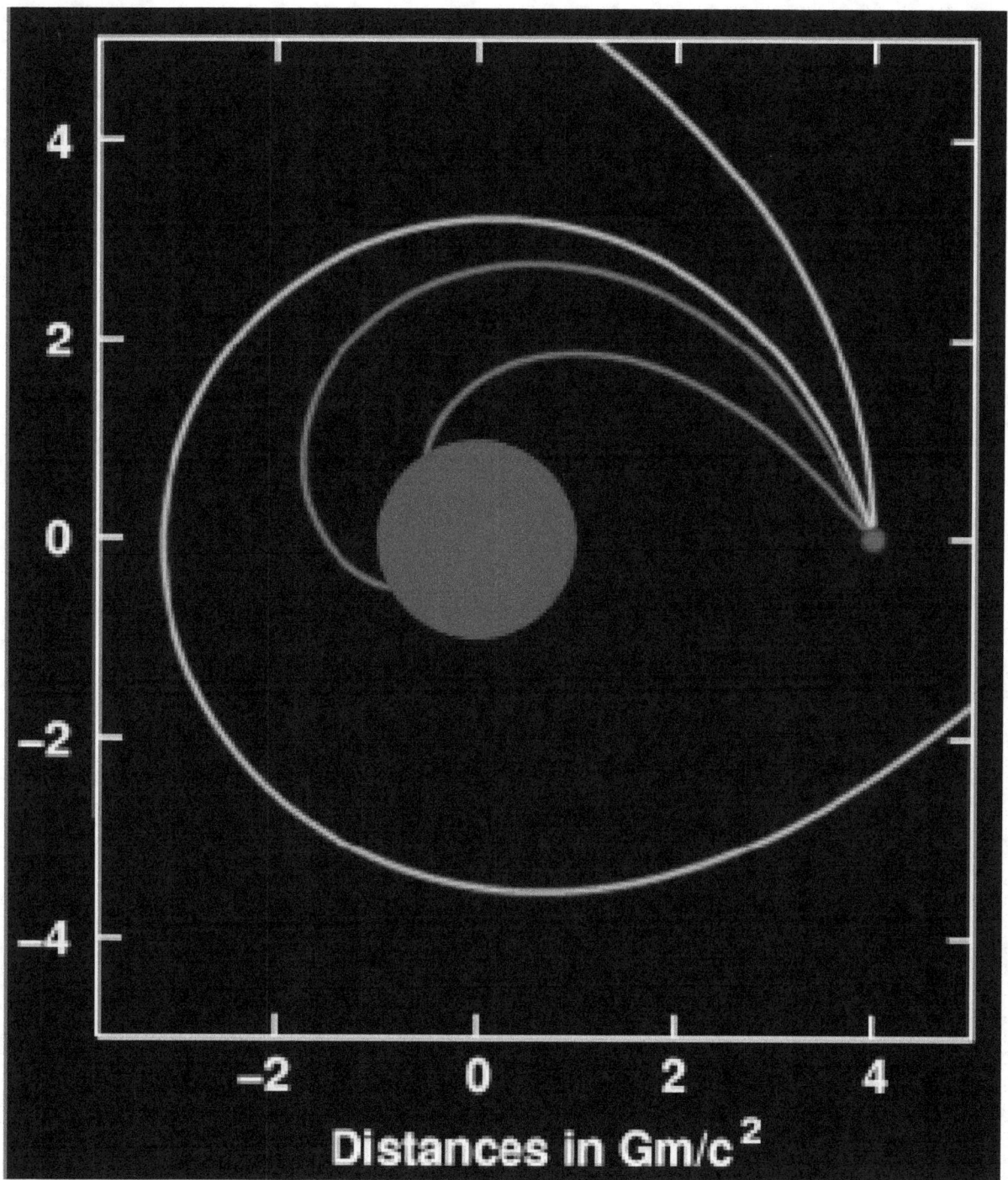

Deflection of light (sent out from the location shown in blue) near a compact body (shown in gray)

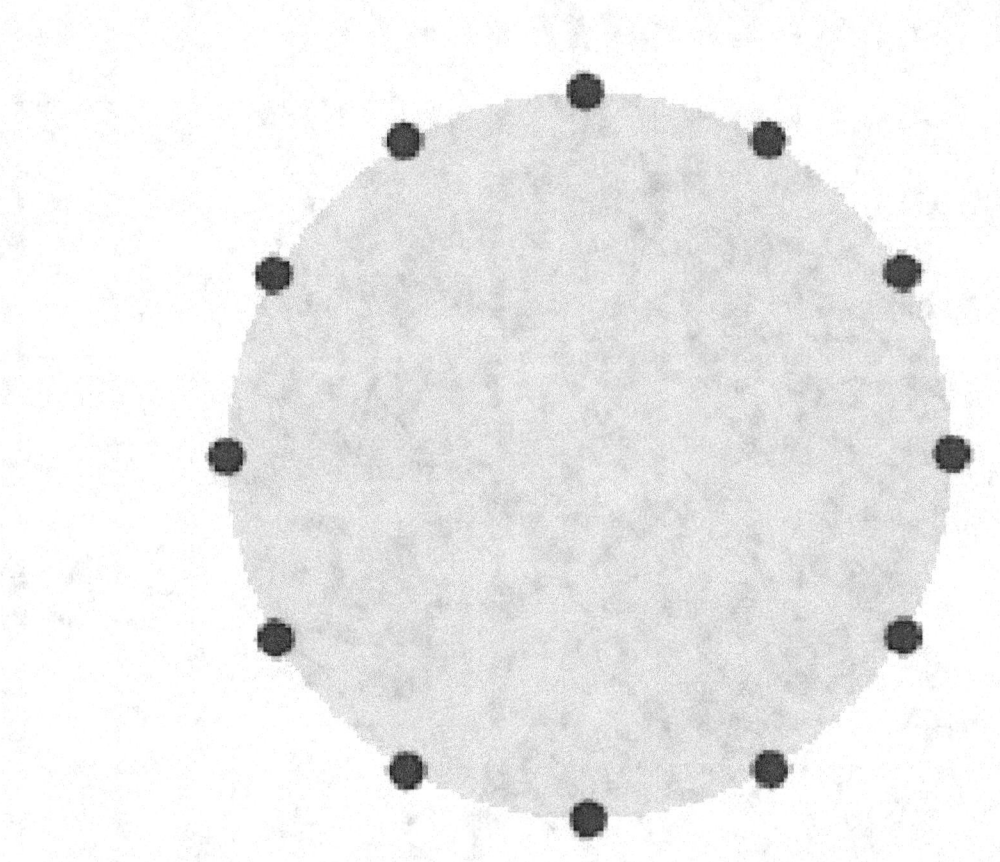

Ring of test particles influenced by gravitational wave

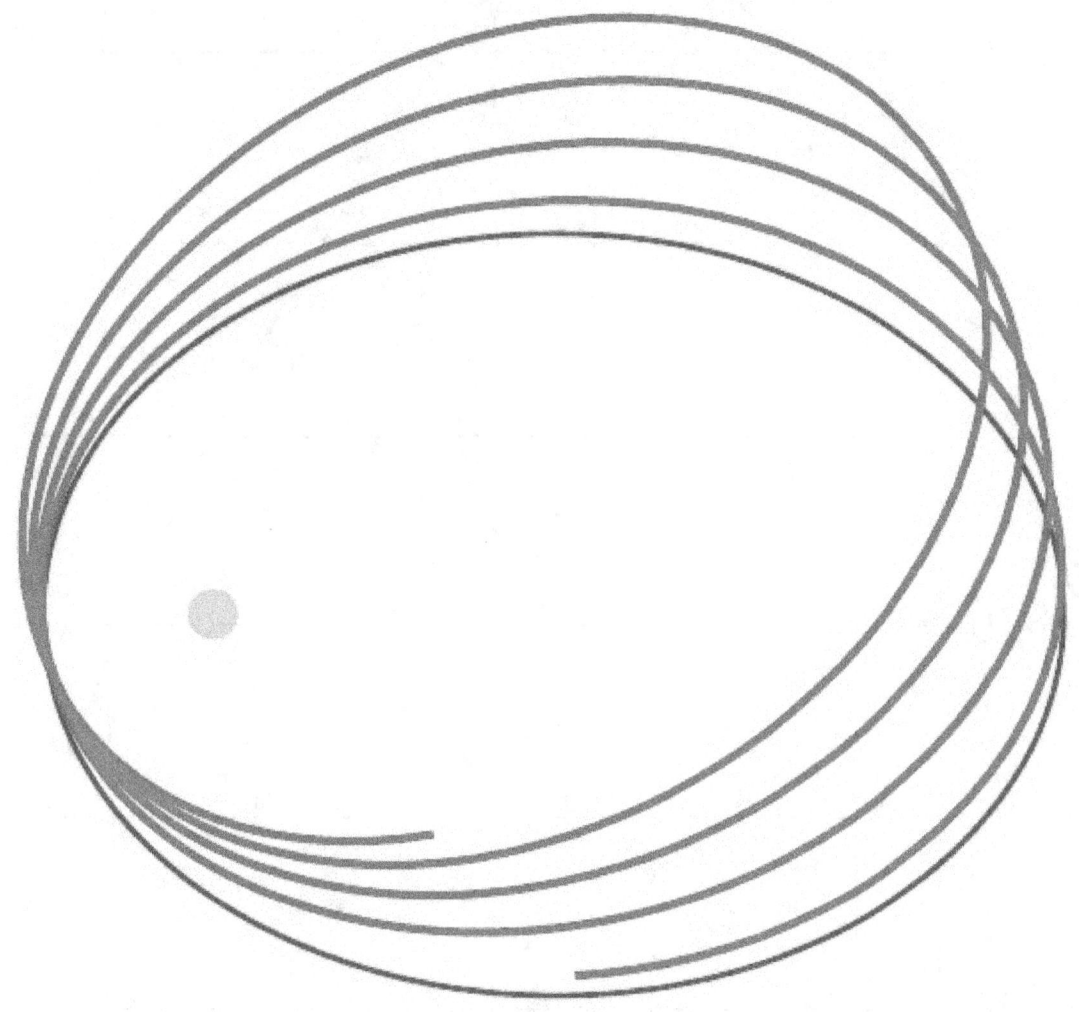

Newtonian (red) vs. Einsteinian orbit (blue) of a lone planet orbiting a star

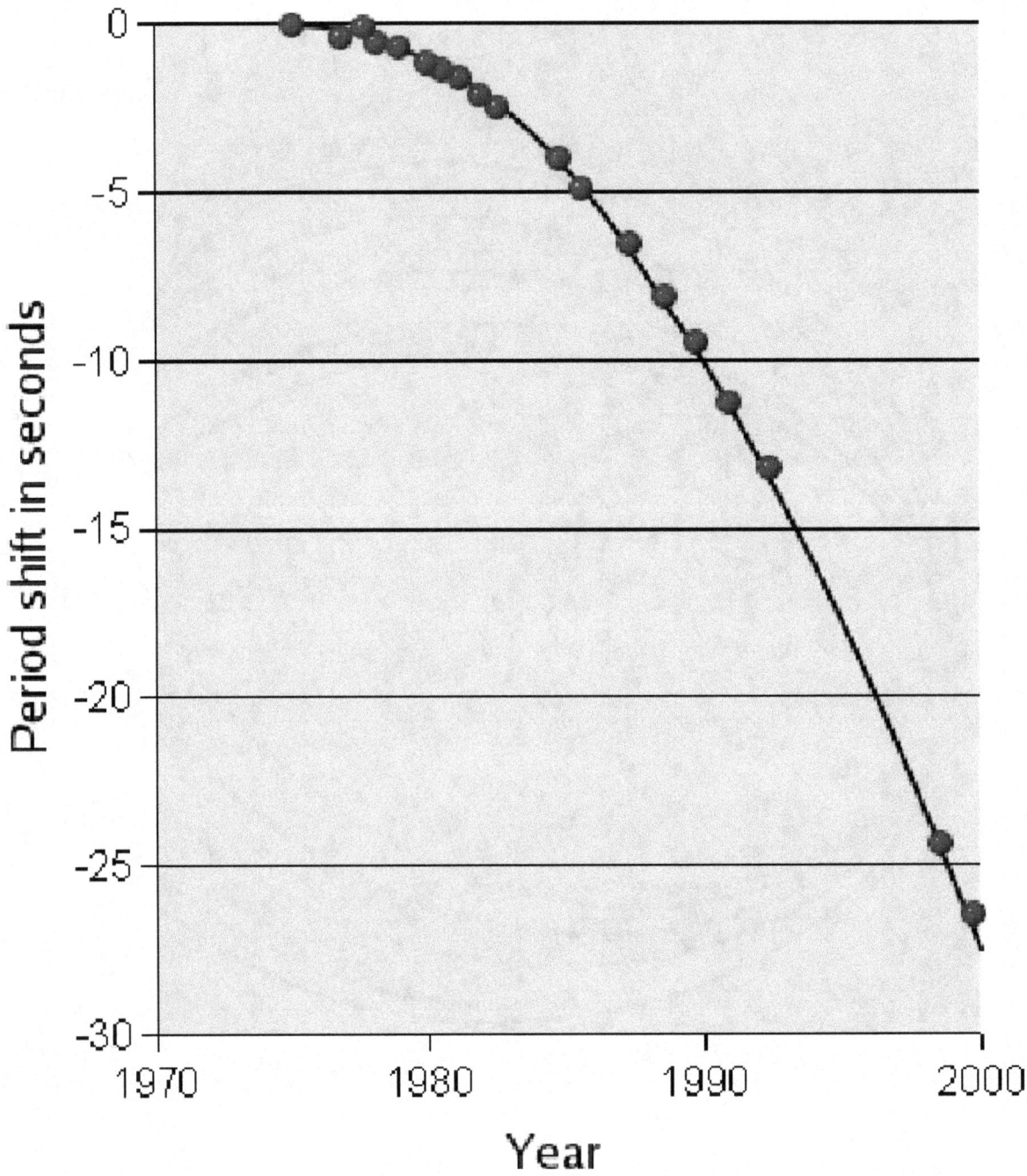

Orbital decay for PSR1913+16: time shift in seconds, tracked over three decades.[78]

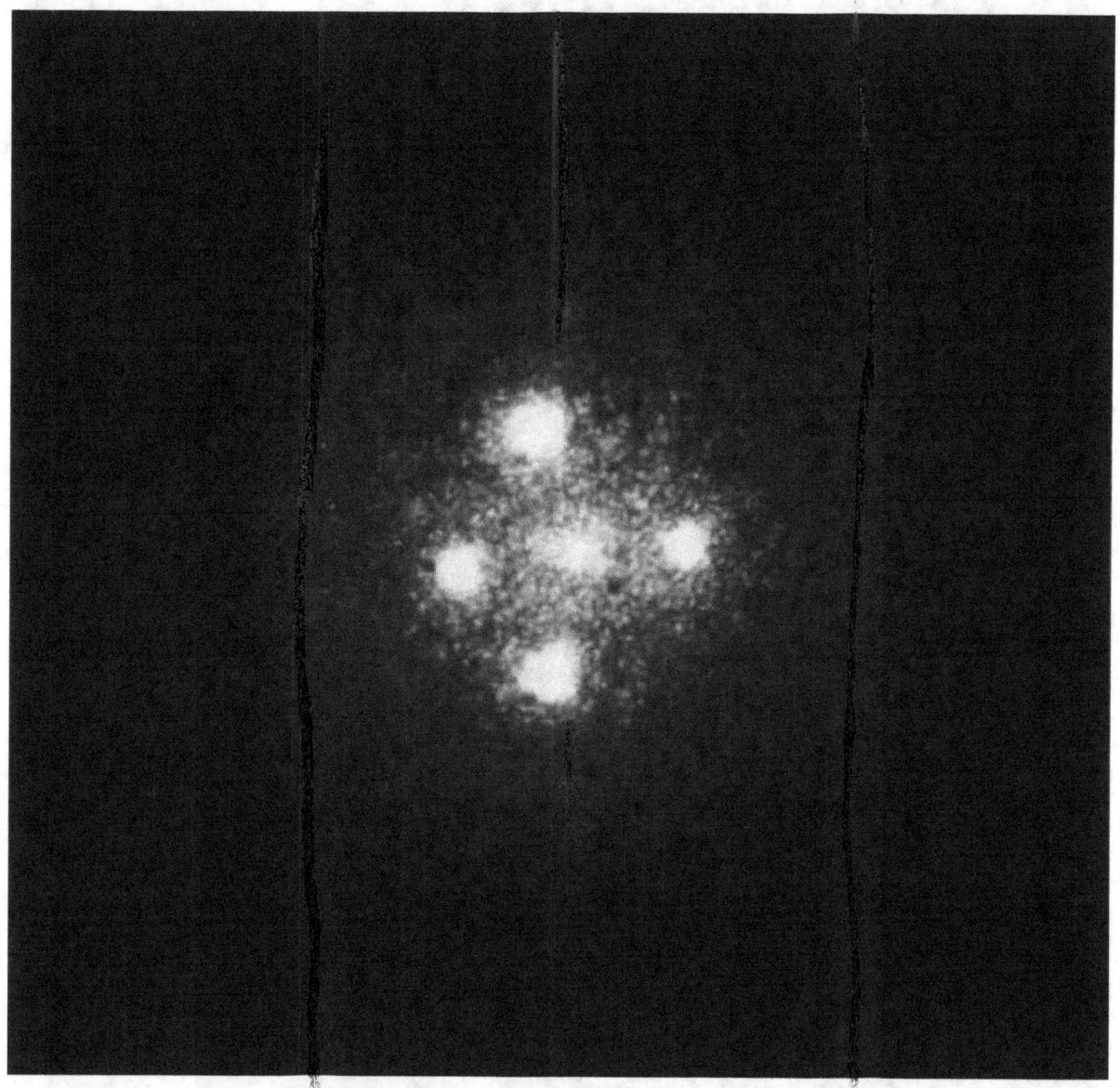

Einstein cross: four images of the same astronomical object, produced by a gravitational lens

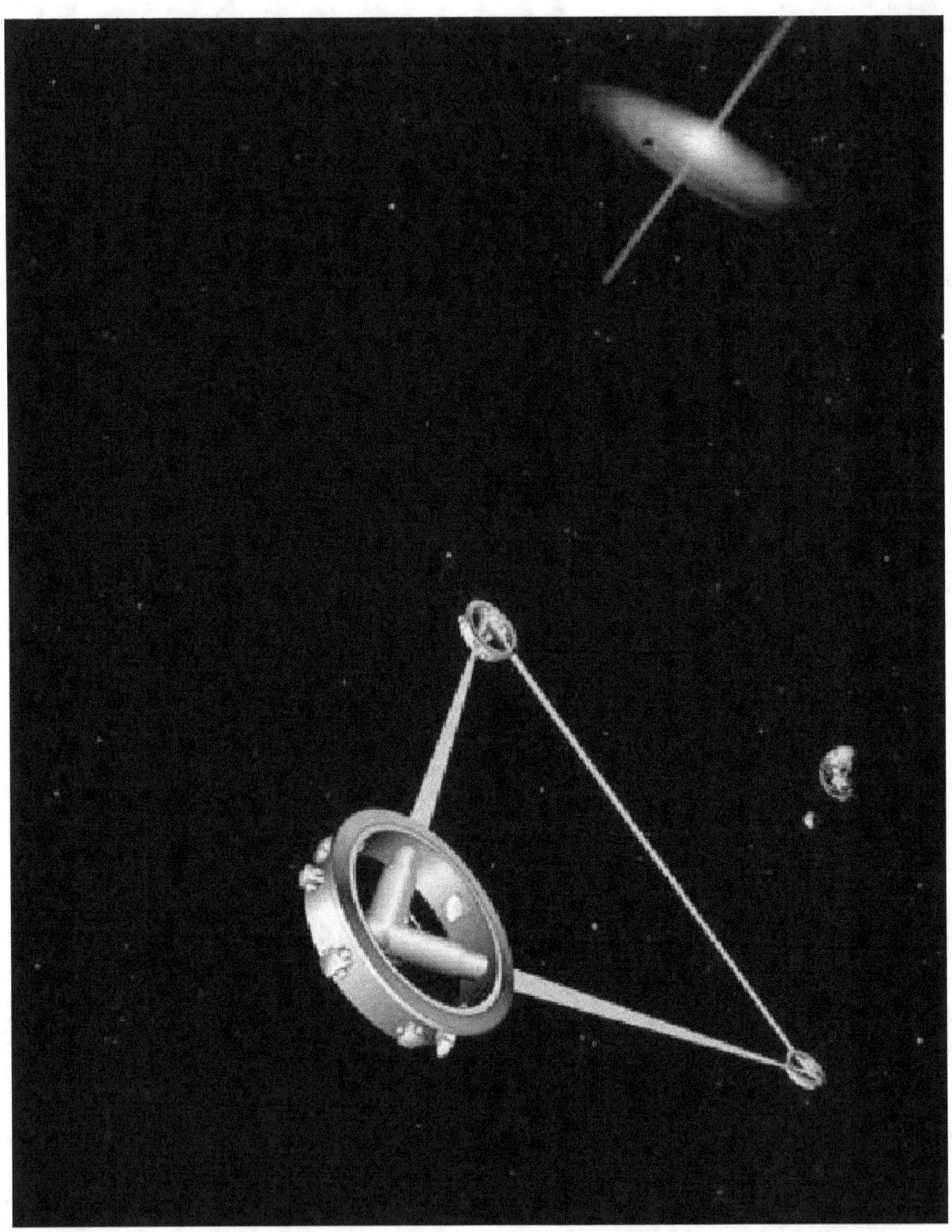

Artist's impression of the space-borne gravitational wave detector LISA

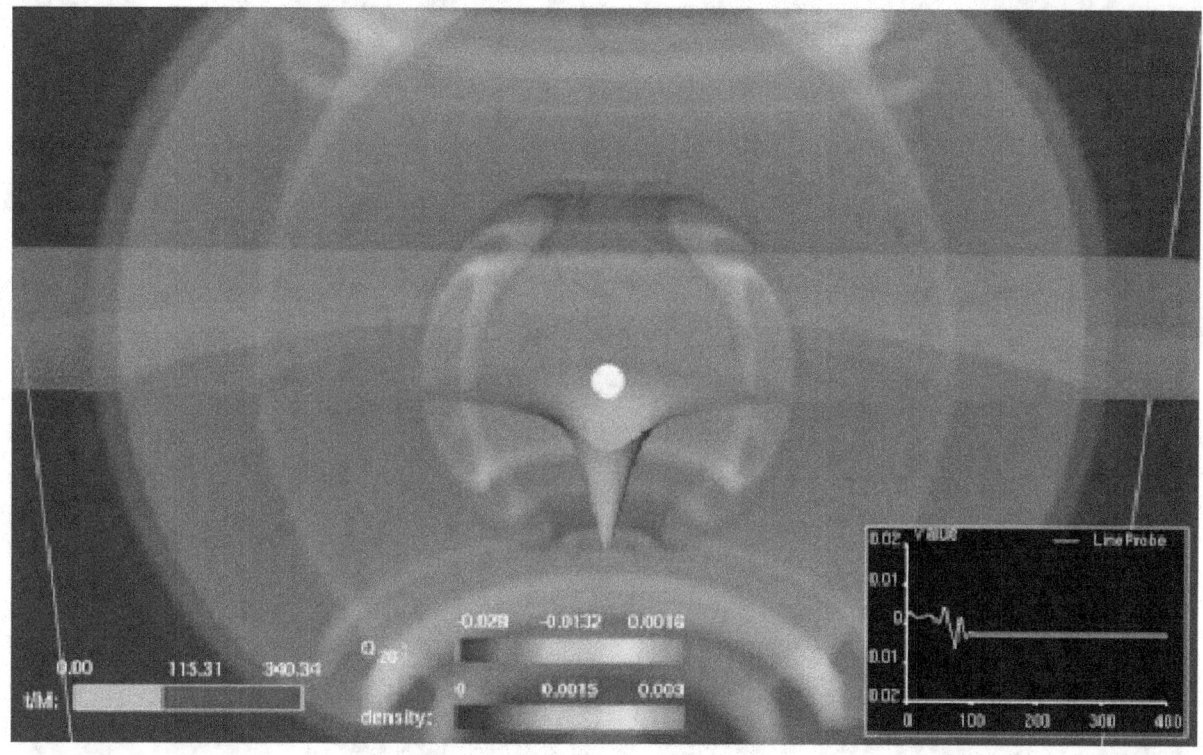

Simulation based on the equations of general relativity: a star collapsing to form a black hole while emitting gravitational waves

This blue horseshoe is a distant galaxy that has been magnified and warped into a nearly complete ring by the strong gravitational pull of the massive foreground luminous red galaxy.

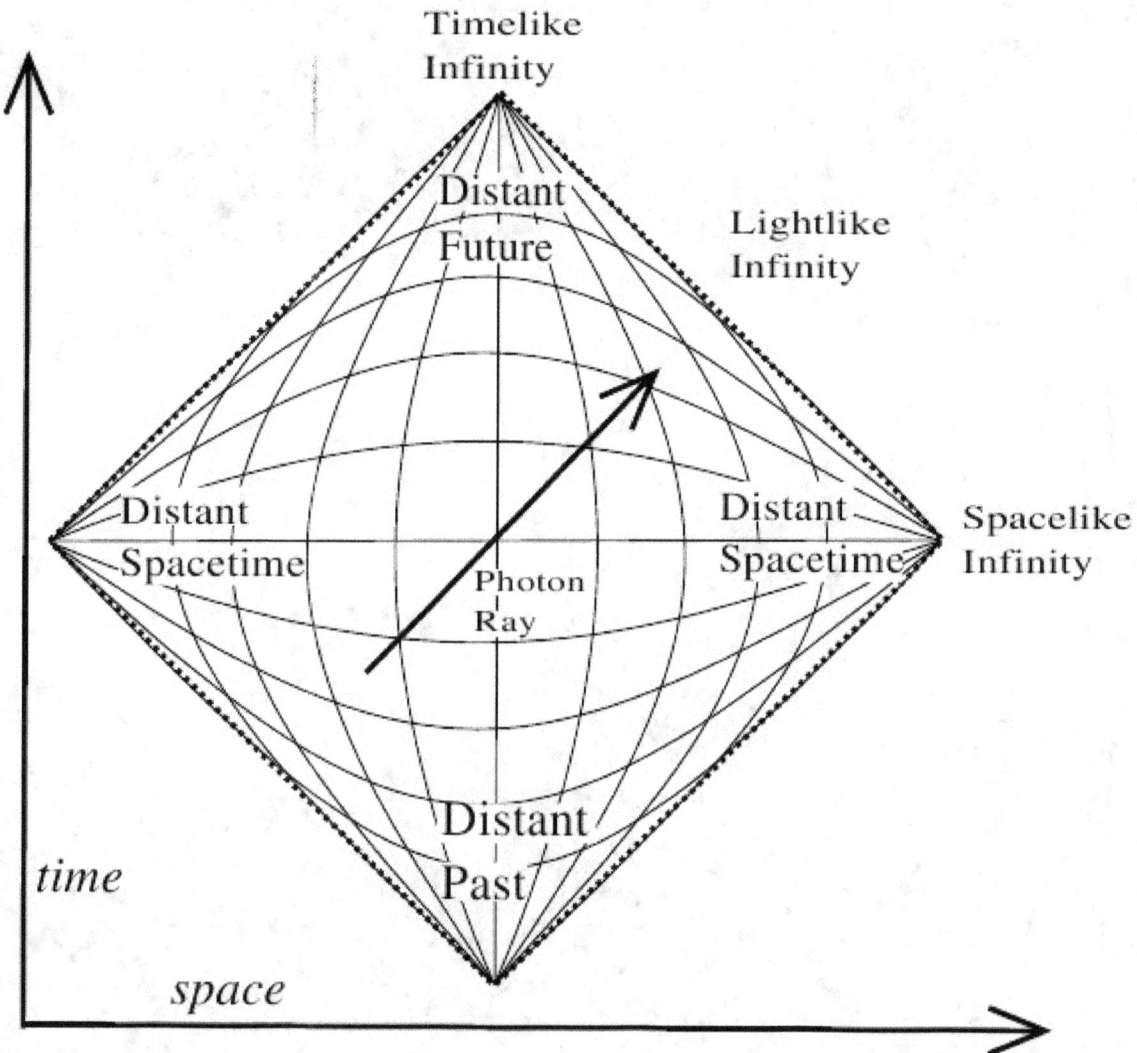

Penrose–Carter diagram of an infinite Minkowski universe

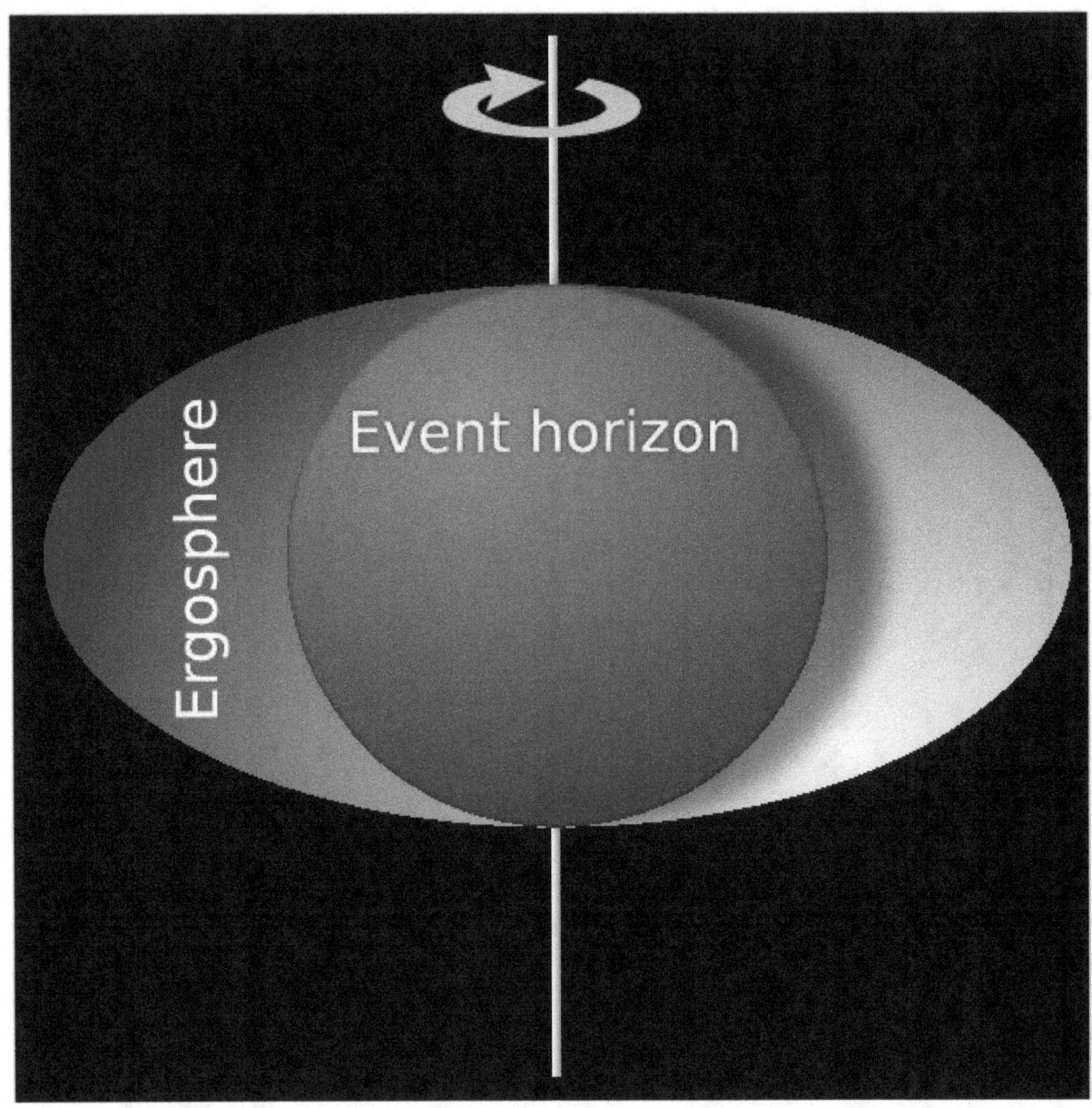

The ergosphere of a rotating black hole, which plays a key role when it comes to extracting energy from such a black hole

Projection of a Calabi–Yau manifold, one of the ways of compactifying the extra dimensions posited by string theory

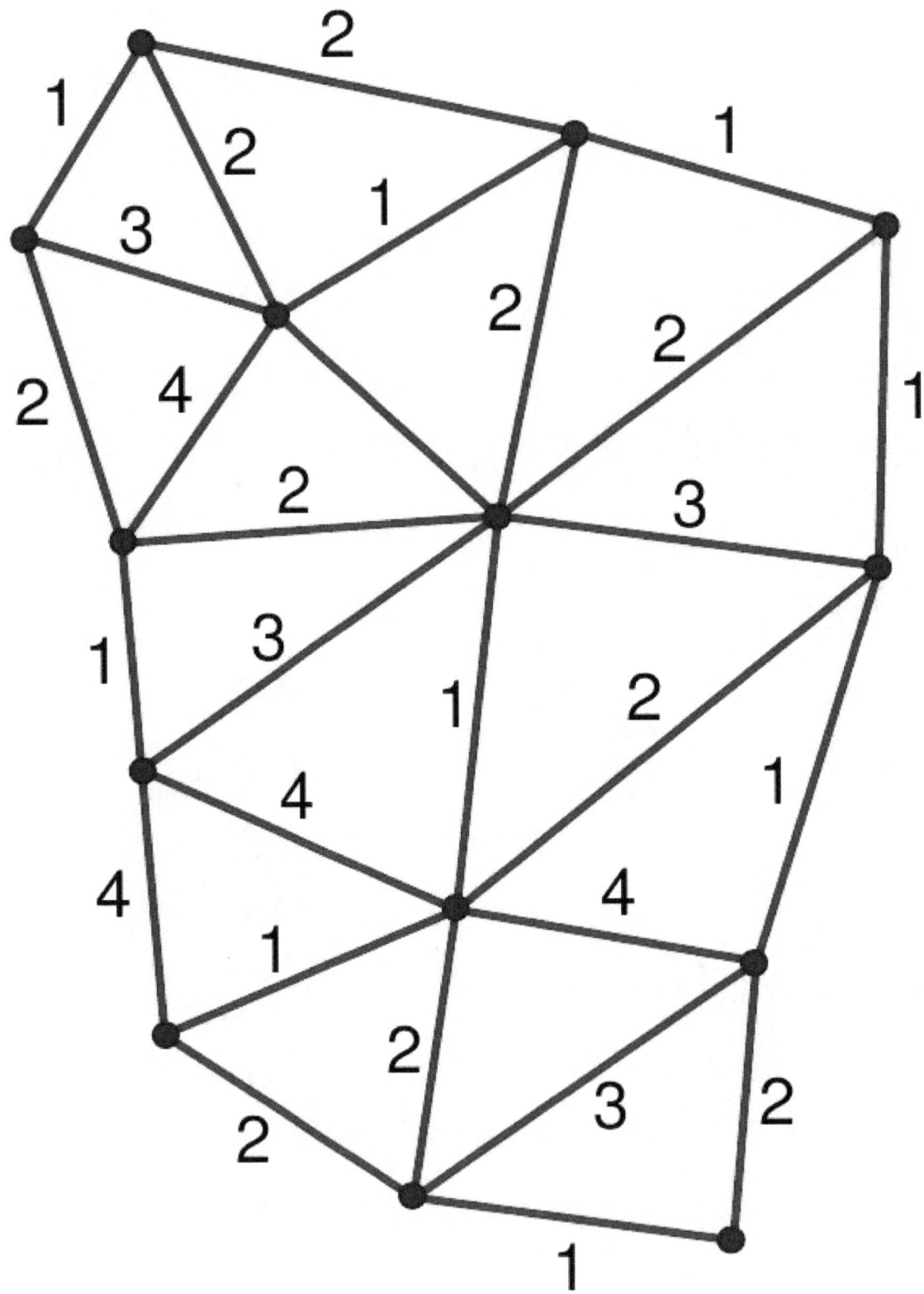

Simple spin network of the type used in loop quantum gravity

Chapter 7

Quantum field theory

"Relativistic quantum field theory" redirects here. For other uses, see Relativity.

In theoretical physics, **quantum field theory** (**QFT**) is a theoretical framework for constructing quantum mechanical models of subatomic particles in particle physics and quasiparticles in condensed matter physics. A QFT treats particles as excited states of an underlying physical field, so these are called field quanta.

In quantum field theory, quantum mechanical interactions between particles are described by interaction terms between the corresponding underlying fields.

7.1 Definition

Quantum electrodynamics (QED) has one electron field and one photon field; quantum chromodynamics (QCD) has one field for each type of quark; and, in condensed matter, there is an atomic displacement field that gives rise to phonon particles. Edward Witten describes QFT as "by far" the most difficult theory in modern physics.[1]

7.1.1 Dynamics

See also: Relativistic dynamics

Ordinary quantum mechanical systems have a fixed number of particles, with each particle having a finite number of degrees of freedom. In contrast, the excited states of a QFT can represent any number of particles. This makes quantum field theories especially useful for describing systems where the particle count/number may change over time, a crucial feature of relativistic dynamics.

7.1.2 States

QFT interaction terms are similar in spirit to those between charges with electric and magnetic fields in Maxwell's equations. However, unlike the classical fields of Maxwell's theory, fields in QFT generally exist in quantum superpositions of states and are subject to the laws of quantum mechanics.

Because the fields are continuous quantities over space, there exist excited states with arbitrarily large numbers of particles in them, providing QFT systems with an effectively infinite number of degrees of freedom. Infinite degrees of freedom can easily lead to divergences of calculated quantities (i.e., the quantities become infinite). Techniques such as renormalization of QFT parameters or discretization of spacetime, as in lattice QCD, are often used to avoid such infinities so as to yield physically meaningful results.

7.1.3 Fields and radiation

The gravitational field and the electromagnetic field are the only two fundamental fields in nature that have infinite range and a corresponding classical low-energy limit, which greatly diminishes and hides their "particle-like" excitations. Albert Einstein in 1905, attributed "particle-like" and discrete exchanges of momenta and energy, characteristic of "field quanta", to the electromagnetic field. Originally, his principal motivation was to explain the thermodynamics of radiation. Although the photoelectric effect and Compton scattering strongly suggest the existence of the photon, it might alternately be explained by a mere quantization of emission; more definitive evidence of the quantum nature of radiation is now taken up into modern quantum optics as in the antibunching effect.[2]

7.2 Theories

There is currently no complete quantum theory of the remaining fundamental force, gravity. Many of the proposed theories to describe gravity as a QFT postulate the existence of a graviton particle that mediates the gravitational force. Presumably, the as yet unknown correct quantum field-theoretic treatment of the gravitational field will behave like Einstein's general theory of relativity in the low-energy limit. Quantum field theory of the fundamental forces itself has been postulated to be the low-energy effective field theory limit of a more fundamental theory such as superstring theory.

Most theories in standard particle physics are formulated as **relativistic quantum field theories**, such as QED, QCD, and the Standard Model. QED, the quantum field-theoretic description of the electromagnetic field, approximately reproduces Maxwell's theory of electrodynamics in the low-energy limit, with small non-linear corrections to the Maxwell equations required due to virtual electron–positron pairs.

In the perturbative approach to quantum field theory, the full field interaction terms are approximated as a perturbative expansion in the number of particles involved. Each term in the expansion can be thought of as forces between particles being mediated by other particles. In QED, the electromagnetic force between two electrons is caused by an exchange of photons. Similarly, intermediate vector bosons mediate the weak force and gluons mediate the strong force in QCD. The notion of a force-mediating particle comes from perturbation theory, and does not make sense in the context of non-perturbative approaches to QFT, such as with bound states.

7.3 History

Main article: History of quantum field theory

7.3.1 Foundations

The early development of the field involved Dirac, Fock, Pauli, Heisenberg and Bogolyubov. This phase of development culminated with the construction of the theory of quantum electrodynamics in the 1950s.

7.3.2 Gauge theory

Gauge theory was formulated and quantized, leading to the **unification of forces** embodied in the standard model of particle physics. This effort started in the 1950s with the work of Yang and Mills, was carried on by Martinus Veltman and a host of others during the 1960s and completed by the 1970s through the work of Gerard 't Hooft, Frank Wilczek, David Gross and David Politzer.

7.3.3 Grand synthesis

Parallel developments in the understanding of phase transitions in condensed matter physics led to the study of the renormalization group. This in turn led to the **grand synthesis** of theoretical physics, which unified theories of particle and condensed matter physics through quantum field theory. This involved the work of Michael Fisher and Leo Kadanoff in the 1970s, which led to the seminal reformulation of quantum field theory by Kenneth G. Wilson in 1975.

7.4 Principles

7.4.1 Classical and quantum fields

Main article: Classical field theory

A classical field is a function defined over some region of space and time.[3] Two physical phenomena which are described by classical fields are Newtonian gravitation, described by Newtonian gravitational field $\mathbf{g}(\mathbf{x}, t)$, and classical electromagnetism, described by the electric and magnetic fields $\mathbf{E}(\mathbf{x}, t)$ and $\mathbf{B}(\mathbf{x}, t)$. Because such fields can in principle take on distinct values at each point in space, they are said to have infinite degrees of freedom.[3]

Classical field theory does not, however, account for the quantum-mechanical aspects of such physical phenomena. For instance, it is known from quantum mechanics that certain aspects of electromagnetism involve discrete particles— photons—rather than continuous fields. The business of *quantum* field theory is to write down a field that is, like a classical field, a function defined over space and time, but which also accommodates the observations of quantum mechanics. This is a *quantum field*.

It is not immediately clear *how* to write down such a quantum field, since quantum mechanics has a structure very unlike a field theory. In its most general formulation, quantum mechanics is a theory of abstract operators (observables) acting on an abstract state space (Hilbert space), where the observables represent physically observable quantities and the state space represents the possible states of the system under study.[4] For instance, the fundamental observables associated with the motion of a single quantum mechanical particle are the position and momentum operators $\hat{x}$ and $\hat{p}$. Field theory, in contrast, treats x as a way to index the field rather than as an operator.[5]

There are two common ways of developing a quantum field: the path integral formalism and canonical quantization.[6] The latter of these is pursued in this article.

Lagrangian formalism

Quantum field theory frequently makes use of the Lagrangian formalism from classical field theory. This formalism is analogous to the Lagrangian formalism used in classical mechanics to solve for the motion of a particle under the influence of a field. In classical field theory, one writes down a Lagrangian density, $\mathcal{L}$, involving a field, $\varphi(\mathbf{x},t)$, and possibly its first derivatives ($\partial\varphi/\partial t$ and $\nabla\varphi$), and then applies a field-theoretic form of the Euler–Lagrange equation. Writing coordinates $(t, \mathbf{x}) = (x^0, x^1, x^2, x^3) = x^\mu$, this form of the Euler–Lagrange equation is[3]

$$\frac{\partial}{\partial x^\mu}\left[\frac{\partial \mathcal{L}}{\partial(\partial\phi/\partial x^\mu)}\right] - \frac{\partial \mathcal{L}}{\partial \phi} = 0,$$

where a sum over μ is performed according to the rules of Einstein notation.

By solving this equation, one arrives at the "equations of motion" of the field.[3] For example, if one begins with the Lagrangian density

$$\mathcal{L}(\phi, \nabla\phi) = -\rho(t, \mathbf{x})\,\phi(t, \mathbf{x}) - \frac{1}{8\pi G}|\nabla\phi|^2,$$

and then applies the Euler–Lagrange equation, one obtains the equation of motion

$$4\pi G\rho(t, \mathbf{x}) = \nabla^2 \phi.$$

This equation is Newton's law of universal gravitation, expressed in differential form in terms of the gravitational potential $\varphi(t, \mathbf{x})$ and the mass density $\rho(t, \mathbf{x})$. Despite the nomenclature, the "field" under study is the gravitational potential, φ, rather than the gravitational field, $\mathbf{g}$. Similarly, when classical field theory is used to study electromagnetism, the "field" of interest is the electromagnetic four-potential $(V/c, \mathbf{A})$, rather than the electric and magnetic fields $\mathbf{E}$ and $\mathbf{B}$.

Quantum field theory uses this same Lagrangian procedure to determine the equations of motion for quantum fields. These equations of motion are then supplemented by commutation relations derived from the canonical quantization procedure described below, thereby incorporating quantum mechanical effects into the behavior of the field.

7.4.2 Single- and many-particle quantum mechanics

Main articles: Quantum mechanics and First quantization

In quantum mechanics, a particle (such as an electron or proton) is described by a complex wavefunction, $\psi(x, t)$, whose time-evolution is governed by the Schrödinger equation:

$$-\frac{\hbar^2}{2m}\frac{\partial^2}{\partial x^2}\psi(x, t) + V(x)\psi(x, t) = i\hbar\frac{\partial}{\partial t}\psi(x, t).$$

Here m is the particle's mass and $V(x)$ is the applied potential. Physical information about the behavior of the particle is extracted from the wavefunction by constructing expected values for various quantities; for example, the expected value of the particle's position is given by integrating $\psi^*(x)\, x\, \psi(x)$ over all space, and the expected value of the particle's momentum is found by integrating $-i\hbar\psi^*(x)\mathrm{d}\psi/\mathrm{d}x$. The quantity $\psi^*(x)\psi(x)$ is itself in the Copenhagen interpretation of quantum mechanics interpreted as a probability density function. This treatment of quantum mechanics, where a particle's wavefunction evolves against a classical background potential $V(x)$, is sometimes called *first quantization*.

This description of quantum mechanics can be extended to describe the behavior of multiple particles, so long as the number and the type of particles remain fixed. The particles are described by a wavefunction $\psi(x_1, x_2, \ldots, xN, t)$, which is governed by an extended version of the Schrödinger equation.

Often one is interested in the case where N particles are all of the same type (for example, the 18 electrons orbiting a neutral argon nucleus). As described in the article on identical particles, this implies that the state of the entire system must be either symmetric (bosons) or antisymmetric (fermions) when the coordinates of its constituent particles are exchanged. This is achieved by using a Slater determinant as the wavefunction of a fermionic system (and a Slater permanent for a bosonic system), which is equivalent to an element of the symmetric or antisymmetric subspace of a tensor product.

For example, the general quantum state of a system of N bosons is written as

$$|\phi_1 \cdots \phi_N\rangle = \sqrt{\frac{\prod_j N_j!}{N!}} \sum_{p \in S_N} |\phi_{p(1)}\rangle \otimes \cdots \otimes |\phi_{p(N)}\rangle,$$

where $|\phi_i\rangle$ are the single-particle states, Nj is the number of particles occupying state j, and the sum is taken over all possible permutations p acting on N elements. In general, this is a sum of $N!$ (N factorial) distinct terms. $\sqrt{\frac{\prod_j N_j!}{N!}}$ is a normalizing factor.

There are several shortcomings to the above description of quantum mechanics, which are addressed by quantum field theory. First, it is unclear how to extend quantum mechanics to include the effects of special relativity.[17] Attempted replacements for the Schrödinger equation, such as the Klein–Gordon equation or the Dirac equation, have many unsatisfactory qualities; for instance, they possess energy eigenvalues that extend to $-\infty$, so that there seems to be no easy definition of a ground state. It turns out that such inconsistencies arise from relativistic wavefunctions not having a well-defined probabilistic interpretation in position space, as probability conservation is not a relativistically covariant concept. The second shortcoming, related to the first, is that in quantum mechanics there is no mechanism to describe particle creation and annihilation;[8] this is crucial for describing phenomena such as pair production, which result from the conversion between mass and energy according to the relativistic relation $E = mc^2$.

7.4.3   Second quantization

Main article: Second quantization

In this section, we will describe a method for constructing a quantum field theory called **second quantization**. This basically involves choosing a way to index the quantum mechanical degrees of freedom in the space of multiple identical-particle states. It is based on the Hamiltonian formulation of quantum mechanics.

Several other approaches exist, such as the Feynman path integral,[9] which uses a Lagrangian formulation. For an overview of some of these approaches, see the article on quantization.

Bosons

For simplicity, we will first discuss second quantization for bosons, which form perfectly symmetric quantum states. Let us denote the mutually orthogonal single-particle states which are possible in the system by $|\phi_1\rangle, |\phi_2\rangle, |\phi_3\rangle$, and so on. For example, the 3-particle state with one particle in state $|\phi_1\rangle$ and two in state $|\phi_2\rangle$ is

$$\frac{1}{\sqrt{3}} \left[|\phi_1\rangle|\phi_2\rangle|\phi_2\rangle + |\phi_2\rangle|\phi_1\rangle|\phi_2\rangle + |\phi_2\rangle|\phi_2\rangle|\phi_1\rangle \right].$$

The first step in second quantization is to express such quantum states in terms of **occupation numbers**, by listing the number of particles occupying each of the single-particle states $|\phi_1\rangle, |\phi_2\rangle$, etc. This is simply another way of labelling the states. For instance, the above 3-particle state is denoted as

$$|1, 2, 0, 0, 0, \dots \rangle.$$

An N-particle state belongs to a space of states describing systems of N particles. The next step is to combine the individual N-particle state spaces into an extended state space, known as Fock space, which can describe systems of any number of particles. This is composed of the state space of a system with no particles (the so-called vacuum state, written as $|0\rangle$), plus the state space of a 1-particle system, plus the state space of a 2-particle system, and so forth. States describing a definite number of particles are known as Fock states: a general element of Fock space will be a linear combination of Fock states. There is a one-to-one correspondence between the occupation number representation and valid boson states in the Fock space.

At this point, the quantum mechanical system has become a quantum field in the sense we described above. The field's elementary degrees of freedom are the occupation numbers, and each occupation number is indexed by a number j indicating which of the single-particle states $|\phi_1\rangle, |\phi_2\rangle, \dots, |\phi_j\rangle, \dots$ it refers to:

$$|N_1, N_2, N_3, \dots, N_j, \dots \rangle.$$

The properties of this quantum field can be explored by defining creation and annihilation operators, which add and subtract particles. They are analogous to ladder operators in the quantum harmonic oscillator problem, which added and subtracted energy quanta. However, these operators literally create and annihilate particles of a given quantum state. The bosonic annihilation operator a_2 and creation operator $a_2^\dagger$ are easily defined in the occupation number representation as having the following effects:

$$a_2 | N_1, N_2, N_3, \dots \rangle = \sqrt{N_2} \, | \, N_1, (N_2 - 1), N_3, \dots \rangle,$$

$$a_2^\dagger | N_1, N_2, N_3, \dots \rangle = \sqrt{N_2 + 1} \, | \, N_1, (N_2 + 1), N_3, \dots \rangle.$$

It can be shown that these are operators in the usual quantum mechanical sense, i.e. linear operators acting on the Fock space. Furthermore, they are indeed Hermitian conjugates, which justifies the way we have written them. They can be shown to obey the commutation relation

$$[a_i, a_j] = 0 \quad , \quad \left[a_i^\dagger, a_j^\dagger\right] = 0 \quad , \quad \left[a_i, a_j^\dagger\right] = \delta_{ij},$$

where δ stands for the Kronecker delta. These are precisely the relations obeyed by the ladder operators for an infinite set of independent quantum harmonic oscillators, one for each single-particle state. Adding or removing bosons from each state is therefore analogous to exciting or de-exciting a quantum of energy in a harmonic oscillator.

Applying an annihilation operator a_k followed by its corresponding creation operator $a_k^\dagger$ returns the number N_k of particles in the kth single-particle eigenstate:

$$a_k^\dagger a_k | \dots, N_k, \dots \rangle = N_k | \dots, N_k, \dots \rangle.$$

The combination of operators $a_k^\dagger a_k$ is known as the number operator for the kth eigenstate.

The Hamiltonian operator of the quantum field (which, through the Schrödinger equation, determines its dynamics) can be written in terms of creation and annihilation operators. For instance, for a field of free (non-interacting) bosons, the total energy of the field is found by summing the energies of the bosons in each energy eigenstate. If the kth single-particle energy eigenstate has energy E_k and there are N_k bosons in this state, then the total energy of these bosons is $E_k N_k$. The energy in the *entire* field is then a sum over k :

$$E_{\text{tot}} = \sum_k E_k N_k$$

This can be turned into the Hamiltonian operator of the field by replacing N_k with the corresponding number operator, $a_k^\dagger a_k$. This yields

$$H = \sum_k E_k \, a_k^\dagger a_k.$$

Fermions

It turns out that a different definition of creation and annihilation must be used for describing fermions. According to the Pauli exclusion principle, fermions cannot share quantum states, so their occupation numbers N_i can only take on the value 0 or 1. The fermionic annihilation operators c and creation operators $c^\dagger$ are defined by their actions on a Fock state thus

$$c_j | N_1, N_2, \dots, N_j = 0, \dots \rangle = 0$$

$$c_j | N_1, N_2, \dots, N_j = 1, \dots \rangle = (-1)^{(N_1 + \dots + N_j - 1)} | N_1, N_2, \dots, N_j = 0, \dots \rangle$$

$$c_j^\dagger | N_1, N_2, \dots, N_j = 0, \dots \rangle = (-1)^{(N_1 + \dots + N_j - 1)} | N_1, N_2, \dots, N_j = 1, \dots \rangle$$

$$c_j^\dagger |N_1, N_2, \ldots, N_j = 1, \ldots\rangle = 0.$$

These obey an anticommutation relation:

$$\{c_i, c_j\} = 0 \quad , \quad \left\{c_i^\dagger, c_j^\dagger\right\} = 0 \quad , \quad \left\{c_i, c_j^\dagger\right\} = \delta_{ij}.$$

One may notice from this that applying a fermionic creation operator twice gives zero, so it is impossible for the particles to share single-particle states, in accordance with the exclusion principle.

Field operators

We have previously mentioned that there can be more than one way of indexing the degrees of freedom in a quantum field. Second quantization indexes the field by enumerating the single-particle quantum states. However, as we have discussed, it is more natural to think about a "field", such as the electromagnetic field, as a set of degrees of freedom indexed by position.

To this end, we can define *field operators* that create or destroy a particle at a particular point in space. In particle physics, these operators turn out to be more convenient to work with, because they make it easier to formulate theories that satisfy the demands of relativity.

Single-particle states are usually enumerated in terms of their momenta (as in the particle in a box problem.) We can construct field operators by applying the Fourier transform to the creation and annihilation operators for these states. For example, the bosonic field annihilation operator $\phi(\mathbf{r})$ is

$$\phi(\mathbf{r}) \overset{\text{def}}{=} \sum_j e^{i\mathbf{k}_j \cdot \mathbf{r}} a_j.$$

The bosonic field operators obey the commutation relation

$$[\phi(\mathbf{r}), \phi(\mathbf{r}')] = 0 \quad , \quad [\phi^\dagger(\mathbf{r}), \phi^\dagger(\mathbf{r}')] = 0 \quad , \quad [\phi(\mathbf{r}), \phi^\dagger(\mathbf{r}')] = \delta^3(\mathbf{r} - \mathbf{r}')$$

where $\delta(x)$ stands for the Dirac delta function. As before, the fermionic relations are the same, with the commutators replaced by anticommutators.

The field operator is not the same thing as a single-particle wavefunction. The former is an operator acting on the Fock space, and the latter is a quantum-mechanical amplitude for finding a particle in some position. However, they are closely related, and are indeed commonly denoted with the same symbol. If we have a Hamiltonian with a space representation, say

$$H = -\frac{\hbar^2}{2m} \sum_i \nabla_i^2 + \sum_{i<j} U(|\mathbf{r}_i - \mathbf{r}_j|)$$

where the indices i and j run over all particles, then the field theory Hamiltonian (in the non-relativistic limit and for negligible self-interactions) is

$$H = -\frac{\hbar^2}{2m} \int d^3r \, \phi^\dagger(\mathbf{r})\nabla^2\phi(\mathbf{r}) + \frac{1}{2} \int d^3r \int d^3r' \, \phi^\dagger(\mathbf{r})\phi^\dagger(\mathbf{r}')U(|\mathbf{r} - \mathbf{r}'|)\phi(\mathbf{r}')\phi(\mathbf{r}).$$

This looks remarkably like an expression for the expectation value of the energy, with ϕ playing the role of the wavefunction. This relationship between the field operators and wavefunctions makes it very easy to formulate field theories starting from space-projected Hamiltonians.

7.4.4   Dynamics

Once the Hamiltonian operator is obtained as part of the canonical quantization process, the time dependence of the state is described with the Schrödinger equation, just as with other quantum theories. Alternatively, the Heisenberg picture can be used where the time dependence is in the operators rather than in the states.

7.4.5   Implications

Unification of fields and particles

The "second quantization" procedure that we have outlined in the previous section takes a set of single-particle quantum states as a starting point. Sometimes, it is impossible to define such single-particle states, and one must proceed directly to quantum field theory. For example, a quantum theory of the electromagnetic field *must* be a quantum field theory, because it is impossible (for various reasons) to define a wavefunction for a single photon.[10] In such situations, the quantum field theory can be constructed by examining the mechanical properties of the classical field and guessing the corresponding quantum theory. For free (non-interacting) quantum fields, the quantum field theories obtained in this way have the same properties as those obtained using second quantization, such as well-defined creation and annihilation operators obeying commutation or anticommutation relations.

Quantum field theory thus provides a unified framework for describing "field-like" objects (such as the electromagnetic field, whose excitations are photons) and "particle-like" objects (such as electrons, which are treated as excitations of an underlying electron field), so long as one can treat interactions as "perturbations" of free fields. There are still unsolved problems relating to the more general case of interacting fields that may or may not be adequately described by perturbation theory. For more on this topic, see Haag's theorem.

Physical meaning of particle indistinguishability

The second quantization procedure relies crucially on the particles being identical. We would not have been able to construct a quantum field theory from a distinguishable many-particle system, because there would have been no way of separating and indexing the degrees of freedom.

Many physicists prefer to take the converse interpretation, which is that *quantum field theory explains what identical particles are*. In ordinary quantum mechanics, there is not much theoretical motivation for using symmetric (bosonic) or antisymmetric (fermionic) states, and the need for such states is simply regarded as an empirical fact. From the point of view of quantum field theory, particles are identical if and only if they are excitations of the same underlying quantum field. Thus, the question "why are all electrons identical?" arises from mistakenly regarding individual electrons as fundamental objects, when in fact it is only the electron field that is fundamental.

Particle conservation and non-conservation

During second quantization, we started with a Hamiltonian and state space describing a fixed number of particles (N), and ended with a Hamiltonian and state space for an arbitrary number of particles. Of course, in many common situations N is an important and perfectly well-defined quantity, e.g. if we are describing a gas of atoms sealed in a box. From the point of view of quantum field theory, such situations are described by quantum states that are eigenstates of the number operator $\hat{N}$, which measures the total number of particles present. As with any quantum mechanical observable, $\hat{N}$ is conserved if it commutes with the Hamiltonian. In that case, the quantum state is trapped in the N-particle subspace of the total Fock space, and the situation could equally well be described by ordinary N-particle quantum mechanics. (Strictly speaking, this is only true in the noninteracting case or in the low energy density limit of renormalized quantum field theories)

For example, we can see that the free-boson Hamiltonian described above conserves particle number. Whenever the Hamiltonian operates on a state, each particle destroyed by an annihilation operator a_k is immediately put back by the creation operator $a_k^\dagger$.

On the other hand, it is possible, and indeed common, to encounter quantum states that are *not* eigenstates of $\hat{N}$, which do not have well-defined particle numbers. Such states are difficult or impossible to handle using ordinary quantum mechanics, but they can be easily described in quantum field theory as quantum superpositions of states having different values of N. For example, suppose we have a bosonic field whose particles can be created or destroyed by interactions with a fermionic field. The Hamiltonian of the combined system would be given by the Hamiltonians of the free boson and free fermion fields, plus a "potential energy" term such as

$$H_I = \sum_{k,q} V_q(a_q + a^\dagger_{-q})c^\dagger_{k+q}c_k,$$

where $a^\dagger_k$ and a_k denotes the bosonic creation and annihilation operators, $c^\dagger_k$ and c_k denotes the fermionic creation and annihilation operators, and V_q is a parameter that describes the strength of the interaction. This "interaction term" describes processes in which a fermion in state k either absorbs or emits a boson, thereby being kicked into a different eigenstate $k + q$. (In fact, this type of Hamiltonian is used to describe interaction between conduction electrons and phonons in metals. The interaction between electrons and photons is treated in a similar way, but is a little more complicated because the role of spin must be taken into account.) One thing to notice here is that even if we start out with a fixed number of bosons, we will typically end up with a superposition of states with different numbers of bosons at later times. The number of fermions, however, is conserved in this case.

In condensed matter physics, states with ill-defined particle numbers are particularly important for describing the various superfluids. Many of the defining characteristics of a superfluid arise from the notion that its quantum state is a superposition of states with different particle numbers. In addition, the concept of a coherent state (used to model the laser and the BCS ground state) refers to a state with an ill-defined particle number but a well-defined phase.

7.4.6 Axiomatic approaches

The preceding description of quantum field theory follows the spirit in which most physicists approach the subject. However, it is not mathematically rigorous. Over the past several decades, there have been many attempts to put quantum field theory on a firm mathematical footing by formulating a set of axioms for it. These attempts fall into two broad classes.

The first class of axioms, first proposed during the 1950s, include the Wightman, Osterwalder–Schrader, and Haag–Kastler systems. They attempted to formalize the physicists' notion of an "operator-valued field" within the context of functional analysis, and enjoyed limited success. It was possible to prove that any quantum field theory satisfying these axioms satisfied certain general theorems, such as the spin-statistics theorem and the CPT theorem. Unfortunately, it proved extraordinarily difficult to show that any realistic field theory, including the Standard Model, satisfied these axioms. Most of the theories that could be treated with these analytic axioms were physically trivial, being restricted to low-dimensions and lacking interesting dynamics. The construction of theories satisfying one of these sets of axioms falls in the field of constructive quantum field theory. Important work was done in this area in the 1970s by Segal, Glimm, Jaffe and others.

During the 1980s, a second set of axioms based on geometric ideas was proposed. This line of investigation, which restricts its attention to a particular class of quantum field theories known as topological quantum field theories, is associated most closely with Michael Atiyah and Graeme Segal, and was notably expanded upon by Edward Witten, Richard Borcherds, and Maxim Kontsevich. However, most of the physically relevant quantum field theories, such as the Standard Model, are not topological quantum field theories; the quantum field theory of the fractional quantum Hall effect is a notable exception. The main impact of axiomatic topological quantum field theory has been on mathematics, with important applications in representation theory, algebraic topology, and differential geometry.

Finding the proper axioms for quantum field theory is still an open and difficult problem in mathematics. One of the Millennium Prize Problems—proving the existence of a mass gap in Yang–Mills theory—is linked to this issue.

7.5 Associated phenomena

In the previous part of the article, we described the most general features of quantum field theories. Some of the quantum field theories studied in various fields of theoretical physics involve additional special ideas, such as renormalizability, gauge symmetry, and supersymmetry. These are described in the following sections.

7.5.1 Renormalization

Main article: Renormalization

Early in the history of quantum field theory, it was found that many seemingly innocuous calculations, such as the perturbative shift in the energy of an electron due to the presence of the electromagnetic field, give infinite results. The reason is that the perturbation theory for the shift in an energy involves a sum over all other energy levels, and there are infinitely many levels at short distances that each give a finite contribution which results in a divergent series.

Many of these problems are related to failures in classical electrodynamics that were identified but unsolved in the 19th century, and they basically stem from the fact that many of the supposedly "intrinsic" properties of an electron are tied to the electromagnetic field that it carries around with it. The energy carried by a single electron—its self energy—is not simply the bare value, but also includes the energy contained in its electromagnetic field, its attendant cloud of photons. The energy in a field of a spherical source diverges in both classical and quantum mechanics, but as discovered by Weisskopf with help from Furry, in quantum mechanics the divergence is much milder, going only as the logarithm of the radius of the sphere.

The solution to the problem, presciently suggested by Stueckelberg, independently by Bethe after the crucial experiment by Lamb, implemented at one loop by Schwinger, and systematically extended to all loops by Feynman and Dyson, with converging work by Tomonaga in isolated postwar Japan, comes from recognizing that all the infinities in the interactions of photons and electrons can be isolated into redefining a finite number of quantities in the equations by replacing them with the observed values: specifically the electron's mass and charge: this is called renormalization. The technique of renormalization recognizes that the problem is essentially purely mathematical, that extremely short distances are at fault. In order to define a theory on a continuum, first place a cutoff on the fields, by postulating that quanta cannot have energies above some extremely high value. This has the effect of replacing continuous space by a structure where very short wavelengths do not exist, as on a lattice. Lattices break rotational symmetry, and one of the crucial contributions made by Feynman, Pauli and Villars, and modernized by 't Hooft and Veltman, is a symmetry-preserving cutoff for perturbation theory (this process is called regularization). There is no known symmetrical cutoff outside of perturbation theory, so for rigorous or numerical work people often use an actual lattice.

On a lattice, every quantity is finite but depends on the spacing. When taking the limit of zero spacing, we make sure that the physically observable quantities like the observed electron mass stay fixed, which means that the constants in the Lagrangian defining the theory depend on the spacing. Hopefully, by allowing the constants to vary with the lattice spacing, all the results at long distances become insensitive to the lattice, defining a continuum limit.

The renormalization procedure only works for a certain class of quantum field theories, called **renormalizable quantum field theories**. A theory is **perturbatively renormalizable** when the constants in the Lagrangian only diverge at worst as logarithms of the lattice spacing for very short spacings. The continuum limit is then well defined in perturbation theory, and even if it is not fully well defined non-perturbatively, the problems only show up at distance scales that are exponentially small in the inverse coupling for weak couplings. The Standard Model of particle physics is perturbatively renormalizable, and so are its component theories (quantum electrodynamics/electroweak theory and quantum chromodynamics). Of the three components, quantum electrodynamics is believed to not have a continuum limit, while the asymptotically free $SU(2)$ and $SU(3)$ weak hypercharge and strong color interactions are nonperturbatively well defined.

The renormalization group describes how renormalizable theories emerge as the long distance low-energy effective field theory for any given high-energy theory. Because of this, renormalizable theories are insensitive to the precise nature of the underlying high-energy short-distance phenomena. This is a blessing because it allows physicists to formulate low energy theories without knowing the details of high energy phenomenon. It is also a curse, because once a renormalizable theory like the standard model is found to work, it gives very few clues to higher energy processes. The only way high energy processes can be seen in the standard model is when they allow otherwise forbidden events, or if they predict quantitative relations between the coupling constants.

7.5.2 Haag's theorem

See also: Haag's theorem

From a mathematically rigorous perspective, there exists no interaction picture in a Lorentz-covariant quantum field theory. This implies that the perturbative approach of Feynman diagrams in QFT is not strictly justified, despite producing vastly precise predictions validated by experiment. This is called Haag's theorem, but most particle physicists relying on QFT largely shrug it off.

7.5.3 Gauge freedom

A gauge theory is a theory that admits a symmetry with a local parameter. For example, in every quantum theory the global phase of the wave function is arbitrary and does not represent something physical. Consequently, the theory is invariant under a global change of phases (adding a constant to the phase of all wave functions, everywhere); this is a global symmetry. In quantum electrodynamics, the theory is also invariant under a *local* change of phase, that is – one may shift the phase of all wave functions so that the shift may be different at every point in space-time. This is a *local* symmetry. However, in order for a well-defined derivative operator to exist, one must introduce a new field, the gauge field, which also transforms in order for the local change of variables (the phase in our example) not to affect the derivative. In quantum electrodynamics this gauge field is the electromagnetic field. The change of local gauge of variables is termed gauge transformation. It is worth noting that by Noether's theorem, for every such symmetry there exists an associated conserved current. The aforementioned symmetry of the wavefunction under global phase changes implies the conservation of electric charge.

In quantum field theory the excitations of fields represent particles. The particle associated with excitations of the gauge field is the gauge boson, which is the photon in the case of quantum electrodynamics.

The degrees of freedom in quantum field theory are local fluctuations of the fields. The existence of a gauge symmetry reduces the number of degrees of freedom, simply because some fluctuations of the fields can be transformed to zero by gauge transformations, so they are equivalent to having no fluctuations at all, and they therefore have no physical meaning. Such fluctuations are usually called "non-physical degrees of freedom" or *gauge artifacts*; usually some of them have a negative norm, making them inadequate for a consistent theory. Therefore, if a classical field theory has a gauge symmetry, then its quantized version (i.e. the corresponding quantum field theory) will have this symmetry as well. In other words, a gauge symmetry cannot have a quantum anomaly. If a gauge symmetry is anomalous (i.e. not kept in the quantum theory) then the theory is non-consistent: for example, in quantum electrodynamics, had there been a gauge anomaly, this would require the appearance of photons with longitudinal polarization and polarization in the time direction, the latter having a negative norm, rendering the theory inconsistent; another possibility would be for these photons to appear only in intermediate processes but not in the final products of any interaction, making the theory non-unitary and again inconsistent (see optical theorem).

In general, the gauge transformations of a theory consist of several different transformations, which may not be commutative. These transformations are together described by a mathematical object known as a gauge group. Infinitesimal gauge transformations are the gauge group generators. Therefore, the number of gauge bosons is the group dimension (i.e. number of generators forming a basis).

All the fundamental interactions in nature are described by gauge theories. These are:

- Quantum chromodynamics, whose gauge group is $SU(3)$. The gauge bosons are eight gluons.

- The electroweak theory, whose gauge group is $U(1) \times SU(2)$, (a direct product of $U(1)$ and $SU(2)$).

- Gravity, whose classical theory is general relativity, admits the equivalence principle, which is a form of gauge symmetry. However, it is explicitly non-renormalizable.

7.5.4 Multivalued gauge transformations

The gauge transformations which leave the theory invariant involve, by definition, only single-valued gauge functions $\Lambda(x_i)$ which satisfy the Schwarz integrability criterion

$$\partial_{x_i x_j} \Lambda = \partial_{x_j x_i} \Lambda.$$

An interesting extension of gauge transformations arises if the gauge functions $\Lambda(x_i)$ are allowed to be multivalued functions which violate the integrability criterion. These are capable of changing the physical field strengths and are therefore not proper symmetry transformations. Nevertheless, the transformed field equations describe correctly the physical laws in the presence of the newly generated field strengths. See the textbook by H. Kleinert cited below for the applications to phenomena in physics.

7.5.5 Supersymmetry

Main article: Supersymmetry

Supersymmetry assumes that every fundamental fermion has a superpartner that is a boson and vice versa. It was introduced in order to solve the so-called Hierarchy Problem, that is, to explain why particles not protected by any symmetry (like the Higgs boson) do not receive radiative corrections to its mass driving it to the larger scales (GUT, Planck...). It was soon realized that supersymmetry has other interesting properties: its gauged version is an extension of general relativity (Supergravity), and it is a key ingredient for the consistency of string theory.

The way supersymmetry protects the hierarchies is the following: since for every particle there is a superpartner with the same mass, any loop in a radiative correction is cancelled by the loop corresponding to its superpartner, rendering the theory UV finite.

Since no superpartners have yet been observed, if supersymmetry exists it must be broken (through a so-called soft term, which breaks supersymmetry without ruining its helpful features). The simplest models of this breaking require that the energy of the superpartners not be too high; in these cases, supersymmetry is expected to be observed by experiments at the Large Hadron Collider. The Higgs particle has been detected at the LHC, and no such superparticles have been discovered.

7.6   See also

- Abraham–Lorentz force

- Basic concepts of quantum mechanics

- Common integrals in quantum field theory

- Einstein–Maxwell–Dirac equations

- Form factor (quantum field theory)

- Green–Kubo relations

- Green's function (many-body theory)

- Invariance mechanics

- List of quantum field theories

- Quantum electrodynamics

- Quantum field theory in curved spacetime

- Quantum flavordynamics

- Quantum hydrodynamics

- Quantum triviality

- Relation between Schrödinger's equation and the path integral formulation of quantum mechanics

- Relationship between string theory and quantum field theory

- Schwinger–Dyson equation

- Static forces and virtual-particle exchange

- Symmetry in quantum mechanics

- Theoretical and experimental justification for the Schrödinger equation

- Ward–Takahashi identity

- Wheeler–Feynman absorber theory

- Wigner's classification

- Wigner's theorem

7.7   Notes

7.8   References

[1] "Beautiful Minds, Vol. 20: Ed Witten". la Repubblica. 2010. Retrieved 22 June 2012. See here.

[2] J. J. Thorn et al. (2004). Observing the quantum behavior of light in an undergraduate laboratory. . J. J. Thorn, M. S. Neel, V. W. Donato, G. S. Bergreen, R. E. Davies, and M. Beck. American Association of Physics Teachers. 2004.DOI: 10.1119/1.1737397.

[3] David Tong, *Lectures on Quantum Field Theory*, chapter 1.

[4] Srednicki, Mark. *Quantum Field Theory* (1st ed.). p. 19.

[5] Srednicki, Mark. *Quantum Field Theory* (1st ed.). pp. 25–6.

[6] Zee, Anthony. *Quantum Field Theory in a Nutshell* (2nd ed.). p. 61.

[7] David Tong, *Lectures on Quantum Field Theory*, Introduction.

[8] Zee, Anthony. *Quantum Field Theory in a Nutshell* (2nd ed.). p. 3.

[9] Abraham Pais, *Inward Bound: Of Matter and Forces in the Physical World* ISBN 0-19-851997-4. Pais recounts how his astonishment at the rapidity with which Feynman could calculate using his method. Feynman's method is now part of the standard methods for physicists.

[10] Newton, T.D.; Wigner, E.P. (1949). "Localized states for elementary systems". *Reviews of Modern Physics* **21** (3): 400–406. Bibcode:1949RvMP...21..400N. doi:10.1103/RevModPhys.21.400.

7.9 Further reading

General readers

- Feynman, R.P. (2001) [1964]. *The Character of Physical Law*. MIT Press. ISBN 0-262-56003-8.

- Feynman, R.P. (2006) [1985]. *QED: The Strange Theory of Light and Matter*. Princeton University Press. ISBN 0-691-12575-9.

- Gribbin, J. (1998). *Q is for Quantum: Particle Physics from A to Z*. Weidenfeld & Nicolson. ISBN 0-297-81752-3.

- Schumm, Bruce A. (2004) *Deep Down Things*. Johns Hopkins Univ. Press. Chpt. 4.

Introductory texts

- McMahon, D. (2008). *Quantum Field Theory*. McGraw-Hill. ISBN 978-0-07-154382-8.

- Bogoliubov, N.; Shirkov, D. (1982). *Quantum Fields*. Benjamin-Cummings. ISBN 0-8053-0983-7.

- Frampton, P.H. (2000). *Gauge Field Theories. Frontiers in Physics (2nd ed.)*. Wiley.

- Greiner, W; Müller, B. (2000). *Gauge Theory of Weak Interactions*. Springer. ISBN 3-540-67672-4.

- Itzykson, C.; Zuber, J.-B. (1980). *Quantum Field Theory*. McGraw-Hill. ISBN 0-07-032071-3.

- Kane, G.L. (1987). *Modern Elementary Particle Physics*. Perseus Books. ISBN 0-201-11749-5.

- Kleinert, H.; Schulte-Frohlinde, Verena (2001). *Critical Properties of φ^4-Theories*. World Scientific. ISBN 981-02-4658-7.

- Kleinert, H. (2008). *Multivalued Fields in Condensed Matter, Electrodynamics, and Gravitation* (PDF). World Scientific. ISBN 978-981-279-170-2.

- Loudon, R (1983). *The Quantum Theory of Light*. Oxford University Press. ISBN 0-19-851155-8.

- Mandl, F.; Shaw, G. (1993). *Quantum Field Theory*. John Wiley & Sons. ISBN 978-0-471-94186-6.

- Peskin, M.; Schroeder, D. (1995). *An Introduction to Quantum Field Theory*. Westview Press. ISBN 0-201-50397-2.

- Ryder, L.H. (1985). *Quantum Field Theory*. Cambridge University Press. ISBN 0-521-33859-X.

- Schwartz, M.D. (2014). *Quantum Field Theory and the Standard Model*. Cambridge University Press. ISBN 978-1107034730.

- Srednicki, Mark (2007) *Quantum Field Theory*. Cambridge Univ. Press.

- Ynduráin, F.J. (1996). *Relativistic Quantum Mechanics and Introduction to Field Theory* (1st ed.). Springer. ISBN 978-3-540-60453-2.

- Zee, A. (2003). *Quantum Field Theory in a Nutshell*. Princeton University Press. ISBN 0-691-01019-6.

Advanced texts

- Brown, Lowell S. (1994). *Quantum Field Theory*. Cambridge University Press. ISBN 978-0-521-46946-3.

- Bogoliubov, N.; Logunov, A.A.; Oksak, A.I.; Todorov, I.T. (1990). *General Principles of Quantum Field Theory*. Kluwer Academic Publishers. ISBN 978-0-7923-0540-8.

- Weinberg, S. (1995). *The Quantum Theory of Fields* **1–3**. Cambridge University Press.

Articles:

- Gerard 't Hooft (2007) "The Conceptual Basis of Quantum Field Theory" in Butterfield, J., and John Earman, eds., *Philosophy of Physics, Part A*. Elsevier: 661–730.

- Frank Wilczek (1999) "Quantum field theory", *Reviews of Modern Physics* 71: S83–S95. Also doi=10.1103/Rev. Mod. Phys. 71.

7.10   External links

- Hazewinkel, Michiel, ed. (2001), "Quantum field theory", *Encyclopedia of Mathematics*, Springer, ISBN 978-1-55608-010-4

- Stanford Encyclopedia of Philosophy: "Quantum Field Theory", by Meinard Kuhlmann.

- Siegel, Warren, 2005. *Fields*. A free text, also available from arXiv:hep-th/9912205.

- Quantum Field Theory by P. J. Mulders

Chapter 8

Standard Model

This article is about the Standard Model of particle physics. For other uses, see Standard model (disambiguation).

This article is a non-mathematical general overview of the Standard Model. For a mathematical description, see the article Standard Model (mathematical formulation).

For the Standard Model of Big Bang cosmology, Lambda-CDM model.

The **Standard Model** of particle physics is a theory concerning the electromagnetic, weak, and strong nuclear inter-

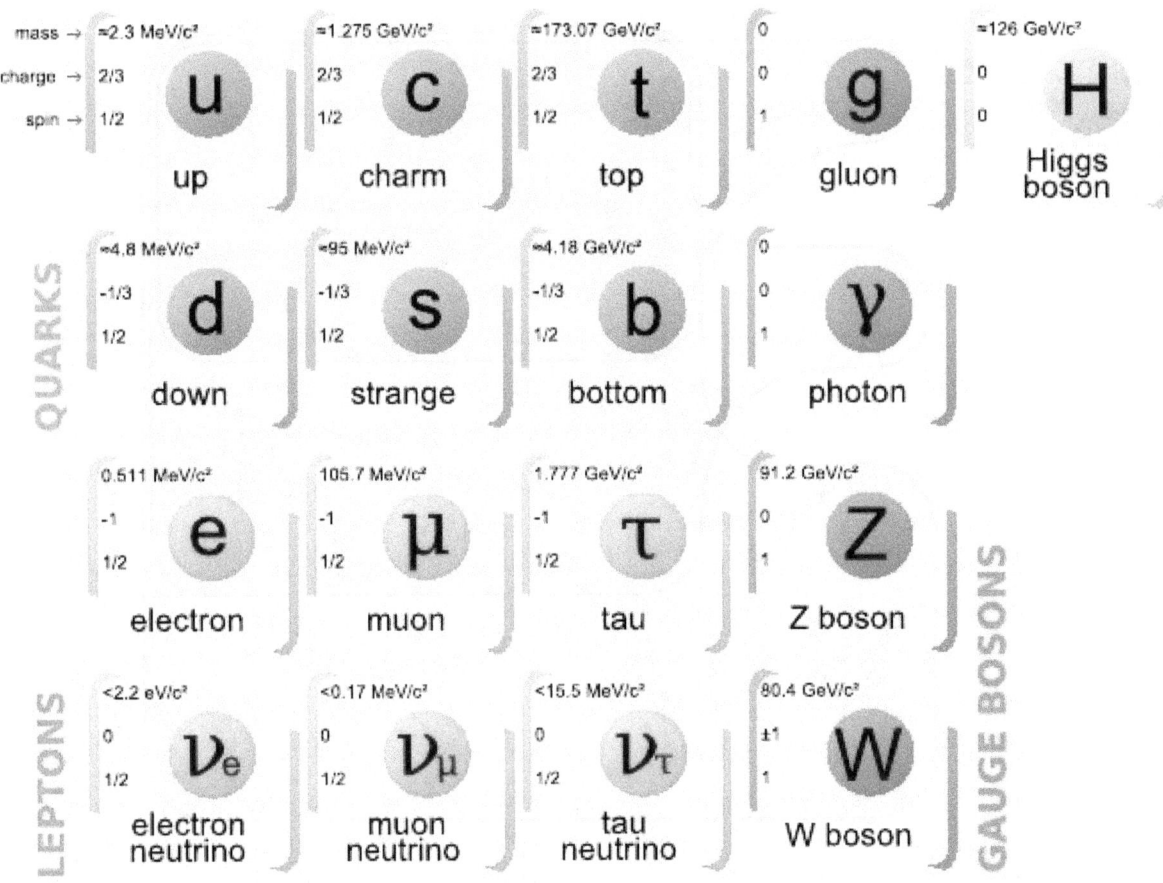

The Standard Model of elementary particles (more schematic depiction), with the three generations of matter, gauge bosons in the fourth column, and the Higgs boson in the fifth.

actions, as well as classifying all the subatomic particles known. It was developed throughout the latter half of the 20th century, as a collaborative effort of scientists around the world.[1] The current formulation was finalized in the mid-1970s upon experimental confirmation of the existence of quarks. Since then, discoveries of the top quark (1995), the tau neutrino (2000), and more recently the Higgs boson (2013), have given further credence to the Standard Model. Because of its success in explaining a wide variety of experimental results, the Standard Model is sometimes regarded as a "theory of almost everything".

Although the Standard Model is believed to be theoretically self-consistent[2] and has demonstrated huge and continued successes in providing experimental predictions, it does leave some phenomena unexplained and it falls short of being a complete theory of fundamental interactions. It does not incorporate the full theory of gravitation[3] as described by general relativity, or account for the accelerating expansion of the universe (as possibly described by dark energy). The model does not contain any viable dark matter particle that possesses all of the required properties deduced from observational cosmology. It also does not incorporate neutrino oscillations (and their non-zero masses).

The development of the Standard Model was driven by theoretical and experimental particle physicists alike. For theorists, the Standard Model is a paradigm of a quantum field theory, which exhibits a wide range of physics including spontaneous symmetry breaking, anomalies, non-perturbative behavior, etc. It is used as a basis for building more exotic models that incorporate hypothetical particles, extra dimensions, and elaborate symmetries (such as supersymmetry) in an attempt to explain experimental results at variance with the Standard Model, such as the existence of dark matter and neutrino oscillations.

8.1   Historical background

The first step towards the Standard Model was Sheldon Glashow's discovery in 1961 of a way to combine the electromagnetic and weak interactions.[4] In 1967 Steven Weinberg[5] and Abdus Salam[6] incorporated the Higgs mechanism[7][8][9] into Glashow's electroweak theory, giving it its modern form.

The Higgs mechanism is believed to give rise to the masses of all the elementary particles in the Standard Model. This includes the masses of the W and Z bosons, and the masses of the fermions, i.e. the quarks and leptons.

After the neutral weak currents caused by Z boson exchange were discovered at CERN in 1973,[10][11][12][13] the electroweak theory became widely accepted and Glashow, Salam, and Weinberg shared the 1979 Nobel Prize in Physics for discovering it. The W and Z bosons were discovered experimentally in 1981, and their masses were found to be as the Standard Model predicted.

The theory of the strong interaction, to which many contributed, acquired its modern form around 1973–74, when experiments confirmed that the hadrons were composed of fractionally charged quarks.

8.2   Overview

At present, matter and energy are best understood in terms of the kinematics and interactions of elementary particles. To date, physics has reduced the laws governing the behavior and interaction of all known forms of matter and energy to a small set of fundamental laws and theories. A major goal of physics is to find the "common ground" that would unite all of these theories into one integrated theory of everything, of which all the other known laws would be special cases, and from which the behavior of all matter and energy could be derived (at least in principle).[14]

8.3   Particle content

The Standard Model includes members of several classes of elementary particles (fermions, gauge bosons, and the Higgs boson), which in turn can be distinguished by other characteristics, such as color charge.

8.3.1 Fermions

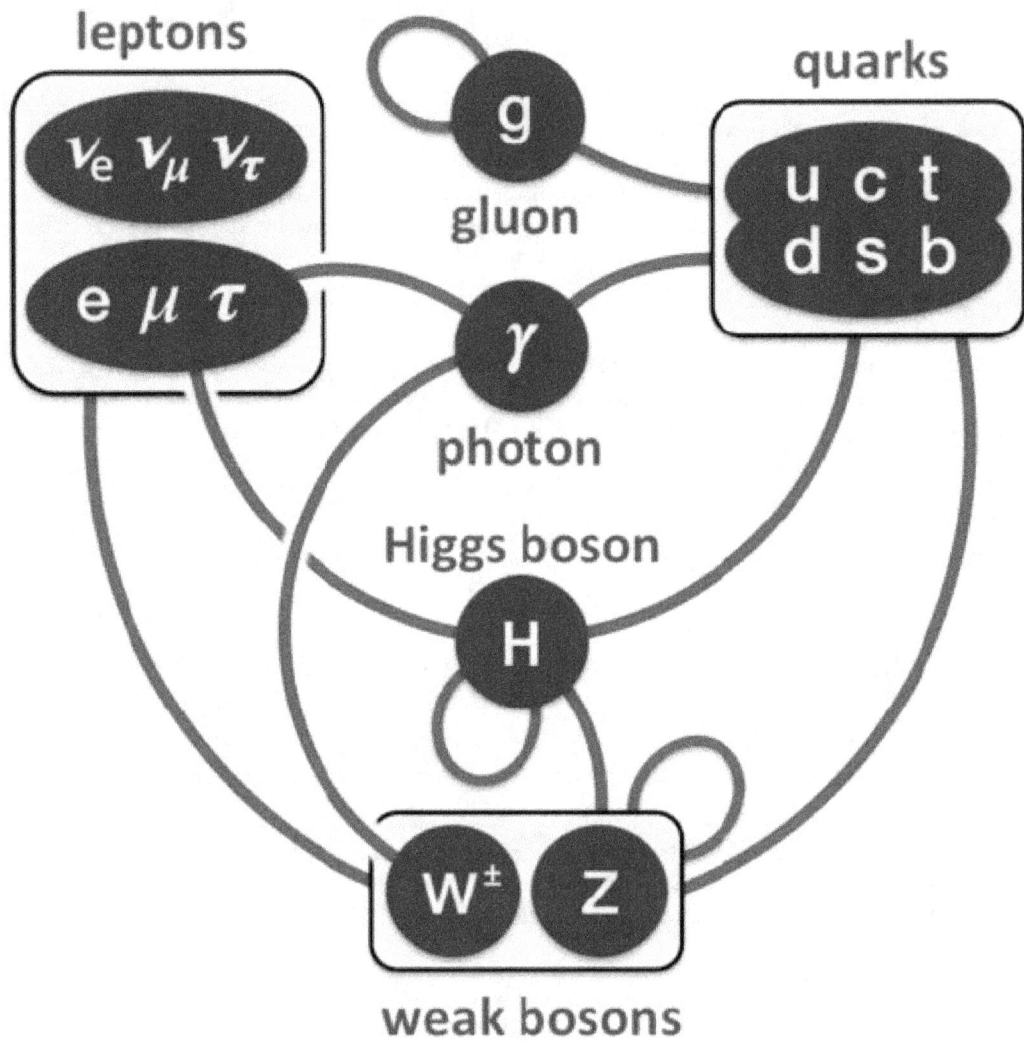

Summary of interactions between particles described by the Standard Model.

The Standard Model includes 12 elementary particles of spin-½ known as fermions. According to the spin-statistics theorem, fermions respect the Pauli exclusion principle. Each fermion has a corresponding antiparticle.

The fermions of the Standard Model are classified according to how they interact (or equivalently, by what charges they carry). There are six quarks (up, down, charm, strange, top, bottom), and six leptons (electron, electron neutrino, muon, muon neutrino, tau, tau neutrino). Pairs from each classification are grouped together to form a generation, with corresponding particles exhibiting similar physical behavior (see table).

The defining property of the quarks is that they carry color charge, and hence, interact via the strong interaction. A phenomenon called color confinement results in quarks being very strongly bound to one another, forming color-neutral composite particles (hadrons) containing either a quark and an antiquark (mesons) or three quarks (baryons). The familiar proton and the neutron are the two baryons having the smallest mass. Quarks also carry electric charge and weak isospin. Hence they interact with other fermions both electromagnetically and via the weak interaction.

The remaining six fermions do not carry colour charge and are called leptons. The three neutrinos do not carry electric charge either, so their motion is directly influenced only by the weak nuclear force, which makes them notoriously difficult to detect. However, by virtue of carrying an electric charge, the electron, muon, and tau all interact electromagnetically.

Each member of a generation has greater mass than the corresponding particles of lower generations. The first generation charged particles do not decay; hence all ordinary (baryonic) matter is made of such particles. Specifically, all atoms consist of electrons orbiting around atomic nuclei, ultimately constituted of up and down quarks. Second and third generations charged particles, on the other hand, decay with very short half lives, and are observed only in very high-energy environments. Neutrinos of all generations also do not decay, and pervade the universe, but rarely interact with baryonic matter.

8.3.2   Gauge bosons

In the Standard Model, gauge bosons are defined as force carriers that mediate the strong, weak, and electromagnetic fundamental interactions.

Interactions in physics are the ways that particles influence other particles. At a macroscopic level, electromagnetism allows particles to interact with one another via electric and magnetic fields, and gravitation allows particles with mass to attract one another in accordance with Einstein's theory of general relativity. The Standard Model explains such forces as resulting from matter particles exchanging other particles, generally referred to as *force mediating particles*. When a force-mediating particle is exchanged, at a macroscopic level the effect is equivalent to a force influencing both of them, and the particle is therefore said to have *mediated* (i.e., been the agent of) that force. The Feynman diagram calculations, which are a graphical representation of the perturbation theory approximation, invoke "force mediating particles", and when applied to analyze high-energy scattering experiments are in reasonable agreement with the data. However, perturbation theory (and with it the concept of a "force-mediating particle") fails in other situations. These include low-energy quantum chromodynamics, bound states, and solitons.

The gauge bosons of the Standard Model all have spin (as do matter particles). The value of the spin is 1, making them bosons. As a result, they do not follow the Pauli exclusion principle that constrains fermions: thus bosons (e.g. photons) do not have a theoretical limit on their spatial density (number per volume). The different types of gauge bosons are described below.

- Photons mediate the electromagnetic force between electrically charged particles. The photon is massless and is well-described by the theory of quantum electrodynamics.

- The W+, W−, and Z gauge bosons mediate the weak interactions between particles of different flavors (all quarks and leptons). They are massive, with the Z being more massive than the W±. The weak interactions involving the W± exclusively act on *left-handed* particles and *right-handed* antiparticles. Furthermore, the W± carries an electric charge of +1 and −1 and couples to the electromagnetic interaction. The electrically neutral Z boson interacts with both left-handed particles and antiparticles. These three gauge bosons along with the photons are grouped together, as collectively mediating the electroweak interaction.

- The eight gluons mediate the strong interactions between color charged particles (the quarks). Gluons are massless. The eightfold multiplicity of gluons is labeled by a combination of color and anticolor charge (e.g. red–antigreen).[nb 1] Because the gluons have an effective color charge, they can also interact among themselves. The gluons and their interactions are described by the theory of quantum chromodynamics.

The interactions between all the particles described by the Standard Model are summarized by the diagrams on the right of this section.

8.3.3   Higgs boson

Main article: Higgs boson

Standard Model Interactions
(Forces Mediated by Gauge Bosons)

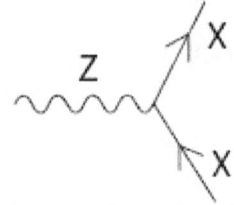

X is any fermion in
the Standard Model.

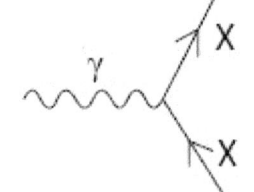

X is electrically charged.

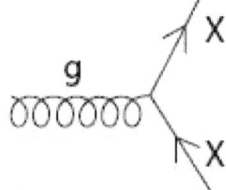

X is any quark.

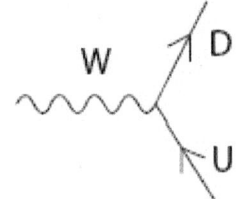

U is a up-type quark;
D is a down-type quark.

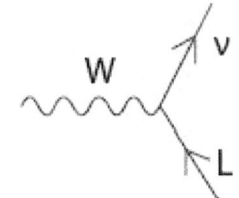

L is a lepton and ν is the
corresponding neutrino.

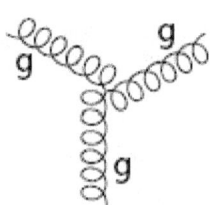

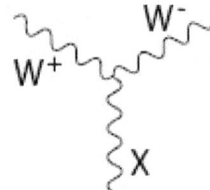

X is a photon or Z-boson.

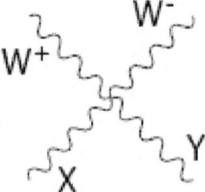

X and Y are any two
electroweak bosons such
that charge is conserved.

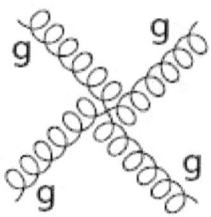

The above interactions form the basis of the standard model. Feynman diagrams in the standard model are built from these vertices. Modifications involving Higgs boson interactions and neutrino oscillations are omitted. The charge of the W bosons is dictated by the fermions they interact with; the conjugate of each listed vertex (i.e. reversing the direction of arrows) is also allowed.

The Higgs particle is a massive scalar elementary particle theorized by Robert Brout, François Englert, Peter Higgs, Gerald Guralnik, C. R. Hagen, and Tom Kibble in 1964 (see 1964 PRL symmetry breaking papers) and is a key building block in the Standard Model.[7][8][9][15] It has no intrinsic spin, and for that reason is classified as a boson (like the gauge bosons, which have integer spin).

The Higgs boson plays a unique role in the Standard Model, by explaining why the other elementary particles, except the photon and gluon, are massive. In particular, the Higgs boson explains why the photon has no mass, while the W and Z bosons are very heavy. Elementary particle masses, and the differences between electromagnetism (mediated by the photon) and the weak force (mediated by the W and Z bosons), are critical to many aspects of the structure of microscopic (and hence macroscopic) matter. In electroweak theory, the Higgs boson generates the masses of the leptons (electron, muon, and tau) and quarks. As the Higgs boson is massive, it must interact with itself.

Because the Higgs boson is a very massive particle and also decays almost immediately when created, only a very high-energy particle accelerator can observe and record it. Experiments to confirm and determine the nature of the Higgs boson using the Large Hadron Collider (LHC) at CERN began in early 2010, and were performed at Fermilab's Tevatron until its closure in late 2011. Mathematical consistency of the Standard Model requires that any mechanism capable of generating the masses of elementary particles become visible at energies above 1.4 TeV;[16] therefore, the LHC (designed to collide two 7 to 8 TeV proton beams) was built to answer the question of whether the Higgs boson actually exists.[17]

On 4 July 2012, the two main experiments at the LHC (ATLAS and CMS) both reported independently that they found a new particle with a mass of about 125 GeV/c^2 (about 133 proton masses, on the order of 10^{-25} kg), which is "consistent with the Higgs boson." Although it has several properties similar to the predicted "simplest" Higgs,[18] they acknowledged that further work would be needed to conclude that it is indeed the Higgs boson, and exactly which version of the Standard Model Higgs is best supported if confirmed.[19][20][21][22][23]

On 14 March 2013 the Higgs Boson was tentatively confirmed to exist.[24]

8.3.4   Total particle count

Counting particles by a rule that distinguishes between particles and their corresponding antiparticles, and among the many color states of quarks and gluons, gives a total of 61 elementary particles.[25]

8.4   Theoretical aspects

Main article: Standard Model (mathematical formulation)

8.4.1   Construction of the Standard Model Lagrangian

Technically, quantum field theory provides the mathematical framework for the Standard Model, in which a Lagrangian controls the dynamics and kinematics of the theory. Each kind of particle is described in terms of a dynamical field that pervades space-time. The construction of the Standard Model proceeds following the modern method of constructing most field theories: by first postulating a set of symmetries of the system, and then by writing down the most general renormalizable Lagrangian from its particle (field) content that observes these symmetries.

The global Poincaré symmetry is postulated for all relativistic quantum field theories. It consists of the familiar translational symmetry, rotational symmetry and the inertial reference frame invariance central to the theory of special relativity. The local SU(3)×SU(2)×U(1) gauge symmetry is an internal symmetry that essentially defines the Standard Model. Roughly, the three factors of the gauge symmetry give rise to the three fundamental interactions. The fields fall into different representations of the various symmetry groups of the Standard Model (see table). Upon writing the most general Lagrangian, one finds that the dynamics depend on 19 parameters, whose numerical values are established by experiment. The parameters are summarized in the table above (note: with the Higgs mass is at 125 GeV, the Higgs self-coupling strength $\lambda \sim 1/8$).

Quantum chromodynamics sector

Main article: Quantum chromodynamics

The quantum chromodynamics (QCD) sector defines the interactions between quarks and gluons, with SU(3) symmetry, generated by T^a. Since leptons do not interact with gluons, they are not affected by this sector. The Dirac Lagrangian of the quarks coupled to the gluon fields is given by

$$\mathcal{L}_{QCD} = i\overline{U}(\partial_\mu - ig_s G_\mu^a T^a)\gamma^\mu U + i\overline{D}(\partial_\mu - ig_s G_\mu^a T^a)\gamma^\mu D.$$

G_μ^a is the SU(3) gauge field containing the gluons, γ^μ are the Dirac matrices, D and U are the Dirac spinors associated with up- and down-type quarks, and g_s is the strong coupling constant.

Electroweak sector

Main article: Electroweak interaction

The electroweak sector is a Yang–Mills gauge theory with the simple symmetry group U(1)×SU(2)L,

$$\mathcal{L}_{EW} = \sum_\psi \bar{\psi}\gamma^\mu \left(i\partial_\mu - g'\frac{1}{2}Y_W B_\mu - g\frac{1}{2}\vec{\tau}_L \vec{W}_\mu \right) \psi$$

where $B\mu$ is the U(1) gauge field; YW is the weak hypercharge—the generator of the U(1) group; $\vec{W}_\mu$ is the three-component SU(2) gauge field; $\vec{\tau}_L$ are the Pauli matrices—infinitesimal generators of the SU(2) group. The subscript L indicates that they only act on left fermions; g' and g are coupling constants.

Higgs sector

Main article: Higgs mechanism

In the Standard Model, the Higgs field is a complex scalar of the group SU(2)L:

$$\varphi = \frac{1}{\sqrt{2}} \begin{pmatrix} \varphi^+ \\ \varphi^0 \end{pmatrix} ,$$

where the indices + and 0 indicate the electric charge (Q) of the components. The weak isospin (YW) of both components is 1.

Before symmetry breaking, the Higgs Lagrangian is:

$$\mathcal{L}_H = \varphi^\dagger \left(\partial^\mu - \frac{i}{2}\left(g'Y_W B^\mu + g\vec{\tau}\vec{W}^\mu \right) \right) \left(\partial_\mu + \frac{i}{2}\left(g'Y_W B_\mu + g\vec{\tau}\vec{W}_\mu \right) \right) \varphi - \frac{\lambda^2}{4}\left(\varphi^\dagger\varphi - v^2 \right)^2 ,$$

which can also be written as:

$$\mathcal{L}_H = \left| \left(\partial_\mu + \frac{i}{2}\left(g'Y_W B_\mu + g\vec{\tau}\vec{W}_\mu \right) \right) \varphi \right|^2 - \frac{\lambda^2}{4}\left(\varphi^\dagger\varphi - v^2 \right)^2 .$$

8.5 Fundamental forces

Main article: Fundamental interaction

The Standard Model classified all four fundamental forces in nature. In the Standard Model, a force is described as an exchange of bosons between the objects affected, such as a photon for the electromagnetic force and a gluon for the strong interaction. Those particles are called force carriers.[26]

8.6   Tests and predictions

The Standard Model (SM) predicted the existence of the W and Z bosons, gluon, and the top and charm quarks before these particles were observed. Their predicted properties were experimentally confirmed with good precision. To give an idea of the success of the SM, the following table compares the measured masses of the W and Z bosons with the masses predicted by the SM:

The SM also makes several predictions about the decay of Z bosons, which have been experimentally confirmed by the Large Electron-Positron Collider at CERN.

In May 2012 BaBar Collaboration reported that their recently analyzed data may suggest possible flaws in the Standard Model of particle physics.[28][29] These data show that a particular type of particle decay called "B to D-star-tau-nu" happens more often than the Standard Model says it should. In this type of decay, a particle called the B-bar meson decays into a D meson, an antineutrino and a tau-lepton. While the level of certainty of the excess (3.4 sigma) is not enough to claim a break from the Standard Model, the results are a potential sign of something amiss and are likely to impact existing theories, including those attempting to deduce the properties of Higgs bosons.[30]

On December 13, 2012, physicists reported the constancy, over space and time, of a basic physical constant of nature that supports the *standard model of physics*. The scientists, studying methanol molecules in a distant galaxy, found the change ($\Delta\mu/\mu$) in the proton-to-electron mass ratio μ to be equal to "$(0.0 \pm 1.0) \times 10^{-7}$ at redshift z = 0.89" and consistent with "a null result".[31][32]

8.7   Challenges

See also: Physics beyond the Standard Model

Self-consistency of the Standard Model (currently formulated as a non-abelian gauge theory quantized through path-integrals) has not been mathematically proven. While regularized versions useful for approximate computations (for example lattice gauge theory) exist, it is not known whether they converge (in the sense of S-matrix elements) in the limit that the regulator is removed. A key question related to the consistency is the Yang–Mills existence and mass gap problem.

Experiments indicate that neutrinos have mass, which the classic Standard Model did not allow.[33] To accommodate this finding, the classic Standard Model can be modified to include neutrino mass.

If one insists on using only Standard Model particles, this can be achieved by adding a non-renormalizable interaction of leptons with the Higgs boson.[34] On a fundamental level, such an interaction emerges in the seesaw mechanism where heavy right-handed neutrinos are added to the theory. This is natural in the left-right symmetric extension of the Standard Model[35][36] and in certain grand unified theories.[37] As long as new physics appears below or around 10^{14} GeV, the neutrino masses can be of the right order of magnitude.

Theoretical and experimental research has attempted to extend the Standard Model into a Unified field theory or a Theory of everything, a complete theory explaining all physical phenomena including constants. Inadequacies of the Standard Model that motivate such research include:

- It does not attempt to explain gravitation, although a theoretical particle known as a graviton would help explain it, and unlike for the strong and electroweak interactions of the Standard Model, there is no known way of describing general relativity, the canonical theory of gravitation, consistently in terms of quantum field theory. The reason for this is, among other things, that quantum field theories of gravity generally break down before reaching the Planck scale. As a consequence, we have no reliable theory for the very early universe;

- Some consider it to be *ad hoc* and inelegant, requiring 19 numerical constants whose values are unrelated and arbitrary. Although the Standard Model, as it now stands, can explain why neutrinos have masses, the specifics of neutrino mass are still unclear. It is believed that explaining neutrino mass will require an additional 7 or 8 constants, which are also arbitrary parameters;

- The Higgs mechanism gives rise to the hierarchy problem if some new physics (coupled to the Higgs) is present at high energy scales. In these cases in order for the weak scale to be much smaller than the Planck scale, severe fine tuning of the parameters is required; there are, however, other scenarios that include quantum gravity in which such fine tuning can be avoided.[38] There are also issues of Quantum triviality, which suggests that it may not be possible to create a consistent quantum field theory involving elementary scalar particles.

- It should be modified so as to be consistent with the emerging "Standard Model of cosmology." In particular, the Standard Model cannot explain the observed amount of cold dark matter (CDM) and gives contributions to dark energy which are many orders of magnitude too large. It is also difficult to accommodate the observed predominance of matter over antimatter (matter/antimatter asymmetry). The isotropy and homogeneity of the visible universe over large distances seems to require a mechanism like cosmic inflation, which would also constitute an extension of the Standard Model.

- The existence of ultra-high-energy cosmic rays are difficult to explain under the Standard Model.

Currently, no proposed Theory of Everything has been widely accepted or verified.

8.8 See also

- Fundamental interaction:

    - Quantum electrodynamics

    - Strong interaction: Color charge, Quantum chromodynamics, Quark model

    - Weak interaction: Electroweak theory, Fermi theory of beta decay, Weak hypercharge, Weak isospin

- Gauge theory: Nontechnical introduction to gauge theory

- Generation

- Higgs mechanism: Higgs boson, Higgsless model

- J. C. Ward

- J. J. Sakurai Prize for Theoretical Particle Physics

- Lagrangian

- Open questions: BTeV experiment, CP violation, Neutrino masses, Quark matter, Quantum triviality

- Penguin diagram

- Quantum field theory

- Standard Model: Mathematical formulation of, Physics beyond the Standard Model

8.9 Notes and references

[1] Technically, there are nine such color–anticolor combinations. However, there is one color-symmetric combination that can be constructed out of a linear superposition of the nine combinations, reducing the count to eight.

8.10   References

[1] R. Oerter (2006). *The Theory of Almost Everything: The Standard Model, the Unsung Triumph of Modern Physics* (Kindle ed.). Penguin Group. p. 2. ISBN 0-13-236678-9.

[2] In fact, there are mathematical issues regarding quantum field theories still under debate (see e.g. Landau pole), but the predictions extracted from the Standard Model by current methods applicable to current experiments are all self-consistent. For a further discussion see e.g. Chapter 25 of R. Mann (2010). *An Introduction to Particle Physics and the Standard Model*. CRC Press. ISBN 978-1-4200-8298-2.

[3] Sean Carroll, Ph.D., Cal Tech, 2007. The Teaching Company, *Dark Matter, Dark Energy: The Dark Side of the Universe*, Guidebook Part 2 page 59, Accessed Oct. 7, 2013, "...Standard Model of Particle Physics: The modern theory of elementary particles and their interactions ... It does not, strictly speaking, include gravity, although it's often convenient to include gravitons among the known particles of nature..."

[4] S.L. Glashow (1961). "Partial-symmetries of weak interactions". *Nuclear Physics* **22**(4): 579–588. Bibcode:1961NucPh..22...G. doi:10.1016/0029-5582(61)90469-2.

[5] S. Weinberg (1967). "A Model of Leptons". *Physical Review Letters* **19** (21): 1264–1266. Bibcode:1967PhRvL..19.1264W. doi:10.1103/PhysRevLett.19.1264.

[6] A. Salam (1968). N. Svartholm, ed. *Elementary Particle Physics: Relativistic Groups and Analyticity*. Eighth Nobel Symposium. Stockholm: Almquvist and Wiksell. p. 367.

[7] F. Englert, R. Brout (1964). "Broken Symmetry and the Mass of Gauge Vector Mesons". *Physical Review Letters* **13** (9): 321–323. Bibcode:1964PhRvL..13..321E. doi:10.1103/PhysRevLett.13.321.

[8] P.W. Higgs (1964). "Broken Symmetries and the Masses of Gauge Bosons". *Physical Review Letters* **13** (16): 508–509. Bibcode:1964PhRvL..13..508H. doi:10.1103/PhysRevLett.13.508.

[9] G.S. Guralnik, C.R. Hagen, T.W.B. Kibble (1964). "Global Conservation Laws and Massless Particles". *Physical Review Letters* **13** (20): 585–587. Bibcode:1964PhRvL..13..585G. doi:10.1103/PhysRevLett.13.585.

[10] F.J. Hasert; et al. (1973). "Search for elastic muon-neutrino electron scattering". *Physics Letters B* **46**(1): 121. Bibcode:1973. doi:10.1016/0370-2693(73)90494-2.

[11] F.J. Hasert; et al. (1973). "Observation of neutrino-like interactions without muon or electron in the Gargamelle neutrino experiment". *Physics Letters B* **46** (1): 138. Bibcode:1973PhLB...46..138H. doi:10.1016/0370-2693(73)90499-1.

[12] F.J. Hasert; et al. (1974). "Observation of neutrino-like interactions without muon or electron in the Gargamelle neutrino experiment". *Nuclear Physics B* **73** (1): 1. Bibcode:1974NuPhB..73....1H. doi:10.1016/0550-3213(74)90038-8.

[13] D. Haidt (4 October 2004). "The discovery of the weak neutral currents". *CERN Courier*. Retrieved 8 May 2008.

[14] "Details can be worked out if the situation is simple enough for us to make an approximation, which is almost never, but often we can understand more or less what is happening." from *The Feynman Lectures on Physics*, Vol 1. pp. 2–7

[15] G.S. Guralnik (2009). "The History of the Guralnik, Hagen and Kibble development of the Theory of Spontaneous Symmetry Breaking and Gauge Particles". *International Journal of Modern Physics A* **24** (14): 2601–2627. arXiv:0907.3466. Bibcode:2009IJMPA..24.2601G. doi:10.1142/S0217751X09045431.

[16] B.W. Lee, C. Quigg, H.B. Thacker (1977). "Weak interactions at very high energies: The role of the Higgs-boson mass". *Physical Review D* **16** (5): 1519–1531. Bibcode:1977PhRvD..16.1519L. doi:10.1103/PhysRevD.16.1519.

[17] "Huge $10 billion collider resumes hunt for 'God particle'". CNN. 11 November 2009. Retrieved 2010-05-04.

[18] M. Strassler (10 July 2012). "Higgs Discovery: Is it a Higgs?". Retrieved 2013-08-06.

[19] "CERN experiments observe particle consistent with long-sought Higgs boson". CERN. 4 July 2012. Retrieved 2012-07-04.

[20] "Observation of a New Particle with a Mass of 125 GeV". CERN. 4 July 2012. Retrieved 2012-07-05.

[21] "ATLAS Experiment". ATLAS. 1 January 2006. Retrieved 2012-07-05.

[22] "Confirmed: CERN discovers new particle likely to be the Higgs boson". *YouTube*. Russia Today. 4 July 2012. Retrieved 2013-08-06.

[23] D. Overbye (4 July 2012). "A New Particle Could Be Physics' Holy Grail". *New York Times*. Retrieved 2012-07-04.

[24] "New results indicate that new particle is a Higgs boson". CERN. 14 March 2013. Retrieved 2013-08-06.

[25] S. Braibant, G. Giacomelli, M. Spurio (2009). *Particles and Fundamental Interactions: An Introduction to Particle Physics*. Springer. pp. 313–314. ISBN 978-94-007-2463-1.

[26] http://home.web.cern.ch/about/physics/standard-model Official CERN website

[27] http://www.pha.jhu.edu/~{}dfehling/particle.gif

[28] "BABAR Data in Tension with the Standard Model". SLAC. 31 May 2012. Retrieved 2013-08-06.

[29] BaBar Collaboration (2012). "Evidence for an excess of $B \rightarrow D^{(*)} \tau^{-} \nu\tau$ decays". *Physical Review Letters* **109** (10): 101802. arXiv:1205.5442. Bibcode:2012PhRvL.109j1802L. doi:10.1103/PhysRevLett.109.101802.

[30] "BaBar data hint at cracks in the Standard Model". *e! Science News*. 18 June 2012. Retrieved 2013-08-06.

[31] J. Bagdonaite; et al. (2012). "A Stringent Limit on a Drifting Proton-to-Electron Mass Ratio from Alcohol in the Early Universe". *Science* **339** (6115): 46. Bibcode:2013Sci...339...46B. doi:10.1126/science.1224898.

[32] C. Moskowitz (13 December 2012). "Phew! Universe's Constant Has Stayed Constant". Space.com. Retrieved 2012-12-14.

[33] "Particle chameleon caught in the act of changing". CERN. 31 May 2010. Retrieved 2012-07-05.

[34] S. Weinberg (1979). "Baryon and Lepton Nonconserving Processes". *Physical Review Letters* **43**(21): 1566. Bibcode:1979PW. doi:10.1103/PhysRevLett.43.1566.

[35] P. Minkowski (1977). "$\mu \rightarrow e \gamma$ at a Rate of One Out of 10^{9} Muon Decays?". *Physics Letters* **B67**(4): 421. Bibcode:1977PhLB...61M. doi:10.1016/0370-2693(77)90435-X.

[36] R. N. Mohapatra, G. Senjanovic (1980). "Neutrino Mass and Spontaneous Parity Nonconservation". *Physical Review Letters* **44** (14): 912–915. Bibcode:1980PhRvL..44..912M. doi:10.1103/PhysRevLett.44.912.

[37] M. Gell-Mann, P. Ramond and R. Slansky (1979). F. van Nieuwenhuizen and D. Z. Freedman, ed. *Supergravity*. North Holland. pp. 315–321. ISBN 0-444-85438-X.

[38] Salvio, Strumia (2014-03-17). "Agravity". *JHEP 1406 (2014) 080*. arXiv:1403.4226. Bibcode:2014JHEP...06..080S. doi:

8.11   Further reading

- R. Oerter (2006). *The Theory of Almost Everything: The Standard Model, the Unsung Triumph of Modern Physics*. Plume.

- B.A. Schumm (2004). *Deep Down Things: The Breathtaking Beauty of Particle Physics*. Johns Hopkins University Press. ISBN 0-8018-7971-X.

- "The Standard Model of Particle Physics Interactive Graphic".

Introductory textbooks

- I. Aitchison, A. Hey (2003). *Gauge Theories in Particle Physics: A Practical Introduction*. Institute of Physics. ISBN 978-0-585-44550-2.

- W. Greiner, B. Müller (2000). *Gauge Theory of Weak Interactions*. Springer. ISBN 3-540-67672-4.

- G.D. Coughlan, J.E. Dodd, B.M. Gripaios (2006). *The Ideas of Particle Physics: An Introduction for Scientists*. Cambridge University Press.

- D.J. Griffiths (1987). *Introduction to Elementary Particles*. John Wiley & Sons. ISBN 0-471-60386-4.

- G.L. Kane (1987). *Modern Elementary Particle Physics*. Perseus Books. ISBN 0-201-11749-5.

Advanced textbooks

- T.P. Cheng, L.F. Li (2006). *Gauge theory of elementary particle physics*. Oxford University Press. ISBN 0-19-851961-3. Highlights the gauge theory aspects of the Standard Model.

- J.F. Donoghue, E. Golowich, B.R. Holstein (1994). *Dynamics of the Standard Model*. Cambridge University Press. ISBN 978-0-521-47652-2. Highlights dynamical and phenomenological aspects of the Standard Model.

- L. O'Raifeartaigh (1988). *Group structure of gauge theories*. Cambridge University Press. ISBN 0-521-34785-8.

- Nagashima Y. Elementary Particle Physics: Foundations of the Standard Model, Volume 2. (Wiley 2013) 920 раппы

- Schwartz, M.D. Quantum Field Theory and the Standard Model (Cambridge University Press 2013) 952 pages

- Langacker P. The standard model and beyond. (CRC Press, 2010) 670 pages Highlights group-theoretical aspects of the Standard Model.

Journal articles

- E.S. Abers, B.W. Lee (1973). "Gauge theories". *Physics Reports* **9**: 1–141. Bibcode:1973PhR.....9....1A.doi:1-1573(73)90027-6.

- M. Baak; et al. (2012). "The Electroweak Fit of the Standard Model after the Discovery of a New Boson at the LHC". *The European Physical Journal C* **72** (11). arXiv:1209.2716. Bibcode:2012EPJC...72.2205B. doi:10.1140/epjc/s10052-012-2205-9.

- Y. Hayato; et al. (1999). "Search for Proton Decay through $p \rightarrow \nu K^+$ in a Large Water Cherenkov Detector". *Physical Review Letters* **83** (8): 1529. arXiv:hep-ex/9904020. Bibcode:1999PhRvL..83.1529H. doi:10.1103.

- S.F. Novaes (2000). "Standard Model: An Introduction". arXiv:hep-ph/0001283 [hep-ph].

- D.P. Roy (1999). "Basic Constituents of Matter and their Interactions — A Progress Report". arXiv:hep-ph/9912523 [hep-ph].

- F. Wilczek (2004). "The Universe Is A Strange Place". *Nuclear Physics B - Proceedings Supplements* **134**: 3. arXiv:astro-ph/0401347. Bibcode:2004NuPhS.134....3W. doi:10.1016/j.nuclphysbps.2004.08.001.

8.12   External links

- "The Standard Model explained in Detail by CERN's John Ellis" omega tau podcast.

- "LHC sees hint of lightweight Higgs boson" "New Scientist".

- "Standard Model may be found incomplete," *New Scientist*.

- "Observation of the Top Quark" at Fermilab.

- "The Standard Model Lagrangian." After electroweak symmetry breaking, with no explicit Higgs boson.

- "Standard Model Lagrangian" with explicit Higgs terms. PDF, PostScript, and LaTeX versions.

- "The particle adventure." Web tutorial.

- Nobes, Matthew (2002) "Introduction to the Standard Model of Particle Physics" on Kuro5hin: Part 1, Part 2, Part 3a, Part 3b.

- "The Standard Model" The Standard Model on the CERN web site explains how the basic building blocks of matter interact, governed by four fundamental forces.

Chapter 9

Physics beyond the Standard Model

Physics beyond the Standard Model (BSM) refers to the theoretical developments needed to explain the deficiencies of the Standard Model, such as the origin of mass, the strong CP problem, neutrino oscillations, matter–antimatter asymmetry, and the nature of dark matter and dark energy.[1] Another problem lies within the mathematical framework of the Standard Model itself—the Standard Model is inconsistent with that of general relativity, to the point that one or both theories break down under certain conditions (for example within known space-time singularities like the Big Bang and black hole event horizons).

Theories that lie beyond the Standard Model include various extensions of the standard model through supersymmetry, such as the Minimal Supersymmetric Standard Model (MSSM) and Next-to-Minimal Supersymmetric Standard Model (NMSSM), or entirely novel explanations, such as string theory, M-theory, and extra dimensions. As these theories tend to reproduce the entirety of current phenomena, the question of which theory is the right one, or at least the "best step" towards a Theory of Everything, can only be settled via experiments, and is one of the most active areas of research in both theoretical and experimental physics.

9.1 Problems with the Standard Model

Despite being the most successful theory of particle physics to date, the Standard Model is not perfect.[2] A large share of the published output of theoretical physicists consists of proposals for various forms of "Beyond the Standard Model" new physics proposals that would modify the Standard Model in ways subtle enough to be consistent with existing data, yet address its imperfections materially enough to predict non-Standard Model outcomes of new experiments that can be proposed.

9.1.1 Phenomena not explained

The Standard Model is inherently an incomplete theory. There are fundamental physical phenomena in nature that the Standard Model does not adequately explain:

- *Gravity.* The standard model does not explain gravity. The approach of simply adding a "graviton" (whose properties are the subject of considerable consensus among physicists if it exists) to the Standard Model does not recreate what is observed experimentally without other modifications, as yet undiscovered, to the Standard Model. Moreover, instead, the Standard Model is widely considered to be incompatible with the most successful theory of gravity to date, general relativity.[3]

- *Dark matter and dark energy.* Cosmological observations tell us the standard model explains about 5% of the energy present in the universe. About 26% should be dark matter, which would behave just like other matter, but which only interacts weakly (if at all) with the Standard Model fields. Yet, the Standard Model does not supply

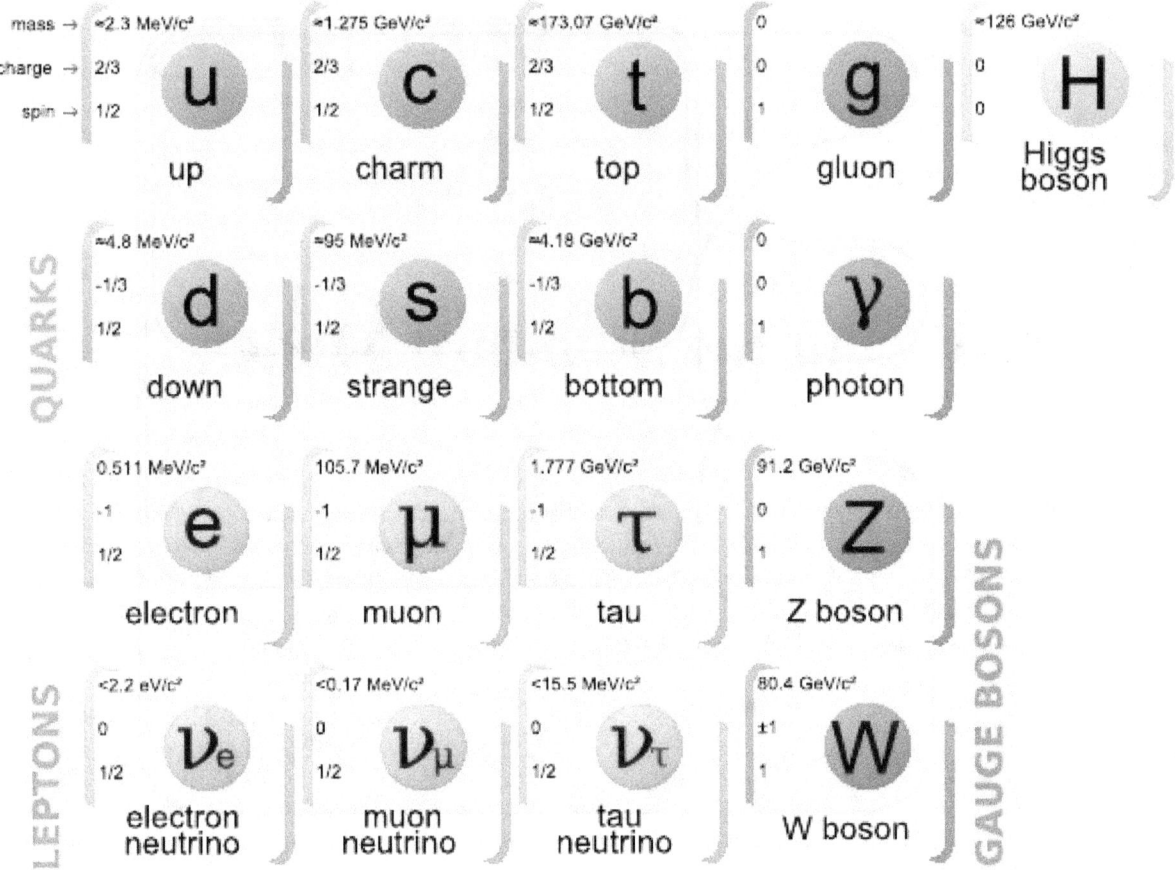

The Standard Model of elementary particles

any fundamental particles that are good dark matter candidates. The rest (69%) should be dark energy, a constant energy density for the vacuum. Attempts to explain dark energy in terms of vacuum energy of the standard model lead to a mismatch of 120 orders of magnitude.[4]

- *Neutrino masses.* According to the standard model, neutrinos are massless particles. However, neutrino oscillation experiments have shown that neutrinos do have mass. Mass terms for the neutrinos can be added to the standard model by hand, but these lead to new theoretical problems. For example, the mass terms need to be extraordinarily small and it is not clear if the neutrino masses would arise in the same way that the masses of other fundamental particles do in the Standard Model.

- *Matter-antimatter asymmetry.* The universe is made out of mostly matter. However, the standard model predicts that matter and antimatter should have been created in (almost) equal amounts if the initial conditions of the universe did not involve disproportionate matter relative to antimatter. Yet, no mechanism sufficient to explain this asymmetry exists in the Standard Model.

9.1.2 Experimental results not explained

No experimental result is widely accepted as definitively contradicting the Standard Model at the "five sigma" level, widely considered to be the threshold of a "discovery" in particle physics. But because every experiment contains some degree of statistical and systemic uncertainty, and the theoretical predictions themselves are also almost never calculated exactly and are subject to uncertainties in measurements of the fundamental constants of the Standard Model (some of which are tiny and others of which are substantial), it is mathematically expected that some of the hundreds of experimental tests of the Standard Model will deviate to some extent from it, even if there were no "new physics" to be discovered.

At any given time there are a number of experimental results that are significantly different from the Standard Model expectation, although many of these have been found to be statistical flukes or experimental errors as more data has been collected. On the other hand, any "beyond the Standard Model" physics would necessarily first manifest experimentally as a statistically significant difference between an experiment and the theoretical prediction.

In each case, physicists seek to determine if a result is a mere statistical fluke or experimental error on the one hand, or a sign of new physics on the other. More statistically significant results cannot be mere statistical flukes but can still result from experimental error or inaccurate estimates of experimental precision. Frequently, experiments are tailored to be more sensitive to experimental results that would distinguish the Standard Model from theoretical alternatives.

Some of the most notable examples include the following:

- *Muonic hydrogen* — the Standard Model makes precise theoretical predictions regarding the atomic radius size of ordinary hydrogen (a proton-electron system) and that of muonic hydrogen (a proton-muon system in which a muon is a "heavy" variant of an electron). However, the measured atomic radius of muonic hydrogen differs significantly from that of the radius predicted by the Standard Model using existing physical constant measurements by what appears to be as many as seven standard deviations.[5] Doubts about the accuracy of the error estimates in earlier experiments, which are still within 4% of each other in measuring a truly tiny distance, and a lack of a well motivated theory that could explain the discrepancy, have caused physicists to be hesitant to describe these results as contradicting the Standard Model despite the apparent statistical significance of the result and a lack of any clearly identified possible source of experimental error in the results.

- *BaBar data suggests possible flaws in the Standard Model* — results from a BaBar experiment may suggest a surplus over Standard Model predictions of a type of particle decay called "B to D-star-tau-nu." In this, an electron and positron collide, resulting in a B meson and an antimatter B-bar meson, which then decays into a D meson and a tau lepton as well as a smaller antineutrino. While the level of certainty of the excess (3.4 sigma in statistical language) is not enough to claim a break from the Standard Model, the results are a potential sign of something amiss and are likely to affect existing theories, including those attempting to deduce the properties of Higgs bosons.[6] In 2015, LHCb reported observing a 2.1 sigma excess in the same ratio of branching fractions.[7]

- Proton radius — the radius measured using electrons is different from the radius measured using muons.[8]

9.1.3 Theoretical predictions not observed

Observation at particle colliders of all of the fundamental particles predicted by the Standard Model has been confirmed. The Higgs boson is predicted by the Standard Model's explanation of the Higgs mechanism, which describes how the weak SU(2) gauge symmetry is broken and how fundamental particles obtain mass; it was the last particle predicted by the Standard Model to be observed. On July 4, 2012, CERN scientists using the Large Hadron Collider announced the discovery of a particle consistent with the Higgs boson, with a mass of about 126 GeV/c^2. A Higgs boson was confirmed to exist on March 14, 2013, although efforts to confirm that it has all of the properties predicted by the Standard Model are ongoing.[9]

A few hadrons (i.e. composite particles made of quarks) whose existence is predicted by the Standard Model, which can be produced only at very high energies in very low frequencies have not yet been definitively observed, and "glueballs"[10] (i.e. composite particles made of gluons) have also not yet been definitively observed. Some very low frequency particle decays predicted by the Standard Model have also not yet been definitively observed because insufficient data is available to make a statistically significant observation.

9.1.4 Theoretical problems

Some features of the standard model are added in an ad hoc way. These are not problems per se (i.e. the theory works fine with these ad hoc features), but they imply a lack of understanding. These ad hoc features have motivated theorists to look for more fundamental theories with fewer parameters. Some of the ad hoc features are:

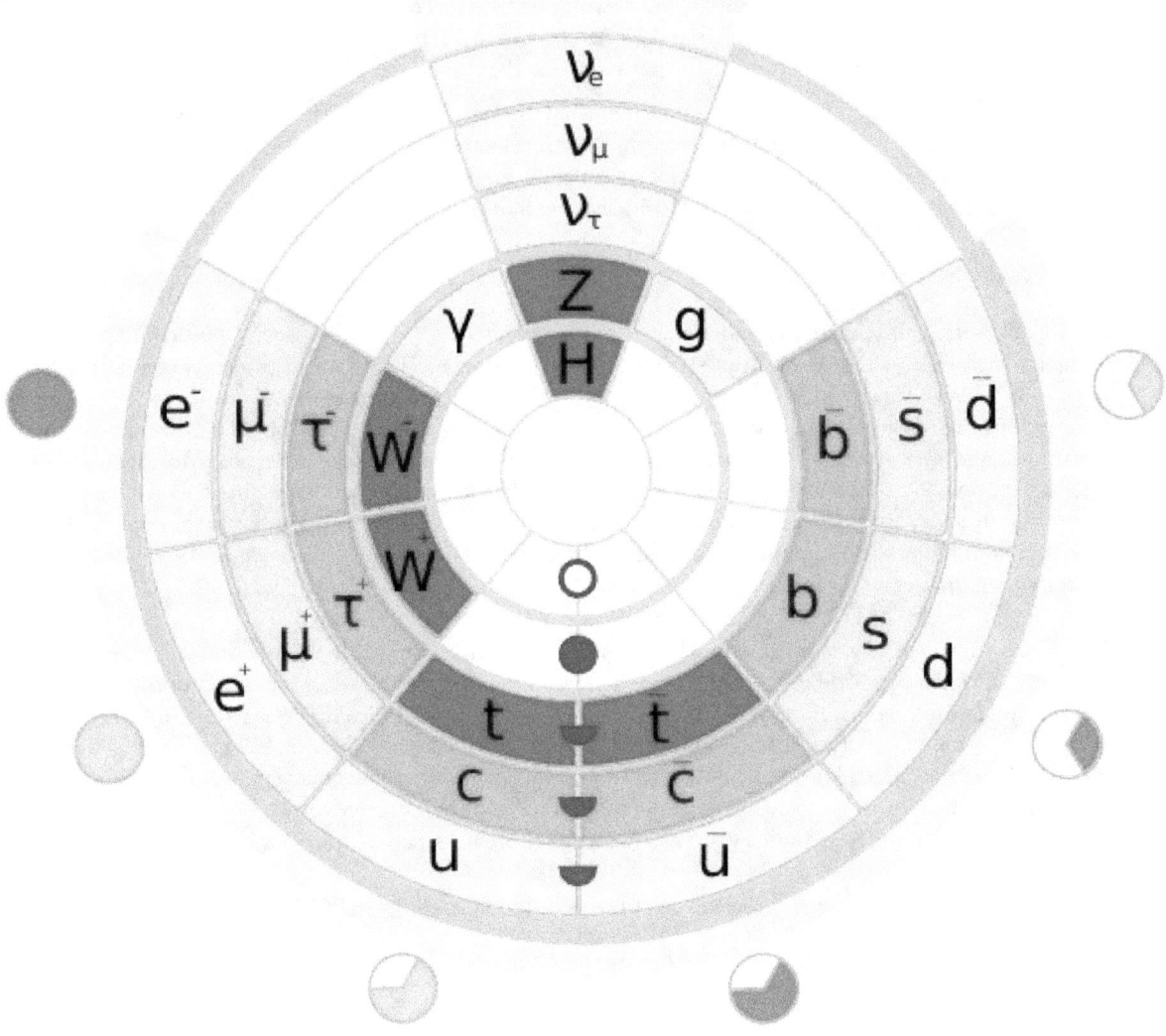

Masses of fundamental particles ----
more than 80 GeV/c²
1-5 GeV/c²
90-110 MeV/c²
less than 16 MeV/c²
Massless

- *Hierarchy problem* — the standard model introduces particle masses through a process known as spontaneous symmetry breaking caused by the Higgs field. Within the standard model, the mass of the Higgs gets some very large quantum corrections due to the presence of virtual particles (mostly virtual top quarks). These corrections are much larger than the actual mass of the Higgs. This means that the bare mass parameter of the Higgs in the standard model must be fine tuned in such a way that almost completely cancels the quantum corrections. This level of fine-tuning is deemed unnatural by many theorists.

- *Number of parameters* — the standard model depends on 19 numerical parameters. Their values are known from experiment, but the origin of the values is unknown. Some theorists have tried to find relations between different parameters, for example, between the masses of particles in different generations.

- *Quantum triviality* — suggests that it may not be possible to create a consistent quantum field theory involving elementary scalar Higgs particles.

- *Strong CP problem* — theoretically it can be argued that the standard model should contain a term that breaks CP symmetry—relating matter to antimatter—in the strong interaction sector. Experimentally, however, no such violation has been found, implying that the coefficient of this term is very close to zero. This fine tuning is also considered unnatural.

9.2   Grand unified theories

Main article: Grand Unified Theory

The standard model has three gauge symmetries; the colour SU(3), the weak isospin SU(2), and the hypercharge U(1) symmetry, corresponding to the three fundamental forces. Due to renormalization the coupling constants of each of these symmetries vary with the energy at which they are measured. Around 10^{16} GeV these couplings become approximately equal. This has led to speculation that above this energy the three gauge symmetries of the standard model are unified in one single gauge symmetry with a simple group gauge group, and just one coupling constant. Below this energy the symmetry is spontaneously broken to the standard model symmetries.[11] Popular choices for the unifying group are the special unitary group in five dimensions SU(5) and the special orthogonal group in ten dimensions SO(10).[12]

Theories that unify the standard model symmetries in this way are called Grand Unified Theories (or GUTs), and the energy scale at which the unified symmetry is broken is called the GUT scale. Generically, grand unified theories predict the creation of magnetic monopoles in the early universe,[13] and instability of the proton.[14] Neither of these have been observed, and this absence of observation puts limits on the possible GUTs.

9.3   Supersymmetry

Main article: Supersymmetry

Supersymmetry extends the Standard Model by adding another class of symmetries to the Lagrangian. These symmetries exchange fermionic particles with bosonic ones. Such a symmetry predicts the existence of *supersymmetric particles*, abbreviated as *sparticles*, which include the sleptons, squarks, neutralinos and charginos. Each particle in the Standard Model would have a superpartner whose spin differs by 1/2 from the ordinary particle. Due to the breaking of supersymmetry, the sparticles are much heavier than their ordinary counterparts; they are so heavy that existing particle colliders may not be powerful enough to produce them.

9.4   Neutrinos

In the standard model, neutrinos have exactly zero mass. This is a consequence of the standard model containing only left-handed neutrinos. With no suitable right-handed partner, it is impossible to add a renormalizable mass term to the standard model.[15] Measurements however indicated that neutrinos spontaneously change flavour, which implies that neutrinos have a mass. These measurements only give the relative masses of the different flavours. The best constraint on the absolute mass of the neutrinos comes from precision measurements of tritium decay, providing an upper limit 2 eV, which makes them at least five orders of magnitude lighter than the other particles in the standard model.[16] This necessitates an extension of the standard model, which not only needs to explain how neutrinos get their mass, but also why the mass is so small.[17]

One approach to add masses to the neutrinos, the so-called seesaw mechanism, is to add right-handed neutrinos and have these couple to left-handed neutrinos with a Dirac mass term. The right-handed neutrinos have to be sterile, meaning that they do not participate in any of the standard model interactions. Because they have no charges, the right-handed neutrinos can act as their own anti-particles, and have a Majorana mass term. Like the other Dirac masses in the standard model, the neutrino Dirac mass is expected to be generated through the Higgs mechanism, and is therefore unpredictable. The standard model fermion masses differ by many orders of magnitude; the Dirac neutrino mass has at least the same uncertainty. On the other hand, the Majorana mass for the right-handed neutrinos does not arise from the Higgs mechanism, and is therefore expected to be tied to some energy scale of new physics beyond the standard model, for example the Planck scale.[18] Therefore, any process involving right-handed neutrinos will be suppressed at low energies. The correction due to these suppressed processes effectively gives the left-handed neutrinos a mass that is inversely proportional to the right-handed Majorana mass, a mechanism known as the see-saw.[19] The presence of heavy right-handed neutrinos thereby explains both the small mass of the left-handed neutrinos and the absence of the right-handed neutrinos in observations. However, due to the uncertainty in the Dirac neutrino masses, the right-handed neutrino masses can lie anywhere. For example, they could be as light as keV and be dark matter,[20] they can have a mass in the LHC energy range[21][22] and lead to observable lepton number violation,[23] or they can be near the GUT scale, linking the right-handed neutrinos to the possibility of a grand unified theory.[24][25]

The mass terms mix neutrinos of different generations. This mixing is parameterized by the PMNS matrix, which is the neutrino analogue of the CKM quark mixing matrix. Unlike the quark mixing, which is almost minimal, the mixing of the neutrinos appears to be almost maximal. This has led to various speculations of symmetries between the various generations that could explain the mixing patterns.[26] The mixing matrix could also contain several complex phases that break CP invariance, although there has been no experimental probe of these. These phases could potentially create a surplus of leptons over anti-leptons in the early universe, a process known as leptogenesis. This asymmetry could then at a later stage be converted in an excess of baryons over anti-baryons, and explain the matter-antimatter asymmetry in the universe.[12]

The light neutrinos are disfavored as an explanation for the observation of dark matter, due to considerations of large-scale structure formation in the early universe. Simulations of structure formation show that they are too hot—i.e. their kinetic energy is large compared to their mass—while formation of structures similar to the galaxies in our universe requires cold dark matter. The simulations show that neutrinos can at best explain a few percent of the missing dark matter. The heavy sterile right-handed neutrinos are however a possible candidate for a dark matter WIMP.[27]

9.5   Preon Models

Several preon models have been proposed to address the unsolved problem concerning the fact that there are three generations of quarks and leptons. Preon models generally postulate some additional new particles which are further postulated to be able to combine to form the quarks and leptons of the standard model. One of the earliest preon models was the Rishon model.[28][29][30]

To date, no preon model is widely accepted or fully verified.

9.6   Theories of everything

9.6.1   Theory of everything

Main article: Theory of everything

Theoretical physics continues to strive toward a theory of everything, a theory that fully explains and links together all known physical phenomena, and predicts the outcome of any experiment that could be carried out in principle. In practical terms the immediate goal in this regard is to develop a theory which would unify the Standard Model with General Relativity in a theory of quantum gravity. Additional features, such as overcoming conceptual flaws in either theory or accurate prediction of particle masses, would be desired. The challenges in putting together such a theory are not just conceptual - they include the experimental aspects of the very high energies needed to probe exotic realms.

Several notable attempts in this direction are supersymmetry, string theory, and loop quantum gravity.

9.6.2   String theory

Main article: String theory

Extensions, revisions, replacements, and reorganizations of the Standard Model exist in attempt to correct for these and other issues. String theory is one such reinvention, and many theoretical physicists think that such theories are the next theoretical step toward a true Theory of Everything. Theories of quantum gravity such as loop quantum gravity and others are thought by some to be promising candidates to the mathematical unification of quantum field theory and general relativity, requiring less drastic changes to existing theories.[31] However recent work places stringent limits on the putative effects of quantum gravity on the speed of light, and disfavours some current models of quantum gravity.[32]

Among the numerous variants of string theory, M-theory, whose mathematical existence was first proposed at a String Conference in 1995, is believed by many to be a proper "ToE" candidate, notably by physicists Brian Greene and Stephen Hawking. Though a full mathematical description is not yet known, solutions to the theory exist for specific cases.[33] Recent works have also proposed alternate string models, some of which lack the various harder-to-test features of M-theory (e.g. the existence of Calabi–Yau manifolds, many extra dimensions, etc.) including works by well-published physicists such as Lisa Randall.[34][35]

9.7   See also

- *A New Kind of Science*
- Antimatter tests of Lorentz violation
- Beyond black holes
- Fundamental physical constants in the standard model
- Higgsless model
- Holographic principle
- Little Higgs
- Lorentz-violating neutrino oscillations
- Minimal Supersymmetric Standard Model
- Peccei–Quinn theory
- Preon
- Standard-Model Extension
- Supergravity
- Seesaw mechanism
- Supersymmetry
- Superfluid vacuum theory
- String theory
- Technicolor (physics)
- Theory of everything
- Unsolved problems in physics
- Unparticle physics

9.8   References

[1] Womersley, J. (February 2005). "Beyond the Standard Model" (PDF). *Symmetry Magazine*. Retrieved 2010-11-23.

[2] Lykken, J. D. (2010). "Beyond the Standard Model". *CERN Yellow Report*. CERN. pp. 101–109. arXiv:1005.1676. CERN-2010-002.

[3] Sushkov, A. O.; Kim, W. J.; Dalvit, D. A. R.; Lamoreaux, S. K. (2011). "New Experimental Limits on Non-Newtonian Forces in the Micrometer Range". *Physical Review Letters* **107** (17): 171101. arXiv:1108.2547. Bibcode:2011PhRvL.107q1101S. doi:10.1103/PhysRevLett.107.171101. It is remarkable that two of the greatest successes of 20th century physics, general relativity and the standard model, appear to be fundamentally incompatible. But see also Donoghue, John F. (2012). "The effective field theory treatment of quantum gravity". *AIP Conference Proceedings* **1473**: 73. arXiv:1209.3511. doi:10.1063/1.4756964. One can find thousands of statements in the literature to the effect that "general relativity and quantum mechanics are incompatible". These are completely outdated and no longer relevant. Effective field theory shows that general relativity and quantum mechanics work together perfectly normally over a range of scales and curvatures, including those relevant for the world that we see around us. However, effective field theories are only valid over some range of scales. General relativity certainly does have problematic issues at extreme scales. There are important problems which the effective field theory does not solve because they are beyond its range of validity. However, this means that the issue of quantum gravity is not what we thought it to be. Rather than a fundamental incompatibility of quantum mechanics and gravity, we are in the more familiar situation of needing a more complete theory beyond the range of their combined applicability. The usual marriage of general relativity and quantum mechanics is fine at ordinary energies, but we now seek to uncover the modifications that must be present in more extreme conditions. This is the modern view of the problem of quantum gravity, and it represents progress over the outdated view of the past."

[4] Krauss, L. (2009). *A Universe from Nothing*. AAI Conference.

[5] Randolf Pohl, Ronald Gilman. Gerald A. Miller, Krzysztof Pachucki, "Muonic hydrogen and the proton radius puzzle" (May 30, 2013) http://arxiv.org/abs/1301.0905 in print Annu. Rev. Nucl. Part. Sci. Vol 63 (2013) 10.1146/annurev-nucl-102212-170627 ("The recent determination of the proton radius using the measurement of the Lamb shift in the muonic hydrogen atom startled the physics world. The obtained value of 0.84087(39) fm differs by about 4% or 7 standard deviations from the CODATA value of 0.8775(51) fm. The latter is composed from the electronic hydrogenate atom value of 0.8758(77) fm and from a similar value with larger uncertainties determined by electron scattering.")

[6] Lees, J. P.; et al. (BaBar Collaboration) (1970). "Evidence for an excess of $B \rightarrow D^{(*)}\tau^{-}\tau\nu$ decays". *Physical Review Letters* **109** (10). arXiv:1205.5442. Bibcode:2012PhRvL.109j1802L. doi:10.1103/PhysRevLett.109.101802.

[7] http://arxiv.org/abs/1506.08614

[8] http://arxiv.org/abd/1502.05314

[9] O'Luanaigh, C. (14 March 2013). "New results indicate that new particle is a Higgs boson". CERN.

[10] Marco Frasca, "What is a Glueball?" (March 31, 2009) http://marcofrasca.wordpress.com/2009/03/31/what-is-a-glueball-2/

[11] Peskin, M. E.; Schroeder, D. V. (1995). *An introduction to quantum field theory*. Addison-Wesley. pp. 786–791. ISBN 978-0-201-50397-5.

[12] Buchmüller, W. (2002). "Neutrinos, Grand Unification and Leptogenesis". arXiv:hep-ph/0204288 [hep-ph].

[13] Milstead, D.; Weinberg, E.J. (2009). "Magnetic Monopoles" (PDF). Particle Data Group. Retrieved 2010-12-20.

[14] P., Nath; P. F., Perez (2006). "Proton stability in grand unified theories, in strings, and in branes". *Physics Reports* **441** (5–6): 191–317. arXiv:hep-ph/0601023. Bibcode:2007PhR...441..191N. doi:10.1016/j.physrep.2007.02.010.

[15] Peskin, M. E.; Schroeder, D. V. (1995). *An introduction to quantum field theory*. Addison-Wesley. pp. 713–715. ISBN 978-0-201-50397-5.

[16] Nakamura, K.; et al. (Particle Data Group) (2010). "Neutrino Properties". Particle Data Group. Retrieved 2010-12-20.

[17] Mohapatra, R. N.; Pal, P. B. (2007). *Massive neutrinos in physics and astrophysics*. Lecture Notes in Physics **72** (3rd ed.). World Scientific. ISBN 978-981-238-071-5.

[18] Senjanovic, G. (2011). "Probing the Origin of Neutrino Mass: from GUT to LHC". arXiv:1107.5322 [hep-ph].

[19] Grossman, Y. (2003). "TASI 2002 lectures on neutrinos". arXiv:hep-ph/0305245v1 [hep-ph].

[20] Dodelson, S.; Widrow, L. M. (1993). "Sterile neutrinos as dark matter". *Physical Review Letters* **72**: 17. arXiv:hep-ph/9303287. Bibcode:1994PhRvL..72...17D. doi:10.1103/PhysRevLett.72.17.

[21] Minkowski, P. (1977). "$\mu \rightarrow e\,\gamma$ at a Rate of One Out of 10₉Muon Decays?". *Physics Letters* B**67**(4): 421.Bibcode. doi:10.1016/0370-2693(77)90435-X.

[22] Mohapatra, R. N.; Senjanovic, G. (1980). "Neutrino mass and spontaneous parity nonconservation". *Physical Review Letters* **44** (14): 912. Bibcode:1980PhRvL..44..912M. doi:10.1103/PhysRevLett.44.912.

[23] Keung, W.-Y.; Senjanovic, G. (1983). "Majorana Neutrinos And The Production Of The Right-handed Charged Gauge Boson". *Physical Review Letters* **50** (19): 1427. Bibcode:1983PhRvL..50.1427K. doi:10.1103/PhysRevLett.50.1427.

[24] Gell-Mann, M.; Ramond, P.; Slansky, R. (1979). P. van Nieuwenhuizen; D. Freedman, eds. *Supergravity*. North Holland.

[25] Glashow, S. L. (1979). M. Levy, ed. *Proceedings of the 1979 Cargèse Summer Institute on Quarks and Leptons*. Plenum Press.

[26] Altarelli, G. (2007). "Lectures on Models of Neutrino Masses and Mixings". arXiv:0711.0161 [hep-ph].

[27] Murayama, H. (2007). "Physics Beyond the Standard Model and Dark Matter". arXiv:0704.2276 [hep-ph].

[28] Harari, H. (1979). "A Schematic Model of Quarks and Leptons". Physics Letters B 86 (1): 83-86.

[29] Shupe, M. A. (1979). "A Composite Model of Leptons and Quarks". Physics Letters B 86 (1): 87-92.

[30] Zenczykowski, P. (2008). "The Harari-Shupe preon model and nonrelativistic quantum phase space". Physics Letters B 660 (5): 567-572.

[31] Smolin, L. (2001). *Three Roads to Quantum Gravity*. Basic Books. ISBN 0-465-07835-4.

[32] Abdo, A. A.; et al. (Fermi GBM/LAT Collaborations) (2009). "A limit on the variation of the speed of light arising from quantum gravity effects". *Nature* **462** (7271): 331–4. arXiv:0908.1832. Bibcode:2009Natur.462..331A. doi:10.1038/nature08574. PMID 19865083.

[33] Maldacena, J.; Strominger, A.; Witten, E. (1997). "Black hole entropy in M-Theory". *Journal of High Energy Physics* **1997** (12): 2. arXiv:hep-th/9711053. Bibcode:1997JHEP...12..002M. doi:10.1088/1126-6708/1997/12/002.

[34] Randall, L.; Sundrum, R. (1999). "Large Mass Hierarchy from a Small Extra Dimension". *Physical Review Letters* **83** (17): 3370. arXiv:hep-ph/9905221. Bibcode:1999PhRvL..83.3370R. doi:10.1103/PhysRevLett.83.3370.

[35] Randall, L.; Sundrum, R. (1999). "An Alternative to Compactification". *Physical Review Letters* **83** (23): 4690. arXiv:hep-th/9906064. Bibcode:1999PhRvL..83.4690R. doi:10.1103/PhysRevLett.83.4690.

9.9 Further reading

- Lisa Randall (2005). *Warped Passages: Unraveling the Mysteries of the Universe's Hidden Dimensions*. HarperCollins. ISBN 0-06-053108-8.

9.10 External resources

- Standard Model Theory @ SLAC

- Scientific American Apr 2006

- LHC. Nature July 2007

- Open Questions

- Working group - schedule

- Les Houches Conference, Summer 2005

Chapter 10

Grand Unified Theory

For the album, see Grand Unification (album).

A **Grand Unified Theory** (**GUT**) is a model in particle physics in which at high energy, the three gauge interactions of the Standard Model which define the electromagnetic, weak, and strong interactions or forces, are merged into one single force. This unified interaction is characterized by one larger gauge symmetry and thus several force carriers, but one unified coupling constant. If Grand Unification is realized in nature, there is the possibility of a grand unification epoch in the early universe in which the fundamental forces are not yet distinct.

Models that do not unify all interactions using one simple Lie group as the gauge symmetry, but do so using semisimple groups, can exhibit similar properties and are sometimes referred to as Grand Unified Theories as well.

Unifying gravity with the other three interactions would provide a theory of everything (TOE), rather than a GUT. Nevertheless, GUTs are often seen as an intermediate step towards a TOE.

The novel particles predicted by GUT models are expected to have energies around the GUT scale—just a few orders of magnitude below the Planck scale—and so will be well beyond the reach of any foreseen particle collider experiments. Therefore, the particles predicted by GUT models will be unable to be observed directly and instead the effects of grand unification might be detected through indirect observations such as proton decay, electric dipole moments of elementary particles, or the properties of neutrinos.[1] Some grand unified theories predict the existence of magnetic monopoles.

As of 2012, all GUT models which aim to be completely realistic are quite complicated, even compared to the Standard Model, because they need to introduce additional fields and interactions, or even additional dimensions of space. The main reason for this complexity lies in the difficulty of reproducing the observed fermion masses and mixing angles. Due to this difficulty, and due to the lack of any observed effect of grand unification so far, there is no generally accepted GUT model.

10.1 History

Historically, the first true GUT which was based on the simple Lie group SU(5), was proposed by Howard Georgi and Sheldon Glashow in 1974.[2] The Georgi–Glashow model was preceded by the Semisimple Lie algebra Pati–Salam model by Abdus Salam and Jogesh Pati,[3] who pioneered the idea to unify gauge interactions.

The acronym GUT was first coined in 1978 by CERN researchers John Ellis, Andrzej Buras, Mary K. Gaillard, and Dimitri Nanopoulos, however in the final version of their paper[4] they opted for the less anatomical *GUM* (Grand Unification Mass). Nanopoulos later that year was the first to use[5] the acronym in a paper.[6]

10.2 Motivation

The fact that the electric charges of electrons and protons seem to cancel each other exactly to extreme precision is essential for the existence of the macroscopic world as we know it, but this important property of elementary particles is not explained in the Standard Model of particle physics. While the description of strong and weak interactions within the Standard Model is based on gauge symmetries governed by the simple symmetry groups SU(3) and SU(2) which allow only discrete charges, the remaining component, the weak hypercharge interaction is described by an abelian symmetry U(1) which in principle allows for arbitrary charge assignments.[note 1] The observed charge quantization, namely the fact that all known elementary particles carry electric charges which appear to be exact multiples of 1/3 of the "elementary" charge, has led to the idea that hypercharge interactions and possibly the strong and weak interactions might be embedded in one Grand Unified interaction described by a single, larger simple symmetry group containing the Standard Model. This would automatically predict the quantized nature and values of all elementary particle charges. Since this also results in a prediction for the relative strengths of the fundamental interactions which we observe, in particular the weak mixing angle, Grand Unification ideally reduces the number of independent input parameters, but is also constrained by observations.

Grand Unification is reminiscent of the unification of electric and magnetic forces by Maxwell's theory of electromagnetism in the 19th century, but its physical implications and mathematical structure are qualitatively different.

10.3 Unification of matter particles

$$
\left[
\begin{array}{ccccc|cccccccccc}
0 & W & X_r & X_g & X_b & 0 & 0 & 0 & 0 & 0 & 0 & 0 & 0 & 0 & 0 \\
\overline{W} & 0 & Y_r & Y_g & Y_b & 0 & 0 & 0 & 0 & 0 & 0 & 0 & 0 & 0 & 0 \\
\overline{X}_r & \overline{Y}_r & 0 & g_{r\bar{g}} & g_{r\bar{b}} & 0 & 0 & 0 & 0 & 0 & 0 & 0 & 0 & 0 & 0 \\
\overline{X}_g & \overline{Y}_g & \bar{g}_{r\bar{g}} & 0 & g_{g\bar{b}} & 0 & 0 & 0 & 0 & 0 & 0 & 0 & 0 & 0 & 0 \\
\overline{X}_b & \overline{Y}_b & \bar{g}_{r\bar{b}} & \bar{g}_{g\bar{b}} & 0 & 0 & 0 & 0 & 0 & 0 & 0 & 0 & 0 & 0 & 0 \\
\hline
0 & 0 & 0 & 0 & 0 & 0 & Y_r & Y_g & Y_b & X_r & X_g & X_b & 0 & 0 & 0 \\
0 & 0 & 0 & 0 & 0 & \overline{Y}_r & 0 & g_{r\bar{g}} & g_{r\bar{b}} & W & 0 & 0 & 0 & X_b & X_g \\
0 & 0 & 0 & 0 & 0 & \overline{Y}_g & \bar{g}_{r\bar{g}} & 0 & g_{g\bar{b}} & 0 & W & 0 & X_b & 0 & X_r \\
0 & 0 & 0 & 0 & 0 & \overline{Y}_b & \bar{g}_{r\bar{b}} & \bar{g}_{g\bar{b}} & 0 & 0 & 0 & W & X_g & X_r & 0 \\
0 & 0 & 0 & 0 & 0 & \overline{X}_r & \overline{W} & 0 & 0 & 0 & g_{r\bar{g}} & g_{r\bar{b}} & 0 & Y_b & Y_g \\
0 & 0 & 0 & 0 & 0 & \overline{X}_g & 0 & \overline{W} & 0 & \bar{g}_{r\bar{g}} & 0 & g_{g\bar{b}} & Y_b & 0 & Y_r \\
0 & 0 & 0 & 0 & 0 & \overline{X}_b & 0 & 0 & \overline{W} & \bar{g}_{r\bar{b}} & \bar{g}_{g\bar{b}} & 0 & Y_g & Y_r & 0 \\
0 & 0 & 0 & 0 & 0 & 0 & 0 & \overline{X}_b & \overline{X}_g & 0 & \overline{Y}_b & \overline{Y}_g & 0 & g_{r\bar{g}} & g_{r\bar{b}} \\
0 & 0 & 0 & 0 & 0 & 0 & \overline{X}_b & 0 & \overline{X}_r & \overline{Y}_b & 0 & \overline{Y}_r & \bar{g}_{r\bar{g}} & 0 & g_{g\bar{b}} \\
0 & 0 & 0 & 0 & 0 & 0 & \overline{X}_g & \overline{X}_r & 0 & \overline{Y}_g & \overline{Y}_r & 0 & \bar{g}_{r\bar{b}} & \bar{g}_{g\bar{b}} & 0
\end{array}
\right]
\left[
\begin{array}{c}
v_e \\
e \\
\bar{d}_r \\
\bar{d}_g \\
\bar{d}_b \\
\bar{e} \\
d_r \\
d_g \\
d_b \\
u_r \\
u_g \\
u_b \\
\bar{u}_r \\
\bar{u}_g \\
\bar{u}_b
\end{array}
\right]
$$

*Schematic representation of fermions and bosons in SU(5) GUT showing **5 + 10** split in the multiplets. Neutral bosons (photon, Z-boson, and neutral gluons) are not shown but occupy the diagonal entries of the matrix in complex superpositions*

For an elementary introduction to how Lie algebras are related to particle physics, see the article Particle physics and representation theory.

10.3.1 SU(5)

Main article: SU(5) (physics)

SU(5) is the simplest GUT. The smallest simple Lie group which contains the standard model, and upon which the first

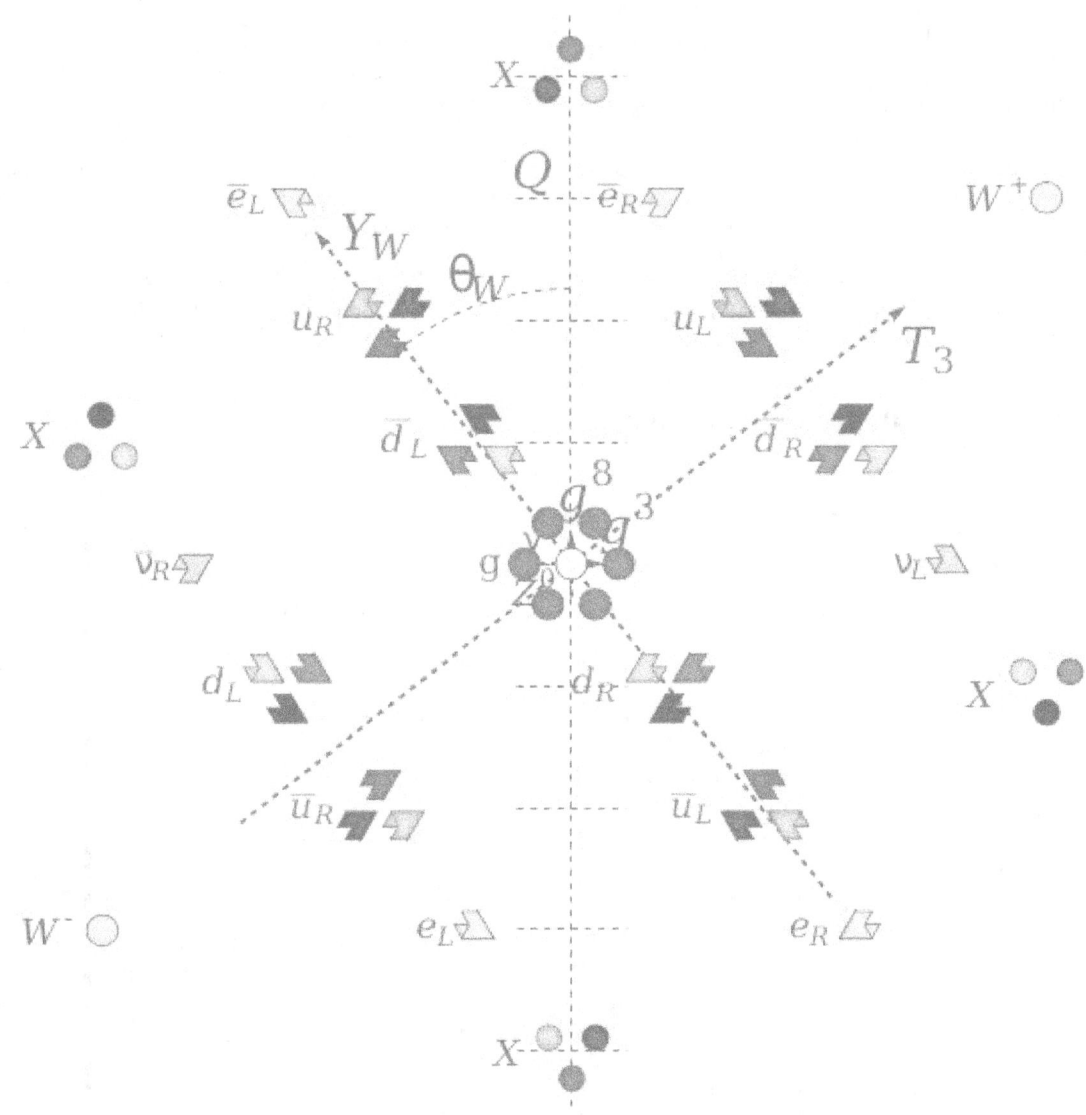

The pattern of weak isospins, weak hypercharges, and strong charges for particles in the SU(5) model, rotated by the predicted weak mixing angle, showing electric charge roughly along the vertical. In addition to Standard Model particles, the theory includes twelve colored X bosons, responsible for proton decay.

Grand Unified Theory was based, is

$$SU(5) \supset SU(3) \times SU(2) \times U(1)$$

Such group symmetries allow the reinterpretation of several known particles as different states of a single particle field. However, it is not obvious that the simplest possible choices for the extended "Grand Unified" symmetry should yield the correct inventory of elementary particles. The fact that all currently known (2009) matter particles fit nicely into three copies of the smallest group representations of SU(5) and immediately carry the correct observed charges, is one of the first and most important reasons why people believe that a Grand Unified Theory might actually be realized in nature.

The two smallest irreducible representations of SU(5) are **5** and **10**. In the standard assignment, the **5** contains the charge conjugates of the right-handed down-type quark color triplet and a left-handed lepton isospin doublet, while the **10** contains the six up-type quark components, the left-handed down-type quark color triplet, and the right-handed electron. This scheme has to be replicated for each of the three known generations of matter. It is notable that the theory is anomaly free with this matter content.

The hypothetical right-handed neutrinos are not contained in any of these representations, which can explain their relative heaviness (see seesaw mechanism).

10.3.2   SO(10)

Main article: SO(10) (physics)
The next simple Lie group which contains the standard model is

$$SO(10) \supset SU(5) \supset SU(3) \times SU(2) \times U(1)$$

Here, the unification of matter is even more complete, since the irreducible spinor representation **16** contains both the **5** and **10** of SU(5) and a right-handed neutrino, and thus the complete particle content of one generation of the extended standard model with neutrino masses. This is already the largest simple group which achieves the unification of matter in a scheme involving only the already known matter particles (apart from the Higgs sector).

Since different standard model fermions are grouped together in larger representations, GUTs specifically predict relations among the fermion masses, such as between the electron and the down quark, the muon and the strange quark, and the tau lepton and the bottom quark for SU(5) and SO(10). Some of these mass relations hold approximately, but most don't (see Georgi-Jarlskog mass relation).

The boson matrix for SO(10) is found by taking the 15 × 15 matrix from the **10** + **5** representation of SU(5) and adding an extra row and column for the right handed neutrino. The bosons are found by adding a partner to each of the 20 charged bosons (2 right-handed W bosons, 6 massive charged gluons and 12 X/Y type bosons) and adding an extra heavy neutral Z-boson to make 5 neutral bosons in total. The boson matrix will have a boson or its new partner in each row and column. These pairs combine to create the familiar 16D Dirac spinor matrices of SO(10).

10.3.3   SU(8)

Assuming 4 generations of fermions instead of 3 makes a total of **64** types of particles. These can be put into **64** = **8** + **56** representations of SU(8). This can be divided into SU(5) × SU(3)F × U(1) which is the SU(5) theory together with some heavy bosons which act on the generation number.

10.3.4   O(16)

Again assuming 4 generations of fermions, the **128** particles and anti-particles can be put into a single spinor representation of O(16).

10.3.5   Symplectic Groups and Quaternion Representations

Symplectic gauge groups could also be considered. For example Sp(8) (which is called Sp(4) in the article symplectic group) has a representation in terms of 4 × 4 quaternion unitary matrices which has a **16** dimensional real representation and so might be considered as a candidate for a gauge group. Sp(8) has 32 charged bosons and 4 neutral bosons. Its

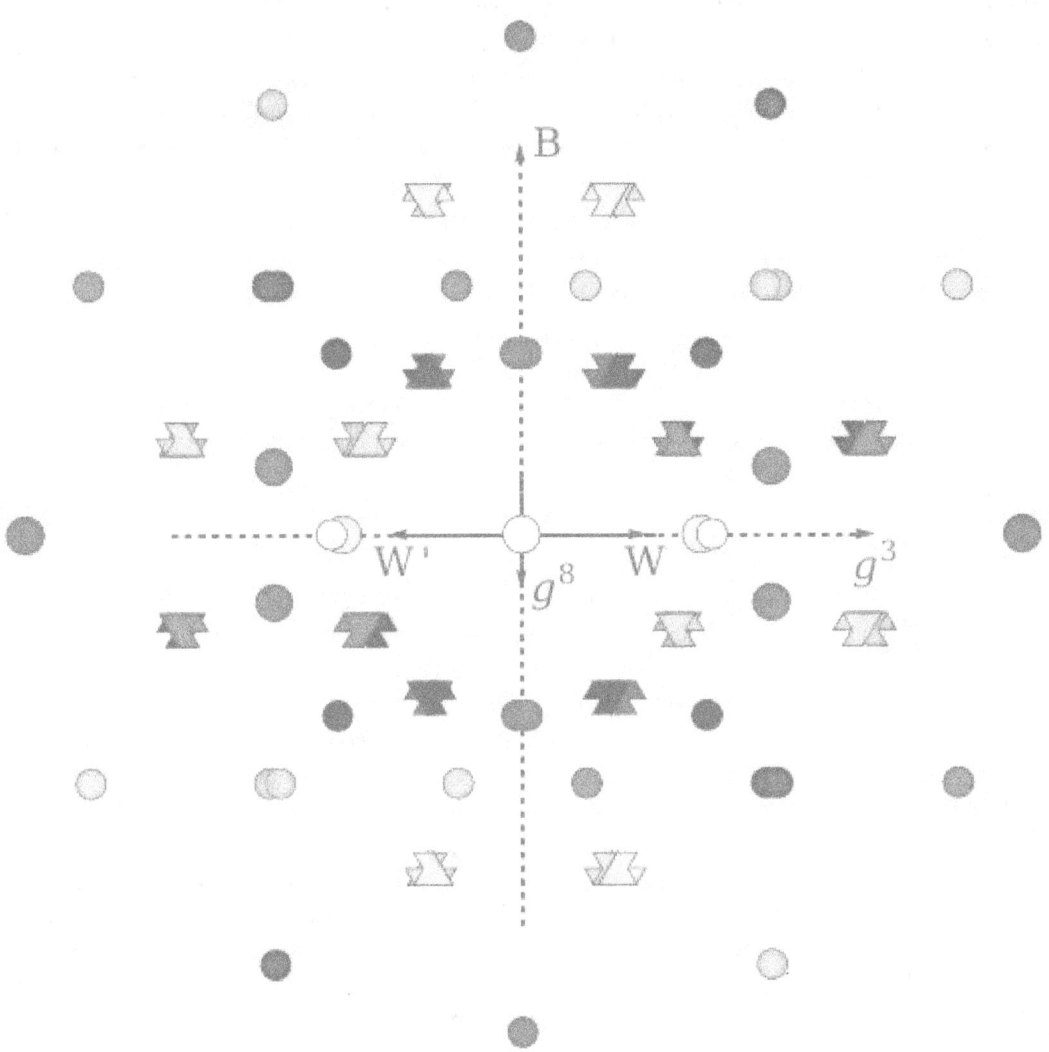

The pattern of weak isospin, W, weaker isospin, W', strong g3 and g8, and baryon minus lepton, B, charges for particles in the SO(10) Grand Unified Theory, rotated to show the embedding in E_6.

subgroups include SU(4) so can at least contain the gluons and photon of SU(3) × U(1). Although it's probably not possible to have weak bosons acting on chiral fermions in this representation. A quaternion representation of the fermions might be:

$$\begin{bmatrix} e + i\overline{e} + jv + k\overline{v} \\ u_r + i\overline{u_r} + jd_r + k\overline{d_r} \\ u_g + i\overline{u_g} + jd_g + k\overline{d_g} \\ u_b + i\overline{u_b} + jd_b + k\overline{d_b} \end{bmatrix}_L$$

A further complication with quaternion representations of fermions is that there are two types of multiplication: left multiplication and right multiplication which must be taken into account. It turns out that including left and right-handed 4 × 4 quaternion matrices is equivalent to including a single right-multiplication by a unit quaternion which adds an extra

SU(2) and so has an extra neutral boson and two more charged bosons. Thus the group of left and right handed 4×4 quaternion matrcies is $Sp(8) \times SU(2)$ which does include the standard model bosons:

$$SU(4, H)_L \times H_R = Sp(8) \times SU(2) \supset SU(4) \times SU(2) \supset SU(3) \times SU(2) \times U(1)$$

If ψ is a quaternion valued spinor, A_μ^{ab} is quaternion hermitian 4×4 matrix coming from $Sp(8)$ and B_μ is a pure imaginary quaternion (both of which are 4-vector bosons) then the interaction term is:

$$\overline{\psi^a} \gamma_\mu \left(A_\mu^{ab} \psi^b + \psi^a B_\mu \right)$$

10.3.6   E8 and Octonion Representations

It can be noted that a generation of 16 fermions can be put into the form of an Octonion with each element of the octonion being an 8-vector. If the 3 generations are then put in a 3x3 hermitian matrix with certain additions for the diagonal elements then these matrices form an exceptional (grassman-) Jordan algebra, which has the symmetry group of one of the exceptional Lie groups (F_4, E_6, E_7 or E_8) depending on the details.

$$\psi = \begin{bmatrix} a & e & \mu \\ \overline{e} & b & \tau \\ \overline{\mu} & \overline{\tau} & c \end{bmatrix}$$

$$[\psi_A, \psi_B] \subset J_3(O)$$

Because they are fermions the anti-commutators of the Jordan algebra become commutators. It is known that E_6 has subgroup O(10) and so is big enough to include the Standard Model. An E_8 gauge group, for example, would have 8 neutral bosons, 120 charged bosons and 120 charged anti-bosons. To account for the 248 fermions in the lowest multiplet of E_8, these would either have to include anti-particles (and so have Baryogenesis), have new undiscovered particles, or have gravity-like (Spin connection) bosons affecting elements of the particles spin direction. Each of these poses theoretical problems.

10.3.7   Beyond Lie Groups

Other structures have been suggested including Lie 3-algebras and Lie superalgebras. Neither of these fit with Yang–Mills theory. In particular Lie superalgebras would introduce bosons with the wrong statistics. Supersymmetry however does fit with Yang–Mills. For example N=4 Super Yang Mills Theory requires an SU(N) gauge group.

10.4   Unification of forces and the role of supersymmetry

The unification of forces is possible due to the energy scale dependence of force coupling parameters in quantum field theory called renormalization group running, which allows parameters with vastly different values at usual energies to converge to a single value at a much higher energy scale.[7]

The renormalization group running of the three gauge couplings in the Standard Model has been found to nearly, but not quite, meet at the same point if the hypercharge is normalized so that it is consistent with SU(5) or SO(10) GUTs, which are precisely the GUT groups which lead to a simple fermion unification. This is a significant result, as other Lie groups lead to different normalizations. However, if the supersymmetric extension MSSM is used instead of the Standard Model, the match becomes much more accurate. In this case, the coupling constants of the strong and electroweak interactions meet at the grand unification energy, also known as the GUT scale:

$\Lambda_{GUT} \approx 10^{16}$ GeV

It is commonly believed that this matching is unlikely to be a coincidence, and is often quoted as one of the main motivations to further investigate supersymmetric theories despite the fact that no supersymmetric partner particles have been experimentally observed (May 2015). Also, most model builders simply assume supersymmetry because it solves the hierarchy problem—i.e., it stabilizes the electroweak Higgs mass against radiative corrections.

10.5   Neutrino masses

Since Majorana masses of the right-handed neutrino are forbidden by SO(10) symmetry, SO(10) GUTs predict the Majorana masses of right-handed neutrinos to be close to the GUT scale where the symmetry is spontaneously broken in those models. In supersymmetric GUTs, this scale tends to be larger than would be desirable to obtain realistic masses of the light, mostly left-handed neutrinos (see neutrino oscillation) via the seesaw mechanism.

10.6   Proposed theories

Several such theories have been proposed, but none is currently universally accepted. An even more ambitious theory that includes *all* fundamental forces, including gravitation, is termed a theory of everything. Some common mainstream GUT models are:

Not quite GUTs:

Note: These models refer to Lie algebras not to Lie groups. The Lie group could be $[SU(4) \times SU(2) \times SU(2)]/\mathbf{Z}_2$, just to take a random example.

The most promising candidate is SO(10). (Minimal) SO(10) does not contain any exotic fermions (i.e. additional fermions besides the Standard Model fermions and the right-handed neutrino), and it unifies each generation into a single irreducible representation. A number of other GUT models are based upon subgroups of SO(10). They are the minimal left-right model, SU(5), flipped SU(5) and the Pati–Salam model. The GUT group E_6 contains SO(10), but models based upon it are significantly more complicated. The primary reason for studying E_6 models comes from $E_8 \times E_8$ heterotic string theory.

GUT models generically predict the existence of topological defects such as monopoles, cosmic strings, domain walls, and others. But none have been observed. Their absence is known as the monopole problem in cosmology. Most GUT models also predict proton decay, although not the Pati–Salam model; current experiments still haven't detected proton decay. This experimental limit on the proton's lifetime pretty much rules out minimal SU(5).

- Proton Decay. These graphics refer to the X bosons and Higgs bosons.
- Dimension 6 proton decay mediated by the X boson in SU(5) GUT
- Dimension 6 proton decay mediated by the X boson in flipped SU(5) GUT
- Dimension 6 proton decay mediated by the triplet Higgs and the anti-triplet Higgs in SU(5) GUT

Some GUT theories like SU(5) and SO(10) suffer from what is called the doublet-triplet problem. These theories predict that for each electroweak Higgs doublet, there is a corresponding colored Higgs triplet field with a very small mass (many orders of magnitude smaller than the GUT scale here). In theory, unifying quarks with leptons, the Higgs doublet would also be unified with a Higgs triplet. Such triplets have not been observed. They would also cause extremely rapid proton decay (far below current experimental limits) and prevent the gauge coupling strengths from running together in the renormalization group.

Most GUT models require a threefold replication of the matter fields. As such, they do not explain why there are three generations of fermions. Most GUT models also fail to explain the little hierarchy between the fermion masses for different generations.

10.7   Ingredients

A GUT model basically consists of a gauge group which is a compact Lie group, a connection form for that Lie group, a Yang–Mills action for that connection given by an invariant symmetric bilinear form over its Lie algebra (which is specified by a coupling constant for each factor), a Higgs sector consisting of a number of scalar fields taking on values within real/complex representations of the Lie group and chiral Weyl fermions taking on values within a complex rep of the Lie group. The Lie group contains the Standard Model group and the Higgs fields acquire VEVs leading to a spontaneous symmetry breaking to the Standard Model. The Weyl fermions represent matter.

10.8   Current status

As of 2012, there is still no hard evidence that nature is described by a Grand Unified Theory. The discovery of neutrino oscillations indicates that the Standard Model is incomplete and has led to renewed interest toward certain GUT such as SO(10). One of the few possible experimental tests of certain GUT is proton decay and also fermion masses. There are a few more special tests for supersymmetric GUT.

The gauge coupling strengths of QCD, the weak interaction and hypercharge seem to meet at a common length scale called the GUT scale and equal approximately to 10^{16} GeV, which is slightly suggestive. This interesting numerical observation is called the **gauge coupling unification**, and it works particularly well if one assumes the existence of superpartners of the Standard Model particles. Still it is possible to achieve the same by postulating, for instance, that ordinary (non supersymmetric) SO(10) models break with an intermediate gauge scale, such as the one of Pati–Salam group

10.9   See also

- Paradigm shift

- Classical unified field theories

- X and Y bosons

- B − L quantum number

10.10   Notes

[1] There are however certain constraints on the choice of particle charges from theoretical consistency, in particular anomaly cancellation.

10.11   References

[1] Ross, G. (1984). *Grand Unified Theories*. Westview Press. ISBN 978-0-8053-6968-7.

[2] Georgi, H.; Glashow, S.L. (1974). "Unity of All Elementary Particle Forces". *Physical Review Letters* **32**: 438–441. Bibcod8G. doi:10.1103/PhysRevLett.32.438.

[3] Pati, J.; Salam, A. (1974). "Lepton Number as the Fourth Color". *Physical Review D* **10**: 275–289. Bibcode:1974PhRvD..10..275P. doi:10.1103/PhysRevD.10.275.

[4] Buras, A.J.; Ellis, J.; Gaillard, M.K.; Nanopoulos, D.V. (1978). "Aspects of the grand unification of strong, weak and electromagnetic interactions" (PDF). *Nuclear Physics B* **135** (1): 66–92. Bibcode:1978NuPhB.135...66B. doi:10.1016/0550-3213(78)90214-6. Retrieved 2011-03-21.

[5] Nanopoulos, D.V. (1979). "Protons Are Not Forever". *Orbis Scientiae* **1**: 91. Harvard Preprint HUTP-78/A062.

[6] Ellis, J. (2002). "Physics gets physical". *Nature* **415** (6875): 957. Bibcode:2002Natur.415..957E. doi:10.1038/415957b.

[7] Ross, G. (1984). *Grand Unified Theories*. Westview Press. ISBN 978-0-8053-6968-7.

10.11.1 Further reading

- Stephen Hawking, A Brief History of Time, includes a brief popular overview.

10.12 External links

- The Algebra of Grand Unified Theories
- Scholarpedia: Grand Unification

Chapter 11

Quantum gravity

Quantum gravity (**QG**) is a field of theoretical physics that seeks to describe the force of gravity according to the principles of quantum mechanics.

The current understanding of gravity is based on Albert Einstein's general theory of relativity, which is formulated within the framework of classical physics. On the other hand, the nongravitational forces are described within the framework of quantum mechanics, a radically different formalism for describing physical phenomena based on probability.[1] The necessity of a quantum mechanical description of gravity follows from the fact that one cannot consistently couple a classical system to a quantum one.[2]

Although a quantum theory of gravity is needed in order to reconcile general relativity with the principles of quantum mechanics, difficulties arise when one attempts to apply the usual prescriptions of quantum field theory to the force of gravity.[3] From a technical point of view, the problem is that the theory one gets in this way is not renormalizable and therefore cannot be used to make meaningful physical predictions. As a result, theorists have taken up more radical approaches to the problem of quantum gravity, the most popular approaches being string theory and loop quantum gravity.[4] A recent development is the theory of causal fermion systems which gives quantum mechanics, general relativity, and quantum field theory as limiting cases.[5][6][7][8][9][10]

Strictly speaking, the aim of quantum gravity is only to describe the quantum behavior of the gravitational field and should not be confused with the objective of unifying all fundamental interactions into a single mathematical framework. While any substantial improvement into the present understanding of gravity would aid further work towards unification, study of quantum gravity is a field in it's own right with various branches having different approaches to unification. Although some quantum gravity theories, such as string theory, try to unify gravity with the other fundamental forces, others, such as loop quantum gravity, make no such attempt; instead, they make an effort to quantize the gravitational field while it is kept separate from the other forces. A theory of quantum gravity that is also a grand unification of all known interactions is sometimes referred to as a theory of everything (TOE).

One of the difficulties of quantum gravity is that quantum gravitational effects are only expected to become apparent near the Planck scale, a scale far smaller in distance (equivalently, far larger in energy) than what is currently accessible at high energy particle accelerators. As a result, quantum gravity is a mainly theoretical enterprise, although there are speculations about how quantum gravity effects might be observed in existing experiments.[11]

11.1 Overview

Much of the difficulty in meshing these theories at all energy scales comes from the different assumptions that these theories make on how the universe works. Quantum field theory depends on particle fields embedded in the flat space-time of special relativity. General relativity models gravity as a curvature within space-time that changes as a gravitational mass moves. Historically, the most obvious way of combining the two (such as treating gravity as simply another particle field) ran quickly into what is known as the renormalization problem. In the old-fashioned understanding of renormalization, gravity particles would attract each other and adding together all of the interactions results in many infinite values which

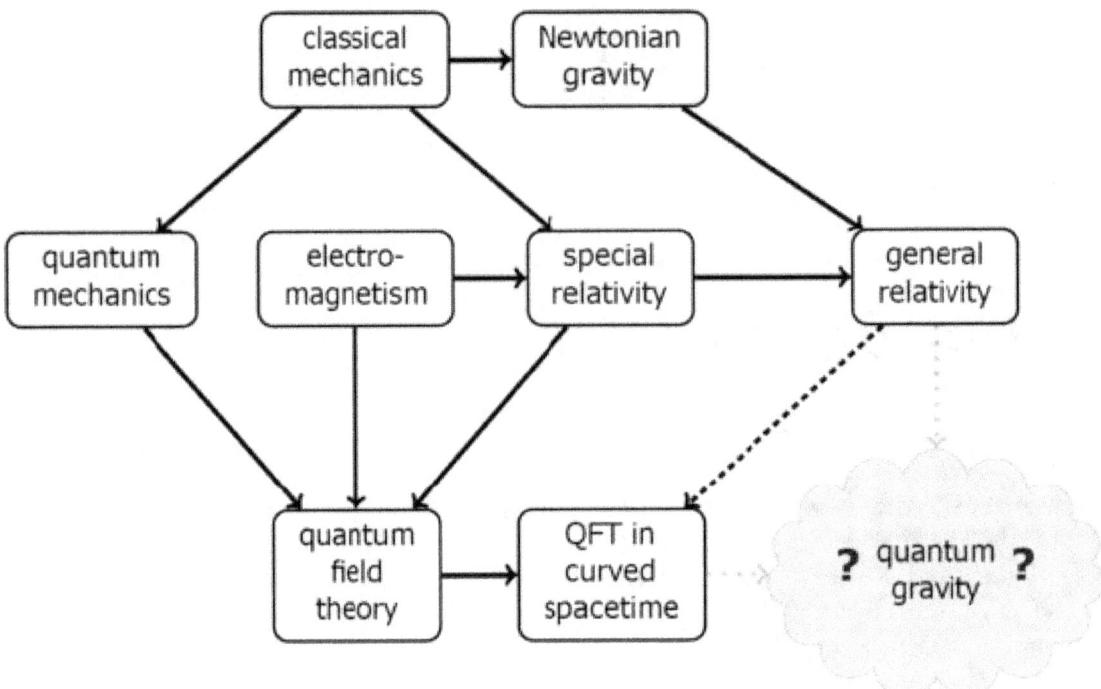

Diagram showing where quantum gravity sits in the hierarchy of physics theories

cannot easily be cancelled out mathematically to yield sensible, finite results. This is in contrast with quantum electrodynamics where, given that the series still do not converge, the interactions sometimes evaluate to infinite results, but those are few enough in number to be removable via renormalization.

11.1.1   Effective field theories

Quantum gravity can be treated as an effective field theory. Effective quantum field theories come with some high-energy cutoff, beyond which we do not expect that the theory provides a good description of nature. The "infinities" then become large but finite quantities depending on this finite cutoff scale, and correspond to processes that involve very high energies near the fundamental cutoff. These quantities can then be absorbed into an infinite collection of coupling constants, and at energies well below the fundamental cutoff of the theory, to any desired precision; only a finite number of these coupling constants need to be measured in order to make legitimate quantum-mechanical predictions. This same logic works just as well for the highly successful theory of low-energy pions as for quantum gravity. Indeed, the first quantum-mechanical corrections to graviton-scattering and Newton's law of gravitation have been explicitly computed[12] (although they are so astronomically small that we may never be able to measure them). In fact, gravity is in many ways a much better quantum field theory than the Standard Model, since it appears to be valid all the way up to its cutoff at the Planck scale.

While confirming that quantum mechanics and gravity are indeed consistent at reasonable energies, it is clear that near or above the fundamental cutoff of our effective quantum theory of gravity (the cutoff is generally assumed to be of the order of the Planck scale), a new model of nature will be needed. Specifically, the problem of combining quantum mechanics and gravity becomes an issue only at very high energies, and may well require a totally new kind of model.

11.1.2 Quantum gravity theory for the highest energy scales

The general approach to deriving a quantum gravity theory that is valid at even the highest energy scales is to assume that such a theory will be simple and elegant and, accordingly, to study symmetries and other clues offered by current theories that might suggest ways to combine them into a comprehensive, unified theory. One problem with this approach is that it is unknown whether quantum gravity will actually conform to a simple and elegant theory, as it should resolve the dual conundrums of special relativity with regard to the uniformity of acceleration and gravity, and general relativity with regard to spacetime curvature.

Such a theory is required in order to understand problems involving the combination of very high energy and very small dimensions of space, such as the behavior of black holes, and the origin of the universe.

11.2 Quantum mechanics and general relativity

11.2.1 The graviton

Main article: Graviton

At present, one of the deepest problems in theoretical physics is harmonizing the theory of general relativity, which describes gravitation, and applications to large-scale structures (stars, planets, galaxies), with quantum mechanics, which describes the other three fundamental forces acting on the atomic scale. This problem must be put in the proper context, however. In particular, contrary to the popular claim that quantum mechanics and general relativity are fundamentally incompatible, one can demonstrate that the structure of general relativity essentially follows inevitably from the quantum mechanics of interacting theoretical spin-2 massless particles (called gravitons).[13][14][15][16][17]

While there is no concrete proof of the existence of gravitons, quantized theories of matter may necessitate their existence. Supporting this theory is the observation that all fundamental forces except gravity have one or more known messenger particles, leading researchers to believe that at least one most likely does exist; they have dubbed this hypothetical particle the *graviton*. The predicted find would result in the classification of the graviton as a "force particle" similar to the photon of the electromagnetic field. Many of the accepted notions of a unified theory of physics since the 1970s assume, and to some degree depend upon, the existence of the graviton. These include string theory, superstring theory, M-theory, and loop quantum gravity. Detection of gravitons is thus vital to the validation of various lines of research to unify quantum mechanics and relativity theory.

11.2.2 The dilaton

Main article: Dilaton

The dilaton made its first appearance in Kaluza–Klein theory, a five-dimensional theory that combined gravitation and electromagnetism. Generally, it appears in string theory. More recently, it has appeared in the lower-dimensional many-bodied gravity problem[18] based on the field theoretic approach of Roman Jackiw. The impetus arose from the fact that complete analytical solutions for the metric of a covariant N-body system have proven elusive in general relativity. To simplify the problem, the number of dimensions was lowered to $(1+1)$, i.e. one spatial dimension and one temporal dimension. This model problem, known as $R=T$ theory[19] (as opposed to the general $G=T$ theory) was amenable to exact solutions in terms of a generalization of the Lambert W function. It was also found that the field equation governing the dilaton (derived from differential geometry) was the Schrödinger equation and consequently amenable to quantization.[20]

Thus, one had a theory which combined gravity, quantization and even the electromagnetic interaction, promising ingredients of a fundamental physical theory. It is worth noting that the outcome revealed a previously unknown and already existing *natural link* between general relativity and quantum mechanics. However, this theory needs to be generalized in $(2+1)$ or $(3+1)$ dimensions although, in principle, the field equations are amenable to such generalization as shown with the inclusion of a one-graviton process[21] and yielding the correct Newtonian limit in d dimensions if a dilaton is included. However, it is not yet clear what the fully generalized field equation governing the dilaton in (3+1) dimensions should be. This is further complicated by the fact that gravitons can propagate in *(3+1)* dimensions and consequently that would imply gravitons and dilatons exist in the real world. Moreover, detection of the dilaton is expected to be even more elusive than the graviton. However, since this approach allows for the combination of gravitational, electromagnetic and quantum effects, their coupling could potentially lead to a means of vindicating the theory, through cosmology and perhaps even *experimentally*.

11.2.3   Nonrenormalizability of gravity

Further information: Renormalization

General relativity, like electromagnetism, is a classical field theory. One might expect that, as with electromagnetism, there should be a corresponding quantum field theory.

However, gravity is perturbatively nonrenormalizable.[22][23] For a quantum field theory to be well-defined according to this understanding of the subject, it must be asymptotically free or asymptotically safe. The theory must be characterized by a choice of *finitely many* parameters, which could, in principle, be set by experiment. For example, in quantum electrodynamics, these parameters are the charge and mass of the electron, as measured at a particular energy scale.

On the other hand, in quantizing gravity, there are *infinitely many independent parameters* (counterterm coefficients) needed to define the theory. For a given choice of those parameters, one could make sense of the theory, but since we can never do infinitely many experiments to fix the values of every parameter, we do not have a meaningful physical theory:

- At low energies, the logic of the renormalization group tells us that, despite the unknown choices of these infinitely many parameters, quantum gravity will reduce to the usual Einstein theory of general relativity.

- On the other hand, if we could probe very high energies where quantum effects take over, then *every one* of the infinitely many unknown parameters would begin to matter, and we could make no predictions at all.

As explained below, there is a way around this problem by treating QG as an effective field theory.

Any meaningful theory of quantum gravity that makes sense and is predictive at all energy scales must have some deep principle that reduces the infinitely many unknown parameters to a finite number that can then be measured.

- One possibility is that normal perturbation theory is not a reliable guide to the renormalizability of the theory, and that there really *is* a UV fixed point for gravity. Since this is a question of non-perturbative quantum field theory, it is difficult to find a reliable answer, but some people still pursue this option.

- Another possibility is that there are new symmetry principles that constrain the parameters and reduce them to a finite set. This is the route taken by string theory, where all of the excitations of the string essentially manifest themselves as new symmetries.

11.2.4   QG as an effective field theory

Main article: Effective field theory

In an effective field theory, all but the first few of the infinite set of parameters in a non-renormalizable theory are suppressed by huge energy scales and hence can be neglected when computing low-energy effects. Thus, at least in the low-energy regime, the model is indeed a predictive quantum field theory.[12] (A very similar situation occurs for the very similar effective field theory of low-energy pions.) Furthermore, many theorists agree that even the Standard Model should really be regarded as an effective field theory as well, with "nonrenormalizable" interactions suppressed by large energy scales and whose effects have consequently not been observed experimentally.

Recent work[12] has shown that by treating general relativity as an effective field theory, one can actually make legitimate predictions for quantum gravity, at least for low-energy phenomena. An example is the well-known calculation of the tiny first-order quantum-mechanical correction to the classical Newtonian gravitational potential between two masses.

11.2.5   Spacetime background dependence

Main article: Background independence

A fundamental lesson of general relativity is that there is no fixed spacetime background, as found in Newtonian mechanics and special relativity; the spacetime geometry is dynamic. While easy to grasp in principle, this is the hardest idea to understand about general relativity, and its consequences are profound and not fully explored, even at the classical level. To a certain extent, general relativity can be seen to be a relational theory,[24] in which the only physically relevant information is the relationship between different events in space-time.

On the other hand, quantum mechanics has depended since its inception on a fixed background (non-dynamic) structure. In the case of quantum mechanics, it is time that is given and not dynamic, just as in Newtonian classical mechanics. In relativistic quantum field theory, just as in classical field theory, Minkowski spacetime is the fixed background of the theory.

String theory

String theory can be seen as a generalization of quantum field theory where instead of point particles, string-like objects propagate in a fixed spacetime background, although the interactions among closed strings give rise to space-time in a dynamical way. Although string theory had its origins in the study of quark confinement and not of quantum gravity, it was soon discovered that the string spectrum contains the graviton, and that "condensation" of certain vibration modes of strings is equivalent to a modification of the original background. In this sense, string perturbation theory exhibits exactly the features one would expect of a perturbation theory that may exhibit a strong dependence on asymptotics (as seen, for example, in the AdS/CFT correspondence) which is a weak form of background dependence.

Background independent theories

Loop quantum gravity is the fruit of an effort to formulate a background-independent quantum theory.

Topological quantum field theory provided an example of background-independent quantum theory, but with no local degrees of freedom, and only finitely many degrees of freedom globally. This is inadequate to describe gravity in 3+1 dimensions, which has local degrees of freedom according to general relativity. In 2+1 dimensions, however, gravity is a topological field theory, and it has been successfully quantized in several different ways, including spin networks.

11.2.6   Semi-classical quantum gravity

Quantum field theory on curved (non-Minkowskian) backgrounds, while not a full quantum theory of gravity, has shown many promising early results. In an analogous way to the development of quantum electrodynamics in the early part of the 20th century (when physicists considered quantum mechanics in classical electromagnetic fields), the consideration of quantum field theory on a curved background has led to predictions such as black hole radiation.

Phenomena such as the Unruh effect, in which particles exist in certain accelerating frames but not in stationary ones, do not pose any difficulty when considered on a curved background (the Unruh effect occurs even in flat Minkowskian backgrounds). The vacuum state is the state with the least energy (and may or may not contain particles). See Quantum field theory in curved spacetime for a more complete discussion.

11.2.7 Points of tension

There are other points of tension between quantum mechanics and general relativity.

- First, classical general relativity breaks down at singularities, and quantum mechanics becomes inconsistent with general relativity in the neighborhood of singularities (however, no one is certain that classical general relativity applies near singularities in the first place).

- Second, it is not clear how to determine the gravitational field of a particle, since under the Heisenberg uncertainty principle of quantum mechanics its location and velocity cannot be known with certainty. The resolution of these points may come from a better understanding of general relativity.[25]

- Third, there is the problem of time in quantum gravity. Time has a different meaning in quantum mechanics and general relativity and hence there are subtle issues to resolve when trying to formulate a theory which combines the two.[26]

11.3 Candidate theories

There are a number of proposed quantum gravity theories.[27] Currently, there is still no complete and consistent quantum theory of gravity, and the candidate models still need to overcome major formal and conceptual problems. They also face the common problem that, as yet, there is no way to put quantum gravity predictions to experimental tests, although there is hope for this to change as future data from cosmological observations and particle physics experiments becomes available.[28][29]

11.3.1 String theory

Main article: String theory
One suggested starting point is ordinary quantum field theories which, after all, are successful in describing the other three basic fundamental forces in the context of the standard model of elementary particle physics. However, while this leads to an acceptable effective (quantum) field theory of gravity at low energies,[30] gravity turns out to be much more problematic at higher energies. Where, for ordinary field theories such as quantum electrodynamics, a technique known as renormalization is an integral part of deriving predictions which take into account higher-energy contributions,[31] gravity turns out to be nonrenormalizable: at high energies, applying the recipes of ordinary quantum field theory yields models that are devoid of all predictive power.[32]

One attempt to overcome these limitations is to replace ordinary quantum field theory, which is based on the classical concept of a point particle, with a quantum theory of one-dimensional extended objects: string theory.[33] At the energies reached in current experiments, these strings are indistinguishable from point-like particles, but, crucially, different modes of oscillation of one and the same type of fundamental string appear as particles with different (electric and other) charges. In this way, string theory promises to be a unified description of all particles and interactions.[34] The theory is successful in that one mode will always correspond to a graviton, the messenger particle of gravity; however, the price to pay are unusual features such as six extra dimensions of space in addition to the usual three for space and one for time.[35]

In what is called the second superstring revolution, it was conjectured that both string theory and a unification of general relativity and supersymmetry known as supergravity[36] form part of a hypothesized eleven-dimensional model known as M-theory, which would constitute a uniquely defined and consistent theory of quantum gravity.[37][38] As presently understood, however, string theory admits a very large number (10^{500} by some estimates) of consistent vacua, comprising the so-called "string landscape". Sorting through this large family of solutions remains a major challenge.

11.3.2 Loop quantum gravity

Main article: Loop quantum gravity
Loop quantum gravity is based first of all on the idea to take seriously the insight of general relativity that spacetime is

a dynamical field and therefore is a quantum object. The second idea is that the quantum discreteness that determines the particle-like behavior of other field theories (for instance, the photons of the electromagnetic field) also affects the structure of space.

The main result of loop quantum gravity is the derivation of a granular structure of space at the Planck length. This is derived as follows. In the case of electromagnetism, the quantum operator representing the energy of each frequency of the field has discrete spectrum. Therefore the energy of each frequency is quantized, and the quanta are the photons. In the case of gravity, the operators representing the area and the volume of each surface or space region have discrete spectrum. Therefore area and volume of any portion of space are quantized, and the quanta are elementary quanta of space. It follows that spacetime has an elementary quantum granular structure at the Planck scale, which cuts-off the ultraviolet infinities of quantum field theory.

The quantum state of spacetime is described in the theory by means of a mathematical structure called spin networks. Spin networks were initially introduced by Roger Penrose in abstract form, and later shown by Carlo Rovelli and Lee Smolin to derive naturally from a non perturbative quantization of general relativity. Spin networks do not represent quantum states of a field in spacetime: they represent directly quantum states of spacetime.

The theory is based on the reformulation of general relativity known as Ashtekar variables, which represent geometric gravity using mathematical analogues of electric and magnetic fields.[39][40] In the quantum theory space is represented by a network structure called a spin network, evolving over time in discrete steps.[41][42][43][44]

The dynamics of the theory is today constructed in several versions. One version starts with the canonical quantization of general relativity. The analogue of the Schrödinger equation is a Wheeler–DeWitt equation, which can be defined in the theory.[45] In the covariant, or spinfoam formulation of the theory, the quantum dynamics is obtained via a sum over discrete versions of spacetime, called spinfoams. These represent histories of spin networks.

11.3.3 Scale Relativity

Main article: Scale relativity

Most quantum gravity theories assume quantum laws as a starting point. However, in the framework of scale relativity, this is not needed.[46] The theory is an extension of special and general relativity, including the relativity of scale transformations. It thus takes a geometrical approach to the problem, where quantum phenomena became a manifestation of the fractality of spacetime. This is similar to the geometrical interpretation of gravitation in general relativity, where gravitation become a manifestation of spacetime curvature instead of a force. Although much remains to be developed, validated predictions have already been obtained in physics, astrophysics and cosmology.

11.3.4 Other approaches

There are a number of other approaches to quantum gravity. The approaches differ depending on which features of general relativity and quantum theory are accepted unchanged, and which features are modified.[47][48] Examples include:

- Acoustic metric and other analog models of gravity

- Asymptotic safety in quantum gravity

- Euclidean quantum gravity

- Causal dynamical triangulation[49]

- Causal fermion systems,[5][6][7][8][9][10] giving quantum mechanics, general relativity and quantum field theory as limiting cases.

- Causal sets[50]

- Covariant Feynman path integral approach

- Group field theory[51]

- Wheeler-DeWitt equation

- Geometrodynamics

- Hořava–Lifshitz gravity

- MacDowell–Mansouri action

- Noncommutative geometry.

- Path-integral based models of quantum cosmology[52]

- Regge calculus

- String-nets giving rise to gapless helicity ±2 excitations with no other gapless excitations[53]

- Superfluid vacuum theory a.k.a. theory of BEC vacuum

- Supergravity

- Twistor theory[54]

- Canonical quantum gravity

- E8 Theory

11.4 Weinberg–Witten theorem

In quantum field theory, the Weinberg–Witten theorem places some constraints on theories of composite gravity/emergent gravity. However, recent developments attempt to show that if locality is only approximate and the holographic principle is correct, the Weinberg–Witten theorem would not be valid.

11.5 Experimental tests

As was emphasized above, quantum gravitational effects are extremely weak and therefore difficult to test. For this reason, the possibility of experimentally testing quantum gravity had not received much attention prior to the late 1990s. However, in the past decade, physicists have realized that evidence for quantum gravitational effects can guide the development of the theory. Since theoretical development has been slow, the field of phenomenological quantum gravity, which studies the possibility of experimental tests, has obtained increased attention.[55][56]

The most widely pursued possibilities for quantum gravity phenomenology include violations of Lorentz invariance, imprints of quantum gravitational effects in the cosmic microwave background (in particular its polarization), and decoherence induced by fluctuations in the space-time foam.

The BICEP2 experiment detected what was initially thought to be primordial B-mode polarization caused by gravitational waves in the early universe. If truly primordial, these waves were born as quantum fluctuations in gravity itself. Cosmologist Ken Olum (Tufts University) stated: "I think this is the only observational evidence that we have that actually shows that gravity is quantized....It's probably the only evidence of this that we will ever have."[57]

11.6 See also

11.7 References

[1] Griffiths, David J. (2004). *Introduction to Quantum Mechanics*. Pearson Prentice Hall. OCLC 803860989.

[2] Wald, Robert M. (1984). *General Relativity*. University of Chicago Press. p. 382. OCLC 471881415.

[3] Zee, Anthony (2010). *Quantum Field Theory in a Nutshell* (2nd ed.). Princeton University Press. p. 172. OCLC 659549695.

[4] Penrose, Roger (2007). *The road to reality : a complete guide to the laws of the universe*. Vintage. p. 1017. OCLC 716437154.

[5] F. Finster, J. Kleiner, Causal Fermion Systems as a Candidate for a Unified Physical Theory, http://arxiv.org/abs/1502.03587

[6] F. Finster, The Principle of the Fermionic Projector, hep-th/0001048, hep-th/0202059, hep- th/0210121, AMS/IP Studies in Advanced Mathematics, vol. **35**, American Mathematical Society, Providence, RI, 2006.

[7] F. Finster, A formulation of quantum field theory realizing a sea of interacting Dirac particles, arXiv:0911.2102 [hep-th], Lett. Math. Phys. **97** (2011), no. 2, 165–183.

[8] F. Finster, An action principle for an interacting fermion system and its analysis in the continuum limit, arXiv:0908.1542 [math-ph] (2009).

[9] F. Finster, The continuum limit of a fermion system involving neutrinos: Weak and gravitational interactions, arXiv:1211.3351 [math-ph] (2012).

[10] F. Finster, Perturbative quantum field theory in the framework of the fermionic projector, arXiv:1310.4121 [math-ph], J. Math. Phys. **55** (2014), no. 4, 042301.

[11] Quantum effects in the early universe might have an observable effect on the structure of the present universe, for example, or gravity might play a role in the unification of the other forces. Cf. the text by Wald cited above.

[12] Donoghue (1995). "Introduction to the Effective Field Theory Description of Gravity". arXiv:gr-qc/9512024. (verify against ISBN 9789810229085)

[13] Kraichnan, R. H. (1955). "Special-Relativistic Derivation of Generally Covariant Gravitation Theory". *Physical Review* **98** (4): 1118–1122. Bibcode:1955PhRv...98.1118K. doi:10.1103/PhysRev.98.1118.

[14] Gupta, S. N. (1954). "Gravitation and Electromagnetism". *Physical Review* **96** (6): 1683–1685. Bibcode:1954PhRv...96.1683G. doi:10.1103/PhysRev.96.1683.

[15] Gupta, S. N. (1957). "Einstein's and Other Theories of Gravitation". *Reviews of Modern Physics* **29** (3): 334–336. BibcodeG. doi:10.1103/RevModPhys.29.334.

[16] Gupta, S. N. (1962). "Quantum Theory of Gravitation". *Recent Developments in General Relativity*. Pergamon Press. pp. 251–258.

[17] Deser, S. (1970). "Self-Interaction and Gauge Invariance". *General Relativity and Gravitation* **1**: 9–18. arXiv:gr-qc/0411023. Bibcode:1970GReGr...1....9D. doi:10.1007/BF00759198.

[18] Ohta, Tadayuki; Mann, Robert (1996). "Canonical reduction of two-dimensional gravity for particle dynamics". *Classical and Quantum Gravity* **13** (9): 2585–2602. arXiv:gr-qc/9605004. Bibcode:1996CQGra..13.2585O.doi:10.1088/0264-9381/13.

[19] Sikkema, A E; Mann, R B (1991). "Gravitation and cosmology in (1+1) dimensions". *Classical and Quantum Gravity* **8**: 219–235. Bibcode:1991CQGra...8..219S. doi:10.1088/0264-9381/8/1/022.

[20] Farrugia; Mann; Scott (2007). "N-body Gravity and the Schroedinger Equation". *Classical and Quantum Gravity* **24** (18): 4647–4659. arXiv:gr-qc/0611144. Bibcode:2007CQGra..24.4647F. doi:10.1088/0264-9381/24/18/006.

[21] Mann, R B; Ohta, T (1997). "Exact solution for the metric and the motion of two bodies in (1+1)-dimensional gravity". *Physical Review D* **55** (8): 4723–4747. arXiv:gr-qc/9611008. Bibcode:1997PhRvD..55.4723M. doi:10.1103/PhysRevD.55.4723.

[22] Feynman, R. P.; Morinigo, F. B.; Wagner, W. G.; Hatfield, B. (1995). *Feynman lectures on gravitation*. Addison-Wesley. ISBN 0-201-62734-5.

[23] Hamber, H. W. (2009). *Quantum Gravitation - The Feynman Path Integral Approach*. Springer Publishing. ISBN 978-3-540-85292-6.

[24] Smolin, Lee (2001). *Three Roads to Quantum Gravity*. Basic Books. pp. 20–25. ISBN 0-465-07835-4. Pages 220–226 are annotated references and guide for further reading.

[25] Hunter Monroe (2005). "Singularity-Free Collapse through Local Inflation". arXiv:astro-ph/0506506.

[26] Edward Anderson (2010). "The Problem of Time in Quantum Gravity". arXiv:1009.2157 [gr-qc]. (also published as chapter 4 of ISBN 9781611229578)

[27] A timeline and overview can be found in Rovelli, Carlo (2000). "Notes for a brief history of quantum gravity". arXiv:gr-qc/0006061. (verify against ISBN 9789812777386)

[28] Ashtekar, Abhay (2007). "Loop Quantum Gravity: Four Recent Advances and a Dozen Frequently Asked Questions". *11th Marcel Grossmann Meeting on Recent Developments in Theoretical and Experimental General Relativity.* p. 126. arXiv:0705.2222. Bibcode:2008mgm..conf..126A. doi:10.1142/9789812834300_0008.

[29] Schwarz, John H. (2007). "String Theory: Progress and Problems". *Progress of Theoretical Physics Supplement* **170**: 214–226. arXiv:hep-th/0702219. Bibcode:2007PThPS.170..214S. doi:10.1143/PTPS.170.214.

[30] Donoghue, John F. (editor) (1995). "Introduction to the Effective Field Theory Description of Gravity". In Cornet, Fernando. *Effective Theories: Proceedings of the Advanced School, Almunecar, Spain, 26 June–1 July 1995.* Singapore: World Scientific. arXiv:gr-qc/9512024. ISBN 981-02-2908-9.

[31] Weinberg, Steven (1996). "Chapters 17–18". *The Quantum Theory of Fields II: Modern Applications.* Cambridge University Press. ISBN 0-521-55002-5.

[32] Goroff, Marc H.; Sagnotti, Augusto; Sagnotti, Augusto (1985). "Quantum gravity at two loops". *Physics Letters B* **160**: 81–86. Bibcode:1985PhLB..160...81G. doi:10.1016/0370-2693(85)91470-4.

[33] An accessible introduction at the undergraduate level can be found in Zwiebach, Barton (2004). *A First Course in String Theory.* Cambridge University Press. ISBN 0-521-83143-1., and more complete overviews in Polchinski, Joseph (1998). *String Theory Vol. I: An Introduction to the Bosonic String.* Cambridge University Press. ISBN 0-521-63303-6. and Polchinski, Joseph (1998b). *String Theory Vol. II: Superstring Theory and Beyond.* Cambridge University Press. ISBN 0-521-63304-4.

[34] Ibanez, L. E. (2000). "The second string (phenomenology) revolution". *Classical & Quantum Gravity* **17** (5): 1117–1128. arXiv:hep-ph/9911499. Bibcode:2000CQGra..17.1117I. doi:10.1088/0264-9381/17/5/321.

[35] For the graviton as part of the string spectrum, e.g. Green, Schwarz & Witten 1987, sec. 2.3 and 5.3; for the extra dimensions, ibid sec. 4.2.

[36] Weinberg, Steven (2000). "Chapter 31". *The Quantum Theory of Fields II: Modern Applications.* Cambridge University Press. ISBN 0-521-55002-5.

[37] Townsend, Paul K. (1996). *Four Lectures on M-Theory.* ICTP Series in Theoretical Physics. p. 385. arXiv:hep-th/9612121. Bibcode:1997hepcbconf..385T.

[38] Duff, Michael (1996). "M-Theory (the Theory Formerly Known as Strings)". *International Journal of Modern Physics A* **11** (32): 5623–5642. arXiv:hep-th/9608117. Bibcode:1996IJMPA..11.5623D. doi:10.1142/S0217751X96002583.

[39] Ashtekar, Abhay (1986). "New variables for classical and quantum gravity". *Physical Review Letters* **57** (18): 2244–2247. Bibcode:1986PhRvL..57.2244A. doi:10.1103/PhysRevLett.57.2244. PMID 10033673.

[40] Ashtekar, Abhay (1987). "New Hamiltonian formulation of general relativity". *Physical Review D* **36** (6): 1587–1602. Bibcode:1987PhRvD..36.1587A. doi:10.1103/PhysRevD.36.1587.

[41] Thiemann, Thomas (2006). "Loop Quantum Gravity: An Inside View". *Approaches to Fundamental Physics.* Lecture Notes in Physics **721**: 185. arXiv:hep-th/0608210. Bibcode:2007LNP...721..185T. doi:10.1007/978-3-540-71117-9_10. ISBN 978-3-540-71115-5.

[42] Rovelli, Carlo (1998). "Loop Quantum Gravity". *Living Reviews in Relativity* **1**. Retrieved 2008-03-13.

[43] Ashtekar, Abhay; Lewandowski, Jerzy (2004). "Background Independent Quantum Gravity: A Status Report". *Classical & Quantum Gravity* **21** (15): R53–R152. arXiv:gr-qc/0404018. Bibcode:2004CQGra..21R..53A. doi:10.1088/0264-9381/211.

[44] Thiemann, Thomas (2003). "Lectures on Loop Quantum Gravity". *Lecture Notes in Physics.* Lecture Notes in Physics **631**: 41–135. arXiv:gr-qc/0210094. Bibcode:2003LNP...631...41T. doi:10.1007/978-3-540-45230-0_3. ISBN 978-3-540-40810-9.

[45] Rovelli, Carlo (2004). *Quantum Gravity.* Cambridge University Press. ISBN 0521715962.

[46] Nottale, L. (2011). *Scale Relativity and Fractal Space-Time: A New Approach to Unifying Relativity and Quantum Mechanics.* World Scientific Publishing Company. ISBN 1848166508.;p. 458

[47] Isham, Christopher J. (1994). "Prima facie questions in quantum gravity". In Ehlers, Jürgen; Friedrich, Helmut. *Canonical Gravity: From Classical to Quantum.* Springer. arXiv:gr-qc/9310031. ISBN 3-540-58339-4.

[48] Sorkin, Rafael D. (1997). "Forks in the Road, on the Way to Quantum Gravity". *International Journal of Theoretical Physics* **36** (12): 2759–2781. arXiv:gr-qc/9706002. Bibcode:1997IJTP...36.2759S. doi:10.1007/BF02435709.

[49] Loll, Renate (1998). "Discrete Approaches to Quantum Gravity in Four Dimensions". *Living Reviews in Relativity* **1**: 13. arXiv:gr-qc/9805049. Bibcode:1998LRR.....1...13L. doi:10.12942/lrr-1998-13. Retrieved 2008-03-09.

[50] Sorkin, Rafael D. (2005). "Causal Sets: Discrete Gravity". In Gomberoff, Andres; Marolf, Donald. *Lectures on Quantum Gravity.* Springer. arXiv:gr-qc/0309009. ISBN 0-387-23995-2.

[51] See Daniele Oriti and references therein.

[52] Hawking, Stephen W. (1987). "Quantum cosmology". In Hawking, Stephen W.; Israel, Werner. *300 Years of Gravitation.* Cambridge University Press. pp. 631–651. ISBN 0-521-37976-8.

[53] Wen 2006

[54] See ch. 33 in Penrose 2004 and references therein.

[55] Hossenfelder, Sabine (2011). "Experimental Search for Quantum Gravity". In V. R. Frignanni. *Classical and Quantum Gravity: Theory, Analysis and Applications.* Chapter 5: Nova Publishers. ISBN 978-1-61122-957-8.

[56] Hossenfelder, Sabine (2010-10-17). V. R. Frignanni, ed. "Experimental Search for Quantum Gravity". *Classical and Quantum Gravity: Theory, Analysis and Applications* (Nova Publishers) **5** (2011). arXiv:1010.3420. Bibcode:2010arXiv1010.3420H. |chapter= ignored (help)

[57] Camille Carlisle. "First Direct Evidence of Big Bang Inflation". SkyandTelescope.com. Retrieved March 18, 2014.

11.8   Further reading

- Ahluwalia, D. V. (2002). "Interface of Gravitational and Quantum Realms". *Modern Physics Letters A* **17** (15–17): 1135. arXiv:gr-qc/0205121. Bibcode:2002MPLA...17.1135A. doi:10.1142/S021773230200765X.

- Ashtekar, Abhay (2005). "The winding road to quantum gravity" (PDF). *Current Science* **89**: 2064–2074.

- Carlip, Steven (2001). "Quantum Gravity: a Progress Report". *Reports on Progress in Physics* **64** (8): 885–942. arXiv:gr-qc/0108040. Bibcode:2001RPPh...64..885C. doi:10.1088/0034-4885/64/8/301.

- Herbert W. Hamber (2009). *Quantum Gravitation.* Springer Publishing. doi:10.1007/978-3-540-85293-3. ISBN 978-3-540-85292-6.

- Kiefer, Claus (2007). *Quantum Gravity.* Oxford University Press. ISBN 0-19-921252-X.

- Kiefer, Claus (2005). "Quantum Gravity: General Introduction and Recent Developments". *Annalen der Physik* **15**: 129–148. arXiv:gr-qc/0508120. Bibcode:2006AnP...518..129K. doi:10.1002/andp.200510175.

- Lämmerzahl, Claus, ed. (2003). *Quantum Gravity: From Theory to Experimental Search.* Lecture Notes in Physics. Springer. ISBN 3-540-40810-X.

- Rovelli, Carlo (2004). *Quantum Gravity.* Cambridge University Press. ISBN 0-521-83733-2.

- Trifonov, Vladimir (2008). "GR-friendly description of quantum systems". *International Journal of Theoretical Physics* **47** (2): 492–510. arXiv:math-ph/0702095. Bibcode:2008IJTP...47..492T. doi:10.1007/s10773-007-9474-3.

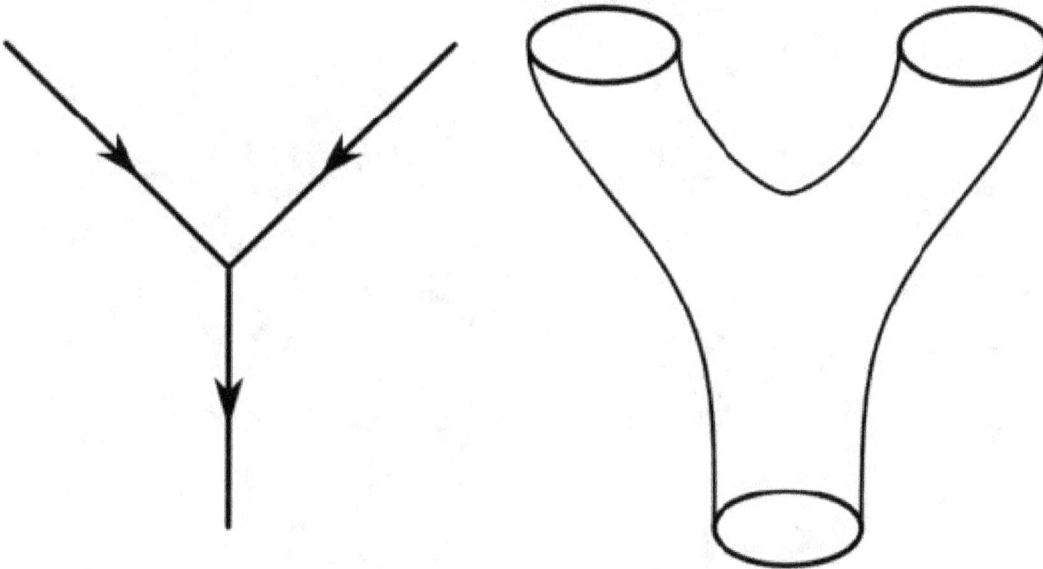

Interaction in the subatomic world: world lines of point-like particles in the Standard Model or a world sheet swept up by closed strings in string theory

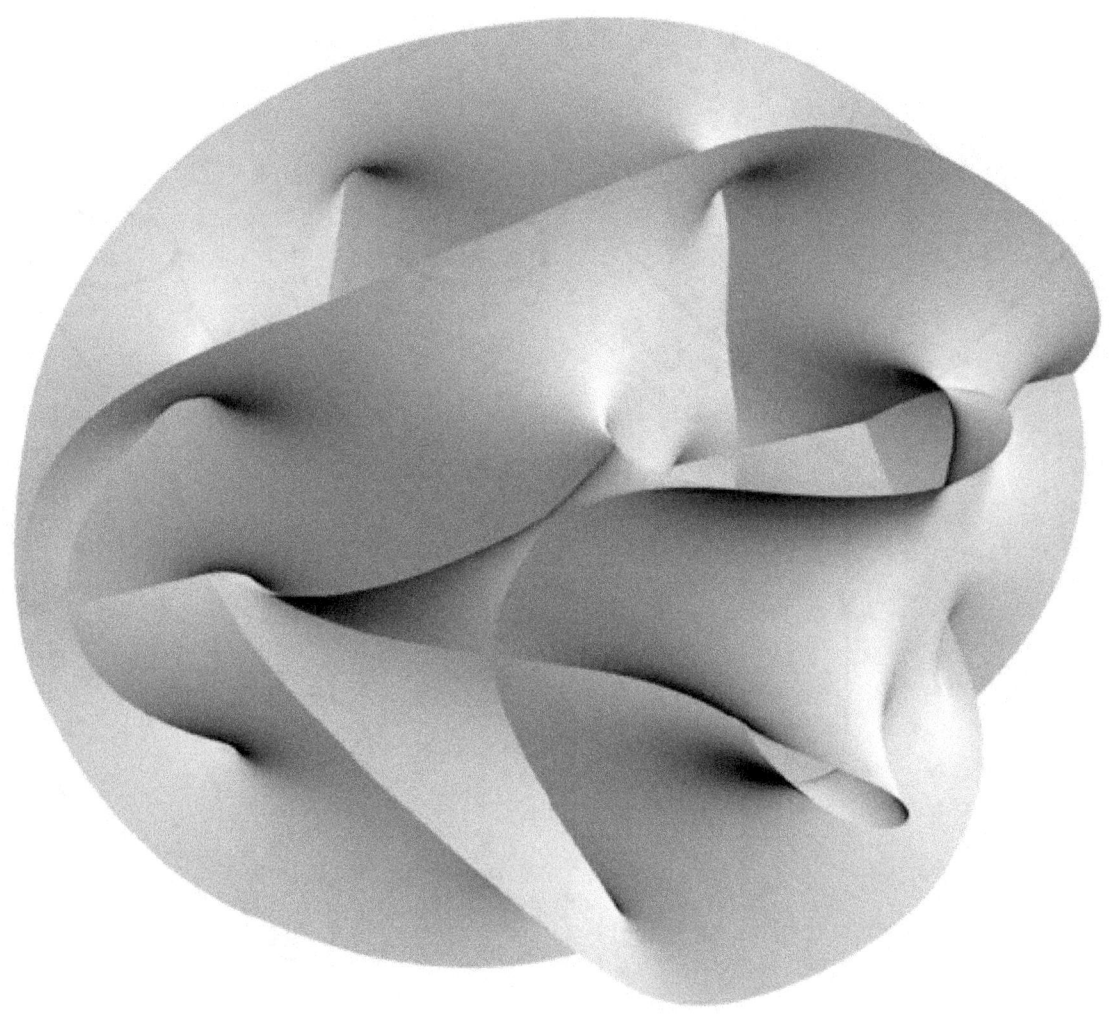

Projection of a Calabi–Yau manifold, one of the ways of compactifying the extra dimensions posited by string theory

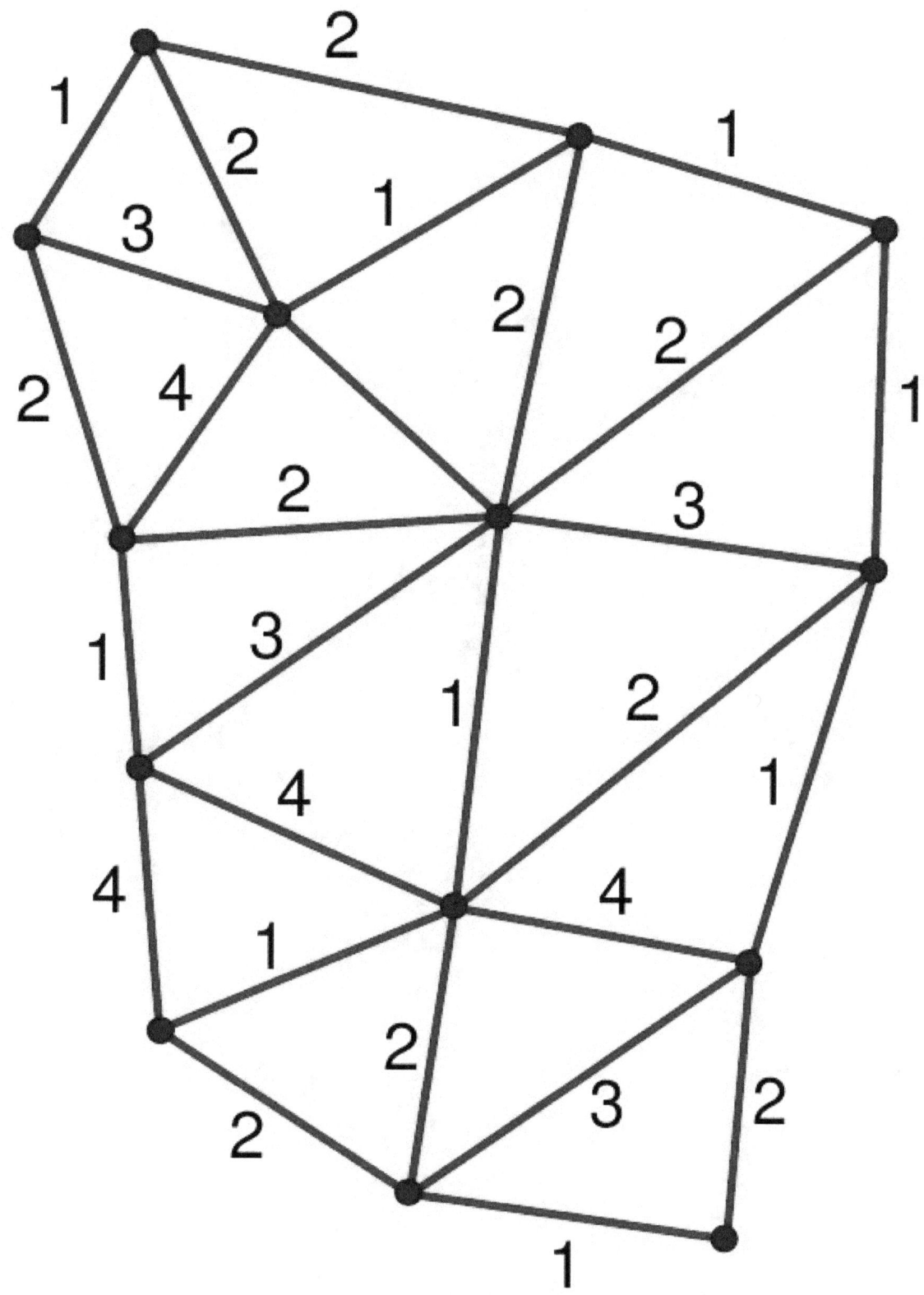

Simple spin network of the type used in loop quantum gravity

Schrödinger's flower. Morphogenesis of a flower-like structure, solution of a growth process equation that takes the form of a Schrödinger equation under fractal conditions.

Chapter 12

String theory

For the study of strings of characters, see Concatenation theory.

For a more accessible and less technical introduction to this topic, see Introduction to M-theory.

In physics, **string theory** is a theoretical framework in which the point-like particles of particle physics are replaced by one-dimensional objects called strings. String theory describes how these strings propagate through space and interact with each other. On distance scales larger than the string scale, a string looks just like an ordinary particle, with its mass, charge, and other properties determined by the vibrational state of the string. In string theory, one of the many vibrational states of the string corresponds to the graviton, a quantum mechanical particle that carries gravitational force. Thus string theory is a theory of quantum gravity.

String theory is a broad and varied subject that attempts to address a number of deep questions of fundamental physics. String theory has been applied to a variety of problems in black hole physics, early universe cosmology, nuclear physics, and condensed matter physics, and it has stimulated a number of major developments in pure mathematics. Because string theory potentially provides a unified description of gravity and particle physics, it is a candidate for a theory of everything, a self-contained mathematical model that describes all fundamental forces and forms of matter. Despite much work on these problems, it is not known to what extent string theory describes the real world or how much freedom the theory allows to choose the details.

String theory was first studied in the late 1960s as a theory of the strong nuclear force, before being abandoned in favor of quantum chromodynamics. Subsequently, it was realized that the very properties that made string theory unsuitable as a theory of nuclear physics made it a promising candidate for a quantum theory of gravity. The earliest version of string theory, bosonic string theory, incorporated only the class of particles known as bosons. It later developed into superstring theory, which posits a connection called supersymmetry between bosons and the class of particles called fermions. Five consistent versions of superstring theory were developed before it was conjectured in the mid-1990s that they were all different limiting cases of a single theory in eleven dimensions known as M-theory. In late 1997, theorists discovered an important relationship called the AdS/CFT correspondence, which relates string theory to another type of physical theory called a quantum field theory.

One of the challenges of string theory is that the full theory does not yet have a satisfactory definition in all circumstances. Another issue is that the theory is thought to describe an enormous landscape of possible universes, and this has complicated efforts to develop theories of particle physics based on string theory. These issues have led some in the community to criticize these approaches to physics and question the value of continued research on string theory unification.

12.1 Fundamentals

In the twentieth century, two theoretical frameworks emerged for formulating the laws of physics. One of these frameworks was Albert Einstein's general theory of relativity, a theory that explains the force of gravity and the structure of space and time. The other was quantum mechanics, a radically different formalism for describing physical phenomena

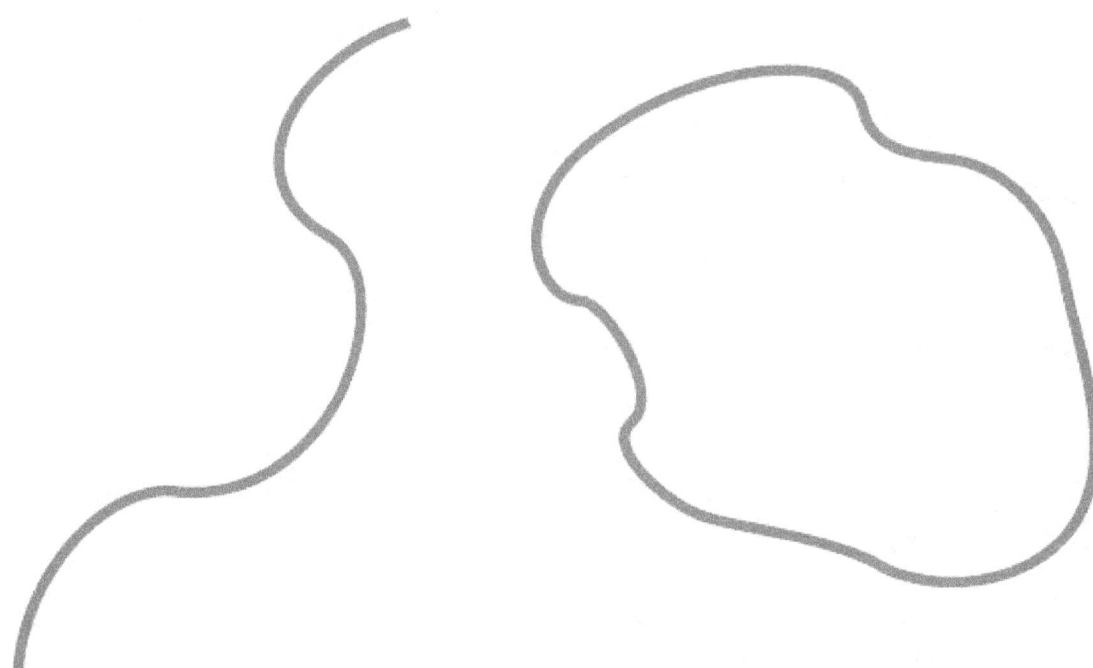

The fundamental objects of string theory are open and closed strings.

using probability. By the late 1970s, these two frameworks had proven to be sufficient to explain most of the observed features of the universe, from elementary particles to atoms to the evolution of stars and the universe as a whole.[1]

In spite of these successes, there are still many problems that remain to be solved. One of the deepest problems in modern physics is the problem of quantum gravity.[2] The general theory of relativity is formulated within the framework of classical physics, whereas the other fundamental forces are described within the framework of quantum mechanics. A quantum theory of gravity is needed in order to reconcile general relativity with the principles of quantum mechanics, but difficulties arise when one attempts to apply the usual prescriptions of quantum theory to the force of gravity.[3] In addition to the problem of developing a consistent theory of quantum gravity, there are many other fundamental problems in the physics of atomic nuclei, black holes, and the early universe.[lower-alpha 1]

String theory is a theoretical framework that attempts to address these questions and many others. The starting point for string theory is the idea that the point-like particles of particle physics can also be modeled as one-dimensional objects called strings. String theory describes how strings propagate through space and interact with each other. In a given version of string theory, there is only one kind of string, which may look like a small loop or segment of ordinary string, and it can vibrate in different ways. On distance scales larger than the string scale, a string will look just like an ordinary particle, with its mass, charge, and other properties determined by the vibrational state of the string. In this way, all of the different elementary particles may be viewed as vibrating strings. In string theory, one of the vibrational states of the string gives rise to the graviton, a quantum mechanical particle that carries gravitational force. Thus string theory is a theory of quantum gravity.[4]

One of the main developments of the past several decades in string theory was the discovery of certain "dualities", mathematical transformations that identify one physical theory with another. Physicists studying string theory have discovered a number of these dualities between different versions of string theory, and this has led to the conjecture that all consistent versions of string theory are subsumed in a single framework known as M-theory.[5]

Studies of string theory have also yielded a number of results on the nature of black holes and the gravitational interaction. There are certain paradoxes that arise when one attempts to understand the quantum aspects of black holes, and work on string theory has attempted to clarify these issues. In late 1997 this line of work culminated in the discovery of the anti-de Sitter/conformal field theory correspondence or AdS/CFT.[6] This is a theoretical result which relates string theory to other physical theories which are better understood theoretically. The AdS/CFT correspondence has implications for the study of black holes and quantum gravity, and it has been applied to other subjects, including nuclear[7] and condensed matter physics.[8][9]

Since string theory incorporates all of the fundamental interactions, including gravity, many physicists hope that it fully describes our universe, making it a theory of everything. One of the goals of current research in string theory is to find a solution of the theory that reproduces the observed spectrum of elementary particles, with a small cosmological constant, containing dark matter and a plausible mechanism for cosmic inflation. While there has been progress toward these goals, it is not known to what extent string theory describes the real world or how much freedom the theory allows to choose the details.[10]

One of the challenges of string theory is that the full theory does not yet have a satisfactory definition in all circumstances. The scattering of strings is most straightforwardly defined using the techniques of perturbation theory, but it is not known in general how to define string theory nonperturbatively.[11] It is also not clear whether there is any principle by which string theory selects its vacuum state, the physical state that determines the properties of our universe.[12] These problems have led some in the community to criticize these approaches to the unification of physics and question the value of continued research on these problems.[13]

12.1.1   Strings

Main article: String (physics)

The application of quantum mechanics to physical objects such as the electromagnetic field, which are extended in space

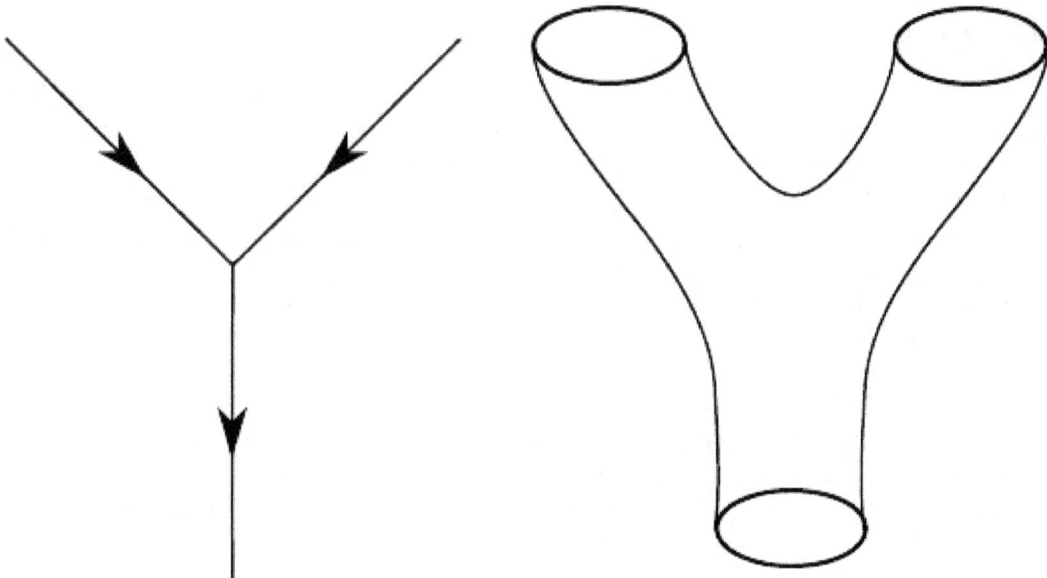

Interaction in the quantum world: worldlines of point-like particles or a worldsheet swept up by closed strings in string theory.

and time, is known as quantum field theory. In particle physics, quantum field theories form the basis for our understanding of elementary particles, which are modeled as excitations in the fundamental fields.[14]

In quantum field theory, one typically computes the probabilities of various physical events using the techniques of perturbation theory. Developed by Richard Feynman and others in the first half of the twentieth century, perturbative quantum field theory uses special diagrams called Feynman diagrams to organize computations. One imagines that these diagrams depict the paths of point-like particles and their interactions.[15]

The starting point for string theory is the idea that the point-like particles of quantum field theory can also be modeled as one-dimensional objects called strings.[16] The interaction of strings is most straightforwardly defined by generalizing the perturbation theory used in ordinary quantum field theory. At the level of Feynman diagrams, this means replacing the one-dimensional diagram representing the path of a point particle by a two-dimensional surface representing the motion of a string.[17] Unlike in quantum field theory, string theory does not yet have a full non-perturbative definition, so many of the theoretical questions that physicists would like to answer remain out of reach.[18]

In theories of particle physics based on string theory, the characteristic length scale of strings is assumed to be on the order of the Planck length, or 10^{-35} meters, the scale at which the effects of quantum gravity are believed to become significant.[19] On much larger length scales, such as the scales visible in physics laboratories, such objects would be indistinguishable from zero-dimensional point particles, and the vibrational state of the string would determine the type of particle. One of the vibrational states of a string corresponds to the graviton, a quantum mechanical particle that carries the gravitational force.[20]

The original version of string theory was bosonic string theory, but this version described only bosons, a class of particles which transmit forces between the matter particles, or fermions. Bosonic string theory was eventually superseded by theories called superstring theories. These theories describe both bosons and fermions, and they incorporate a theoretical idea called supersymmetry. This is a mathematical relation that exists in certain physical theories between the bosons and fermions. In theories with supersymmetry, each boson has a counterpart which is a fermion, and vice versa.[21]

There are several versions of superstring theory: type I, type IIA, type IIB, and two flavors of heterotic string theory ($SO(32)$ and $E_8 \times E_8$). The different theories allow different types of strings, and the particles that arise at low energies exhibit different symmetries. For example, the type I theory includes both open strings (which are segments with endpoints) and closed strings (which form closed loops), while types IIA and IIB include only closed strings.[22]

12.1.2 Extra dimensions

In everyday life, there are three familiar dimensions of space: height, width and length. Einstein's general theory of relativity treats time as a dimension on par with the three spatial dimensions; in general relativity, space and time are not modeled as separate entities but are instead unified to a four-dimensional spacetime. In this framework, the phenomenon of gravity is viewed as a consequence of the geometry of spacetime.[23]

In spite of the fact that the universe is well described by four-dimensional spacetime, there are several reasons why physicists consider theories in other dimensions. In some cases, by modeling spacetime in a different number of dimensions, a theory becomes more mathematically tractable, and one can perform calculations and gain general insights more easily.[lower-alpha 2] There are also situations where theories in two or three spacetime dimensions are useful for describing phenomena in condensed matter physics.[24] Finally, there exist scenarios in which there could actually be more than four dimensions of spacetime which have nonetheless managed to escape detection.[25]

One notable feature of string theories is that these theories require extra dimensions of spacetime for their mathematical consistency. In bosonic string theory, spacetime is 26-dimensional, while in superstring theory it is ten-dimensional. In order to describe real physical phenomena using string theory, one must therefore imagine scenarios in which these extra dimensions would not be observed in experiments.[26]

Compactification is one way of modifying the number of dimensions in a physical theory. In compactification, some of the extra dimensions are assumed to "close up" on themselves to form circles.[27] In the limit where these curled up dimensions become very small, one obtains a theory in which spacetime has effectively a lower number of dimensions. A standard analogy for this is to consider a multidimensional object such as a garden hose. If the hose is viewed from a sufficient distance, it appears to have only one dimension, its length. However, as one approaches the hose, one discovers that it contains a second dimension, its circumference. Thus, an ant crawling on the surface of the hose would move in two dimensions.[28]

Compactification can be used to construct models in which spacetime is effectively four-dimensional. However, not every way of compactifying the extra dimensions produces a model with the right properties to describe nature. In a viable model of particle physics, the compact extra dimensions must be shaped like a Calabi–Yau manifold.[29] A Calabi–Yau manifold is a special space which is typically taken to be six-dimensional in applications to string theory. It is named after

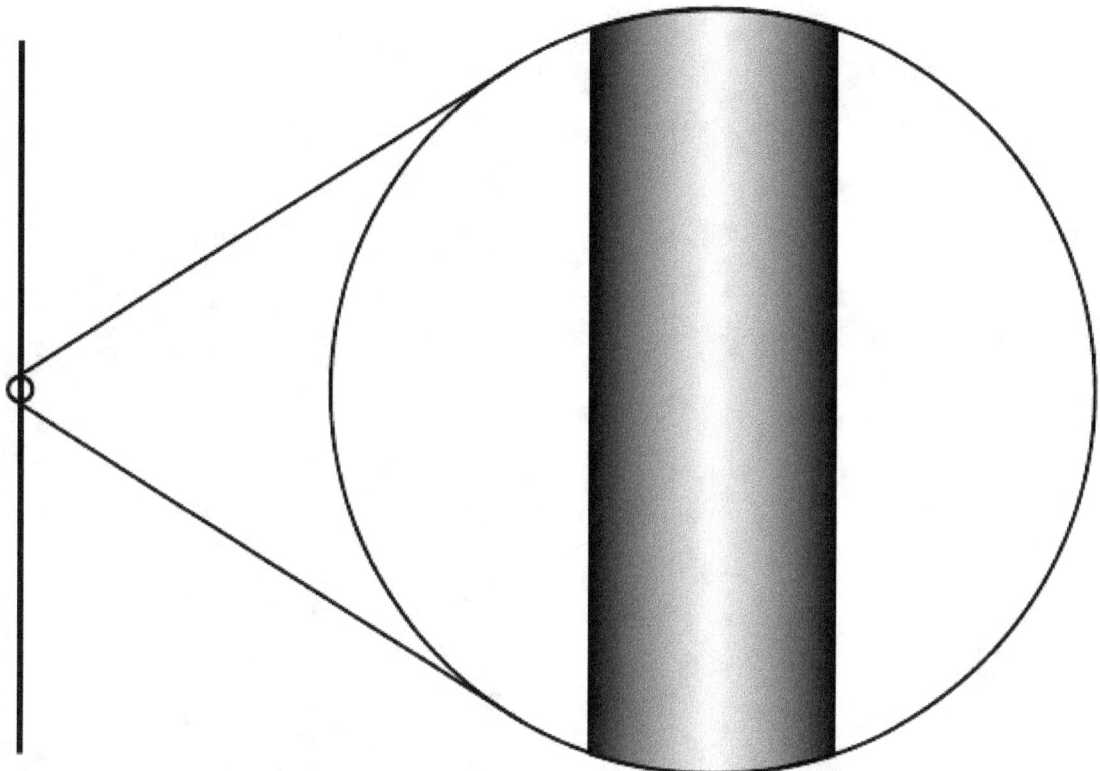

An example of compactification: At large distances, a two dimensional surface with one circular dimension looks one-dimensional.

mathematicians Eugenio Calabi and Shing-Tung Yau.[30]

Another approach to reducing the number of dimensions is the so called brane-world scenario. In this approach, physicists assume that the observable universe is a four-dimensional subspace of a higher dimensional space. In such models, the force-carrying bosons of particle physics arise from open strings with endpoints attached to the four-dimensional subspace, while gravity arises from closed strings propagating through the larger ambient space. This idea plays an important role in attempts to develop models of real world physics based on string theory, and it provides a natural explanation for the weakness of gravity compared to the other fundamental forces.[31]

12.1.3   Dualities

Main articles: S-duality and T-duality

One notable fact about string theory is that the different versions of the theory all turn out to be related in highly nontrivial ways. One of the relationships that can exist between different string theories is called S-duality. This is a relationship which says that a collection of strongly interacting particles in one theory can, in some cases, be viewed as a collection of weakly interacting particles in a completely different theory. Roughly speaking, a collection of particles is said to be strongly interacting if they combine and decay often and weakly interacting if they do so infrequently. Type I string theory turns out to be equivalent by S-duality to the $SO(32)$ heterotic string theory. Similarly, type IIB string theory is related to itself in a nontrivial way by S-duality.[32]

Another relationship between different string theories is T-duality. Here one considers strings propagating around a circular extra dimension. T-duality states that a string propagating around a circle of radius R is equivalent to a string propagating around a circle of radius $1/R$ in the sense that all observable quantities in one description are identified with

A cross section of a quintic Calabi–Yau manifold

quantities in the dual description. For example, a string has momentum as it propagates around a circle, and it can also wind around the circle one or more times. The number of times the string winds around a circle is called the winding number. If a string has momentum p and winding number n in one description, it will have momentum n and winding number p in the dual description. For example, type IIA string theory is equivalent to type IIB string theory via T-duality, and the two versions of heterotic string theory are also related by T-duality.[33]

In general, the term *duality* refers to a situation where two seemingly different physical systems turn out to be equivalent in a nontrivial way. Two theories related by a duality need not be string theories. For example, Montonen–Olive duality is example of an S-duality relationship between quantum field theories. The AdS/CFT correspondence is example of a duality which relates string theory to a quantum field theory. If two theories are related by a duality, it means that one theory can be transformed in some way so that it ends up looking just like the other theory. The two theories are then said to be *dual* to one another under the transformation. Put differently, the two theories are mathematically different descriptions of the same phenomena.[34]

12.1.4   Branes

Main article: Brane

In string theory and related theories, a brane is a physical object that generalizes the notion of a point particle to higher dimensions. For example, a point particle can be viewed as a brane of dimension zero, while a string can be viewed as a brane of dimension one. It is also possible to consider higher-dimensional branes. In dimension p, these are called p-branes. The word brane comes from the word "membrane" which refers to a two-dimensional brane.[35]

Branes are dynamical objects which can propagate through spacetime according to the rules of quantum mechanics. They have mass and can have other attributes such as charge. A p-brane sweeps out a $(p+1)$-dimensional volume in spacetime called its *worldvolume*. Physicists often study fields analogous to the electromagnetic field which live on the worldvolume of a brane.[36]

In string theory, D-branes are an important class of branes that arise when one considers open strings. As an open string propagates through spacetime, its endpoints are required to lie on a D-brane. The letter "D" in D-brane refers to a certain mathematical condition on the system known as the Dirichlet boundary condition. The study of D-branes in string theory has led to important results such as the AdS/CFT correspondence, which has shed light on many problems in quantum field theory.[37]

Branes are also frequently studied from a purely mathematical point of view. Mathematically, branes can be described as objects of certain categories, such as the derived category of coherent sheaves on a complex algebraic variety, or the Fukaya category of a symplectic manifold.[38] The connection between the physical notion of a brane and the mathematical notion of a category has led to important mathematical insights in the fields of algebraic and symplectic geometry[39] and representation theory.[40]

12.2   M-theory

Main article: M-theory

Prior to 1995, theorists believed that there were five consistent versions of superstring theory (type I, type IIA, type IIB, and two versions of heterotic string theory). This understanding changed in 1995 when Edward Witten suggested that the five theories were just special limiting cases of an eleven-dimensional theory called M-theory. Witten's conjecture was based on the work of a number of other physicists, including Ashoke Sen, Chris Hull, Paul Townsend, and Michael Duff. His announcement led to a flurry of research activity now known as the second superstring revolution.[41]

12.2.1   Unification of superstring theories

In the 1970s, many physicists became interested in supergravity theories, which combine general relativity with supersymmetry. Whereas general relativity makes sense in any number of dimensions, supergravity places an upper limit on the number of dimensions.[42] In 1978, work by Werner Nahm showed that the maximum spacetime dimension in which one can formulate a consistent supersymmetric theory is eleven.[43] In the same year, Eugene Cremmer, Bernard Julia, and Joel Scherk of the École Normale Supérieure showed that supergravity not only permits up to eleven dimensions but is in fact most elegant in this maximal number of dimensions.[44][45]

Initially, many physicists hoped that by compactifying eleven-dimensional supergravity, it might be possible to construct realistic models of our four-dimensional world. The hope was that such models would provide a unified description of the four fundamental forces of nature: electromagnetism, the strong and weak nuclear forces, and gravity. Interest in eleven-dimensional supergravity soon waned as various flaws in this scheme were discovered. One of the problems was that the laws of physics appear to distinguish between clockwise and counterclockwise, a phenomenon known as chirality. Edward Witten and others observed this chirality property cannot be readily derived by compactifying from eleven dimensions.[46]

In the first superstring revolution in 1984, many physicists turned to string theory as a unified theory of particle physics and quantum gravity. Unlike supergravity theory, string theory was able to accommodate the chirality of the standard model, and it provided a theory of gravity consistent with quantum effects.[47] Another feature of string theory that many physicists were drawn to in the 1980s and 1990s was its high degree of uniqueness. In ordinary particle theories, one can consider any collection of elementary particles whose classical behavior is described by an arbitrary Lagrangian. In string theory, the possibilities are much more constrained: by the 1990s, physicists had argued that there were only five consistent supersymmetric versions of the theory.[48]

Although there were only a handful of consistent superstring theories, it remained a mystery why there was not just one consistent formulation.[49] However, as physicists began to examine string theory more closely, they realized that these theories are related in intricate and nontrivial ways. They found that a system of strongly interacting strings can, in some cases, be viewed as a system of weakly interacting strings. This phenomenon is known as S-duality. It was studied by Ashoke Sen in the context of heterotic strings in four dimensions[50][51] and by Chris Hull and Paul Townsend in the context of the type IIB theory.[52] Theorists also found that different string theories may be related by T-duality. This duality implies that strings propagating on completely different spacetime geometries may be physically equivalent.[53]

At around the same time, as many physicists were studying the properties of strings, a small group of physicists was examining the possible applications of higher dimensional objects. In 1987, Eric Bergshoeff, Ergin Sezgin, and Paul Townsend showed that eleven-dimensional supergravity includes two-dimensional branes.[54] Intuitively, these objects look like sheets or membranes propagating through the eleven-dimensional spacetime. Shortly after this discovery, Michael Duff, Paul Howe, Takeo Inami, and Kellogg Stelle considered a particular compactification of eleven-dimensional supergravity with one of the dimensions curled up into a circle.[55] In this setting, one can imagine the membrane wrapping around the circular dimension. If the radius of the circle is sufficiently small, then this membrane looks just like a string in ten-dimensional spacetime. In fact, Duff and his collaborators showed that this construction reproduces exactly the strings appearing in type IIA superstring theory.[56]

Speaking at a string theory conference in 1995, Edward Witten made the surprising suggestion that all five superstring theories were in fact just different limiting cases of a single theory in eleven spacetime dimensions. Witten's announcement drew together all of the previous results on S- and T-duality and the appearance of higher dimensional branes in string theory.[57] In the months following Witten's announcement, hundreds of new papers appeared on the Internet confirming different parts of his proposal.[58] Today this flurry of work is known as the second superstring revolution.[59]

Initially, some physicists suggested that the new theory was a fundamental theory of membranes, but Witten was skeptical of the role of membranes in the theory. In a paper from 1996, Hořava and Witten wrote "As it has been proposed that the eleven-dimensional theory is a supermembrane theory but there are some reasons to doubt that interpretation, we will non-committally call it the M-theory, leaving to the future the relation of M to membranes."[60] In the absence of an understanding of the true meaning and structure of M-theory, Witten has suggested that the *M* should stand for "magic", "mystery", or "membrane" according to taste, and the true meaning of the title should be decided when a more fundamental formulation of the theory is known.[61]

12.2.2 Matrix theory

Main article: Matrix theory (physics)

In mathematics, a matrix is a rectangular array of numbers or other data. In physics, a matrix model is a particular kind of physical theory whose mathematical formulation involves the notion of a matrix in an important way. A matrix model describes the behavior of a set of matrices within the framework of quantum mechanics.[62]

One important example of a matrix model is the BFSS matrix model proposed by Tom Banks, Willy Fischler, Stephen Shenker, and Leonard Susskind in 1997. This theory describes the behavior of a set of nine large matrices. In their original paper, these authors showed, among other things, that the low energy limit of this matrix model is described by eleven-dimensional supergravity. These calculations led them to propose that the BFSS matrix model is exactly equivalent to M-theory. The BFSS matrix model can therefore be used as a prototype for a correct formulation of M-theory and a tool for investigating the properties of M-theory in a relatively simple setting.[63]

The development of the matrix model formulation of M-theory has led physicists to consider various connections between string theory and a branch of mathematics called noncommutative geometry. This subject is a generalization of ordinary geometry in which mathematicians define new geometric notions using tools from noncommutative algebra.[64] In a paper from 1998, Alain Connes, Michael R. Douglas, and Albert Schwarz showed that some aspects of matrix models and M-theory are described by a noncommutative quantum field theory, a special kind of physical theory in which spacetime is described mathematically using noncommutative geometry.[65] This established a link between matrix models and M-theory on the one hand, and noncommutative geometry on the other hand. It quickly led to the discovery of other important links between noncommutative geometry and various physical theories.[66][67]

12.3   Black holes

In general relativity, a black hole is defined as a region of spacetime in which the gravitational field is so strong that no particle or radiation can escape. In the currently accepted models of stellar evolution, black holes are thought to arise when massive stars undergo gravitational collapse, and many galaxies are thought to contain supermassive black holes at their centers. Black holes are also important for theoretical reasons, as they present profound challenges for theorists attempting to understand the quantum aspects of gravity. String theory has proved to be an important tool for investigating the theoretical properties of black holes because it provides a framework in which theorists can study their thermodynamics.[68]

12.3.1   Bekenstein–Hawking formula

In the branch of physics called statistical mechanics, entropy is a measure of the randomness or disorder of a physical system. This concept was studied in the 1870s by the Austrian physicist Ludwig Boltzmann, who showed that the thermodynamic properties of a gas could be derived from the combined properties of its many constituent molecules. Boltzmann argued that by averaging the behaviors of all the different molecules in a gas, one can understand macroscopic properties such as volume, temperature, and pressure. In addition, this perspective led him to give a precise definition of entropy as the natural logarithm of the number of different states of the molecules (also called *microstates*) that give rise to the same macroscopic features.[69]

In the twentieth century, physicists began to apply the same concepts to black holes. In most systems such as gases, the entropy scales with the volume. In the 1970s, the physicist Jacob Bekenstein suggested that the entropy of a black hole is instead proportional to the *surface area* of its event horizon, the boundary beyond which matter and radiation is lost to its gravitational attraction.[70] When combined with ideas of the physicist Stephen Hawking,[71] Bekenstein's work yielded a precise formula for the entropy of a black hole. The formula expresses the entropy S as

$$S = \frac{c^3 kA}{4\hbar G}$$

where c is the speed of light, k is Boltzmann's constant, $\hbar$ is the reduced Planck constant, G is Newton's constant, and A is the surface area of the event horizon.[72]

Like any physical system, a black hole has an entropy defined in terms of the number of different microstates that lead to the same macroscopic features. The Bekenstein–Hawking entropy formula gives the expected value of the entropy of a black hole, but by the 1990s, physicists still lacked a derivation of this formula by counting microstates in a theory of quantum gravity. Finding such a derivation of this formula was considered an important test of the viability of any theory of quantum gravity such as string theory.[73]

12.3.2   Derivation within string theory

In a paper from 1996, Andrew Strominger and Cumrun Vafa showed how to derive the Beckenstein–Hawking formula for certain black holes in string theory.[74] Their calculation was based on the observation that D-branes—which look like fluctuating membranes when they are weakly interacting—become dense, massive objects with event horizons when the interactions are strong. In other words, a system of strongly interacting D-branes in string theory is indistinguishable from a black hole. Strominger and Vafa analyzed such D-brane systems and calculated the number of different ways of placing D-branes in spacetime so that their combined mass and charge is equal to a given mass and charge for the resulting black hole. Their calculation reproduced the Bekenstein–Hawking formula exactly, including the factor of 1/4.[75] Subsequent work by Strominger, Vafa, and others refined the original calculations and gave the precise values of the "quantum corrections" needed to describe very small black holes.[76][77]

The black holes that Strominger and Vafa considered in their original work were quite different from real astrophysical black holes. One difference was that Strominger and Vafa considered only extremal black holes in order to make the calculation tractable. These are defined as black holes with the lowest possible mass compatible with a given charge.[78] Strominger and Vafa also restricted attention to black holes in five-dimensional spacetime with unphysical supersymmetry.[79]

Although it was originally developed in this very particular and physically unrealistic context in string theory, the entropy calculation of Strominger and Vafa has led to a qualitative understanding of how black hole entropy can be accounted for in any theory of quantum gravity. Indeed, in 1998, Strominger argued that the original result could be generalized to an arbitrary consistent theory of quantum gravity without relying on strings or supersymmetry.[80] In collaboration with several other authors in 2010, he showed that some results on black hole entropy could be extended to non-extremal astrophysical black holes.[81][82]

12.4   AdS/CFT correspondence

Main article: AdS/CFT correspondence

One approach to formulating string theory and studying its properties is provided by the anti-de Sitter/conformal field theory (AdS/CFT) correspondence. This is a theoretical result which implies that string theory is in some cases equivalent to a quantum field theory. In addition to providing insights into the mathematical structure of string theory, the AdS/CFT correspondence has shed light on many aspects of quantum field theory in regimes where traditional calculational techniques are ineffective.[83] The AdS/CFT correspondence was first proposed by Juan Maldacena in late 1997.[84] Important aspects of the correspondence were elaborated in articles by Steven Gubser, Igor Klebanov, and Alexander Markovich Polyakov,[85] and by Edward Witten.[86] By 2010, Maldacena's article had over 7000 citations, becoming the most highly cited article in the field of high energy physics.[lower-alpha 3]

12.4.1   Overview of the correspondence

In the AdS/CFT correspondence, the geometry of spacetime is described in terms of a certain vacuum solution of Einstein's equation called anti-de Sitter space.[87] In very elementary terms, anti-de Sitter space is a mathematical model of spacetime in which the notion of distance between points (the metric) is different from the notion of distance in ordinary Euclidean geometry. It is closely related to hyperbolic space, which can be viewed as a disk as illustrated on the left.[88] This image shows a tessellation of a disk by triangles and squares. One can define the distance between points of this disk in such a way that all the triangles and squares are the same size and the circular outer boundary is infinitely far from any point in the interior.[89]

One can imagine a stack of hyperbolic disks where each disk represents the state of the universe at a given time. The resulting geometric object is three-dimensional anti-de Sitter space.[90] It looks like a solid cylinder in which any cross section is a copy of the hyperbolic disk. Time runs along the vertical direction in this picture. The surface of this cylinder plays an important role in the AdS/CFT correspondence. As with the hyperbolic plane, anti-de Sitter space is curved in such a way that any point in the interior is actually infinitely far from this boundary surface.[91]

This construction describes a hypothetical universe with only two space dimensions and one time dimension, but it can be generalized to any number of dimensions. Indeed, hyperbolic space can have more than two dimensions and one can "stack up" copies of hyperbolic space to get higher-dimensional models of anti-de Sitter space.[92]

An important feature of anti-de Sitter space is its boundary (which looks like a cylinder in the case of three-dimensional anti-de Sitter space). One property of this boundary is that, within a small region on the surface around any given point, it looks just like Minkowski space, the model of spacetime used in nongravitational physics.[93] One can therefore consider an auxiliary theory in which "spacetime" is given by the boundary of anti-de Sitter space. This observation is the starting point for AdS/CFT correspondence, which states that the boundary of anti-de Sitter space can be regarded as the "spacetime" for a quantum field theory. The claim is that this quantum field theory is equivalent to a gravitational theory, such as string theory, in the bulk anti-de Sitter space in the sense that there is a "dictionary" for translating entities and calculations in one theory into their counterparts in the other theory. For example, a single particle in the gravitational theory might correspond to some collection of particles in the boundary theory. In addition, the predictions in the two theories are quantitatively identical so that if two particles have a 40 percent chance of colliding in the gravitational theory, then the corresponding collections in the boundary theory would also have a 40 percent chance of colliding.[94]

12.4.2 Applications to quantum gravity

The discovery of the AdS/CFT correspondence was a major advance in physicists' understanding of string theory and quantum gravity. One reason for this is that the correspondence provides a formulation of string theory in terms of quantum field theory, which is well understood by comparison. Another reason is that it provides a general framework in which physicists can study and attempt to resolve the paradoxes of black holes.[95]

In 1975, Stephen Hawking published a calculation which suggested that black holes are not completely black but emit a dim radiation due to quantum effects near the event horizon.[96] At first, Hawking's result posed a problem for theorists because it suggested that black holes destroy information. More precisely, Hawking's calculation seemed to conflict with one of the basic postulates of quantum mechanics, which states that physical systems evolve in time according to the Schrödinger equation. This property is usually referred to as unitarity of time evolution. The apparent contradiction between Hawking's calculation and the unitarity postulate of quantum mechanics came to be known as the black hole information paradox.[97]

The AdS/CFT correspondence resolves the black hole information paradox, at least to some extent, because it shows how a black hole can evolve in a manner consistent with quantum mechanics in some contexts. Indeed, one can consider black holes in the context of the AdS/CFT correspondence, and any such black hole corresponds to a configuration of particles on the boundary of anti-de Sitter space.[98] These particles obey the usual rules of quantum mechanics and in particular evolve in a unitary fashion, so the black hole must also evolve in a unitary fashion, respecting the principles of quantum mechanics.[99] In 2005, Hawking announced that the paradox had been settled in favor of information conservation by the AdS/CFT correspondence, and he suggested a concrete mechanism by which black holes might preserve information.[100]

12.4.3 Applications to quantum field theory

Main articles: AdS/QCD correspondence and AdS/CMT correspondence

In addition to its applications to theoretical problems in quantum gravity, the AdS/CFT correspondence has been applied to a variety of problems in quantum field theory. One physical system that has been studied using the AdS/CFT correspondence is the quark–gluon plasma, an exotic state of matter produced in particle accelerators. This state of matter arises for brief instants when heavy ions such as gold or lead nuclei are collided at high energies. Such collisions cause the quarks that make up atomic nuclei to deconfine at temperatures of approximately two trillion kelvins, conditions similar to those present at around 10^{-11} seconds after the Big Bang.[102]

The physics of the quark–gluon plasma is governed by a theory called quantum chromodynamics, but this theory is mathematically intractable in problems involving the quark–gluon plasma.[lower-alpha 4] In an article appearing in 2005, Đàm Thanh Sơn and his collaborators showed that the AdS/CFT correspondence could be used to understand some aspects of the quark–gluon plasma by describing it in the language of string theory.[103] By applying the AdS/CFT correspondence, Sơn and his collaborators were able to describe the quark gluon plasma in terms of black holes in five-dimensional spacetime. The calculation showed that the ratio of two quantities associated with the quark–gluon plasma, the shear viscosity and volume density of entropy, should be approximately equal to a certain universal constant. In 2008, the predicted value of this ratio for the quark–gluon plasma was confirmed at the Relativistic Heavy Ion Collider at Brookhaven National Laboratory.[104][105]

The AdS/CFT correspondence has also been used to study aspects of condensed matter physics. Over the decades, experimental condensed matter physicists have discovered a number of exotic states of matter, including superconductors and superfluids. These states are described using the formalism of quantum field theory, but some phenomena are difficult to explain using standard field theoretic techniques. Some condensed matter theorists including Subir Sachdev hope that the AdS/CFT correspondence will make it possible to describe these systems in the language of string theory and learn more about their behavior.[106]

So far some success has been achieved in using string theory methods to describe the transition of a superfluid to an insulator. A superfluid is a system of electrically neutral atoms that flows without any friction. Such systems are often produced in the laboratory using liquid helium, but recently experimentalists have developed new ways of producing artificial superfluids by pouring trillions of cold atoms into a lattice of criss-crossing lasers. These atoms initially behave as a superfluid, but as experimentalists increase the intensity of the lasers, they become less mobile and then suddenly transition to an insulating state. During the transition, the atoms behave in an unusual way. For example, the atoms slow to a halt at a rate that depends on the temperature and on Planck's constant, the fundamental parameter of quantum mechanics, which does not enter into the description of the other phases. This behavior has recently been understood by considering a dual description where properties of the fluid are described in terms of a higher dimensional black hole.[107]

12.5 Phenomenology

Main article: String phenomenology

In addition to being an idea of considerable theoretical interest, string theory provides a framework for constructing models of real world physics that combine general relativity and particle physics. Phenomenology is the branch of theoretical physics in which physicists construct realistic models of nature from more abstract theoretical ideas. String phenomenology is the part of string theory that attempts to construct realistic models based on string theory.

Partly because of theoretical and mathematical difficulties and partly because of the extremely high energies needed to test these theories experimentally, there is so far no experimental evidence that would unambiguously point to any of these models being a correct fundamental description of nature. This has led some in the community to criticize these approaches to unification and question the value of continued research on these problems.[108]

12.5.1 Particle physics

The currently accepted theory describing elementary particles and their interactions is known as the standard model of particle physics. This theory provides a unified description of three of the fundamental forces of nature: electromagnetism and the strong and weak nuclear forces. Despite its remarkable success in explaining a wide range of physical phenomena, the standard model cannot be a complete description of reality. This is because the standard model fails to incorporate the force of gravity and because of problems such as the hierarchy problem and the inability to explain the structure of fermion masses or dark matter.

String theory has been used to construct a variety of models of particle physics going beyond the standard model. Typically, such models are based on the idea of compactification. Starting with the ten- or eleven-dimensional spacetime of string or M-theory, physicists postulate a shape for the extra dimensions. By choosing this shape appropriately, they can construct models roughly similar to the standard model of particle physics, together with additional undiscovered particles.[109] One popular way of deriving realistic physics from string theory is to start with the heterotic theory in ten dimensions and assume that the six extra dimensions of spacetime are shaped like a six-dimensional Calabi–Yau manifold. Such compactifications offer many ways of extracting realistic physics from string theory. Other similar methods can be used to construct realistic models of our four-dimensional world based on M-theory.[110]

12.5.2 Cosmology

Main article: String cosmology

The Big Bang theory is the prevailing cosmological model for the universe from the earliest known periods through its subsequent large-scale evolution. Despite its success in explaining many observed features of the universe including galactic redshifts, the relative abundance of light elements such as hydrogen and helium, and the existence of a cosmic microwave background, there are several questions that remain unanswered. For example, the standard Big Bang model does not explain why the universe appears to be same in all directions, why it appears flat on very large distance scales, or why certain hypothesized particles such as magnetic monopoles are not observed in experiments.[111]

Currently, the leading candidate for a theory going beyond the Big Bang is the theory of cosmic inflation. Developed by Alan Guth and others in the 1980s, inflation postulates a period of extremely rapid accelerated expansion of the universe prior to the expansion described by the standard Big Bang theory. The theory of cosmic inflation preserves the successes of the Big Bang while providing a natural explanation for some of the mysterious features of the universe.[112] The theory has also received striking support from observations of the cosmic microwave background, the radiation that has filled the sky since around 380,000 years after the Big Bang.[113]

In the theory of inflation, the rapid initial expansion of the universe is caused by a hypothetical particle called the inflaton. The exact properties of this particle are not fixed by the theory but should ultimately be derived from a more fundamental theory such as string theory.[114] Indeed, there have been a number of attempts to identify an inflation within the spectrum of particles described by string theory and to study inflation using string theory. While these approaches might eventually find support in observational data such as measurements of the cosmic microwave background, the application of string theory to cosmology is still in its early stages.[115]

12.6 Connections to mathematics

In addition to influencing research in theoretical physics, string theory has stimulated a number of major developments in pure mathematics. Like many developing ideas in theoretical physics, string theory does not at present have a mathematically rigorous formulation in which all of its concepts can be defined precisely. As a result, physicists who study string theory are often guided by physical intuition to conjecture relationships between the seemingly different mathematical structures that are used to formalize different parts of the theory. These conjectures are later proved by mathematicians, and in this way, string theory serves as a source of new ideas in pure mathematics.[116]

12.6.1 Mirror symmetry

Main article: Mirror symmetry (string theory)

After Calabi–Yau manifolds had entered physics as a way to compactify extra dimensions in string theory, many physicists began studying these manifolds. In the late 1980s, several physicists noticed that given such a compactification of string theory, it is not possible to reconstruct uniquely a corresponding Calabi–Yau manifold.[117] Instead, two different versions of string theory, type IIA and type IIB, can be compactified on completely different Calabi–Yau manifolds giving rise to the same physics. In this situation, the manifolds are called mirror manifolds, and the relationship between the two physical theories is called mirror symmetry.[118]

Regardless of whether Calabi–Yau compactifications of string theory provide a correct description of nature, the existence of the mirror duality between different string theories has significant mathematical consequences. The Calabi–Yau manifolds used in string theory are of interest in pure mathematics, and mirror symmetry allows mathematicians to solve problems in enumerative geometry, a branch of mathematics concerned with counting the numbers of solutions to geometric questions.[119][120]

Enumerative geometry studies a class of geometric objects called algebraic varieties which are defined by the vanishing of polynomials. For example, the Clebsch cubic illustrated on the right is an algebraic variety defined using a certain polynomial of degree three in four variables. A celebrated result of nineteenth-century mathematicians Arthur Cayley and George Salmon states that there are exactly 27 straight lines that lie entirely on such a surface.[121]

Generalizing this problem, one can ask how many lines can be drawn on a quintic Calabi–Yau manifold, such as the one illustrated above, which is defined by a polynomial of degree five. This problem was solved by the nineteenth-century German mathematician Hermann Schubert, who found that there are exactly 2,875 such lines. In 1986, geometer Sheldon Katz proved that the number of curves, such as circles, that are defined by polynomials of degree two and lie entirely in the quintic is 609,250.[122]

By the year 1991, most of the classical problems of enumerative geometry had been solved and interest in enumerative geometry had begun to diminish.[123] The field was reinvigorated in May 1991 when physicists Philip Candelas, Xenia de la Ossa, Paul Green, and Linda Parks showed that mirror symmetry could be used to translate difficult mathematical questions about one Calabi–Yau manifold into easier questions about its mirror.[124] In particular, they used mirror symmetry to show that a six-dimensional Calabi–Yau manifold can contain exactly 317,206,375 curves of degree three.[125] In addition to counting degree-three curves, Candelas and his collaborators obtained a number of more general results for counting rational curves which went far beyond the results obtained by mathematicians.[126]

Originally, these results of Candelas were justified on physical grounds. However, mathematicians generally prefer rigorous proofs that do not require an appeal to physical intuition. Inspired by physicists' work on mirror symmetry, mathematicians have therefore constructed their own arguments proving the enumerative predictions of mirror symmetry.[lower-alpha 5] Today mirror symmetry is an active area of research in mathematics, and mathematicians are working to develop a more complete mathematical understanding of mirror symmetry based on physicists' intuition.[127] Major approaches to mirror symmetry include the homological mirror symmetry program of Maxim Kontsevich[128] and the SYZ conjecture of Andrew Strominger, Shing-Tung Yau, and Eric Zaslow.[129]

12.6.2 Monstrous moonshine

Main article: Monstrous moonshine

Group theory is the branch of mathematics that studies the concept of symmetry. For example, one can consider a geometric shape such as an equilateral triangle. There are various operations that one can perform on this triangle without changing its shape. One can rotate it through $120°$, $240°$, or $360°$, or one can reflect in any of the lines labeled S_0, S_1, or S_2 in the picture. Each of these operations is called a *symmetry*, and the collection of these symmetries satisfies certain technical properties making it into what mathematicians call a group. In this particular example, the group is known as the dihedral group of order 6 because it has six elements. A general group may describe finitely many or infinitely many symmetries; if there are only finitely many symmetries, it is called a finite group.[130]

Mathematicians often strive for a classification (or list) of all mathematical objects of a given type. It is generally believed that finite groups are too diverse to admit a useful classification. A more modest but still challenging problem is to classify all finite *simple* groups. These are finite groups which may be used as building blocks for constructing arbitrary finite groups in the same way that prime numbers can be used to construct arbitrary whole numbers by taking products.[lower-alpha 6] One of the major achievements of contemporary group theory is the classification of finite simple groups, a mathematical theorem which provides a list of all possible finite simple groups.[131]

This classification theorem identifies several infinite families of groups as well as 26 additional groups which do not fit into any family. The latter groups are called the "sporadic" groups, and each one owes its existence to a remarkable combination of circumstances. The largest sporadic group, the so called monster group, has over 10^{53} elements, more than a thousand times the number of atoms in the Earth.[132]

A seemingly unrelated construction is the j-function of number theory. This object belongs to a special class of functions called modular functions, whose graphs form a certain kind of repeating pattern.[133] Although this function appears in a branch of mathematics which seems very different from the theory of finite groups, the two subjects turn out to be intimately related. In the late 1970s, mathematicians John McKay and John Thompson noticed that certain numbers arising in the analysis of the monster group (namely, the dimensions of its irreducible representations) are related to numbers that appear in a formula for the j-function (namely, the coefficients of its Fourier series).[134] This relationship was further developed by John Horton Conway and Simon Norton[135] who called it monstrous moonshine because it seemed so far fetched.[136]

In 1992, Richard Borcherds constructed a bridge between the theory of modular functions and finite groups and, in the process, explained the observations of McKay and Thompson.[137][138] Borcherds' work used ideas from string theory in an essential way, extending earlier results of Igor Frenkel, James Lepowsky, and Arne Meurman, who had realized the monster group as the symmetries of a particular version of string theory.[139] In 1998, Borcherds was awarded the Fields medal for his work.[140]

Since the 1990s, the connection between string theory and moonshine has led to further results in mathematics and physics.[141] In 2010, physicists Tohru Eguchi, Hirosi Ooguri, and Yuji Tachikawa discovered connections between a different sporadic group, the Mathieu group M_{24}, and a certain version of string theory.[142] Miranda Cheng, John Duncan, and Jeffrey A. Harvey proposed a generalization of this moonshine phenomenon called umbral moonshine,[143] and their conjecture was proved mathematically by Duncan, Michael Griffin, and Ken Ono.[144] Witten has also speculated that the version of string theory appearing in monstrous moonshine might be related to a certain simplified model of gravity in three spacetime dimensions.[145]

12.7 History

Main article: History of string theory

12.7.1 Early results

Some of the structures reintroduced by string theory arose for the first time much earlier as part of the program of classical unification started by Albert Einstein. The first person to add a fifth dimension to a theory of gravity was Gunnar Nordström in 1914, who noted that gravity in five dimensions describes both gravity and electromagnetism in four. Nordström attempted to unify electromagnetism with his theory of gravitation, which was however superseded by Einstein's general relativity in 1919. Thereafter, German mathematician Theodor Kaluza combined the fifth dimension with general relativity, and only Kaluza is usually credited with the idea. In 1926, the Swedish physicist Oskar Klein gave a physical interpretation of the unobservable extra dimension—it is wrapped into a small circle. Einstein introduced a non-symmetric metric tensor, while much later Brans and Dicke added a scalar component to gravity. These ideas would be revived within string theory, where they are demanded by consistency conditions.

String theory was originally developed during the late 1960s and early 1970s as a never completely successful theory of hadrons, the subatomic particles like the proton and neutron that feel the strong interaction. In the 1960s, Geoffrey Chew and Steven Frautschi discovered that the mesons make families called Regge trajectories with masses related to spins in a way that was later understood by Yoichiro Nambu, Holger Bech Nielsen and Leonard Susskind to be the relationship expected from rotating strings. Chew advocated making a theory for the interactions of these trajectories that did not presume that they were composed of any fundamental particles, but would construct their interactions from self-consistency conditions on the S-matrix. The S-matrix approach was started by Werner Heisenberg in the 1940s as a way of constructing a theory that did not rely on the local notions of space and time, which Heisenberg believed break down at the nuclear scale. While the scale was off by many orders of magnitude, the approach he advocated was ideally suited for a theory of quantum gravity.

Working with experimental data, R. Dolen, D. Horn and C. Schmid developed some sum rules for hadron exchange. When a particle and antiparticle scatter, virtual particles can be exchanged in two qualitatively different ways. In the s-channel, the two particles annihilate to make temporary intermediate states that fall apart into the final state particles. In the t-channel, the particles exchange intermediate states by emission and absorption. In field theory, the two contributions add together, one giving a continuous background contribution, the other giving peaks at certain energies. In the data, it was clear that the peaks were stealing from the background—the authors interpreted this as saying that the t-channel contribution was dual to the s-channel one, meaning both described the whole amplitude and included the other.

The result was widely advertised by Murray Gell-Mann, leading Gabriele Veneziano to construct a scattering amplitude that had the property of Dolen-Horn-Schmid duality, later renamed world-sheet duality. The amplitude needed poles where the particles appear, on straight line trajectories, and there is a special mathematical function whose poles are evenly spaced on half the real line— the Gamma function— which was widely used in Regge theory. By manipulating combinations of Gamma functions, Veneziano was able to find a consistent scattering amplitude with poles on straight lines, with mostly positive residues, which obeyed duality and had the appropriate Regge scaling at high energy. The amplitude could fit near-beam scattering data as well as other Regge type fits, and had a suggestive integral representation that could be used for generalization.

Over the next years, hundreds of physicists worked to complete the bootstrap program for this model, with many surprises. Veneziano himself discovered that for the scattering amplitude to describe the scattering of a particle that appears in the theory, an obvious self-consistency condition, the lightest particle must be a tachyon. Miguel Virasoro and Joel Shapiro found a different amplitude now understood to be that of closed strings, while Ziro Koba and Holger Nielsen generalized Veneziano's integral representation to multiparticle scattering. Veneziano and Sergio Fubini introduced an operator formalism for computing the scattering amplitudes that was a forerunner of world-sheet conformal theory, while Virasoro understood how to remove the poles with wrong-sign residues using a constraint on the states. Claud Lovelace calculated a loop amplitude, and noted that there is an inconsistency unless the dimension of the theory is 26. Charles Thorn, Peter Goddard and Richard Brower went on to prove that there are no wrong-sign propagating states in dimensions less than or equal to 26.

In 1969, Yoichiro Nambu, Holger Bech Nielsen, and Leonard Susskind recognized that the theory could be given a description in space and time in terms of strings. The scattering amplitudes were derived systematically from the action principle by Peter Goddard, Jeffrey Goldstone, Claudio Rebbi, and Charles Thorn, giving a space-time picture to the vertex operators introduced by Veneziano and Fubini and a geometrical interpretation to the Virasoro conditions.

In 1970, Pierre Ramond added fermions to the model, which led him to formulate a two-dimensional supersymmetry to cancel the wrong-sign states. John Schwarz and André Neveu added another sector to the fermi theory a short time later. In the fermion theories, the critical dimension was 10. Stanley Mandelstam formulated a world sheet conformal theory for both the bose and fermi case, giving a two-dimensional field theoretic path-integral to generate the operator formalism. Michio Kaku and Keiji Kikkawa gave a different formulation of the bosonic string, as a string field theory, with infinitely many particle types and with fields taking values not on points, but on loops and curves.

In 1974, Tamiaki Yoneya discovered that all the known string theories included a massless spin-two particle that obeyed the correct Ward identities to be a graviton. John Schwarz and Joel Scherk came to the same conclusion and made the bold leap to suggest that string theory was a theory of gravity, not a theory of hadrons. They reintroduced Kaluza–Klein theory as a way of making sense of the extra dimensions. At the same time, quantum chromodynamics was recognized as the correct theory of hadrons, shifting the attention of physicists and apparently leaving the bootstrap program in the dustbin of history.

String theory eventually made it out of the dustbin, but for the following decade all work on the theory was completely ignored. Still, the theory continued to develop at a steady pace thanks to the work of a handful of devotees. Ferdinando Gliozzi, Joel Scherk, and David Olive realized in 1976 that the original Ramond and Neveu Schwarz-strings were separately inconsistent and needed to be combined. The resulting theory did not have a tachyon, and was proven to have space-time supersymmetry by John Schwarz and Michael Green in 1981. The same year, Alexander Polyakov gave the theory a modern path integral formulation, and went on to develop conformal field theory extensively. In 1979, Daniel Friedan showed that the equations of motions of string theory, which are generalizations of the Einstein equations of General Relativity, emerge from the Renormalization group equations for the two-dimensional field theory. Schwarz and Green discovered T-duality, and constructed two superstring theories—IIA and IIB related by T-duality, and type I theories with open strings. The consistency conditions had been so strong, that the entire theory was nearly uniquely determined, with only a few discrete choices.

12.7.2 First superstring revolution

In the early 1980s, Edward Witten discovered that most theories of quantum gravity could not accommodate chiral fermions like the neutrino. This led him, in collaboration with Luis Alvarez-Gaumé to study violations of the conservation laws in gravity theories with anomalies, concluding that type I string theories were inconsistent. Green and Schwarz discovered a contribution to the anomaly that Witten and Alvarez-Gaumé had missed, which restricted the gauge group of the type I string theory to be SO(32). In coming to understand this calculation, Edward Witten became convinced that string theory was truly a consistent theory of gravity, and he became a high-profile advocate. Following Witten's lead, between 1984 and 1986, hundreds of physicists started to work in this field, and this is sometimes called the first superstring revolution.

During this period, David Gross, Jeffrey Harvey, Emil Martinec, and Ryan Rohm discovered heterotic strings. The gauge group of these closed strings was two copies of E8, and either copy could easily and naturally include the standard model. Philip Candelas, Gary Horowitz, Andrew Strominger and Edward Witten found that the Calabi–Yau manifolds are the compactifications that preserve a realistic amount of supersymmetry, while Lance Dixon and others worked out the physical properties of orbifolds, distinctive geometrical singularities allowed in string theory. Cumrun Vafa generalized T-duality from circles to arbitrary manifolds, creating the mathematical field of mirror symmetry. Daniel Friedan, Emil Martinec and Stephen Shenker further developed the covariant quantization of the superstring using conformal field theory techniques. David Gross and Vipul Periwal discovered that string perturbation theory was divergent. Stephen Shenker showed it diverged much faster than in field theory suggesting that new non-perturbative objects were missing.

In the 1990s, Joseph Polchinski discovered that the theory requires higher-dimensional objects, called D-branes and identified these with the black-hole solutions of supergravity. These were understood to be the new objects suggested by the perturbative divergences, and they opened up a new field with rich mathematical structure. It quickly became clear that D-branes and other p-branes, not just strings, formed the matter content of the string theories, and the physical interpretation of the strings and branes was revealed—they are a type of black hole. Leonard Susskind had incorporated the holographic principle of Gerardus 't Hooft into string theory, identifying the long highly excited string states with ordinary thermal black hole states. As suggested by 't Hooft, the fluctuations of the black hole horizon, the world-sheet or world-volume theory, describes not only the degrees of freedom of the black hole, but all nearby objects too.

12.7.3   Second superstring revolution

In 1995, at the annual conference of string theorists at the University of Southern California (USC), Edward Witten gave a speech on string theory that in essence united the five string theories that existed at the time, and giving birth to a new 11-dimensional theory called M-theory. M-theory was also foreshadowed in the work of Paul Townsend at approximately the same time. The flurry of activity that began at this time is sometimes called the second superstring revolution.[146]

During this period, Tom Banks, Willy Fischler, Stephen Shenker and Leonard Susskind formulated matrix theory, a full holographic description of M-theory using IIA D0 branes.[147] This was the first definition of string theory that was fully non-perturbative and a concrete mathematical realization of the holographic principle. It is an example of a gauge-gravity duality and is now understood to be a special case of the AdS/CFT correspondence. Andrew Strominger and Cumrun Vafa calculated the entropy of certain configurations of D-branes and found agreement with the semi-classical answer for extreme charged black holes.[148] Petr Hořava and Witten found the eleven-dimensional formulation of the heterotic string theories, showing that orbifolds solve the chirality problem. Witten noted that the effective description of the physics of D-branes at low energies is by a supersymmetric gauge theory, and found geometrical interpretations of mathematical structures in gauge theory that he and Nathan Seiberg had earlier discovered in terms of the location of the branes.

In 1997, Juan Maldacena noted that the low energy excitations of a theory near a black hole consist of objects close to the horizon, which for extreme charged black holes looks like an anti-de Sitter space.[149] He noted that in this limit the gauge theory describes the string excitations near the branes. So he hypothesized that string theory on a near-horizon extreme-charged black-hole geometry, an anti-deSitter space times a sphere with flux, is equally well described by the low-energy limiting gauge theory, the N = 4 supersymmetric Yang–Mills theory. This hypothesis, which is called the AdS/CFT correspondence, was further developed by Steven Gubser, Igor Klebanov and Alexander Polyakov,[150] and by Edward Witten,[151] and it is now well-accepted. It is a concrete realization of the holographic principle, which has far-reaching implications for black holes, locality and information in physics, as well as the nature of the gravitational interaction.[152] Through this relationship, string theory has been shown to be related to gauge theories like quantum chromodynamics and this has led to more quantitative understanding of the behavior of hadrons, bringing string theory back to its roots.[153]

12.8   Criticism

12.8.1   Number of solutions

To construct models of particle physics based on string theory, physicists typically begin by specifying a shape for the extra dimensions of spacetime. Each of these different shapes corresponds to a different possible universe, or "vacuum state", with a different collection of particles and forces. String theory as it is currently understood has an enormous number of vacuum states, typically estimated to be around 10^{500}, and these might be sufficiently diverse to accommodate almost any phenomena that might be observed at low energies.[154]

Many critics of string theory have expressed concerns about the large number of possible universes described by string theory. In his book *Not Even Wrong*, Peter Woit, a lecturer in the mathematics department at Columbia University, has argued that the large number of different physical scenarios renders string theory vacuous as a framework for constructing models of particle physics. According to Woit,

The possible existence of, say, 10^{500} consistent different vacuum states for superstring theory probably destroys the hope of using the theory to predict anything. If one picks among this large set just those states whose properties agree with present experimental observations, it is likely there still will be such a large number of these that one can get just about whatever value one wants for the results of any new observation.[155]

Some physicists believe this large number of solutions is actually a virtue because it may allow a natural anthropic explanation of the observed values of physical constants, in particular the small value of the cosmological constant.[156] The anthropic principle is the idea that some of the numbers appearing in the laws of physics are not fixed by any fundamental principle but must be compatible with the evolution of intelligent life. In 1987, Steven Weinberg published an article in which he argued that the cosmological constant could not have been too large, or else galaxies and intelligent life would not have been able to develop.[157] Weinberg suggested that there might be a huge number of possible consistent universes, each with a different value of the cosmological constant, and observations indicate a small value of the cosmological constant only because humans happen to live in a universe that has allowed intelligent life, and hence observers, to exist.[158]

String theorist Leonard Susskind has argued that string theory provides a natural anthropic explanation of the small value of the cosmological constant.[159] According to Susskind, the different vacuum states of string theory might be realized as different universes within a larger multiverse. The fact that the observed universe has a small cosmological constant is just a tautological consequence of the fact that a small value is required for life to exist.[160] Many prominent theorists and critics have disagreed with Susskind's conclusions.[161] According to Woit, "in this case [anthropic reasoning] is nothing more than an excuse for failure. Speculative scientific ideas fail not just when they make incorrect predictions, but also when they turn out to be vacuous and incapable of predicting anything."[162]

12.8.2 Background independence

Main article: Background independence

One of the fundamental principles of Einstein's general theory of relativity is the idea that the laws of physics should be background independent. This means that the geometry of spacetime is not specified from the outset but is instead determined dynamically by the theory. In general relativity, the geometry of spacetime can evolve in time, responding to whatever matter is present.[163]

One of the older criticisms of string theory is that it is not manifestly background independent. In string theory, one must typically specify a fixed reference geometry for spacetime, and all other possible geometries are described as perturbations of this fixed one. In his book *The Trouble With Physics*, physicist Lee Smolin of the Perimeter Institute for Theoretical Physics claims that this is the principal weakness of string theory as a theory of quantum gravity, saying that string theory has failed to incorporate this important insight from general relativity.[164]

Others have disagreed with Smolin's characterization of string theory. In a review of Smolin's book, string theorist Joseph Polchinski writes

> [Smolin] is mistaking an aspect of the mathematical language being used for one of the physics being described. New physical theories are often discovered using a mathematical language that is not the most suitable for them... In string theory it has always been clear that the physics is background-independent even if the language being used is not, and the search for more suitable language continues. Indeed, as Smolin belatedly notes, [AdS/CFT] provides a solution to this problem, one that is unexpected and powerful.[165]

Polchinski notes that an important open problem in quantum gravity is to develop holographic descriptions of gravity which do not require the gravitational field to be asymptotically anti-de Sitter.[166]

12.8.3 Sociological issues

Since the superstring revolutions of the 1980s and 1990s, string theory has become the dominant paradigm of high energy theoretical physics.[167] Some string theorists have expressed the view that there does not exist an equally successful alternative theory addressing the deep questions of fundamental physics. In an interview from 1987, Nobel laureate David Gross made the following controversial comments about the reasons for the popularity of string theory:

> The most important [reason] is that there are no other good ideas around. That's what gets most people into it. When people started to get interested in string theory they didn't know anything about it. In fact, the first reaction of most people is that the theory is extremely ugly and unpleasant, at least that was the case a few years ago when the understanding of string theory was much less developed. It was difficult for people to learn about it and to be turned on. So I think the real reason why people have got attracted by it is because there is no other game in town. All other approaches of constructing grand unified theories, which were more conservative to begin with, and only gradually became more and more radical, have failed, and this game hasn't failed yet.[168]

Several other high profile theorists and commentators have expressed similar views, suggesting that there are no viable alternatives to string theory.[169]

Many critics of string theory have commented on this state of affairs. In his book criticizing string theory, Peter Woit views the status of string theory research as unhealthy and detrimental to the future of fundamental physics. He argues that the extreme popularity of string theory among theoretical physicists is partly a consequence of the financial structure of academia and the fierce competition for scarce resources.[170] In his book *The Road to Reality*, mathematical physicist Roger Penrose expresses similar views, stating "The often frantic competitiveness that this ease of communication engenders leads to 'bandwagon' effects, where researchers fear to be left behind if they do not join in."[171] Penrose also claims that the technical difficulty of modern physics forces young scientists to rely on the preferences of established researchers, rather than forging new paths of their own.[172] Lee Smolin expresses a slightly different position in his critique, claiming that string theory grew out of a tradition of particle physics which discourages speculation about the foundations of physics, while his preferred approach, loop quantum gravity, encourages more radical thinking. According to Smolin,

> String theory is a powerful, well-motivated idea and deserves much of the work that has been devoted to it. If it has so far failed, the principal reason is that its intrinsic flaws are closely tied to its strengths—and, of course, the story is unfinished, since string theory may well turn out to be part of the truth. The real question is not why we have expended so much energy on string theory but why we haven't expended nearly enough on alternative approaches.[173]

Smolin goes on to offer a number of prescriptions for how scientists might encourage a greater diversity of approaches to quantum gravity research.[174]

12.9   References

12.9.1   Notes

[1] For example, physicists are still working to understand the phenomenon of quark confinement, the paradoxes of black holes, and the origin of dark energy.

[2] For example, in the context of the AdS/CFT correspondence, theorists often formulate and study theories of gravity in unphysical numbers of spacetime dimensions.

[3] "Top Cited Articles during 2010 in hep-th". Retrieved 25 July 2013.

[4] More precisely, one cannot apply the methods of perturbative quantum field theory.

[5] Two independent mathematical proofs of mirror symmetry were given by Givental 1996, 1998 and Lian, Liu, Yau 1997, 1999, 2000.

[6] More precisely, a nontrivial group is called *simple* if its only normal subgroups are the trivial group and the group itself. The Jordan–Hölder theorem exhibits finite simple groups as the building blocks for all finite groups.

12.9.2 Citations

[1] Becker, Becker, and Schwarz 2007, p. 1

[2] Becker, Becker, and Schwarz 2007, p. 1

[3] Zwiebach 2009, p. 6

[4] Becker, Becker, and Schwarz 2007, pp. 2–3

[5] Becker, Becker, and Schwarz 2007, pp. 9–12

[6] Becker, Becker, and Schwarz 2007, pp. 14–15

[7] Klebanov and Maldacena 2009

[8] Merali 2011

[9] Sachdev 2013

[10] Becker, Becker, and Schwarz 2007, pp. 3, 15–16

[11] Becker, Becker, and Schwarz 2007, p. 8

[12] Becker, Becker, and Schwarz 13–14

[13] Woit 2006

[14] Zee 2010

[15] Zee 2010

[16] Becker, Becker, and Schwarz 2007, p. 2

[17] Becker, Becker, and Schwarz 2007, p. 6

[18] Zwiebach 2009, p. 12

[19] Becker, Becker, and Schwarz 2007, p. 6

[20] Becker, Becker, and Schwarz 2007, pp. 2–3

[21] Becker, Becker, and Schwarz 2007, p. 4

[22] Zwiebach 2009, p. 324

[23] Wald 1984, p. 4

[24] Zee 2010, Parts V and VI

[25] Zwiebach 2009, p. 9

[26] Zwiebach 2009, p. 8

[27] Yau and Nadis 2010, Ch. 6

[28] Greene 2000, p. 186

[29] Yau and Nadis 2010, Ch. 6

[30] Yau and Nadis 2010, p. ix

[31] Randall and Sundrum 1999

[32] Becker, Becker, and Schwarz 2007

[33] Becker, Becker, and Schwarz 2007

[34] Zwiebach 2009, p. 376

[35] Moore 2005, p. 214

[36] Moore 2005, p. 214

[37] Moore 2005, p. 215

[38] Aspinwall et al. 2009

[39] Kontsevich 1995

[40] Kapustin and Witten 2007

[41] Duff 1998

[42] Duff 1998, p. 64

[43] Nahm 1978

[44] Cremmer, Julia, and Scherk 1978

[45] Duff 1998, p. 65

[46] Duff 1998, p. 65

[47] Duff 1998, p. 65

[48] Duff 1998, p. 65

[49] Duff 1998, p. 65

[50] Sen 1994a

[51] Sen 1994b

[52] Hull and Townsend 1995

[53] Duff 1998, p. 67

[54] Bergshoeff, Sezgin, and Townsend 1987

[55] Duff et al. 1987

[56] Duff 1998, p. 66

[57] Witten 1995

[58] Duff 1998, pp. 67–68

[59] Becker, Becker, and Schwarz 2007, p. 296

[60] Hořava and Witten 1996

[61] Duff 1996, sec. 1

[62] Banks et al. 1997

[63] Banks et al. 1997

[64] Connes 1994

[65] Connes, Douglas, and Schwarz 1998

[66] Nekrasov and Schwarz 1998

[67] Seiberg and Witten 1999

[68] de Haro et al. 2013, p. 2

[69] Yau and Nadis 2010, p. 187–188

[70] Bekenstein 1973

[71] Hawking 1975

[72] Wald 1984, p. 417

[73] Yau and Nadis 2010, p. 189

[74] Strominger and Vafa 1996

[75] Yau and Nadis 2010, pp. 190–192

[76] Maldacena, Strominger, and Witten 1997

[77] Ooguri, Strominger, and Vafa 2004

[78] Yau and Nadis 2010, pp. 192–193

[79] Yau and Nadis 2010, pp. 194–195

[80] Strominger 1998

[81] Guica et al. 2009

[82] Castro, Maloney, and Strominger 2010

[83] Klebanov and Maldacena 2009

[84] Maldacena 1998

[85] Gubser, Klebanov, and Polyakov 1998

[86] Witten 1998

[87] Klebanov and Maldacena 2009, p. 28

[88] Maldacena 2005, p. 60

[89] Maldacena 2005, p. 61

[90] Maldacena 2005, p. 60

[91] Maldacena 2005, p. 61

[92] Maldacena 2005, p. 60

[93] Zwiebach 2009, p. 552

[94] Maldacena 2005, pp. 61–62

[95] de Haro et al. 2013, p. 2

[96] Hawking 1975

[97] Susskind 2008

[98] Zwiebach 2009, p. 554

[99] Maldacena 2005, p. 63

[100] Hawking 2005

[101] Merali 2011

[102] Zwiebach 2009, p. 559

[103] Kovtun, Son, and Starinets 2001

[104] Merali 2011, p. 303

[105] Luzum and Romatschke 2008

[106] Merali 2011, p. 303

[107] Sachdev 2013, p. 51

[108] Woit 2006

[109] Candelas et al. 1985

[110] Yau and Nadis 2010, pp. 147–150

[111] Becker, Becker, and Schwarz 2007, pp. 530–531

[112] Becker, Becker, and Schwarz 2007, p. 531

[113] Becker, Becker, and Schwarz 2007, p. 538

[114] Becker, Becker, and Schwarz 2007, p. 533

[115] Becker, Becker, and Schwarz 2007, pp. 539–543

[116] Deligne et al. 1999, p. 1

[117] Hori et al. 2003, p. xvii

[118] Aspinwall et al. 2009, p. 13

[119] Hori et al. 2003

[120] Aspinwall et al. 2009

[121] Yau and Nadis 2010, p. 167

[122] Yau and Nadis 2010, p. 166

[123] Yau and Nadis 2010, p. 169

[124] Candelas et al. 1991

[125] Yau and Nadis 2010, p. 169

[126] Yau and Nadis 2010, p. 171

[127] Hori et al. 2003, p. xix

[128] Kontsevich 1995

[129] Strominger, Yau, and Zaslow 1996

[130] Dummit and Foote 2004

[131] Dummit and Foote 2004, pp. 102–103

[132] Klarreich 2015

[133] Gannon 2006, p. 2

[134] Gannon 2006, p. 4

[135] Conway and Norton 1979

[136] Gannon 2006, p. 5

[137] Gannon 2006, p. 8

[138] Borcherds 1992

[139] Frenkel, Lepowsky, and Meurman 1988

[140] Gannon 2006, p. 11

[141] Klarreich 2015

[142] Eguchi, Ooguri, and Tachikawa 2010

[143] Cheng, Duncan, and Harvey 2013

[144] Duncan, Griffin, and Ono 2015

[145] Witten 2007

[146] Duff 1998

[147] Banks et al. 1997

[148] Strominger and Vafa 1996

[149] Maldacena 1998

[150] Gubser, Klebanov, and Polyakov 1998

[151] Witten 1998

[152] de Haro et al. 2013, p. 2

[153] Kovtun, Son, and Starinets 2001

[154] Woit 2006, pp. 240–242

[155] Woit 2006, p. 242

[156] Woit 2006, p. 242

[157] Weinberg 1987

[158] Woit 2006, p. 243

[159] Susskind 2005

[160] Woit 2006, pp. 242–243

[161] Woit 2006, p. 240

[162] Woit 2006, p. 249

[163] Smolin 2006, p. 81

[164] Smolin 2006, p. 184

[165] Polchinski 2007

[166] Polchinski 2007

[167] Penrose 2004, p. 1017

[168] Woit 2006, pp. 224–225

[169] Woit 2006, Ch. 16

[170] Woit 2006, p. 239

[171] Penrose 2004, p. 1018

[172] Penrose 2004, pp. 1019–1020

[173] Smolin 2006, p. 349

[174] Smolin 2006, Ch. 20

12.9.3   Bibliography

- Aspinwall, Paul; Bridgeland, Tom; Craw, Alastair; Douglas, Michael; Gross, Mark; Kapustin, Anton; Moore, Gregory; Segal, Graeme; Szendröi, Balázs; Wilson, P.M.H., eds. (2009). *Dirichlet Branes and Mirror Symmetry*. American Mathematical Society. ISBN 978-0-8218-3848-8.

- Banks, Tom; Fischler, Willy; Schenker, Stephen; Susskind, Leonard (1997). "M theory as a matrix model: A conjecture". *Physical Review D* **55** (8): 5112. arXiv:hep-th/9610043. Bibcode:1997PhRvD..55.5112B. doi:10.1.

- Becker, Katrin; Becker, Melanie; Schwarz, John (2007). *String theory and M-theory: A modern introduction*. Cambridge University Press. ISBN 978-0-521-86069-7.

- Bekenstein, Jacob (1973). "Black holes and entropy". *Physical Review D* **7** (8): 2333. Bibcode:1973PhRvD...7.2333B. doi:10.1103/PhysRevD.7.2333.

- Bergshoeff, Eric; Sezgin, Ergin; Townsend, Paul (1987). "Supermembranes and eleven-dimensional supergravity". *Physics Letters B* **189** (1): 75–78. Bibcode:1987PhLB..189...75B. doi:10.1016/0370-2693(87)91272-X.

- Borcherds, Richard (1992). "Monstrous moonshine and Lie superalgebras". *Inventiones mathematicae* **109** (1): 405–444. Bibcode:1992InMat.109..405B. doi:10.1007/BF01232032.

- Candelas, Philip; de la Ossa, Xenia; Green, Paul; Parks, Linda (1991). "A pair of Calabi–Yau manifolds as an exactly soluble superconformal field theory". *Nuclear Physics B* **359** (1): 21–74. Bibcode:1991NuPhB.359...21C. doi:10.1016/0550-3213(91)90292-6.

- Candelas, Philip; Horowitz, Gary; Strominger, Andrew; Witten, Edward (1985). "Vacuum configurations for superstrings". *Nuclear Physics B* **258**: 46–74. Bibcode:1985NuPhB.258...46C. doi:10.1016/0550-3213(85)90602-9.

- Castro, Alejandra; Maloney, Alexander; Strominger, Andrew (2010). "Hidden conformal symmetry of the Kerr black hole". *Physical Review D* **82** (2). arXiv:1004.0996. Bibcode:2010PhRvD..82b4008C. doi:10.1103/PhysRe.

- Cheng, Miranda; Duncan, John; Harvey, Jeffrey (2013). "Umbral Moonshine". arXiv:1204.2779.

- Connes, Alain (1994). *Noncommutative Geometry*. Academic Press. ISBN 978-0-12-185860-5.

- Connes, Alain; Douglas, Michael; Schwarz, Albert (1998). "Noncommutative geometry and matrix theory". *Journal of High Energy Physics*. 19981 (2): 003. arXiv:hep-th/9711162.Bibcode:1998JHEP...02..003C.doi:10.106708/1998/02/003.

- Conway, John; Norton, Simon (1979). "Monstrous moonshine". *Bull. London. Math. Soc.* **11** (3): 308–339.

- Cremmer, Eugene; Julia, Bernard; Scherk, Joel (1978). "Supergravity theory in eleven dimensions". *Physics Letters B* **76** (4): 409–412. Bibcode:1978PhLB...76..409C. doi:10.1016/0370-2693(78)90894-8.

- de Haro, Sebastian; Dieks, Dennis; 't Hooft, Gerard; Verlinde, Erik (2013). "Forty Years of String Theory Reflecting on the Foundations". *Foundations of Physics* **43** (1): 1–7. Bibcode:2013FoPh...43....1D. doi:10.1007/s10701-012-9691-3.

- Deligne, Pierre; Etingof, Pavel; Freed, Daniel; Jeffery, Lisa; Kazhdan, David; Morgan, John; Morrison, David; Witten, Edward, eds. (1999). *Quantum Fields and Strings: A Course for Mathematicians* **1**. American Mathematical Society. ISBN 978-0821820124.

- Duff, Michael (1996). "M-theory (the theory formerly known as strings)". *International Journal of Modern Physics A* **11** (32): 6523–41. arXiv:hep-th/9608117. Bibcode:1996IJMPA..11.5623D. doi:10.1142/S0217751X96002583.

- Duff, Michael (1998). "The theory formerly known as strings".*Scientific American***278**(2): 64–9.doi:10.1038/98-64.

- Duff, Michael; Howe, Paul; Inami, Takeo; Stelle, Kellogg (1987). "Superstrings in D=10 from supermembranes in D=11". *Nuclear Physics B* **191** (1): 70–74. Bibcode:1987PhLB..191...70D. doi:10.1016/0370-2693(87)91323-2.

- Dummit, David; Foote, Richard (2004). *Abstract Algebra*. Wiley. ISBN 978-0-471-43334-7.

- Duncan, John; Griffin, Michael; Ono, Ken (2015). "Proof of the Umbral Moonshine Conjecture". arXiv:1503.01472.

- Eguchi, Tohru; Ooguri, Hirosi; Tachikawa, Yuji (2011). "Notes on the K3 surface and the Mathieu group M_{24}". *Experimental Mathematics* **20** (1): 91–96. doi:10.1080/10586458.2011.544585.

- Frenkel, Igor; Lepowsky, James; Meurman, Arne (1988). *Vertex Operator Algebras and the Monster*. Pure and Applied Mathematics **134**. Academic Press. ISBN 0-12-267065-5.

- Gannon, Terry. *Moonshine Beyond the Monster: The Bridge Connecting Algebra, Modular Forms, and Physics*. Cambridge University Press.

- Givental, Alexander (1996). "Equivariant Gromov-Witten invariants". *International Mathematics Research Notices* **1996** (13): 613–663. doi:10.1155/S1073792896000414.

- Givental, Alexander (1998). "A mirror theorem for toric complete intersections". *Topological field theory, primitive forms and related topics*: 141–175. doi:10.1007/978-1-4612-0705-4_5. ISBN 978-1-4612-6874-1.

- Gubser, Steven; Klebanov, Igor; Polyakov, Alexander (1998). "Gauge theory correlators from non-critical string theory". *Physics Letters B* **428**: 105–114. arXiv:hep-th/9802109.Bibcode:1998PhLB..428..105G.doi:10.101-2693(98)00377-3.

- Guica, Monica; Hartman, Thomas; Song, Wei; Strominger, Andrew (2009). "The Kerr/CFT Correspondence". *Physical Review D* **80** (12). arXiv:0809.4266. Bibcode:2009PhRvD..80l4008G. doi:10.1103/PhysRevD.80.1240.

- Hawking, Stephen (1975). "Particle creation by black holes". *Communications in mathematical physics* **43** (3): 199–220. Bibcode:1975CMaPh..43..199H. doi:10.1007/BF02345020.

- Hawking, Stephen (2005). "Information loss in black holes". *Physical Review D* **72** (8). arXiv:hep-th/0507171. Bibcode:2005PhRvD..72h4013H. doi:10.1103/PhysRevD.72.084013.

- Hořava, Petr; Witten, Edward (1996). "Heterotic and Type I string dynamics from eleven dimensions". *Nuclear Physics B* **460** (3): 506–524. arXiv:hep-th/9510209. Bibcode:1996NuPhB.460..506H. doi:10.1016/0550-3213(95)00621-4.

- Hori, Kentaro; Katz, Sheldon; Klemm, Albrecht; Pandharipande, Rahul; Thomas, Richard; Vafa, Cumrun; Vakil, Ravi; Zaslow, Eric, eds. (2003). *Mirror Symmetry* (PDF). American Mathematical Society. ISBN 0-8218-2955-6.

- Hull, Chris; Townsend, Paul (1995). "Unity of superstring dualities". *Nuclear Physics B* **4381** (1): 109–137. arXiv:hep-th/9410167. Bibcode:1995NuPhB.438..109H. doi:10.1016/0550-3213(94)00559-W.

- Kapustin, Anton; Witten, Edward (2007). "Electric-magnetic duality and the geometric Langlands program". *Communications in Number Theory and Physics* **1** (1): 1–236. arXiv:hep-th/0604151. Bibcode:2007CNTP....1....1K. doi:10.4310/cntp.2007.v1.n1.a1.

- Klarreich, Erica. "Mathematicians chase moonshine's shadow". *Quanta Magazine*. Retrieved March 2015.

- Klebanov, Igor; Maldacena, Juan (2009). "Solving Quantum Field Theories via Curved Spacetimes" (PDF). *Physics Today* **62**: 28. Bibcode:2009PhT....62a..28K. doi:10.1063/1.3074260. Retrieved May 2013.

- Kontsevich, Maxim (1995). "Homological algebra of mirror symmetry". *Proceedings of the International Congress of Mathematicians*: 120–139. arXiv:alg-geom/9411018. Bibcode:1994alg.geom.11018K.

- Kovtun, P. K.; Son, Dam T.; Starinets, A. O. (2001). "Viscosity in strongly interacting quantum field theories from black hole physics". *Physical review letters* **94** (11): 111601. arXiv:hep-th/0405231.Bibcode:2005PhRvL..94k. doi:10.1103/PhysRevLett.94.111601. PMID 15903845.

- Lian, Bong; Liu, Kefeng; Yau, Shing-Tung (1997). "Mirror principle, I". *Asian Journal of Math* **1**: 729–763. arXiv:alg-geom/9712011. Bibcode:1997alg.geom.12011L.

- Lian, Bong; Liu, Kefeng; Yau, Shing-Tung (1999a). "Mirror principle, II". *Asian Journal of Math* **3**: 109–146. arXiv:math/9905006. Bibcode:1999math......5006L.

- Lian, Bong; Liu, Kefeng; Yau, Shing-Tung (1999b). "Mirror principle, III". *Asian Journal of Math* **3**: 771–800. arXiv:math/9912038. Bibcode:1999math.....12038L.

- Lian, Bong; Liu, Kefeng; Yau, Shing-Tung (2000). "Mirror principle, IV". *Surveys in Differential Geometry*: 475–496. arXiv:math/0007104. Bibcode:2000math......7104L.

- Luzum, Matthew; Romatschke, Paul (2008). "Conformal relativistic viscous hydrodynamics: Applications to RHIC results at $\sqrt{s_{NN}}$=200 GeV". *Physical Review C* **78** (3). arXiv:0804.4015. doi:10.1103/PhysRevC.78.034915.

- Maldacena, Juan (1998). "The Large N limit of superconformal field theories and supergravity". *Advances in Theoretical and Mathematical Physics* **2**: 231–252. arXiv:hep-th/9711200. Bibcode:1998AdTMP...2..231M. doi:10.1063/1.59653.

- Maldacena, Juan (2005)."The Illusion of Gravity"(PDF).*Scientific American***293**(5): 56–63.Bibcode:2005S. doi:10.1038/scientificamerican1105-56. PMID 16318027. Retrieved July 2013.

- Maldacena, Juan; Strominger, Andrew; Witten, Edward (1997). "Black hole entropy in M-theory". *Journal of High Energy Physics* **1997** (12).

- Merali, Zeeya (2011). "Collaborative physics: string theory finds a bench mate". *Nature* **478** (7369): 302–304. Bibcode:2011Natur.478..302M. doi:10.1038/478302a. PMID 22012369.

- Moore, Gregory (2005). "What is ... a Brane?" (PDF). *Notices of the AMS* **52**: 214. Retrieved June 2013.

- Nahm, Walter (1978). "Supersymmetries and their representations".*Nuclear Physics B***135**(1): 149–166.B..149N. doi:10.1016/0550-3213(78)90218-3.

- Nekrasov, Nikita; Schwarz, Albert (1998). "Instantons on noncommutative $\mathbf{R}^4$ and (2,0) superconformal six dimensional theory". *Communications in Mathematical Physics* **198** (3): 689–703. arXiv:hep-th/9802.689N. doi:10.1007/s002200050490.

- Ooguri, Hirosi; Strominger, Andrew; Vafa, Cumrun (2004). "Black hole attractors and the topological string". *Physical Review D* **70** (10).

- Polchinski, Joseph (2007). "All Strung Out?". *American Scientist*. Retrieved April 2015.

- Penrose, Roger (2005). *The Road to Reality: A Complete Guide to the Laws of the Universe*. Knopf. ISBN 0-679-45443-8.

- Randall, Lisa; Sundrum, Raman (1999). "An alternative to compactification". *Physical Review Letters* **83** (23): 4690. arXiv:hep-th/9906064. Bibcode:1999PhRvL..83.4690R. doi:10.1103/PhysRevLett.83.4690.

- Sachdev, Subir (2013). "Strange and stringy". *Scientific American* **308** (44): 44. Bibcode:2012SciAm.308a..44S. doi:10.1038/scientificamerican0113-44.

- Seiberg, Nathan; Witten, Edward (1999). "String Theory and Noncommutative Geometry". *Journal of High Energy Physics* **1999** (9): 032. arXiv:hep-th/9908142.Bibcode:1999JHEP...09..032S.doi:10.1088/1126-6708/192.

- Sen, Ashoke (1994a). "Strong-weak coupling duality in four-dimensional string theory". *International Journal of Modern Physics A* **9** (21): 3707–3750. arXiv:hep-th/9402002.Bibcode:1994IJMPA...9.3707S.doi:10.1142/S7.

- Sen, Ashoke (1994b). "Dyon-monopole bound states, self-dual harmonic forms on the multi-monopole moduli space, and *SL*(2,**Z**) invariance in string theory". *Physics Letters B* **329** (2): 217–221. arXiv:hep-th/9402032. Bibcode:1994PhLB..329..217S. doi:10.1016/0370-2693(94)90763-3.

- Smolin, Lee (2006). *The Trouble with Physics: The Rise of String Theory, the Fall of a Science, and What Comes Next*. New York: Houghton Mifflin Co. ISBN 0-618-55105-0.

- Strominger, Andrew (1998). "Black hole entropy from near-horizon microstates". *Journal of High Energy Physics* **1998** (2). arXiv:hep-th/9712251. Bibcode:1998JHEP...02..009S. doi:10.1088/1126-6708/1998/02/009.

- Strominger, Andrew; Vafa, Cumrun (1996). "Microscopic origin of the Bekenstein–Hawking entropy". *Physics Letters B* **379** (1): 99–104. arXiv:hep-th/9601029. Bibcode:1996PhLB..379...99S. doi:10.1016/0370-2693(-0.

- Strominger, Andrew; Yau, Shing-Tung; Zaslow, Eric (1996). "Mirror symmetry is T-duality". *Nuclear Physics B* **479** (1): 243–259. arXiv:hep-th/9606040. Bibcode:1996NuPhB.479..243S. doi:10.1016/0550-3213(96)00434-8.

- Susskind, Leonard (2005). *The Cosmic Landscape: String Theory and the Illusion of Intelligent Design*. Back Bay Books. ISBN 978-0316013338.

- Susskind, Leonard (2008). *The Black Hole War: My Battle with Stephen Hawking to Make the World Safe for Quantum Mechanics*. Little, Brown and Company. ISBN 978-0-316-01641-4.

- Wald, Robert (1984). *General Relativity*. University of Chicago Press. ISBN 978-0-226-87033-5.

- Weinberg, Steven (1987). *Anthropic bound on the cosmological constant* **59** (22). Physical Review Letters. p. 2607.

- Witten, Edward (1995). "String theory dynamics in various dimensions". *Nuclear Physics B* **443** (1): 85–126. arXiv:hep-th/9503124. Bibcode:1995NuPhB.443...85W. doi:10.1016/0550-3213(95)00158-O.

- Witten, Edward (1998). "Anti-de Sitter space and holography". *Advances in Theoretical and Mathematical Physics* **2**: 253–291. arXiv:hep-th/9802150. Bibcode:1998AdTMP...2..253W.

- Witten, Edward (2007). "Three-dimensional gravity revisited". arXiv:0706.3359 [hep-th].

- Woit, Peter (2006). *Not Even Wrong: The Failure of String Theory and the Search for Unity in Physical Law*. Basic Books. p. 105. ISBN 0-465-09275-6.

- Yau, Shing-Tung; Nadis, Steve (2010). *The Shape of Inner Space: String Theory and the Geometry of the Universe's Hidden Dimensions*. Basic Books. ISBN 978-0-465-02023-2.

- Zee, Anthony (2010). *Quantum Field Theory in a Nutshell* (2nd ed.). Princeton University Press. ISBN 978-0-691-14034-6.

- Zwiebach, Barton (2009). *A First Course in String Theory*. Cambridge University Press. ISBN 978-0-521-88032-9.

12.10   Further reading

12.10.1   Popularizations

General

- Greene, Brian (2003). *The Elegant Universe: Superstrings, Hidden Dimensions, and the Quest for the Ultimate Theory*. New York: W.W. Norton & Company. ISBN 0-393-05858-1.

- Greene, Brian (2004). *The Fabric of the Cosmos: Space, Time, and the Texture of Reality*. New York: Alfred A. Knopf. ISBN 0-375-41288-3.

Critical

- Penrose, Roger (2005). *The Road to Reality: A Complete Guide to the Laws of the Universe*. Knopf. ISBN 0-679-45443-8.

- Smolin, Lee (2006). *The Trouble with Physics: The Rise of String Theory, the Fall of a Science, and What Comes Next*. New York: Houghton Mifflin Co. ISBN 0-618-55105-0.

- Woit, Peter (2006). *Not Even Wrong: The Failure of String Theory And the Search for Unity in Physical Law*. London: Jonathan Cape &: New York: Basic Books. ISBN 978-0-465-09275-8.

12.10.2   Textbooks

For physicists

- Becker, Katrin; Becker, Melanie; Schwarz, John (2007). *String Theory and M-theory: A Modern Introduction*. Cambridge University Press. ISBN 978-0-521-86069-7.

- Green, Michael; Schwarz, John; Witten, Edward (2012). *Superstring theory. Vol. 1: Introduction*. Cambridge University Press. ISBN 978-1107029118.

- Green, Michael; Schwarz, John; Witten, Edward (2012). *Superstring theory. Vol. 2: Loop amplitudes, anomalies and phenomenology*. Cambridge University Press. ISBN 978-1107029132.

- Polchinski, Joseph (1998). *String Theory Vol. 1: An Introduction to the Bosonic String*. Cambridge University Press. ISBN 0-521-63303-6.

- Polchinski, Joseph (1998). *String Theory Vol. 2: Superstring Theory and Beyond*. Cambridge University Press. ISBN 0-521-63304-4.

- Zwiebach, Barton (2009). *A First Course in String Theory*. Cambridge University Press. ISBN 978-0-521-88032-9.

For mathematicians

- Deligne, Pierre; Etingof, Pavel; Freed, Daniel; Jeffery, Lisa; Kazhdan, David; Morgan, John; Morrison, David; Witten, Edward, eds. (1999). *Quantum Fields and Strings: A Course for Mathematicians, Vol. 2*. American Mathematical Society. ISBN 978-0821819883.

12.11 External links

- *The Elegant Universe*—A three-hour miniseries with Brian Greene by *NOVA* (original PBS Broadcast Dates: October 28, 8–10 p.m. and November 4, 8–9 p.m., 2003). Various images, texts, videos and animations explaining string theory.

- Not Even Wrong—A blog critical of string theory

- The Official String Theory Web Site

- Why String Theory—An introduction to string theory.

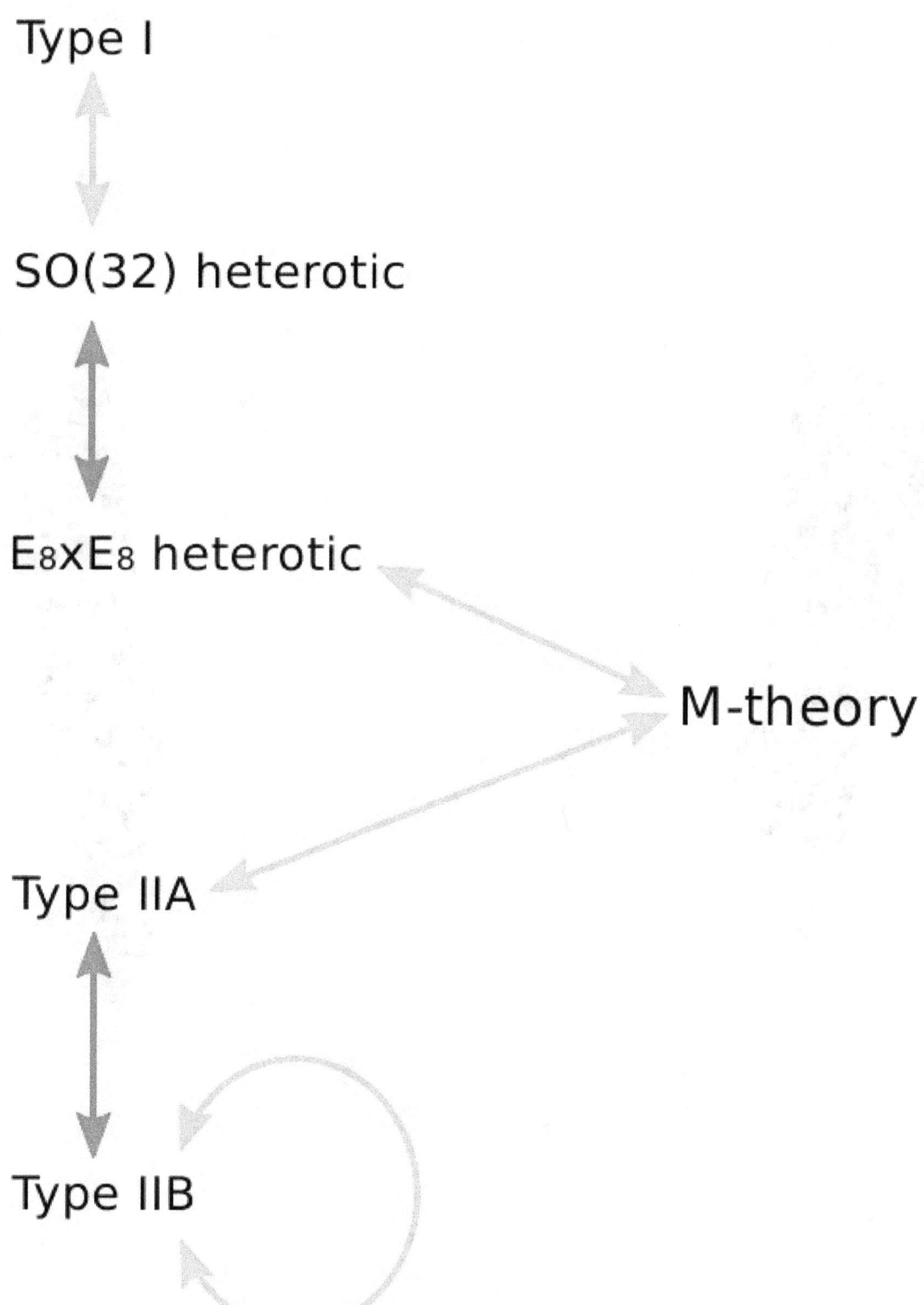

A diagram of string theory dualities. Yellow arrows indicate S-duality. Blue arrows indicate T-duality.

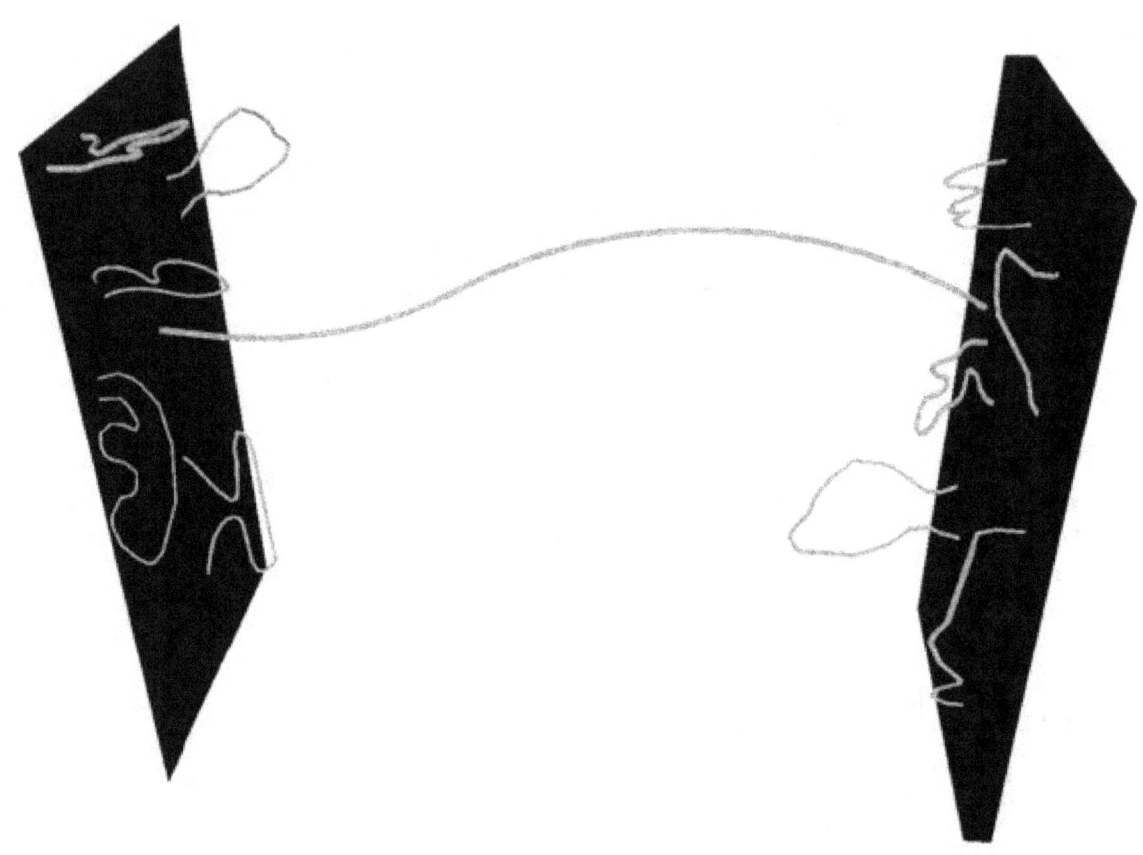

Open strings attached to a pair of D-branes

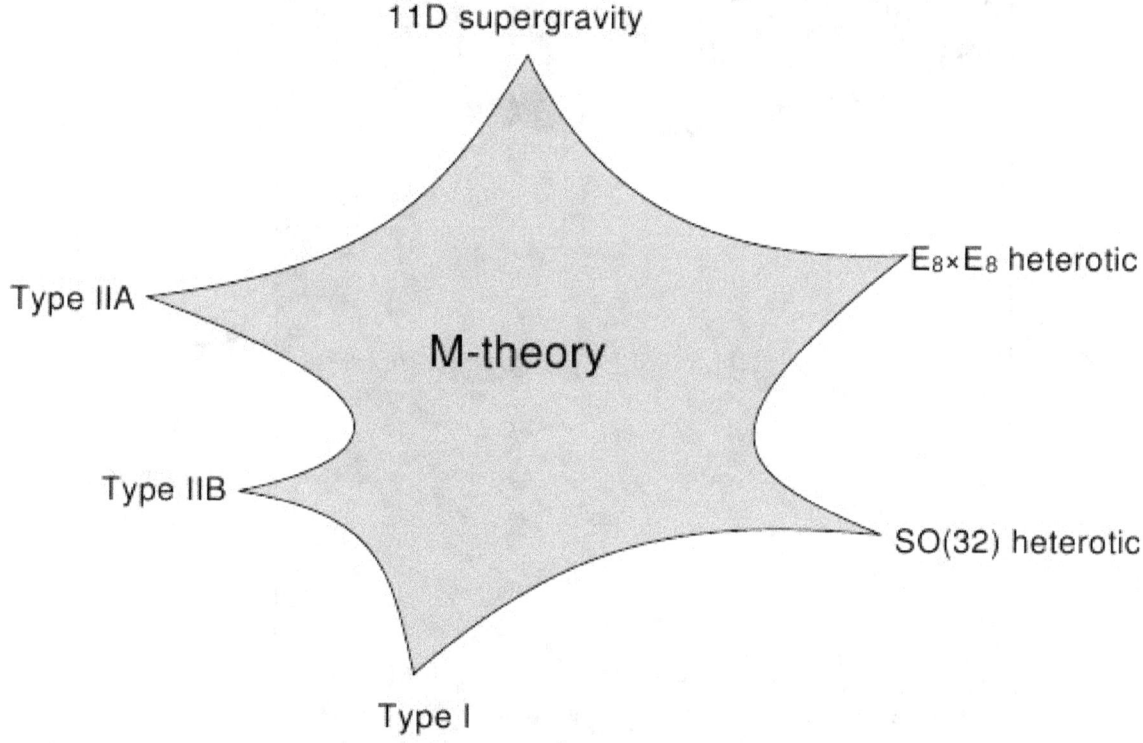

A schematic illustration of the relationship between M-theory, the five superstring theories, and eleven-dimensional supergravity. The shaded region represents a family of different physical scenarios that are possible in M-theory. In certain limiting cases corresponding to the cusps, it is natural to describe the physics using one of the six theories labeled there.

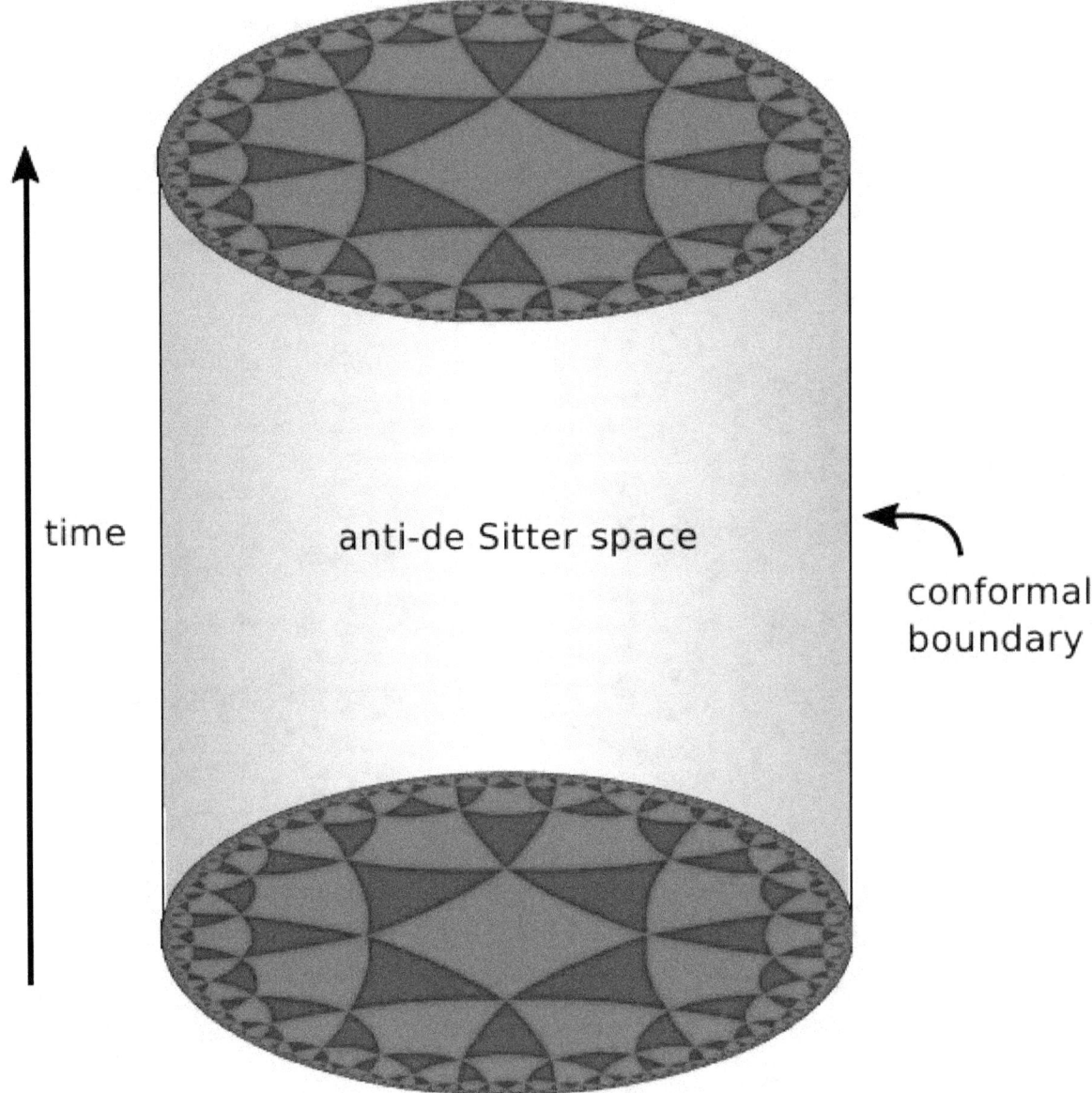

Three-dimensional anti-de Sitter space is like a stack of hyperbolic disks, each one representing the state of the universe at a given time. The resulting spacetime looks like a solid cylinder.

A magnet levitating above a high-temperature superconductor. Today some physicists are working to understand high-temperature super-conductivity using the AdS/CFT correspondence.[101]

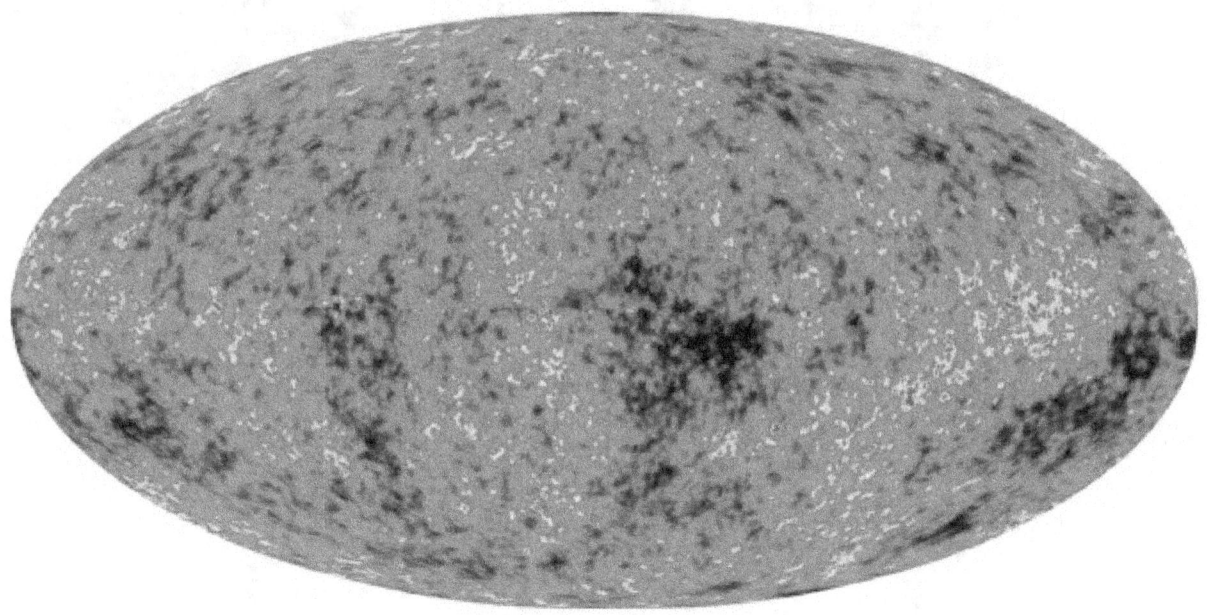

A map of the cosmic microwave background produced by the Wilkinson Microwave Anisotropy Probe

The Clebsch cubic is an example of a kind of geometric object called an algebraic variety. A classical result of enumerative geometry states that there are exactly 27 straight lines that lie entirely on this surface.

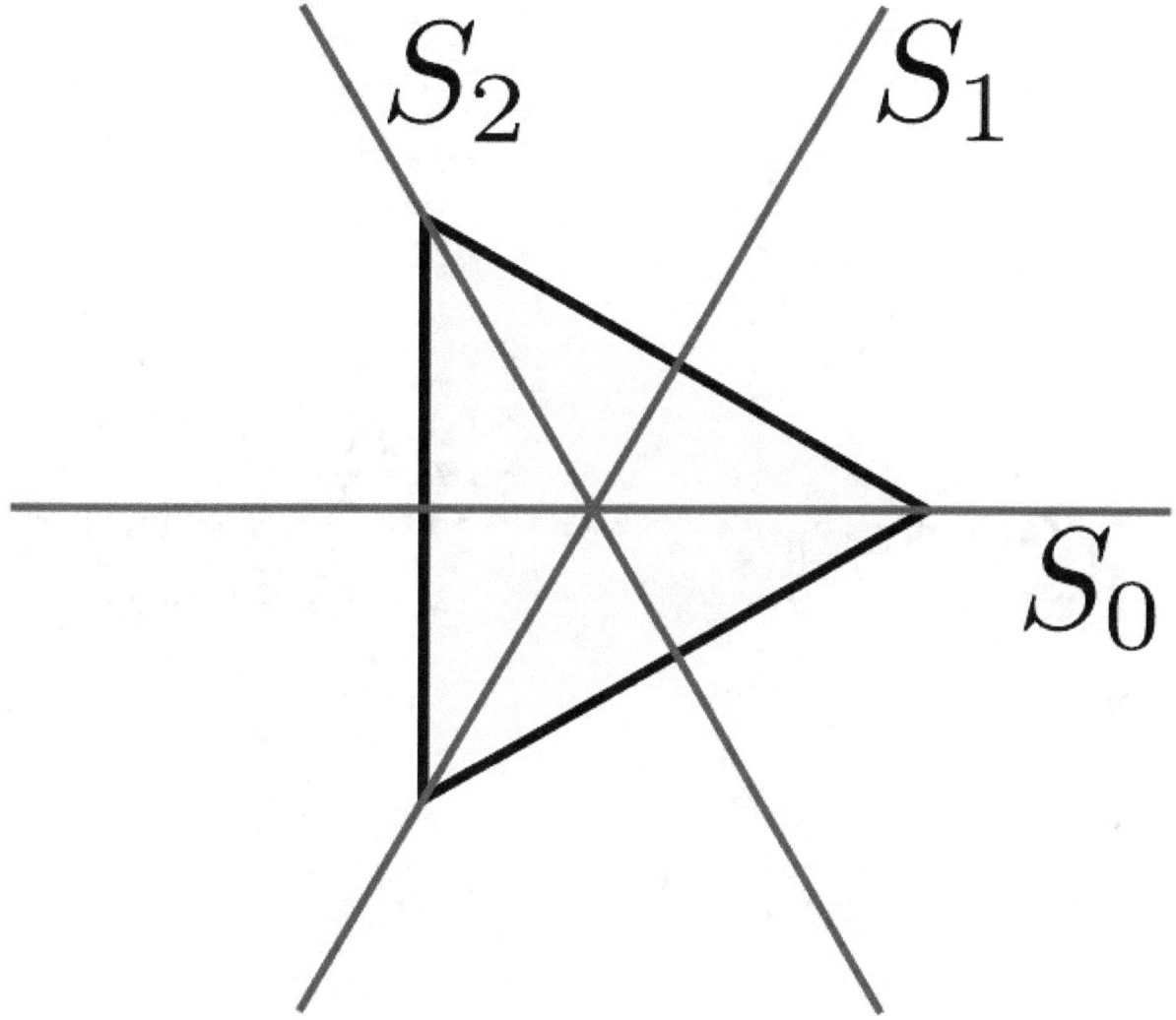

An equilateral triangle can be rotated through 120°, 240°, or 360°, or reflected in any of the three lines pictured without changing its shape.

A graph of the j-function in the complex plane

Leonard Susskind

Gabriele Veneziano

John Schwarz

Edward Witten

Joseph Polchinski

Juan Maldacena

Chapter 13

Planck length

In physics, the **Planck length**, denoted ℓP, is a unit of length, equal to $1.616199(97) \times 10^{-35}$ metres. It is a base unit in the system of Planck units, developed by physicist Max Planck. The Planck length can be defined from three fundamental physical constants: the speed of light in a vacuum, the Planck constant, and the gravitational constant.

13.1 Value

The Planck length ℓP is defined as

$$\ell_{\mathrm{P}} = \sqrt{\frac{\hbar G}{c^3}} \approx 1.616\ 199(97) \times 10^{-35}\ \mathrm{m}$$

where c is the speed of light in a vacuum, G is the gravitational constant, and $\hbar$ is the reduced Planck constant. The two digits enclosed by parentheses are the estimated standard error associated with the reported numerical value.[1][2]

The Planck length is about 10^{-20} times the diameter of a proton, and thus is exceedingly small.

13.2 Theoretical significance

There is currently no proven physical significance of the Planck length; it is, however, a topic of theoretical research. Since the Planck length is so many orders of magnitude smaller than any current instrument could possibly measure, there is no way of examining it directly. According to the generalized uncertainty principle (a concept from speculative models of quantum gravity), the Planck length is, in principle, within a factor of 10, the shortest measurable length – and no theoretically known improvement in measurement instruments could change that.

In some forms of quantum gravity, the Planck length is the length scale at which the structure of spacetime becomes dominated by quantum effects, and it is impossible to determine the difference between two locations less than one Planck length apart. The precise effects of quantum gravity are unknown; it is often guessed that spacetime might have a discrete or foamy structure at a Planck length scale.

The Planck area, equal to the square of the Planck length, plays a role in black hole entropy. The value of this entropy, in units of the Boltzmann constant, is known to be given by $\frac{A}{4\ell_p^2}$, where A is the area of the event horizon. The Planck area is the area by which the surface of a spherical black hole increases when the black hole swallows one bit of information, as was proven by Jacob Bekenstein.[3]

If large extra dimensions exist, the measured strength of gravity may be much smaller than its true (small-scale) value. In this case the Planck length would have no fundamental physical significance, and quantum gravitational effects would appear at other scales.

In string theory, the Planck length is the order of magnitude of the oscillating strings that form elementary particles, and shorter lengths do not make physical sense.[4] The string scale l_s is related to the Planck scale by $\ell P = g_s^{1/4} l_s$, where g_s is the string coupling constant. Contrary to what the name suggests, the string coupling constant is not constant, but depends on the value of a scalar field known as the dilaton.

In loop quantum gravity, area is quantized, and the Planck area is, within a factor of 10, the smallest possible area value.

In doubly special relativity, the Planck length is observer-invariant.

The search for the laws of physics valid at the Planck length is a part of the search for the theory of everything.

13.3 Visualization

The size of the Planck length can be visualized as follows: if a particle or dot about 0.1 mm in size (which is approximately the smallest the unaided human eye can see) were magnified in size to be as large as the observable universe, then inside that universe-sized "dot", the Planck length would be roughly the size of an actual 0.1 mm dot. In other words, a 0.1 mm dot is halfway between the Planck length and the size of the observable universe on a logarithmic scale.

13.4 See also

- Fock–Lorentz symmetry

- Orders of magnitude (length)

- Planck energy

- Planck mass

- Planck epoch

- Planck scale

- Planck temperature

- Planck time

13.5 Notes and references

[1] John Baez, The Planck Length

[2] NIST, "Planck length", NIST's published CODATA constants

[3] "Phys. Rev. D 7, 2333 (1973): Black Holes and Entropy". Prd.aps.org. Retrieved 2013-10-21.

[4] Cliff Burgess; Fernando Quevedo (November 2007). "The Great Cosmic Roller-Coaster Ride". *Scientific American* (print) (Scientific American, Inc.). p. 55.

13.6 Bibliography

- Garay, Luis J. (January 1995). "Quantum gravity and minimum length". *International Journal of Modern Physics A* **10** (2): 145 ff. arXiv:gr-qc/9403008v2. Bibcode:1995IJMPA..10..145G. doi:10.1142/S0217751X95000085.

13.7   External links

- Bowley, Roger; Eaves, Laurence (2010). "Planck Length". *Sixty Symbols*. Brady Haran for the University of Nottingham.

Chapter 14

Unified field theory

"Unified theory" redirects here. For the band, see Unified Theory (band).

In physics, a **unified field theory** (**UFT**), occasionally referred to as a **uniform field theory**,[1] is a type of field theory that allows all that is usually thought of as fundamental forces and elementary particles to be written in terms of a single field. There is no accepted unified field theory, and thus it remains an open line of research. The term was coined by Einstein, who attempted to unify the general theory of relativity with electromagnetism. The "theory of everything" and Grand Unified Theory are closely related to unified field theory, but differ by not requiring the basis of nature to be fields, and often by attempting to explain physical constants of nature.

This article describes unified field theory as it is currently understood in connection with quantum theory. Earlier attempts based on classical physics are described in the article on classical unified field theories.

There may be no *a priori* reason why the correct description of nature has to be a unified field theory. However, this goal has led to a great deal of progress in modern theoretical physics and continues to motivate research.

14.1 Introduction

According to the current understanding of physics, forces are not transmitted directly between interacting objects, but instead are described by intermediary entities called fields. All four of the known fundamental forces are mediated by fields, which in the Standard Model of particle physics result from exchange of gauge bosons. Specifically the four fundamental interactions to be unified are:

- Strong interaction: the interaction responsible for holding quarks together to form hadrons, and holding neutrons and also protons together to form atomic nuclei. The exchange particle that mediates this force is the gluon.

- Electromagnetic interaction: the familiar interaction that acts on electrically charged particles. The photon is the exchange particle for this force.

- Weak interaction: a short-range interaction responsible for some forms of radioactivity, that acts on electrons, neutrinos, and quarks. It is mediated by the W and Z bosons.

- Gravitational interaction: a long-range attractive interaction that acts on *all* particles. The postulated exchange particle has been named the graviton.

Modern unified field theory attempts to bring these four interactions together into a single framework.

14.2   History

The first successful classical unified field theory was developed by James Clerk Maxwell. In 1820 Hans Christian Ørsted discovered that electric currents exerted forces on magnets, while in 1831, Michael Faraday made the observation that time-varying magnetic fields could induce electric currents. Until then, electricity and magnetism had been thought of as unrelated phenomena. In 1864, Maxwell published his famous paper on a dynamical theory of the electromagnetic field. This was the first example of a theory that was able to encompass previously separate field theories (namely electricity and magnetism) to provide a unifying theory of electromagnetism. By 1905, Albert Einstein had used the constancy of the speed of light in Maxwell's theory to unify our notions of space and time into an entity we now call spacetime and in 1915 he expanded this theory of special relativity to a description of gravity, General Relativity, using a field to describe the curving geometry of four-dimensional spacetime.

In the years following the creation of the general theory, a large number of physicists and mathematicians enthusiastically participated in the attempt to unify the then-known fundamental interactions.[2] In view of later developments in this domain, of particular interest are the theories of Hermann Weyl of 1919, who introduced the concept of an (electromagnetic) gauge field in a classical field theory[3] and, two years later, that of Theodor Kaluza, who extended General Relativity to five dimensions.[4] Continuing in this latter direction, Oscar Klein proposed in 1926 that the fourth spatial dimension be curled up into a small, unobserved circle. In Kaluza–Klein theory, the gravitational curvature of the extra spatial direction behaves as an additional force similar to electromagnetism. These and other models of electromagnetism and gravity were pursued by Albert Einstein in his attempts at a classical unified field theory. By 1930 Einstein had already considered the Einstein–Maxwell–Dirac System [Dongen]. This system is (heuristically) the super-classical [Varadarajan] limit of (the not mathematically well-defined) Quantum Electrodynamics. One can extend this system to include the weak and strong nuclear forces to get the Einstein–Yang–Mills–Dirac System.

14.3   Modern progress

In 1963 American physicist Sheldon Glashow proposed that the weak nuclear force and electricity and magnetism could arise from a partially unified electroweak theory. In 1967, Pakistani Abdus Salam and American Steven Weinberg independently revised Glashow's theory by having the masses for the W particle and Z particle arise through spontaneous symmetry breaking with the Higgs mechanism. This unified theory modeled the electroweak interaction as a force mediated by four particles: the photon for the electromagnetic aspect, and a neutral Z particle and two charged W particles for weak aspect. As a result of the spontaneous symmetry breaking, the weak force becomes short-range and the Z and W bosons acquire masses of 80.4 and 91.2 GeV/c^2, respectively. Their theory was first given experimental support by the discovery of weak neutral currents in 1973. In 1983, the Z and W bosons were first produced at CERN by Carlo Rubbia's team. For their insights, Glashow, Salam, and Weinberg were awarded the Nobel Prize in Physics in 1979. Carlo Rubbia and Simon van der Meer received the Prize in 1984.

After Gerardus 't Hooft showed the Glashow–Weinberg–Salam electroweak interactions to be mathematically consistent, the electroweak theory became a template for further attempts at unifying forces. In 1974, Sheldon Glashow and Howard Georgi proposed unifying the strong and electroweak interactions into Georgi–Glashow model, the first Grand Unified Theory, which would have observable effects for energies much above 100 GeV.

Since then there have been several proposals for Grand Unified Theories, e.g. the Pati–Salam model, although none is currently universally accepted. A major problem for experimental tests of such theories is the energy scale involved, which is well beyond the reach of current accelerators. Grand Unified Theories make predictions for the relative strengths of the strong, weak, and electromagnetic forces, and in 1991 LEP determined that supersymmetric theories have the correct ratio of couplings for a Georgi–Glashow Grand Unified Theory. Many Grand Unified Theories (but not Pati–Salam) predict that the proton can decay, and if this were to be seen, details of the decay products could give hints at more aspects of the Grand Unified Theory. It is at present unknown if the proton can decay, although experiments have determined a lower bound of 10^{35} years for its lifetime.

14.4 Current status

Gravity has yet to be successfully included in a theory of everything. Simply trying to combine the graviton with the strong and electroweak interactions runs into fundamental difficulties since the resulting theory is not renormalizable. Theoretical physicists have not yet formulated a widely accepted, consistent theory that combines general relativity and quantum mechanics. The incompatibility of the two theories remains an outstanding problem in the field of physics. Some theoretical physicists currently believe that a quantum theory of general relativity may require frameworks other than field theory itself, such as string theory or loop quantum gravity. Some models in string theory that are promising by way of realizing our familiar standard model are the perturbative heterotic string models, 11-dimensional M-theory, Singular geometries (e.g. orbifold and orientifold), D-branes and other branes, flux compactification and warped geometry, and non-perturbative type IIB superstring solutions (F-theory).[5]

14.5 Notes

[1] *See, e.g., Beyond Art: A Third Culture* page 199. *Compare* Uniform field theory.

[2] See Catherine Goldstein & Jim Ritter (2003) "The varieties of unity: sounding unified theories 1920-1930" in A. Ashtekar, et al. (eds.), *Revisiting the Foundations of Relativistic Physics*, Dordrecht, Kluwer, p. 93-149; Vladimir Vizgin (1994), *Unified Field Theories in the First Third of the 20th Century*, Basel, Birkhäuser; Hubert Goenner On the History of Unified Field Theories.

[3] Erhard Scholtz (ed) (2001), *Hermann Weyl's* Raum - Zeit- Materie *and a General Introduction to His Scientific Work*, Basel, Birkhäuser.

[4] Daniela Wuensch (2003), "The fifth dimension: Theodor Kaluza's ground-breaking idea", *Annalen der Physik*, vol. 12, p. 519–542.

[5] http://arxiv.org/abs/0812.1372

14.6 References

- Shushi Tomer, *A Possible Connection between Quantum and General Relativity Theories*, SSRN: http://papers.ssrn.com/sol3/papers.cfm?abstract_id=2538728 (December 15, 2014)

- Jeroen van van Dongen *Einstein's Unification*, Cambridge University Press (July 26, 2010)

- Varadarajan, V.S. *Supersymmetry for Mathematicians: An Introduction (Courant Lecture Notes)*, American Mathematical Society (July 2004)

14.7 External links

- On the History of Unified Field Theories, by Hubert F. M. Goenner

Chapter 15

Symmetry breaking

This article is about the concept in physics. For the concept in biology, see Symmetry breaking and cortical rotation. For the concept in mathematics, see Symmetry Breaking Constraints.

Symmetry breaking in physics describes a phenomenon where (infinitesimally) small fluctuations acting on a system which is crossing a critical point decide the system's fate, by determining which branch of a bifurcation is taken. To an outside observer unaware of the fluctuations (or "noise"), the choice will appear arbitrary. This process is called symmetry "breaking", because such transitions usually bring the system from a symmetric but disorderly state into one or more definite states. Symmetry breaking is supposed to play a major role in pattern formation.

In 1972, Nobel laureate P.W. Anderson used the idea of symmetry breaking to show some of the drawbacks of Reductionism in his paper titled "More is different" in *Science*.[1]

Symmetry breaking can be distinguished into two types, Explicit symmetry breaking and Spontaneous symmetry breaking, characterized by whether the equations of motion fail to be invariant or the ground state fails to be invariant.

15.1 Explicit symmetry breaking

Main article: Explicit symmetry breaking

In explicit symmetry breaking, the equations of motion describing a system are not invariant under the broken symmetry.

15.2 Spontaneous symmetry breaking

Main article: Spontaneous symmetry breaking

In spontaneous symmetry breaking, the equations of motion of the system are invariant, but the system is not because the background (spacetime) of the system, its vacuum, is non-invariant. Such a symmetry breaking is parametrized by an order parameter. A special case of this type of symmetry breaking is dynamical symmetry breaking.

15.3 Examples

Symmetry breaking can cover any of the following scenarios:[2]

216

- The breaking of an exact symmetry of the underlying laws of physics by the random formation of some structure;

- A situation in physics in which a minimal energy state has less symmetry than the system itself;

- Situations where the actual state of the system does not reflect the underlying symmetries of the dynamics because the manifestly symmetric state is unstable (stability is gained at the cost of local asymmetry);

- Situations where the equations of a theory may have certain symmetries, though their solutions may not (the symmetries are "hidden").

One of the first cases of broken symmetry discussed in the physics literature is related to the form taken by a uniformly rotating body of incompressible fluid in gravitational and hydrostatic equilibrium. Jacobi[3] and soon later Liouville,[4] in 1834, discussed the fact that a tri-axial ellipsoid was an equilibrium solution for this problem when the kinetic energy compared to the gravitational energy of the rotating body exceeded a certain critical value. The axial symmetry presented by the McLaurin spheroids is broken at this bifurcation point. Furthermore, above this bifurcation point, and for constant angular momentum, the solutions that minimize the kinetic energy are the *non*-axially symmetric Jacobi ellipsoids instead of the Maclaurin spheroids.

15.4 See also

- Anomalous Symmetry Breaking

- Higgs mechanism

- QCD vacuum

- Goldstone boson

- 1964 PRL symmetry breaking papers

- J. J. Sakurai Prize for Theoretical Particle Physics

15.5 References

[1]Anderson, P.W. (1972)."More is Different"(PDF).*Science*177(4047): 393–396.Bibcode:1972Sci...177..393A.doi:10.193. PMID 17796623.

[2] http://www.angelfire.com/stars5/astroinfo/gloss/s.html

[3] Jacobi, C.G.J. (1834). "Über die figur des gleichgewichts". *Annalen der Physik und Chemie* (33): 229–238.

[4] Liouville, J. (1834). "Sur la figure d'une masse fluide homogène, en équilibre et douée d'un mouvement de rotation". *Journal de l'École Polytechnique* (14): 289–296.

Chapter 16

Fundamental interaction

Fundamental interactions, also known as **fundamental forces**, are the interactions in physical systems that don't appear to be reducible to more basic interactions. There are four conventionally accepted fundamental interactions—gravitational, electromagnetic, strong nuclear, and weak nuclear. Each one is understood as the dynamics of a *field*. The gravitational force is modeled as a continuous classical field. The other three are each modeled as discrete quantum fields, and exhibit a measurable unit or *elementary particle*.

Gravitation and electromagnetism act over a potentially infinite distance across the universe. They mediate macroscopic phenomena every day. The other two fields act over minuscule, subatomic distances. The strong nuclear interaction is responsible for the binding of atomic nuclei. The weak nuclear interaction also acts on the nucleus, mediating radioactive decay.

Theoretical physicists working beyond the Standard Model seek to quantize the gravitational field toward predictions that particle physicists can experimentally confirm, thus yielding acceptance to a theory of quantum gravity (QG). (Phenomena suitable to model as a fifth force—perhaps an added gravitational effect—remain widely disputed). Other theorists seek to unite the electroweak and strong fields within a Grand Unified Theory (GUT). While all four fundamental interactions are widely thought to align at an extremely minuscule scale, particle accelerators cannot produce the massive energy levels required to experimentally probe at that Planck scale (which would experimentally confirm such theories). Yet some theories, such as the string theory, seek both QG and GUT within one framework, unifying all four fundamental interactions along with mass generation within a theory of everything (ToE).

16.1 General relativity

In his 1687 theory, Isaac Newton postulated space as an infinite and unalterable physical structure existing before, within, and around all objects while their states and relations unfold at a constant pace everywhere, thus absolute space and time. Inferring that all objects bearing mass approach at a constant rate, but collide by impact proportional to their masses, Newton inferred that matter exhibits an attractive force. His law of universal gravitation mathematically stated it to span the entire universe instantly (despite absolute time), or, if not actually a force, to be instant interaction among all objects (despite absolute space). As conventionally interpreted, Newton's theory of motion modeled a *central force* without a communicating medium.[2] Thus Newton's theory violated the first principle of mechanical philosophy, as stated by Descartes, *No action at a distance*. Conversely, during the 1820s, when explaining magnetism, Michael Faraday inferred a *field* filling space and transmitting that force. Faraday conjectured that ultimately, all forces unified into one.

In the early 1870s, James Clerk Maxwell unified electricity and magnetism as effects of an electromagnetic-field whose third consequence was light, traveling at constant speed in a vacuum. The electromagnetic field theory contradicted predictions of Newton's theory of motion, unless physical states of the luminiferous aether—presumed to fill all space whether within matter or in a vacuum and to manifest the electromagnetic field—aligned all phenomena and thereby held valid the Newtonian principle relativity or invariance. Disfavoring hypotheses at unobservables, Albert Einstein discarded the aether, and aligned electrodynamics with relativity by denying absolute space and time, and stating relative space and time. The two phenomena altered in the vicinity of an object measured to be in motion—length contraction and time dilation for the object experienced to be in relative motion—Einstein's principle special relativity, published in 1905.

Special relativity was accepted as a theory, too. It rendered Newton's theory of motion apparently untenable, especially since Newtonian physics postulated an object's mass to be constant. A consequence of special relativity is mass being a variant form of energy, condensed into an object. By the equivalence principle, published by Einstein in 1907, gravitation is indistinguishable from acceleration, perhaps two phenomena sharing a mechanism. That year, Hermann Minkowski modeled special relativity to a unification of space and time, 4D spacetime. So stretching the three spatial dimensions onto the single dimension of time's arrow, Einstein arrived at general theory of relativity in 1915.[3] Einstein interpreted space as a substance, *Einstein aether*, whose physical properties receive motion from an object and transmit it to other objects while modulating events' unfolding. Equivalent to energy, mass contracts space, which dilates time—events unfold more slowly—establishing local tension. The object relieves it in the likeness of a free fall at light speed along the pathway of least resistance, a straight line's equivalent on the curved surface of 4D spacetime, a pathway termed *worldline*.

Einstein abolished *action at a distance* by theorizing a gravitational field—4D spacetime—that waves while transmitting motion across the universe at light speed. All objects always travel at light speed in 4D spacetime. At zero relative speed, an object is observed to travel none through space, but age most rapidly. That is, an object at relative rest in 3D space exhibits its constant energy to an observer by exhibiting top speed along 1D time flow. Conversely, at highest relative speed, an object traverses 3D space at light speed, yet is ageless, none of its constant energy available to internal motion as flow along 1D time. Whereas Newtonian inertia is an idealized case of an object either keeping rest or holding constant velocity by hypothetical existence in a universe otherwise devoid of matter, Einsteinian inertia is indistinguishable from an object experiencing no acceleration by existing in a gravitational field possibly full of matter distributed uniformly. Conversely, even massless energy manifests gravitation—which is acceleration—on local objects by "curving" the surface of 4D spacetime. Physicists renounced belief that motion must be mediated by a *force*.

16.2 Standard Model

Main article: Standard Model
See also: Lambda-CDM model

The electromagnetic, strong, and weak interactions associate with elementary particles, whose behaviors are modeled in quantum mechanics (QM). For predictive success with QM's probabilistic outcomes, particle physics conventionally models QM events across a field set to special relativity, altogether relativistic quantum field theory (QFT).[4] Force particles, called gauge bosons—*force carriers* or *messenger particles* of underlying fields—interact with matter particles, called fermions. Everyday matter is atoms, composed of three fermion types: up-quarks and down-quarks constituting, as well as electrons orbiting, the atom's nucleus. Atoms interact, form molecules, and manifest further properties through electromagnetic interactions among their electrons absorbing and emitting photons, the electromagnetic field's force carrier, which if unimpeded traverse potentially infinite distance. Electromagnetism's QFT is quantum electrodynamics (QED).

The electromagnetic interaction was modeled with the weak interaction, whose force carriers are W and Z bosons, traversing minuscule distance, in electroweak theory (EWT). Electroweak interaction would operate at such high temperatures as soon after the presumed Big Bang, but, as the early universe cooled, split into electromagnetic and weak interactions. The strong interaction, whose force carrier is the gluon, traversing minuscule distance among quarks, is modeled in quantum chromodynamics (QCD). EWT, QCD, and the Higgs mechanism, whereby the Higgs field manifests Higgs bosons that interact with some quantum particles and thereby endow those particles with mass, comprise particle physics' Standard Model (SM). Predictions are usually made using calculational approximation methods, although such perturbation theory is inadequate to model some experimental observations (for instance bound states and solitons). Still, physicists widely accept the Standard Model as science's most experimentally confirmed theory.

Beyond the Standard Model, some theorists work to unite the electroweak and strong interactions within a Grand Unified Theory (GUT). Some attempts at GUTs hypothesize "shadow" particles, such that every known matter particle associates with an undiscovered force particle, and vice versa, altogether supersymmetry (SUSY). Other theorists seek to quantize the gravitational field by modeling behavior of its hypothetical force carrier, the graviton and achieve quantum gravity (QG). One approach to QG is loop quantum gravity (LQG). Still other theorists seek both QG and GUT within one framework, reducing all four fundamental interactions to a Theory of Everything (ToE). The most prevalent aim at a ToE is string theory, although to model matter particles, it added SUSY to force particles—and so, strictly speaking, became

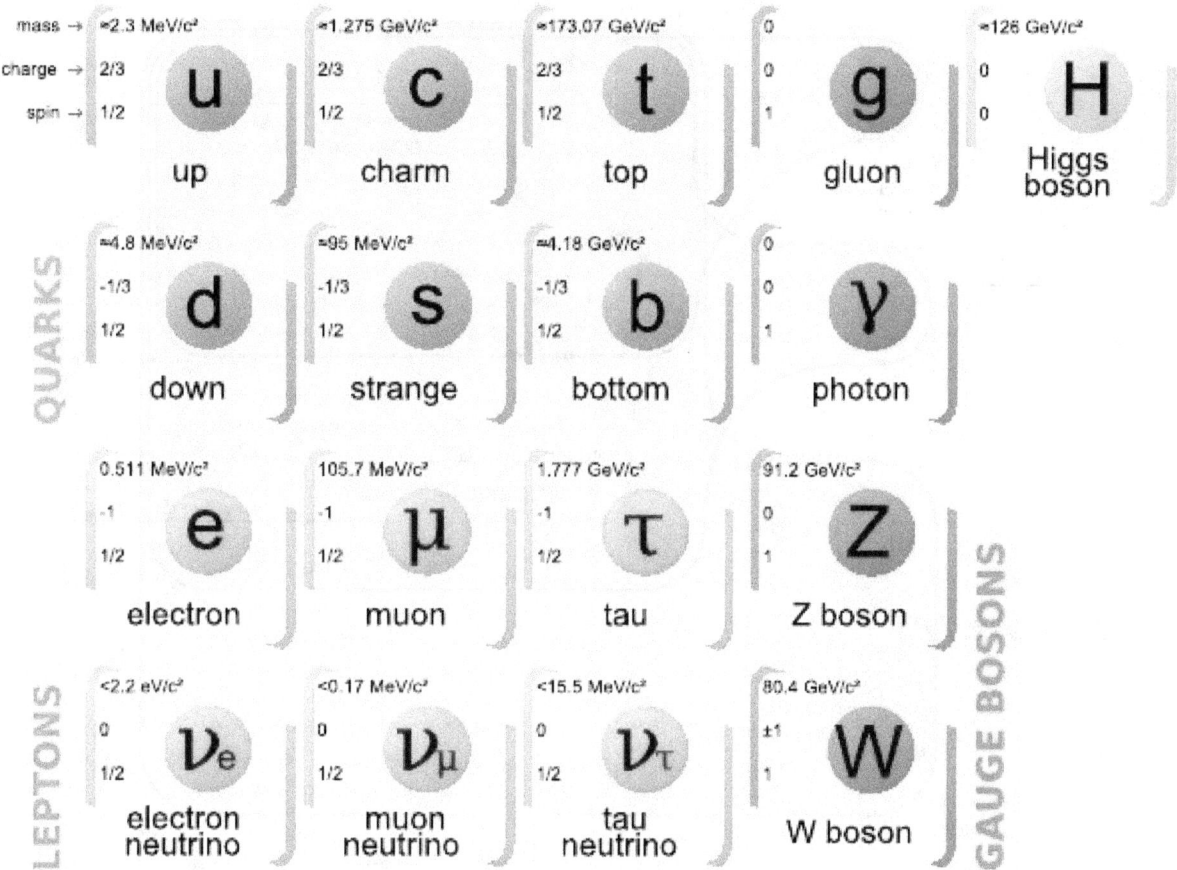

The Standard Model of elementary particles, with the fermions in the first three columns, the gauge bosons in the fourth column, and the Higgs boson in the fifth column

superstring theory. Multiple, seemingly disparate superstring theories were unified on a backbone, M theory. Theories beyond the Standard Model remain highly speculative, lacking great experimental support.

16.3   Overview of the fundamental interactions

In the conceptual model of fundamental interactions, matter consists of fermions, which carry properties called charges and spin $\pm\frac{1}{2}$ (intrinsic angular momentum $\pm\frac{\hbar}{2}$, where ħ is the reduced Planck constant). They attract or repel each other by exchanging bosons.

The interaction of any pair of fermions in perturbation theory can then be modeled thus:

> Two fermions go in → *interaction* by boson exchange → Two changed fermions go out.

The exchange of bosons always carries energy and momentum between the fermions, thereby changing their speed and direction. The exchange may also transport a charge between the fermions, changing the charges of the fermions in the process (e.g., turn them from one type of fermion to another). Since bosons carry one unit of angular momentum, the fermion's spin direction will flip from $+\frac{1}{2}$ to $-\frac{1}{2}$ (or vice versa) during such an exchange (in units of the reduced Planck's constant).

Because an interaction results in fermions attracting and repelling each other, an older term for "interaction" is force.

According to the present understanding, there are four fundamental interactions or forces: gravitation, electromagnetism,

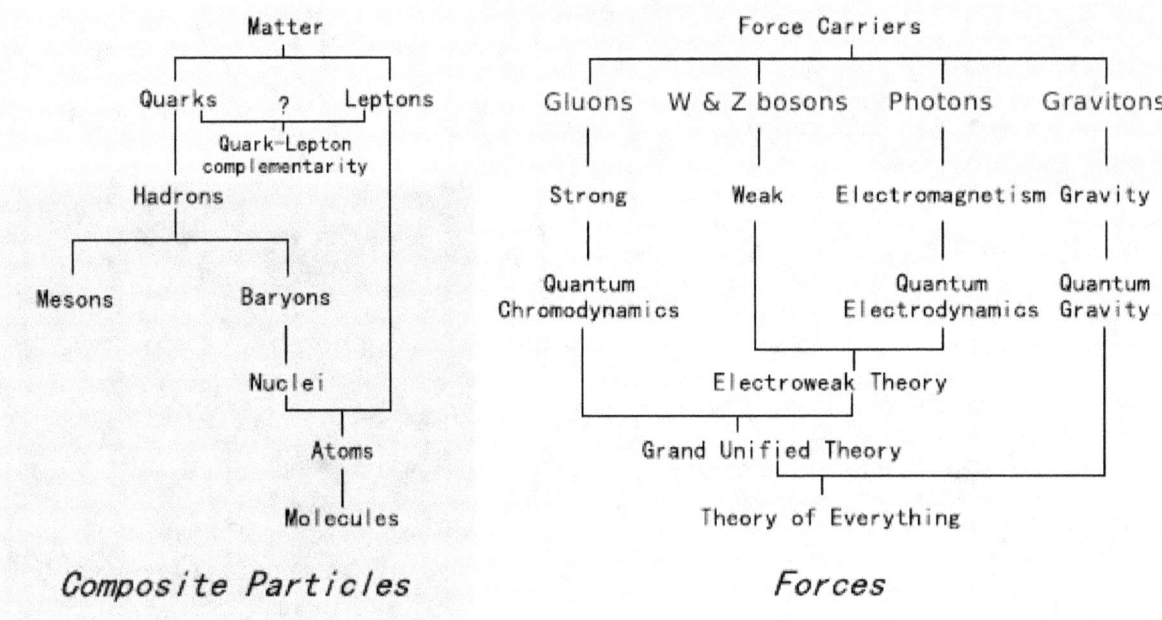

An overview of the various families of elementary and composite particles, and the theories describing their interactions. Fermions are on the left, and Bosons are on the right.

the weak interaction, and the strong interaction. Their magnitude and behavior vary greatly, as described in the table below. Modern physics attempts to explain every observed physical phenomenon by these fundamental interactions. Moreover, reducing the number of different interaction types is seen as desirable. Two cases in point are the unification of:

- Electric and magnetic force into electromagnetism;

- The electromagnetic interaction and the weak interaction into the electroweak interaction; see below.

Both magnitude ("relative strength") and "range", as given in the table, are meaningful only within a rather complex theoretical framework. It should also be noted that the table below lists properties of a conceptual scheme that is still the subject of ongoing research.

The modern (perturbative) quantum mechanical view of the fundamental forces other than gravity is that particles of matter (fermions) do not directly interact with each other, but rather carry a charge, and exchange virtual particles (gauge bosons), which are the interaction carriers or force mediators. For example, photons mediate the interaction of electric charges, and gluons mediate the interaction of color charges.

16.4 The interactions

16.4.1 Gravitation

Main article: Gravitation

Gravitation is by far the weakest of the four interactions. The weakness of gravity can easily be demonstrated by suspending a pin using a simple magnet (such as a refrigerator magnet). The magnet is able to hold the pin against the gravitational pull of the entire Earth.

Yet gravitation is very important for macroscopic objects and over macroscopic distances for the following reasons. Gravitation:

- is the only interaction that acts on all particles having mass, energy and/or momentum;

- has an infinite range, like electromagnetism but unlike strong and weak interaction;

- cannot be absorbed, transformed, or shielded against;

- always attracts and never repels.

Even though electromagnetism is far stronger than gravitation, electrostatic attraction is not relevant for large celestial bodies, such as planets, stars, and galaxies, simply because such bodies contain equal numbers of protons and electrons and so have a net electric charge of zero. Nothing "cancels" gravity, since it is only attractive, unlike electric forces which can be attractive or repulsive. On the other hand, all objects having mass are subject to the gravitational force, which only attracts. Therefore, only gravitation matters on the large scale structure of the universe.

The long range of gravitation makes it responsible for such large-scale phenomena as the structure of galaxies, black holes, and it retards the expansion of the universe. Gravitation also explains astronomical phenomena on more modest scales, such as planetary orbits, as well as everyday experience: objects fall; heavy objects act as if they were glued to the ground; and animals can only jump so high.

Gravitation was the first interaction to be described mathematically. In ancient times, Aristotle hypothesized that objects of different masses fall at different rates. During the Scientific Revolution, Galileo Galilei experimentally determined that this was not the case — neglecting the friction due to air resistance, and buoyancy forces if an atmosphere is present (e.g. the case of a dropped air-filled balloon vs a water-filled balloon) all objects accelerate toward the Earth at the same rate. Isaac Newton's law of Universal Gravitation (1687) was a good approximation of the behaviour of gravitation. Our present-day understanding of gravitation stems from Albert Einstein's General Theory of Relativity of 1915, a more accurate (especially for cosmological masses and distances) description of gravitation in terms of the geometry of space-time.

Merging general relativity and quantum mechanics (or quantum field theory) into a more general theory of quantum gravity is an area of active research. It is hypothesized that gravitation is mediated by a massless spin-2 particle called the graviton.

Although general relativity has been experimentally confirmed (at least, in the weak field or Post-Newtonian case) on all but the smallest scales, there are rival theories of gravitation. Those taken seriously by the physics community all reduce to general relativity in some limit, and the focus of observational work is to establish limitations on what deviations from general relativity are possible.

Proposed extra dimensions could explain why the gravity force is so weak.[6]

16.4.2 Electroweak interaction

Main article: Electroweak interaction

Electromagnetism and weak interaction appear to be very different at everyday low energies. They can be modeled using two different theories. However, above unification energy, on the order of 100 GeV, they would merge into a single electroweak force.

Electroweak theory is very important for modern cosmology, particularly on how the universe evolved. This is because shortly after the Big Bang, the temperature was approximately above 10^{15} K. Electromagnetic force and weak force were merged into a combined electroweak force.

For contributions to the unification of the weak and electromagnetic interaction between elementary particles, Abdus Salam, Sheldon Glashow and Steven Weinberg were awarded the Nobel Prize in Physics in 1979.[7][8]

Electromagnetism

Main article: Electromagnetism

Electromagnetism is the force that acts between electrically charged particles. This phenomenon includes the electrostatic force acting between charged particles at rest, and the combined effect of electric and magnetic forces acting between charged particles moving relative to each other.

Electromagnetism is infinite-ranged like gravity, but vastly stronger, and therefore describes a number of macroscopic phenomena of everyday experience such as friction, rainbows, lightning, and all human-made devices using electric current, such as television, lasers, and computers. Electromagnetism fundamentally determines all macroscopic, and many atomic level, properties of the chemical elements, including all chemical bonding.

In a four kilogram (~1 gallon) jug of water there are

$$4000 \text{ g } H_2O \cdot \frac{1 \text{ mol } H_2O}{18 \text{ g } H_2O} \cdot \frac{10 \text{ mol } e^-}{1 \text{ mol } H_2O} \cdot \frac{96,000 \text{ C}}{1 \text{ mol } e^-} = 2.1 \times 10^8 C$$

of total electron charge. Thus, if we place two such jugs a meter apart, the electrons in one of the jugs repel those in the other jug with a force of

$$\frac{1}{4\pi\varepsilon_0} \frac{(2.1 \times 10^8 C)^2}{(1m)^2} = 4.1 \times 10^{26} N.$$

This is larger than the planet Earth would weigh if weighed on another Earth. The atomic nuclei in one jug also repel those in the other with the same force. However, these repulsive forces are cancelled by the attraction of the electrons in jug A with the nuclei in jug B and the attraction of the nuclei in jug A with the electrons in jug B, resulting in no net force. Electromagnetic forces are tremendously stronger than gravity but cancel out so that for large bodies gravity dominates.

Electrical and magnetic phenomena have been observed since ancient times, but it was only in the 19th century that it was discovered that electricity and magnetism are two aspects of the same fundamental interaction. By 1864, Maxwell's equations had rigorously quantified this unified interaction. Maxwell's theory, restated using vector calculus, is the classical theory of electromagnetism, suitable for most technological purposes.

The constant speed of light in a vacuum (customarily described with the letter "c") can be derived from Maxwell's equations, which are consistent with the theory of special relativity. Einstein's 1905 theory of special relativity, however, which flows from the observation that the speed of light is constant no matter how fast the observer is moving, showed that the theoretical result implied by Maxwell's equations has profound implications far beyond electro-magnetism on the very nature of time and space.

In other work that departed from classical electro-magnetism, Einstein also explained the photoelectric effect by hypothesizing that light was transmitted in quanta, which we now call photons. Starting around 1927, Paul Dirac combined quantum mechanics with the relativistic theory of electromagnetism. Further work in the 1940s, by Richard Feynman, Freeman Dyson, Julian Schwinger, and Sin-Itiro Tomonaga, completed this theory, which is now called quantum electrodynamics, the revised theory of electromagnetism. Quantum electrodynamics and quantum mechanics provide a theoretical basis for electromagnetic behavior such as quantum tunneling, in which a certain percentage of electrically charged particles move in ways that would be impossible under classical electromagnetic theory, that is necessary for everyday electronic devices such as transistors to function.

Weak interaction

Main article: Weak interaction

The *weak interaction* or *weak nuclear force* is responsible for some nuclear phenomena such as beta decay. Electromagnetism and the weak force are now understood to be two aspects of a unified electroweak interaction — this discovery was the first step toward the unified theory known as the Standard Model. In the theory of the electroweak interaction, the carriers of the weak force are the massive gauge bosons called the W and Z bosons. The weak interaction is the only known interaction which does not conserve parity; it is left-right asymmetric. The weak interaction even violates CP symmetry but does conserve CPT.

16.4.3   Strong interaction

Main article: Strong interaction

The *strong interaction*, or *strong nuclear force*, is the most complicated interaction, mainly because of the way it varies with distance. At distances greater than 10 femtometers, the strong force is practically unobservable. Moreover, it holds only inside the atomic nucleus.

After the nucleus was discovered in 1908, it was clear that a new force was needed to overcome the electrostatic repulsion, a manifestation of electromagnetism, of the positively charged protons. Otherwise the nucleus could not exist. Moreover, the force had to be strong enough to squeeze the protons into a volume that is 10^{-15} of that of the entire atom. From the short range of this force, Hideki Yukawa predicted that it was associated with a massive particle, whose mass is approximately 100 MeV.

The 1947 discovery of the pion ushered in the modern era of particle physics. Hundreds of hadrons were discovered from the 1940s to 1960s, and an extremely complicated theory of hadrons as strongly interacting particles was developed. Most notably:

- The pions were understood to be oscillations of vacuum condensates;

- Jun John Sakurai proposed the rho and omega vector bosons to be force carrying particles for approximate symmetries of isospin and hypercharge;

- Geoffrey Chew, Edward K. Burdett and Steven Frautschi grouped the heavier hadrons into families that could be understood as vibrational and rotational excitations of strings.

While each of these approaches offered deep insights, no approach led directly to a fundamental theory.

Murray Gell-Mann along with George Zweig first proposed fractionally charged quarks in 1961. Throughout the 1960s, different authors considered theories similar to the modern fundamental theory of quantum chromodynamics (QCD) as simple models for the interactions of quarks. The first to hypothesize the gluons of QCD were Moo-Young Han and Yoichiro Nambu, who introduced the quark color charge and hypothesized that it might be associated with a force-carrying field. At that time, however, it was difficult to see how such a model could permanently confine quarks. Han and Nambu also assigned each quark color an integer electrical charge, so that the quarks were fractionally charged only on average, and they did not expect the quarks in their model to be permanently confined.

In 1971, Murray Gell-Mann and Harald Fritzsch proposed that the Han/Nambu color gauge field was the correct theory of the short-distance interactions of fractionally charged quarks. A little later, David Gross, Frank Wilczek, and David Politzer discovered that this theory had the property of asymptotic freedom, allowing them to make contact with experimental evidence. They concluded that QCD was the complete theory of the strong interactions, correct at all distance scales. The discovery of asymptotic freedom led most physicists to accept QCD, since it became clear that even the long-distance properties of the strong interactions could be consistent with experiment, if the quarks are permanently confined.

Assuming that quarks are confined, Mikhail Shifman, Arkady Vainshtein, and Valentine Zakharov were able to compute the properties of many low-lying hadrons directly from QCD, with only a few extra parameters to describe the vacuum. In 1980, Kenneth G. Wilson published computer calculations based on the first principles of QCD, establishing, to a level of confidence tantamount to certainty, that QCD will confine quarks. Since then, QCD has been the established theory of the strong interactions.

QCD is a theory of fractionally charged quarks interacting by means of 8 photon-like particles called gluons. The gluons interact with each other, not just with the quarks, and at long distances the lines of force collimate into strings. In this way, the mathematical theory of QCD not only explains how quarks interact over short distances, but also the string-like behavior, discovered by Chew and Frautschi, which they manifest over longer distances.

16.4.4 Beyond the Standard Model

Main article: Physics beyond the Standard Model
See also: Elementary particle § Beyond the Standard Model

Numerous theoretical efforts have been made to systematize the existing four fundamental interactions on the model of electro-weak unification.

Grand Unified Theories (GUTs) are proposals to show that all of the fundamental interactions, other than gravity, arise from a single interaction with symmetries that break down at low energy levels. GUTs predict relationships among constants of nature that are unrelated in the SM. GUTs also predict gauge coupling unification for the relative strengths of the electromagnetic, weak, and strong forces, a prediction verified at the Large Electron–Positron Collider in 1991 for supersymmetric theories.

Theories of everything, which integrate GUTs with a quantum gravity theory face a greater barrier, because no quantum gravity theories, which include string theory, loop quantum gravity, and twistor theory, have secured wide acceptance. Some theories look for a graviton to complete the Standard Model list of force carrying particles, while others, like loop quantum gravity, emphasize the possibility that time-space itself may have a quantum aspect to it.

Some theories beyond the Standard Model include a hypothetical fifth force, and the search for such a force is an ongoing line of experimental research in physics. In supersymmetric theories, there are particles that acquire their masses only through supersymmetry breaking effects and these particles, known as moduli can mediate new forces. Another reason to look for new forces is the recent discovery that the expansion of the universe is accelerating (also known as dark energy), giving rise to a need to explain a nonzero cosmological constant, and possibly to other modifications of general relativity. Fifth forces have also been suggested to explain phenomena such as CP violations, dark matter, and dark flow.

16.5 See also

- Standard Model

  - Strong interaction
  - Electroweak interaction
  - Weak interaction
  - Gravity
    - Quantum gravity
    - String Theory
    - Theory of Everything

- Grand Unified Theory

  - Gauge coupling unification
  - Unified Field Theory

- Quintessence, a hypothesized fifth force.

- *People*: Isaac Newton, James Clerk Maxwell, Albert Einstein, Richard Feynman, Sheldon Glashow, Abdus Salam, Steven Weinberg, Gerardus 't Hooft, David Gross, Edward Witten, Howard Georgi.

16.6   References

[1] http://www.pha.jhu.edu/~{}dfehling/particle.gif

[2] Newton's absolute space was a medium, but not one transmitting gravitation.

[3] Special relativity holds for objects at vast speed but of negligible mass, for instance elementary particles. Yet by yielding gravitation, which is a manner of acceleration, notable mass breaks inertia—that is, constant speed and direction—and thereby violates special relativity. Special relativity could approximately predict a massive object's motion during barely an instant, however, and thus is a temporally limited case of general relativity.

[4] Meinard Kuhlmann, "Physicists debate whether the world is made of particles or fields—or something else entirely", *Scientific American*, 24 Jul 2013.

[5] Approximate. See Coupling constant for more exact strengths, depending on the particles and energies involved.

[6] CERN (20 January 2012), "Extra dimensions, gravitons, and tiny black holes".

[7] Bais, Sander (2005), *The Equations. Icons of knowledge*, ISBN 0-674-01967-9 p.84

[8] "The Nobel Prize in Physics 1979". The Nobel Foundation. Retrieved 2008-12-16.

Bibliography General:

- Davies, Paul (1986), *The Forces of Nature*, Cambridge Univ. Press 2nd ed.

- Feynman, Richard (1967), *The Character of Physical Law*, MIT Press, ISBN 0-262-56003-8

- Schumm, Bruce A. (2004), *Deep Down Things*, Johns Hopkins University Press While all interactions are discussed, discussion is especially thorough on the weak.

- Weinberg, Steven (1993), *The First Three Minutes: A Modern View of the Origin of the Universe*, Basic Books, ISBN 0-465-02437-8

- Weinberg, Steven (1994), *Dreams of a Final Theory*, Basic Books, ISBN 0-679-74408-8

Texts:

- Padmanabhan, T. (1998), *After The First Three Minutes: The Story of Our Universe*, Cambridge Univ. Press, ISBN 0-521-62972-1

- Perkins, Donald H. (2000), *Introduction to High Energy Physics*, Cambridge Univ. Press, ISBN 0-521-62196-8

- Riazuddin (December 29, 2009). "Non-standard interactions" (PDF). *NCP 5th Particle Physics Sypnoisis* (Islamabad: Riazuddin, Head of High-Energy Theory Group at National Center for Physics) **1** (1): 1–25. Retrieved March 19, 2011.

Chapter 17

Gravity

For other uses, see Gravity (disambiguation).
"Gravitation" and "Law of Gravity" redirect here. For other uses, see Gravitation (disambiguation) and Law of Gravity (disambiguation).

Gravity or **gravitation** is a natural phenomenon by which all things with mass are brought towards (or 'gravitate'

Hammer and feather drop: Apollo 15 astronaut David Scott on the Moon enacting the legend of Galileo's gravity experiment. (1.38 MB. ogg/Theora format).

towards) one another including stars, planets, galaxies and even light and sub-atomic particles. Gravity is responsible for the complexity in the universe, by creating spheres of hydrogen, igniting them under pressure to form stars and grouping them into galaxies. Without gravity, the universe would be an uncomplicated one, existing without thermal energy and composed only of equally spaced particles. On Earth, gravity gives weight to physical objects and causes the tides. Gravity has an infinite range, and it cannot be absorbed, transformed, or shielded against.

Gravity is most accurately described by the general theory of relativity (proposed by Albert Einstein in 1915) which describes gravity, not as a force, but as a consequence of the curvature of spacetime caused by the uneven distribution of mass/energy; and resulting in time dilation, where time lapses more slowly in strong gravitation. However, for most applications, gravity is well approximated by Newton's law of universal gravitation, which postulates that gravity is a force where two bodies of mass are directly drawn (or 'attracted') to each other according to a mathematical relationship, where the attractive force is proportional to the product of their masses and inversely proportional to the square of the distance between them. This is considered to occur over an infinite range, such that all bodies (with mass) in the universe are drawn to each other no matter how far they are apart.

Gravity is the weakest of the four fundamental interactions of nature. The gravitational attraction is approximately 10^{-38} times the strength of the strong force (i.e. gravity is 38 orders of magnitude weaker), 10^{-36} times the strength of the electromagnetic force, and 10^{-29} times the strength of the weak force. As a consequence, gravity has a negligible influence on the behavior of sub-atomic particles, and plays no role in determining the internal properties of everyday matter (but see quantum gravity). On the other hand, gravity is the dominant force at the macroscopic scale, that is the cause of the formation, shape, and trajectory (orbit) of astronomical bodies, including those of asteroids, comets, planets, stars, and galaxies. It is responsible for causing the Earth and the other planets to orbit the Sun; for causing the Moon to orbit the Earth; for the formation of tides; for natural convection, by which fluid flow occurs under the influence of a density gradient and gravity; for heating the interiors of forming stars and planets to very high temperatures; for solar system, galaxy, stellar formation and evolution; and for various other phenomena observed on Earth and throughout the universe.

In pursuit of a theory of everything, the merging of general relativity and quantum mechanics (or quantum field theory) into a more general theory of quantum gravity has become an area of research.

17.1   History of gravitational theory

Main article: History of gravitational theory

17.1.1   Scientific revolution

Modern work on gravitational theory began with the work of Galileo Galilei in the late 16th and early 17th centuries. In his famous (though possibly apocryphal[1]) experiment dropping balls from the Tower of Pisa, and later with careful measurements of balls rolling down inclines, Galileo showed that gravity accelerates all objects at the same rate. This was a major departure from Aristotle's belief that heavier objects accelerate faster.[2] Galileo postulated air resistance as the reason that lighter objects may fall more slowly in an atmosphere. Galileo's work set the stage for the formulation of Newton's theory of gravity.

17.1.2   Newton's theory of gravitation

Main article: Newton's law of universal gravitation

In 1687, English mathematician Sir Isaac Newton published *Principia*, which hypothesizes the inverse-square law of universal gravitation. In his own words, "I deduced that the forces which keep the planets in their orbs must [be] reciprocally as the squares of their distances from the centers about which they revolve: and thereby compared the force requisite to keep the Moon in her Orb with the force of gravity at the surface of the Earth; and found them answer pretty nearly."[3] The equation is the following:

$F = G \frac{m_1 m_2}{r^2}$

Sir Isaac Newton, an English physicist who lived from 1642 to 1727

Where F is the force, m_1 and m_2 are the masses of the objects interacting, r is the distance between the centers of the masses and G is the gravitational constant.

Newton's theory enjoyed its greatest success when it was used to predict the existence of Neptune based on motions of Uranus that could not be accounted for by the actions of the other planets. Calculations by both John Couch Adams and

Urbain Le Verrier predicted the general position of the planet, and Le Verrier's calculations are what led Johann Gottfried Galle to the discovery of Neptune.

A discrepancy in Mercury's orbit pointed out flaws in Newton's theory. By the end of the 19th century, it was known that its orbit showed slight perturbations that could not be accounted for entirely under Newton's theory, but all searches for another perturbing body (such as a planet orbiting the Sun even closer than Mercury) had been fruitless. The issue was resolved in 1915 by Albert Einstein's new theory of general relativity, which accounted for the small discrepancy in Mercury's orbit.

Although Newton's theory has been superseded by the Einstein's general relativity, most modern non-relativistic gravitational calculations are still made using the Newton's theory because it is simpler to work with and it gives sufficiently accurate results for most applications involving sufficiently small masses, speeds and energies.

17.1.3   Equivalence principle

The equivalence principle, explored by a succession of researchers including Galileo, Loránd Eötvös, and Einstein, expresses the idea that all objects fall in the same way. The simplest way to test the weak equivalence principle is to drop two objects of different masses or compositions in a vacuum and see whether they hit the ground at the same time. Such experiments demonstrate that all objects fall at the same rate when other forces (such as air resistance and electromagnetic effects) are negligible. More sophisticated tests use a torsion balance of a type invented by Eötvös. Satellite experiments, for example STEP, are planned for more accurate experiments in space.[4]

Formulations of the equivalence principle include:

- The weak equivalence principle: *The trajectory of a point mass in a gravitational field depends only on its initial position and velocity, and is independent of its composition.*[5]

- The Einsteinian equivalence principle: *The outcome of any local non-gravitational experiment in a freely falling laboratory is independent of the velocity of the laboratory and its location in spacetime.*[6]

- The strong equivalence principle requiring both of the above.

17.1.4   General relativity

See also: Introduction to general relativity

In general relativity, the effects of gravitation are ascribed to spacetime curvature instead of a force. The starting point for general relativity is the equivalence principle, which equates free fall with inertial motion and describes free-falling inertial objects as being accelerated relative to non-inertial observers on the ground.[7][8] In Newtonian physics, however, no such acceleration can occur unless at least one of the objects is being operated on by a force.

Einstein proposed that spacetime is curved by matter, and that free-falling objects are moving along locally straight paths in curved spacetime. These straight paths are called geodesics. Like Newton's first law of motion, Einstein's theory states that if a force is applied on an object, it would deviate from a geodesic. For instance, we are no longer following geodesics while standing because the mechanical resistance of the Earth exerts an upward force on us, and we are non-inertial on the ground as a result. This explains why moving along the geodesics in spacetime is considered inertial.

Einstein discovered the field equations of general relativity, which relate the presence of matter and the curvature of spacetime and are named after him. The Einstein field equations are a set of 10 simultaneous, non-linear, differential equations. The solutions of the field equations are the components of the metric tensor of spacetime. A metric tensor describes a geometry of spacetime. The geodesic paths for a spacetime are calculated from the metric tensor.

Notable solutions of the Einstein field equations include:

- The Schwarzschild solution, which describes spacetime surrounding a spherically symmetric non-rotating uncharged massive object. For compact enough objects, this solution generated a black hole with a central singularity. For radial distances from the center which are much greater than the Schwarzschild radius, the accelerations predicted by the Schwarzschild solution are practically identical to those predicted by Newton's theory of gravity.

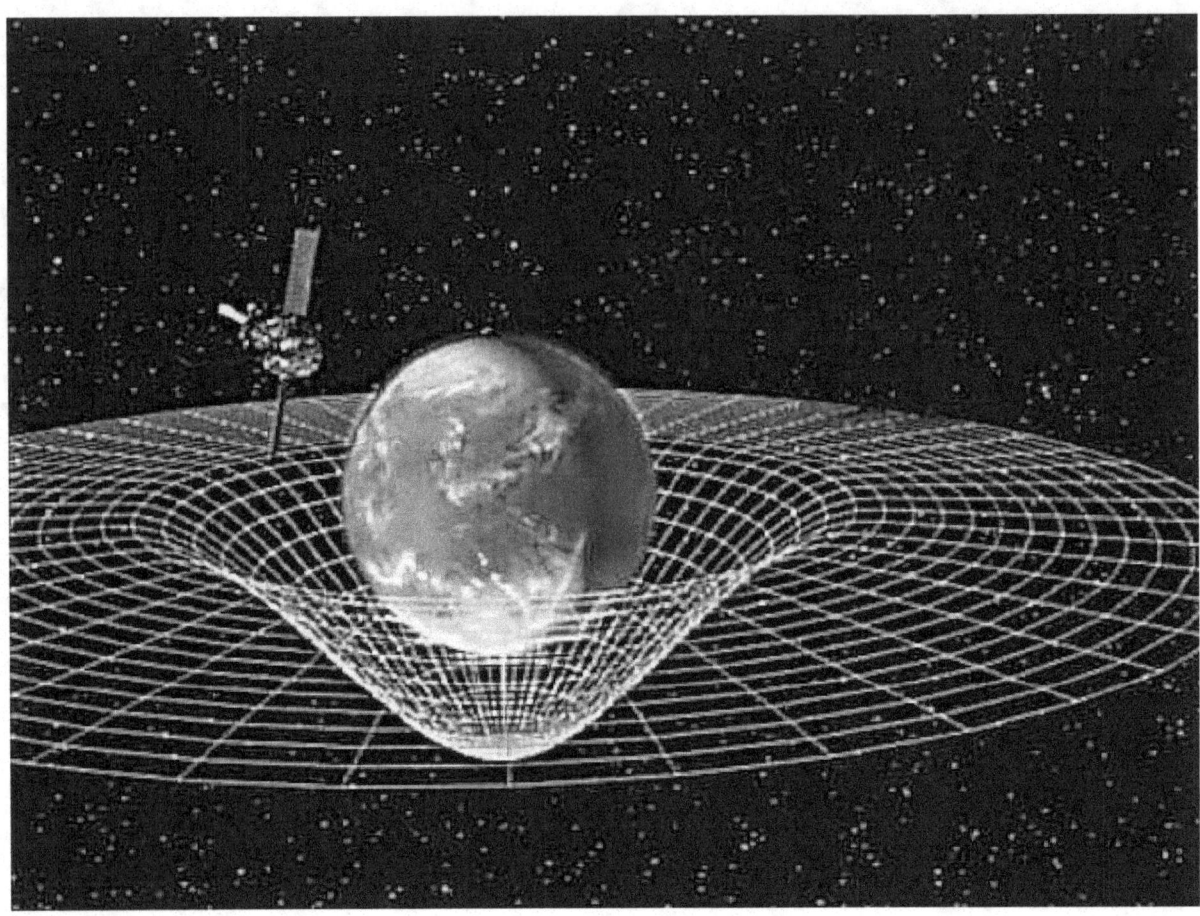

Two-dimensional analogy of spacetime distortion generated by the mass of an object. Matter changes the geometry of spacetime, this (curved) geometry being interpreted as gravity. White lines do not represent the curvature of space but instead represent the coordinate system imposed on the curved spacetime, which would be rectilinear in a flat spacetime.

- The Reissner-Nordström solution, in which the central object has an electrical charge. For charges with a geometrized length which are less than the geometrized length of the mass of the object, this solution produces black holes with two event horizons.

- The Kerr solution for rotating massive objects. This solution also produces black holes with multiple event horizons.

- The Kerr-Newman solution for charged, rotating massive objects. This solution also produces black holes with multiple event horizons.

- The cosmological Friedmann-Lemaître-Robertson-Walker solution, which predicts the expansion of the universe.

The tests of general relativity included the following:[9]

- General relativity accounts for the anomalous perihelion precession of Mercury.[10]

- The prediction that time runs slower at lower potentials has been confirmed by the Pound–Rebka experiment, the Hafele–Keating experiment, and the GPS.

- The prediction of the deflection of light was first confirmed by Arthur Stanley Eddington from his observations during the Solar eclipse of May 29, 1919.[11][12] Eddington measured starlight deflections twice those predicted by Newtonian corpuscular theory, in accordance with the predictions of general relativity. However, his interpretation of the results was later disputed.[13] More recent tests using radio interferometric measurements of quasars passing behind the Sun have more accurately and consistently confirmed the deflection of light to the degree predicted by general relativity.[14] See also gravitational lens.

- The time delay of light passing close to a massive object was first identified by Irwin I. Shapiro in 1964 in inter-planetary spacecraft signals.

- Gravitational radiation has been indirectly confirmed through studies of binary pulsars.

- Alexander Friedmann in 1922 found that Einstein equations have non-stationary solutions (even in the presence of the cosmological constant). In 1927 Georges Lemaître showed that static solutions of the Einstein equations, which are possible in the presence of the cosmological constant, are unstable, and therefore the static universe envisioned by Einstein could not exist. Later, in 1931, Einstein himself agreed with the results of Friedmann and Lemaître. Thus general relativity predicted that the Universe had to be non-static—it had to either expand or contract. The expansion of the universe discovered by Edwin Hubble in 1929 confirmed this prediction.[15]

- The theory's prediction of frame dragging was consistent with the recent Gravity Probe B results.[16]

- General relativity predicts that light should lose its energy when travelling away from the massive bodies. The group of Radek Wojtak of the Niels Bohr Institute at the University of Copenhagen collected data from 8000 galaxy clusters and found that the light coming from the cluster centers tended to be red-shifted compared to the cluster edges, confirming the energy loss due to gravity.[17]

17.1.5   Gravity and quantum mechanics

Main articles: Graviton and Quantum gravity

In the decades after the discovery of general relativity, it was realized that general relativity is incompatible with quantum mechanics.[18] It is possible to describe gravity in the framework of quantum field theory like the other fundamental forces, such that the attractive force of gravity arises due to exchange of virtual gravitons, in the same way as the electromagnetic force arises from exchange of virtual photons.[19][20] This reproduces general relativity in the classical limit. However, this approach fails at short distances of the order of the Planck length,[18] where a more complete theory of quantum gravity (or a new approach to quantum mechanics) is required.

17.2   Specifics

17.2.1   Earth's gravity

Main article: Earth's gravity

Every planetary body (including the Earth) is surrounded by its own gravitational field, which exerts an attractive force on all objects. Assuming a spherically symmetrical planet, the strength of this field at any given point above the surface is proportional to the planetary body's mass and inversely proportional to the square of the distance from the center of the body.

The strength of the gravitational field is numerically equal to the acceleration of objects under its influence. The rate of acceleration of falling objects near the Earth's surface varies very slightly depending on elevation, latitude, and other factors. For purposes of weights and measures, a standard gravity value is defined by the International Bureau of Weights and Measures, under the International System of Units (SI).

That value, denoted g, is $g = 9.80665$ m/s^2 (32.1740 ft/s^2).[21][22]

The standard value of 9.80665 m/s^2 is the one originally adopted by the International Committee on Weights and Measures in 1901 for 45° latitude, even though it has been shown to be too high by about five parts in ten thousand.[23] This value has persisted in meteorology and in some standard atmospheres as the value for 45° latitude even though it applies more precisely to latitude of 45°32'33".[24]

Assuming the standardized value for g and ignoring air resistance, this means that an object falling freely near the Earth's surface increases its velocity by 9.80665 m/s (32.1740 ft/s or 22 mph) for each second of its descent. Thus, an object starting from rest will attain a velocity of 9.80665 m/s (32.1740 ft/s) after one second, approximately 19.62 m/s (64.4 ft/s) after two seconds, and so on, adding 9.80665 m/s (32.1740 ft/s) to each resulting velocity. Also, again ignoring air resistance, any and all objects, when dropped from the same height, will hit the ground at the same time. It is relevant to note that Earth's gravity doesn't have exactly the same value in all regions. There are slight variations in different parts of the globe due to latitude, surface features such as mountains and ridges, and perhaps unusually high or low sub-surface densities.[25]

  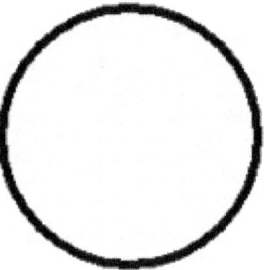

If an object with comparable mass to that of the Earth were to fall towards it, then the corresponding acceleration of the Earth would be observable.

According to Newton's 3rd Law, the Earth itself experiences a force equal in magnitude and opposite in direction to that which it exerts on a falling object. This means that the Earth also accelerates towards the object until they collide. Because the mass of the Earth is huge, however, the acceleration imparted to the Earth by this opposite force is negligible in comparison to the object's. If the object doesn't bounce after it has collided with the Earth, each of them then exerts a repulsive contact force on the other which effectively balances the attractive force of gravity and prevents further acceleration.

The force of gravity on Earth is the resultant (vector sum) of two forces: (a) The gravitational attraction in accordance with Newton's universal law of gravitation, and (b) the centrifugal force, which results from the choice of an earthbound, rotating frame of reference. At the equator, the force of gravity is the weakest due to the centrifugal force caused by the Earth's rotation. The force of gravity varies with latitude and increases from about 9.780 m/s^2 at the Equator to about 9.832 m/s^2 at the poles.

17.2.2 Equations for a falling body near the surface of the Earth

Main article: Equations for a falling body

Under an assumption of constant gravitational attraction, Newton's law of universal gravitation simplifies to $F = mg$, where m is the mass of the body and g is a constant vector with an average magnitude of 9.81 m/s^2 on Earth. This resulting force is the object's weight. The acceleration due to gravity is equal to this g. An initially stationary object which is allowed to fall freely under gravity drops a distance which is proportional to the square of the elapsed time. The image on the right, spanning half a second, was captured with a stroboscopic flash at 20 flashes per second. During the first $\frac{1}{20}$ of a second the ball drops one unit of distance (here, a unit is about 12 mm); by $\frac{2}{20}$ it has dropped at total of 4 units; by $\frac{3}{20}$, 9 units and so on.

Under the same constant gravity assumptions, the potential energy, Ep, of a body at height h is given by $Ep = mgh$ (or $Ep = Wh$, with W meaning weight). This expression is valid only over small distances h from the surface of the Earth. Similarly, the expression $h = \frac{v^2}{2g}$ for the maximum height reached by a vertically projected body with initial velocity v is useful for small heights and small initial velocities only.

17.2.3 Gravity and astronomy

The application of Newton's law of gravity has enabled the acquisition of much of the detailed information we have about the planets in our solar system, the mass of the Sun, and details of quasars; even the existence of dark matter is inferred using Newton's law of gravity. Although we have not traveled to all the planets nor to the Sun, we know their masses. These masses are obtained by applying the laws of gravity to the measured characteristics of the orbit. In space an object maintains its orbit because of the force of gravity acting upon it. Planets orbit stars, stars orbit Galactic Centers, galaxies orbit a center of mass in clusters, and clusters orbit in superclusters. The force of gravity exerted on one object by another is directly proportional to the product of those objects' masses and inversely proportional to the square of the distance between them.

17.2.4 Gravitational radiation

Main article: Gravitational wave

In general relativity, gravitational radiation is generated in situations where the curvature of spacetime is oscillating, such as is the case with co-orbiting objects. The gravitational radiation emitted by the Solar System is far too small to measure. However, gravitational radiation has been indirectly observed as an energy loss over time in binary pulsar systems such as PSR B1913+16. It is believed that neutron star mergers and black hole formation may create detectable amounts of gravitational radiation. Gravitational radiation observatories such as the Laser Interferometer Gravitational Wave Observatory (LIGO) have been created to study the problem. No confirmed detections have been made of this hypothetical radiation.

17.2.5 Speed of gravity

Main article: Speed of gravity

In December 2012, a research team in China announced that it had produced measurements of the phase lag of Earth tides during full and new moons which seem to prove that the speed of gravity is equal to the speed of light.[27] This means that if the Sun suddenly disappeared, the Earth would keep orbiting it normally for 8 minutes, which is the time light takes to travel that distance. The team's findings were released in the Chinese Science Bulletin in February 2013.[28]

17.3 Anomalies and discrepancies

There are some observations that are not adequately accounted for, which may point to the need for better theories of gravity or perhaps be explained in other ways.

- **Extra-fast stars**: Stars in galaxies follow a distribution of velocities where stars on the outskirts are moving faster than they should according to the observed distributions of normal matter. Galaxies within galaxy clusters show a similar pattern. Dark matter, which would interact gravitationally but not electromagnetically, would account for the discrepancy. Various modifications to Newtonian dynamics have also been proposed.

- **Flyby anomaly**: Various spacecraft have experienced greater acceleration than expected during gravity assist maneuvers.

- **Accelerating expansion**: The metric expansion of space seems to be speeding up. Dark energy has been proposed to explain this. A recent alternative explanation is that the geometry of space is not homogeneous (due to clusters of galaxies) and that when the data are reinterpreted to take this into account, the expansion is not speeding up after all,[29] however this conclusion is disputed.[30]

- **Anomalous increase of the astronomical unit**: Recent measurements indicate that planetary orbits are widening faster than if this were solely through the sun losing mass by radiating energy.

- **Extra energetic photons**: Photons travelling through galaxy clusters should gain energy and then lose it again on the way out. The accelerating expansion of the universe should stop the photons returning all the energy, but even taking this into account photons from the cosmic microwave background radiation gain twice as much energy as expected. This may indicate that gravity falls off *faster* than inverse-squared at certain distance scales.[31]

- **Extra massive hydrogen clouds**: The spectral lines of the Lyman-alpha forest suggest that hydrogen clouds are more clumped together at certain scales than expected and, like dark flow, may indicate that gravity falls off *slower* than inverse-squared at certain distance scales.[31]

- **Power**: Proposed extra dimensions could explain why the gravity force is so weak.[32]

17.4 Alternative theories

Main article: Alternatives to general relativity

17.4.1 Historical alternative theories

- Aristotelian theory of gravity

- Le Sage's theory of gravitation (1784) also called LeSage gravity, proposed by Georges-Louis Le Sage, based on a fluid-based explanation where a light gas fills the entire universe.

- Ritz's theory of gravitation, *Ann. Chem. Phys.* 13, 145, (1908) pp. 267–271, Weber-Gauss electrodynamics applied to gravitation. Classical advancement of perihelia.

- Nordström's theory of gravitation (1912, 1913), an early competitor of general relativity.

- Kaluza Klein theory (1921)

- Whitehead's theory of gravitation (1922), another early competitor of general relativity.

17.4.2 Recent alternative theories

- Brans–Dicke theory of gravity (1961) [33]

- Induced gravity (1967), a proposal by Andrei Sakharov according to which general relativity might arise from quantum field theories of matter

- $f(R)$ gravity (1970)

- Horndeski theory (1974) [34]

- Supergravity (1976)

- String theory

- In the modified Newtonian dynamics (MOND) (1981), Mordehai Milgrom proposes a modification of Newton's Second Law of motion for small accelerations [35]

- The self-creation cosmology theory of gravity (1982) by G.A. Barber in which the Brans-Dicke theory is modified to allow mass creation

- Loop quantum gravity (1988) by Carlo Rovelli, Lee Smolin, and Abhay Ashtekar

- Nonsymmetric gravitational theory (NGT) (1994) by John Moffat

- Tensor–vector–scalar gravity (TeVeS) (2004), a relativistic modification of MOND by Jacob Bekenstein

- Gravity as an entropic force, gravity arising as an emergent phenomenon from the thermodynamic concept of entropy.

- In the superfluid vacuum theory the gravity and curved space-time arise as a collective excitation mode of non-relativistic background superfluid.

- Chameleon theory (2004) by Justin Khoury and Amanda Weltman.

- Pressuron theory (2013) by Olivier Minazzoli and Aurélien Hees.

17.5   See also

- Angular momentum

- Anti-gravity, the idea of neutralizing or repelling gravity

- Artificial gravity

- Birkeland current

- Gravitational wave

- Gravitational wave background

- Cosmic gravitational wave background

- Einstein–Infeld–Hoffmann equations

- Escape velocity, the minimum velocity needed to escape from a gravity well

- g-force, a measure of acceleration

- Gauge gravitation theory

- Gauss's law for gravity

- Gravitational binding energy

- Gravity assist

- Gravity gradiometry

- Gravity Recovery and Climate Experiment

- Gravity Research Foundation

- Jovian–Plutonian gravitational effect

- Kepler's third law of planetary motion

- Lagrangian point

- Micro-g environment, also called microgravity

- Mixmaster dynamics

- n-body problem

- Newton's laws of motion

- Pioneer anomaly

- Scalar theories of gravitation

- Speed of gravity

- Standard gravitational parameter

- Standard gravity

- Weightlessness

17.6 Footnotes

[1] Ball, Phil (June 2005). "Tall Tales". *Nature News*. doi:10.1038/news050613-10.

[2] Galileo (1638), *Two New Sciences*, First Day Salviati speaks: "If this were what Aristotle meant you would burden him with another error which would amount to a falsehood; because, since there is no such sheer height available on earth, it is clear that Aristotle could not have made the experiment; yet he wishes to give us the impression of his having performed it when he speaks of such an effect as one which we see."

[3]   • Chandrasekhar, Subrahmanyan (2003). *Newton's Principia for the common reader*. Oxford: Oxford University Press. (pp.1–2). The quotation comes from a memorandum thought to have been written about 1714. As early as 1645 Ismaël Bullialdus had argued that any force exerted by the Sun on distant objects would have to follow an inverse-square law. However, he also dismissed the idea that any such force did exist. See, for example,

Linton, Christopher M. (2004). *From Eudoxus to Einstein—A History of Mathematical Astronomy*. Cambridge: Cambridge University Press. p. 225. ISBN 978-0-521-82750-8.

[4] M.C.W.Sandford (2008). "STEP: Satellite Test of the Equivalence Principle". Rutherford Appleton Laboratory. Retrieved 2011-10-14.

[5] Paul S Wesson (2006). *Five-dimensional Physics*. World Scientific. p. 82. ISBN 981-256-661-9.

[6] Haugen, Mark P.; C. Lämmerzahl (2001). *Principles of Equivalence: Their Role in Gravitation Physics and Experiments that Test Them*. Springer. arXiv:gr-qc/0103067. ISBN 978-3-540-41236-6.

[7] "Gravity and Warped Spacetime". black-holes.org. Retrieved 2010-10-16.

[8] Dmitri Pogosyan. "Lecture 20: Black Holes—The Einstein Equivalence Principle". University of Alberta. Retrieved 2011-10-14.

[9] Pauli, Wolfgang Ernst (1958). "Part IV. General Theory of Relativity". *Theory of Relativity*. Courier Dover Publications. ISBN 978-0-486-64152-2.

[10] Max Born (1924), *Einstein's Theory of Relativity* (The 1962 Dover edition, page 348 lists a table documenting the observed and calculated values for the precession of the perihelion of Mercury, Venus, and Earth.)

[11] Dyson, F.W.; Eddington, A.S.; Davidson, C.R. (1920). "A Determination of the Deflection of Light by the Sun's Gravitational Field, from Observations Made at the Total Eclipse of May 29, 1919". *Phil. Trans. Roy. Soc. A* **220** (571–581): 291–333. Bibcode:1920RSPTA.220..291D. doi:10.1098/rsta.1920.0009.. Quote, p. 332: "Thus the results of the expeditions to Sobral and Principe can leave little doubt that a deflection of light takes place in the neighbourhood of the sun and that it is of the amount demanded by Einstein's generalised theory of relativity, as attributable to the sun's gravitational field."

[12] Weinberg, Steven (1972). *Gravitation and cosmology*. John Wiley & Sons.. Quote, p. 192: "About a dozen stars in all were studied, and yielded values 1.98 ± 0.11" and 1.61 ± 0.31", in substantial agreement with Einstein's prediction $\theta\odot = 1.75$"."

[13] Earman, John; Glymour, Clark (1980). "Relativity and Eclipses: The British eclipse expeditions of 1919 and their predecessors". *Historical Studies in the Physical Sciences* **11**: 49–85. doi:10.2307/27757471.

[14] Weinberg, Steven (1972). *Gravitation and cosmology*. John Wiley & Sons. p. 194.

[15] See W.Pauli, 1958, pp.219–220

[16] "NASA's Gravity Probe B Confirms Two Einstein Space-Time Theories". Nasa.gov. Retrieved 2013-07-23.

[17] Bhattacharjee, Yudhijit. "Galaxy Clusters Validate Einstein's Theory". News.sciencemag.org. Retrieved 2013-07-23.

[18] Randall, Lisa (2005). *Warped Passages: Unraveling the Universe's Hidden Dimensions*. Ecco. ISBN 0-06-053108-8.

[19] Feynman, R. P.; Morinigo, F. B.; Wagner, W. G.; Hatfield, B. (1995). *Feynman lectures on gravitation*. Addison-Wesley. ISBN 0-201-62734-5.

[20] Zee, A. (2003). *Quantum Field Theory in a Nutshell*. Princeton University Press. ISBN 0-691-01019-6.

[21] Bureau International des Poids et Mesures (2006). "The International System of Units (SI)" (PDF) (8th ed.). p. 131. Retrieved 2009-11-25. Unit names are normally printed in Roman (upright) type ... Symbols for quantities are generally single letters set in an italic font, although they may be qualified by further information in subscripts or superscripts or in brackets.

[22] "SI Unit rules and style conventions". National Institute For Standards and Technology (USA). September 2004. Retrieved 2009-11-25. Variables and quantity symbols are in italic type. Unit symbols are in Roman type.

[23] List, R. J. editor, 1968, Acceleration of Gravity, *Smithsonian Meteorological Tables*, Sixth Ed. Smithsonian Institution, Washington, D.C., p. 68.

[24] U.S. Standard Atmosphere, 1976, U.S. Government Printing Office, Washington, D.C., 1976. (Linked file is very large.)

[25] "Astronomy Picture of the Day".

[26] "Milky Way Emerges as Sun Sets over Paranal". *www.eso.org*. European Southern Obseevatory. Retrieved 29 April 2015.

[27] Chinese scientists find evidence for speed of gravity, astrowatch.com, 12/28/12.

[28] TANG, Ke Yun; HUA ChangCai; WEN Wu; CHI ShunLiang; YOU QingYu; YU Dan (February 2013). "Observational evidences for the speed of the gravity based on the Earth tide" (PDF). *Chinese Science Bulletin* **58** (4-5): 474–477. doi:10.-012-5603-3.Retrieved 12 June 2013.

[29] Dark energy may just be a cosmic illusion, *New Scientist*, issue 2646, 7 March 2008.

[30] Swiss-cheese model of the cosmos is full of holes, *New Scientist*, issue 2678, 18 October 2008.

[31] Chown, Marcus (16 March 2009). "Gravity may venture where matter fears to tread". *New Scientist* (2699). Retrieved 4 August 2013.

[32] CERN (20 January 2012). "Extra dimensions, gravitons, and tiny black holes".

[33] Brans, C.H. (Mar 2014). "Jordan-Brans-Dicke Theory". *Scholarpedia* **9**: 31358. Bibcode:2014Schpj...931358B. doi: 358.

[34] Horndeski, G.W. (Sep 1974). "Second-Order Scalar-Tensor Field Equations in a Four-Dimensional Space". *International Journal of Theoretical Physics* **88** (10): 363–384. Bibcode:1974IJTP...10..363H. doi:10.1007/BF01807638.

[35] Milgrom, M. (Jun 2014). "The MOND paradigm of modified dynamics". *Scholarpedia* **9**: 31410. Bibcode:2014SchpJ...931410M. doi:10.4249/scholarpedia.31410.

17.7   References

- Halliday, David; Robert Resnick; Kenneth S. Krane (2001). *Physics v. 1*. New York: John Wiley & Sons. ISBN 0-471-32057-9.

- Serway, Raymond A.; Jewett, John W. (2004). *Physics for Scientists and Engineers* (6th ed.). Brooks/Cole. ISBN 0-534-40842-7.

- Tipler, Paul (2004). *Physics for Scientists and Engineers: Mechanics, Oscillations and Waves, Thermodynamics* (5th ed.). W. H. Freeman. ISBN 0-7167-0809-4.

17.8 Further reading

- Thorne, Kip S.; Misner, Charles W.; Wheeler, John Archibald (1973). *Gravitation*. W.H. Freeman. ISBN 0-7167-0344-0.

17.9 External links

- Hazewinkel, Michiel, ed. (2001), "Gravitation", *Encyclopedia of Mathematics*, Springer, ISBN 978-1-55608-010-4

- Hazewinkel, Michiel, ed. (2001), "Gravitation, theory of", *Encyclopedia of Mathematics*, Springer, ISBN 978-1-55608-010-4

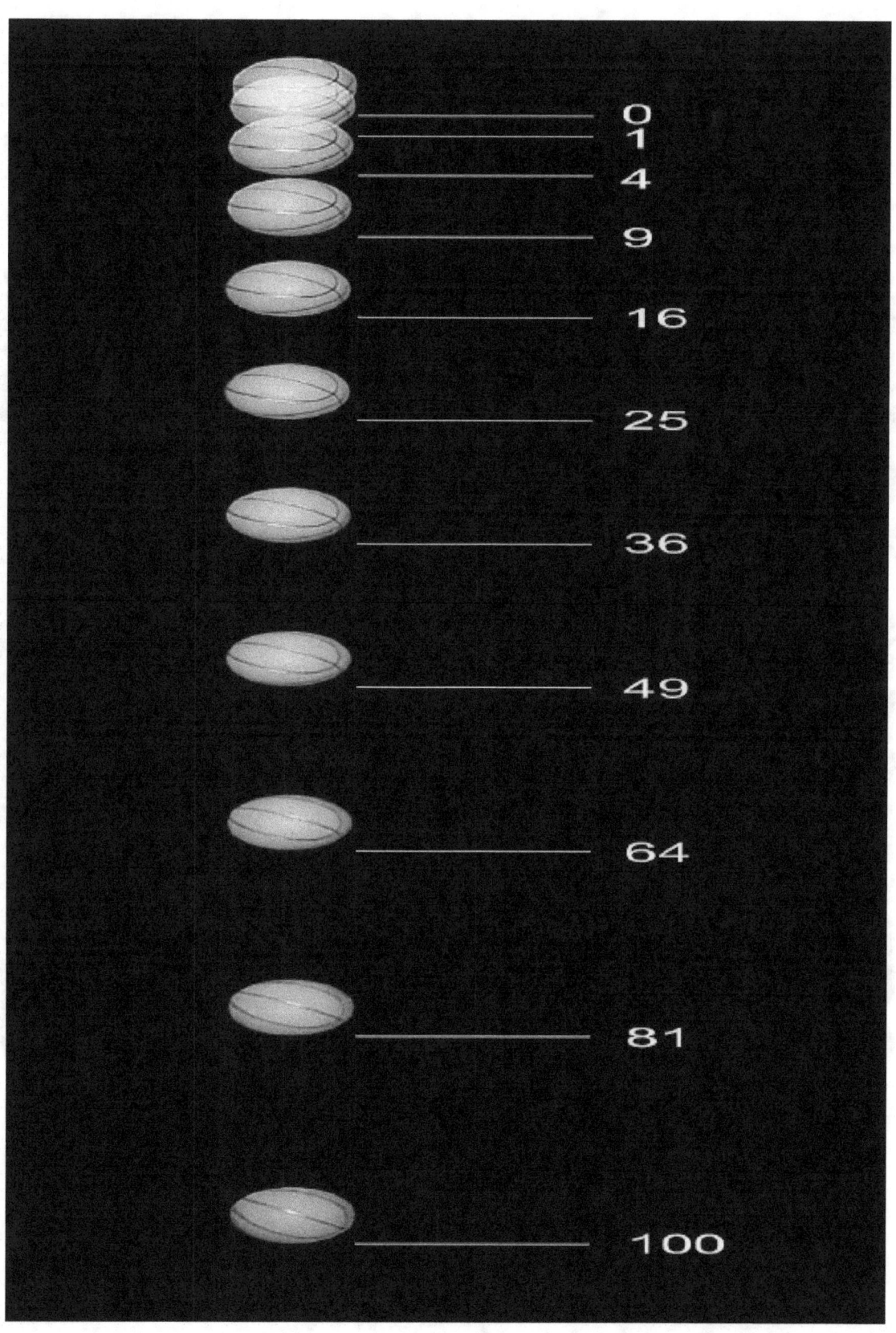

Gravity acts on stars that conform our Milky Way.[26]

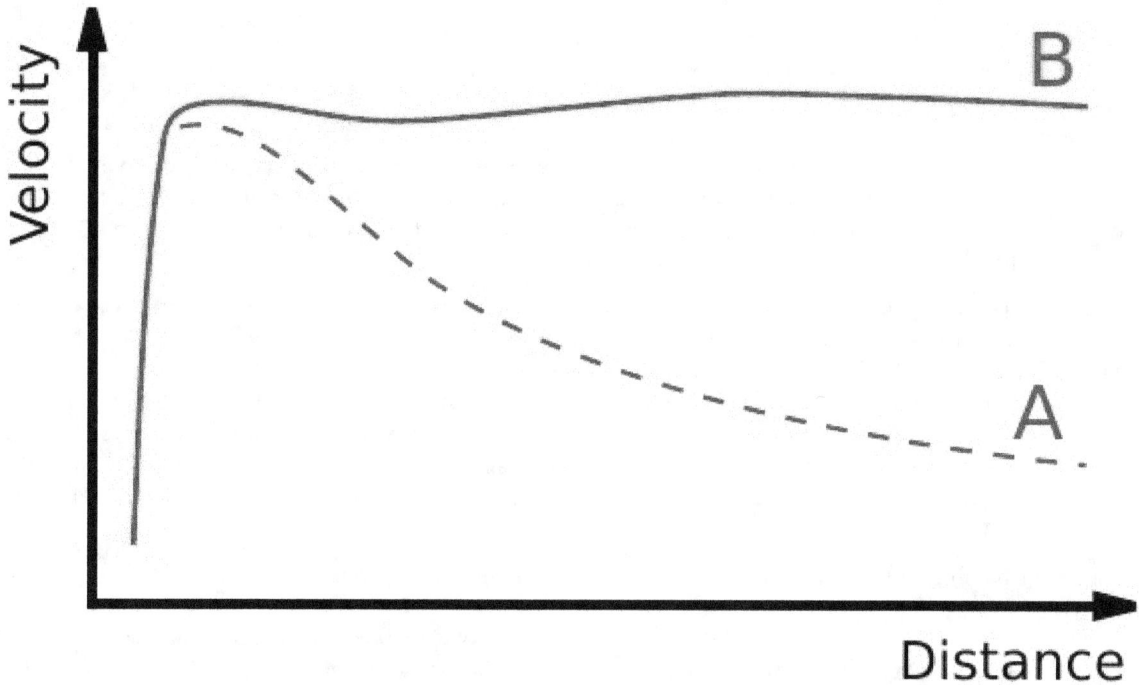

Rotation curve of a typical spiral galaxy: predicted (A) and observed (B). The discrepancy between the curves is attributed to dark matter.

Chapter 18

Strong interaction

In particle physics, the **strong interaction** is the mechanism responsible for the strong nuclear force (also called the **strong force, nuclear strong force** or **colour force**), one of the four fundamental interactions of nature, the others being electromagnetism, the weak interaction and gravitation. Effective only at a distance of a femtometer, it is approximately 100 times stronger than electromagnetism, a million times stronger than the weak force interaction and 10^{38} times stronger than gravitation at that range.[1] It ensures the stability of ordinary matter, as it confines the quark elementary particles into hadron particles, such as the proton and neutron, the largest components of the mass of ordinary matter. Furthermore, most of the mass-energy of a common proton or neutron is in the form of the strong force field energy; the individual quarks provide only about 1% of the mass-energy of a proton.

The strong interaction is observable in two areas: on a larger scale (about 1 to 3 femtometers (fm)), it is the force that binds protons and neutrons (nucleons) together to form the nucleus of an atom. On the smaller scale (less than about 0.8 fm, the radius of a nucleon), it is the force (carried by gluons) that holds quarks together to form protons, neutrons, and other hadron particles. The strong force inherently has so high a strength that the energy of an object bound by the strong force (a hadron) is high enough to produce new massive particles. Thus, if hadrons are struck by high-energy particles, they give rise to new hadrons instead of emitting freely moving radiation (gluons). This property of the strong force is called colour confinement, and it prevents the free "emission" of the strong force: instead, in practice, jets of massive particles are observed.

In the context of binding protons and neutrons together to form atomic nuclei, the strong interaction is called the nuclear force (or *residual strong force*). In this case, it is the residuum of the strong interaction between the quarks that make up the protons and neutrons. As such, the residual strong interaction obeys a quite different distance-dependent behavior between nucleons, from when it is acting to bind quarks within nucleons. The binding energy that is partly released on the breakup of a nucleus is related to the residual strong force and is harnessed in nuclear power and fission-type nuclear weapons.[2][3]

The strong interaction is thought to be mediated by massless particles called gluons, that are exchanged between quarks, antiquarks, and other gluons. Gluons, in turn, are thought to interact with quarks and gluons as all carry a type of charge called colour charge. Colour charge is analogous to electromagnetic charge, but it comes in three types rather than one (+/- red, +/- green, +/- blue) that results in a different type of force, with different rules of behavior. These rules are detailed in the theory of quantum chromodynamics (QCD), which is the theory of quark-gluon interactions.

Just after the Big Bang, and during the electroweak epoch, the electroweak force separated from the strong force. Although it is expected that a Grand Unified Theory exists to describe this, no such theory has been successfully formulated, and the unification remains an unsolved problem in physics.

18.1   History

Before the 1970s, physicists were uncertain about the binding mechanism of the atomic nucleus. It was known that the nucleus was composed of protons and neutrons and that protons possessed positive electric charge, while neutrons

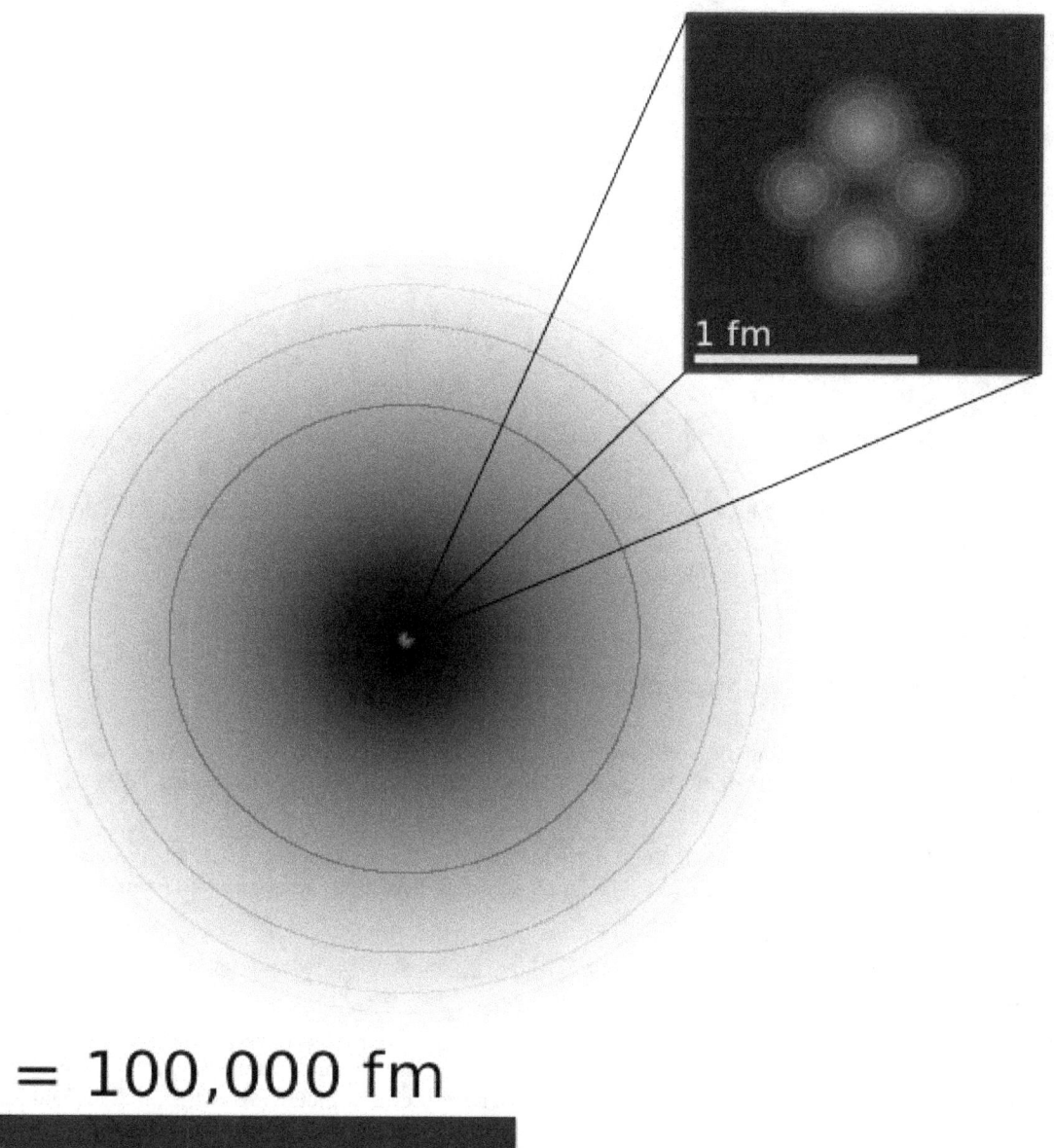

1 Å = 100,000 fm

The nucleus of a helium atom. The two protons have the same charge, but still stay together due to the residual nuclear force

were electrically neutral. However, these facts seemed to contradict one another. By physical understanding at that time, positive charges would repel one another and the nucleus should therefore fly apart. However, this was never observed. New physics was needed to explain this phenomenon.

A stronger attractive force was postulated to explain how the atomic nucleus was bound together despite the protons' mutual electromagnetic repulsion. This hypothesized force was called the *strong force*, which was believed to be a fundamental force that acted on the protons and neutrons that make up the nucleus.

It was later discovered that protons and neutrons were not fundamental particles, but were made up of constituent particles called quarks. The strong attraction between nucleons was the side-effect of a more fundamental force that bound the quarks together in the protons and neutrons. The theory of quantum chromodynamics explains that quarks carry what is called a colour charge, although it has no relation to visible colour.[4] Quarks with unlike colour charge attract one another as a result of the **strong interaction**, which is mediated by particles called gluons.

18.2 Details

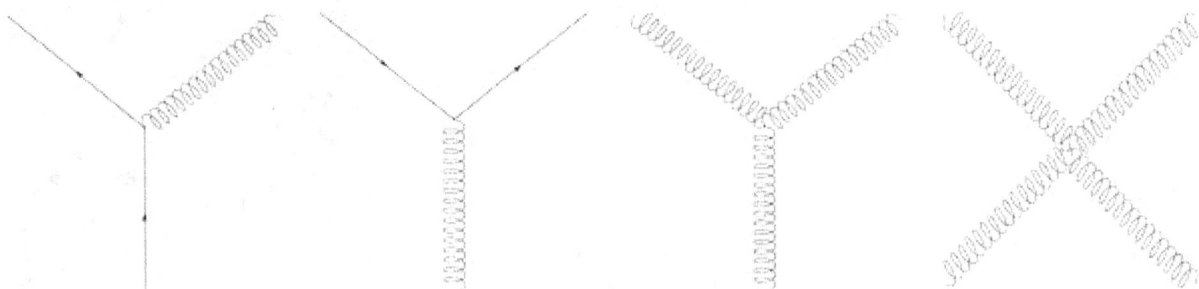

The fundamental couplings of the strong interaction, from left to right: gluon radiation, gluon splitting and gluon self-coupling.

The word *strong* is used since the strong interaction is the "strongest" of the four fundamental forces; its strength is around 10^2 times that of the electromagnetic force, some 10^6 times as great as that of the weak force, and about 10^{39} times that of gravitation, at a distance of a femtometer or less.

18.2.1 Behaviour of the strong force

The contemporary understanding of strong force is described by quantum chromodynamics (QCD), a part of the standard model of particle physics. Mathematically, QCD is a non-Abelian gauge theory based on a local (gauge) symmetry group called SU(3).

Quarks and gluons are the only fundamental particles that carry non-vanishing colour charge, and hence participate in strong interactions. The strong force itself acts directly only on elementary quark and gluon particles.

All quarks and gluons in QCD interact with each other through the strong force. The strength of interaction is parametrized by the strong coupling constant. This strength is modified by the gauge colour charge of the particle, a group theoretical property.

The strong force acts between quarks. Unlike all other forces (electromagnetic, weak, and gravitational), the strong force does not diminish in strength with increasing distance. After a limiting distance (about the size of a hadron) has been reached, it remains at a strength of about 10,000 newtons, no matter how much farther the distance between the quarks.[5] In QCD, this phenomenon is called colour confinement; it implies that only hadrons, not individual free quarks, can be observed. The explanation is that the amount of work done against a force of 10,000 newtons (about the weight of a one-metric ton mass on the surface of the Earth) is enough to create particle-antiparticle pairs within a very short distance of an interaction. In simple terms, the very energy applied to pull two quarks apart will create a pair of new quarks that will pair up with the original ones. The failure of all experiments that have searched for free quarks is considered to be evidence for this phenomenon.

The elementary quark and gluon particles affected are unobservable directly, but they instead emerge as jets of newly created hadrons, whenever energy is deposited into a quark-quark bond, as when a quark in a proton is struck by a very fast quark (in an impacting proton) during a particle accelerator experiment. However, quark–gluon plasmas have been observed.

Every quark in the universe does not attract every other quark in the above distance independent manner, since colour-confinement implies that the strong force acts without distance-diminishment only between pairs of single quarks, and that in collections of bound quarks (i.e., hadrons), the net colour-charge of the quarks cancels out, as seen from far away. Collections of quarks (hadrons) therefore appear (nearly) without colour-charge, and the strong force is therefore nearly absent between these hadrons (i.e., between baryons or mesons). However, the cancellation is not quite perfect. A small residual force remains (described below) known as the **residual strong force**. This residual force *does* diminish rapidly with distance, and is thus very short-range (effectively a few femtometers). It manifests as a force between the "colourless" hadrons, and is therefore sometimes known as the **strong nuclear force** or simply nuclear force.

18.2.2   Residual strong force

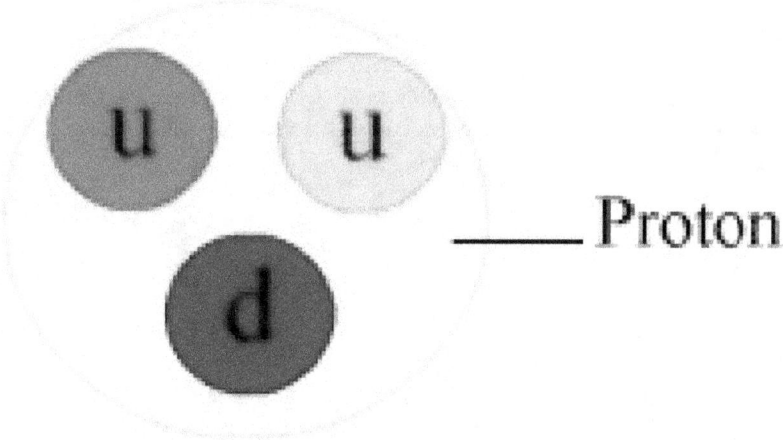

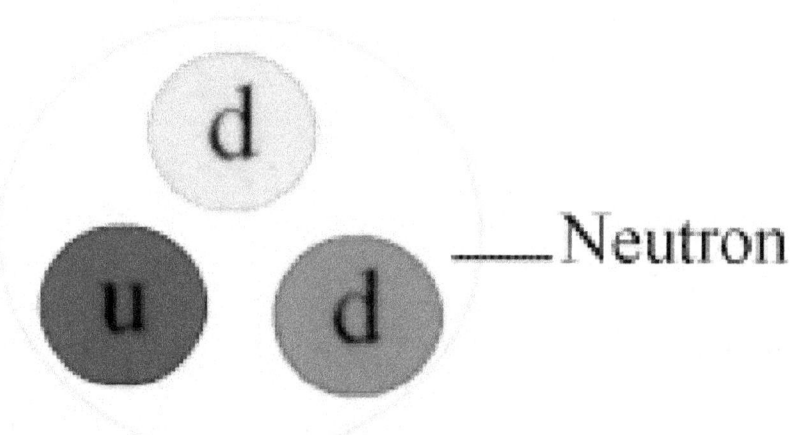

An animation of the nuclear force (or residual strong force) interaction between a proton and a neutron. The small coloured double circles are gluons, which can be seen binding the proton and neutron together. These gluons also hold the quark-antiquark combination called the pion together, and thus help transmit a residual part of the strong force even between colourless hadrons. Anticolours are shown as per this diagram. For a larger version, click here

The residual effect of the strong force is called the nuclear force. The nuclear force acts between hadrons, such as mesons or the nucleons in atomic nuclei. This "residual strong force", acting indirectly, transmits gluons that form part of the virtual pi and rho mesons, which, in turn, transmit the nuclear force between nucleons.

The residual strong force is thus a minor residuum of the strong force that binds quarks together into protons and neutrons.

This same force is much weaker *between* neutrons and protons, because it is mostly neutralized *within* them, in the same way that electromagnetic forces between neutral atoms (van der Waals forces) are much weaker than the electromagnetic forces that hold the atoms internally together.[6]

Unlike the strong force itself, the nuclear force, or residual strong force, *does* diminish in strength, and in fact diminishes rapidly with distance. The decrease is approximately as a negative exponential power of distance, though there is no simple expression known for this; see Yukawa potential. This fact, together with the less-rapid decrease of the disruptive electromagnetic force between protons with distance, causes the instability of larger atomic nuclei, such as all those with atomic numbers larger than 82 (the element lead).

18.3 See also

- Nuclear binding energy

- Colour charge

- Coupling constant

- Nuclear physics

- QCD matter

- Quantum field theory and Gauge theory

- Standard model of particle physics and Standard Model (mathematical formulation)

- Weak interaction, electromagnetism and gravity

- Intermolecular force

- Vortex

- Yukawa interaction

18.4 References

[1] Relative strength of interaction varies with distance. See for instance Matt Strassler's essay, "The strength of the known forces".

[2] on Binding energy: see Binding Energy, Mass Defect, Furry Elephant physics educational site, retr 2012 7 1

[3] on Binding energy: see Chapter 4 NUCLEAR PROCESSES, THE STRONG FORCE, M. Ragheb 1/27/2012, University of Illinois

[4] Feynman, R. P. (1985). *QED: The Strange Theory of Light and Matter*. Princeton University Press. p. 136. ISBN 0-691-08388-6. The idiot physicists, unable to come up with any wonderful Greek words anymore, call this type of polarization by the unfortunate name of 'colour,' which has nothing to do with colour in the normal sense.

[5] Fritzsch, op. cite. p. 164. The author states that the force between differently coloured quarks remains constant at any distance after they travel only a tiny distance from each other, and is equal to that need to raise one ton, which is 1000 kg x 9.8 m/s^2 = ~10,000 N.

[6] Fritzsch, H. (1983). *Quarks: The Stuff of Matter*. Basic Books. pp. 167–168. ISBN 978-0-465-06781-7.

18.5   Further reading

- Christman, J. R. (2001). "MISN-0-280: *The Strong Interaction*" (PDF). *Project PHYSNET*. External link in |work= (help)

- Griffiths, David (1987). *Introduction to Elementary Particles*. John Wiley & Sons. ISBN 0-471-60386-4.

- Halzen, F.; Martin, A. D. (1984). *Quarks and Leptons: An Introductory Course in Modern Particle Physics*. John Wiley & Sons. ISBN 0-471-88741-2.

- Kane, G. L. (1987). *Modern Elementary Particle Physics*. Perseus Books. ISBN 0-201-11749-5.

- Morris, R. (2003). *The Last Sorcerers: The Path from Alchemy to the Periodic Table*. Joseph Henry Press. ISBN 0-309-50593-3.

18.6   External links

- Strong force at *Encyclopædia Britannica*

Chapter 19

Weak interaction

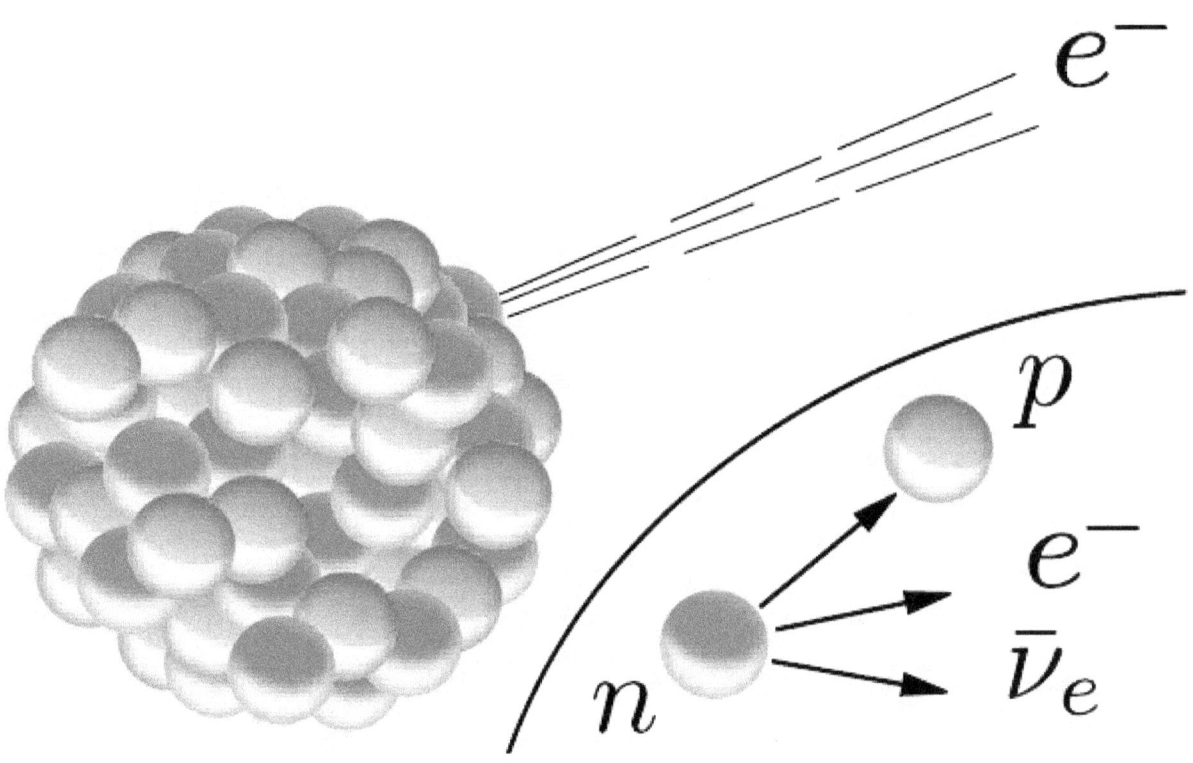

The radioactive beta decay is possible due to the weak interaction, which transforms a neutron into: a proton, an electron, and an electron antineutrino.

In particle physics, the **weak interaction** is the mechanism responsible for the **weak force** or **weak nuclear force**, one of the four known fundamental interactions of nature, alongside the strong interaction, electromagnetism, and gravitation. The weak interaction is responsible for the radioactive decay of subatomic particles, and it plays an essential role in nuclear fission. The theory of the weak interaction is sometimes called **quantum flavordynamics (QFD)**, in analogy with the terms QCD and QED, but the term is rarely used because the weak force is best understood in terms of electro-weak theory (EWT).[1]

In the Standard Model of particle physics, the weak interaction is caused by the emission or absorption of W and Z bosons. All known fermions interact through the weak interaction. Fermions are particles that have half-integer spin (one of the fundamental properties of particles). A fermion can be an elementary particle, such as the electron, or it can be a composite particle, such as the proton. The masses of W^+, W^-, and Z bosons are each far greater than that of protons or neutrons, consistent with the short range of the weak force. The force is termed *weak* because its field strength over a given distance is typically several orders of magnitude less than that of the strong nuclear force and electromagnetic force.

During the quark epoch, the electroweak force split into the electromagnetic and weak forces. Important examples of weak interaction include beta decay, and the production, from hydrogen, of deuterium needed to power the sun's thermonuclear process. Most fermions will decay by a weak interaction over time. Such decay also makes radiocarbon dating possible, as carbon-14 decays through the weak interaction to nitrogen-14. It can also create radioluminescence, commonly used in tritium illumination, and in the related field of betavoltaics.[2]

Quarks, which make up composite particles like neutrons and protons, come in six "flavours" – up, down, strange, charm, top and bottom – which give those composite particles their properties. The weak interaction is unique in that it allows for quarks to swap their flavour for another. For example, during beta minus decay, a down quark decays into an up quark, converting a neutron to a proton. Also the weak interaction is the only fundamental interaction that breaks parity-symmetry, and similarly, the only one to break CP-symmetry.

19.1   History

In 1933, Enrico Fermi proposed the first theory of the weak interaction, known as Fermi's interaction. He suggested that beta decay could be explained by a four-fermion interaction, involving a contact force with no range.[3][4]

However, it is better described as a non-contact force field having a finite range, albeit very short. In 1968, Sheldon Glashow, Abdus Salam and Steven Weinberg unified the electromagnetic force and the weak interaction by showing them to be two aspects of a single force, now termed the electro-weak force.

The existence of the W and Z bosons was not directly confirmed until 1983.

19.2   Properties

The weak interaction is unique in a number of respects:

1. It is the only interaction capable of changing the flavor of quarks (i.e., of changing one type of quark into another).

2. It is the only interaction that violates **P** or parity-symmetry. It is also the only one that violates **CP** symmetry.

3. It is propagated by carrier particles (known as gauge bosons) that have significant masses, an unusual feature which is explained in the Standard Model by the Higgs mechanism.

Due to their large mass (approximately 90 GeV/c^2[5]) these carrier particles, termed the W and Z bosons, are short-lived: they have a lifetime of under 1×10^{-24} seconds.[6] The weak interaction has a coupling constant (an indicator of interaction strength) of between 10^{-7} and 10^{-6}, compared to the strong interaction's coupling constant of about 1 and the electromagnetic coupling constant of about 10^{-2};[7] consequently the weak interaction is weak in terms of strength.[8] The weak interaction has a very short range (around 10^{-17}–10^{-16} m[8]).[7] At distances around 10^{-18} meters, the weak interaction has a strength of a similar magnitude to the electromagnetic force, but this starts to decrease exponentially with increasing distance. At distances of around 3×10^{-17} m, the weak interaction is 10,000 times weaker than the electromagnetic.[9]

The weak interaction affects all the fermions of the Standard Model, as well as the Higgs boson; neutrinos interact through gravity and the weak interaction only, and neutrinos were the original reason for the name *weak force*.[8] The weak interaction does not produce bound states (nor does it involve binding energy) – something that gravity does on an astronomical scale, that the electromagnetic force does at the atomic level, and that the strong nuclear force does inside nuclei.[10]

Its most noticeable effect is due to its first unique feature: flavor changing. A neutron, for example, is heavier than a proton (its sister nucleon), but it cannot decay into a proton without changing the flavor (type) of one of its two *down*

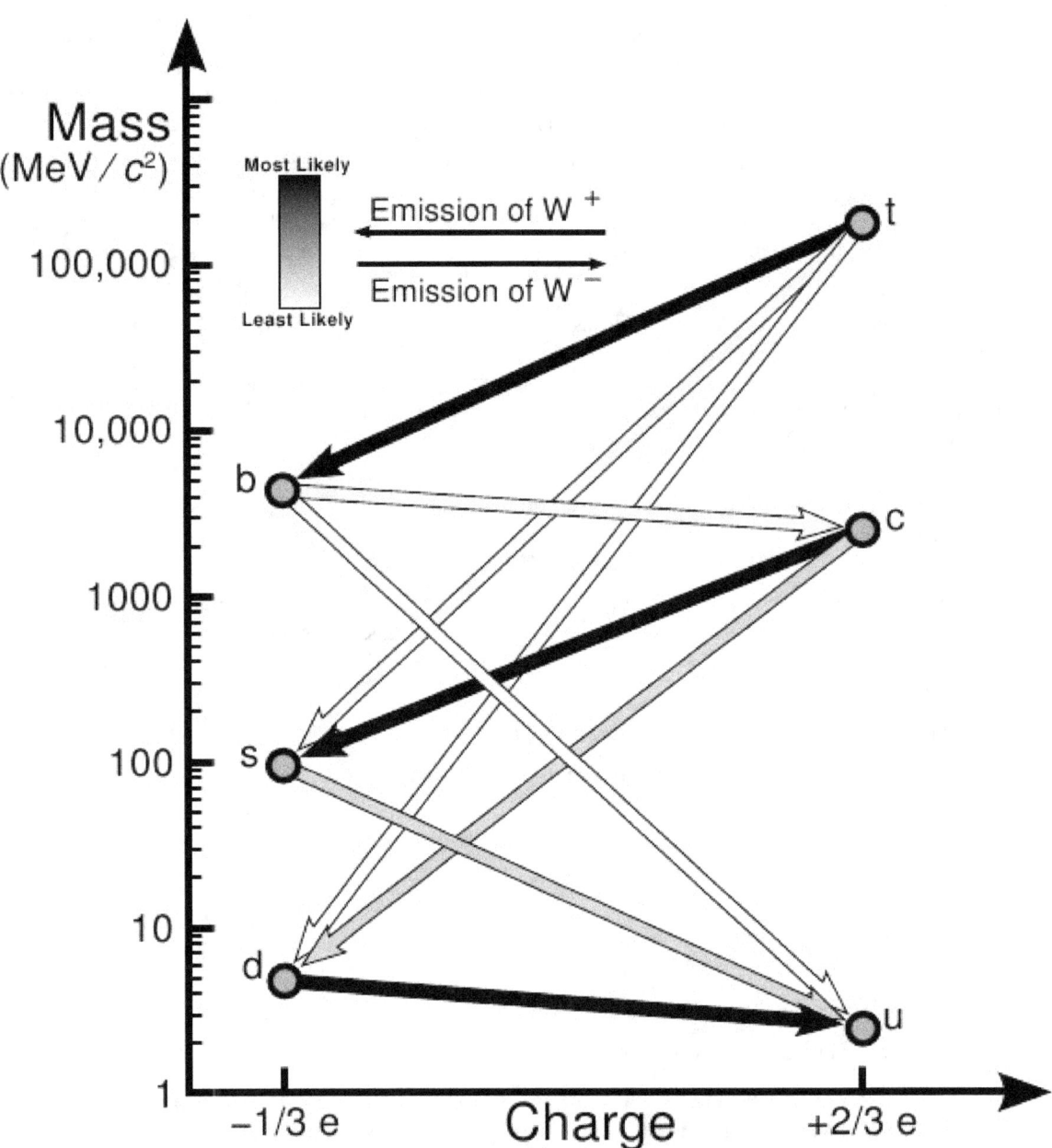

A diagram depicting the various decay routes due to the weak interaction and some indication of their likelihood. The intensity of the lines are given by the CKM parameters.

quarks to *up*. Neither the strong interaction nor electromagnetism permit flavour changing, so this must proceed by **weak decay**; without weak decay, quark properties such as strangeness and charm (associated with the quarks of the same name) would also be conserved across all interactions. All mesons are unstable because of weak decay.[11] In the process known as beta decay, a *down* quark in the neutron can change into an *up* quark by emitting a virtual W− boson which is then converted into an electron and an electron antineutrino.[12] Another example is the electron capture, a common variant of radioactive decay, where a proton (up quark) and an electron within an atom interact, and are changed to a neutron (down quark) and an electron neutrino.

Due to the large mass of a boson, weak decay is much more unlikely than strong or electromagnetic decay, and hence occurs less rapidly. For example, a neutral pion (which decays electromagnetically) has a life of about 10^{-16} seconds, while a charged pion (which decays through the weak interaction) lives about 10^{-8} seconds, a hundred million times longer.[13] In contrast, a free neutron (which also decays through the weak interaction) lives about 15 minutes.[12]

19.2.1   Weak isospin and weak hypercharge

Main article: Weak isospin

All particles have a property called weak isospin (T_3), which serves as a quantum number and governs how that particle interacts in the weak interaction. Weak isospin therefore plays the same role in the weak interaction as electric charge does in electromagnetism, and color charge in the strong interaction. All fermions have a weak isospin value of either $+\frac{1}{2}$ or $-\frac{1}{2}$. For example, the up quark has a T_3 of $+\frac{1}{2}$ and the down quark $-\frac{1}{2}$. A quark never decays through the weak interaction into a quark of the same T_3: quarks with a T_3 of $+\frac{1}{2}$ decay into quarks with a T_3 of $-\frac{1}{2}$ and vice versa.

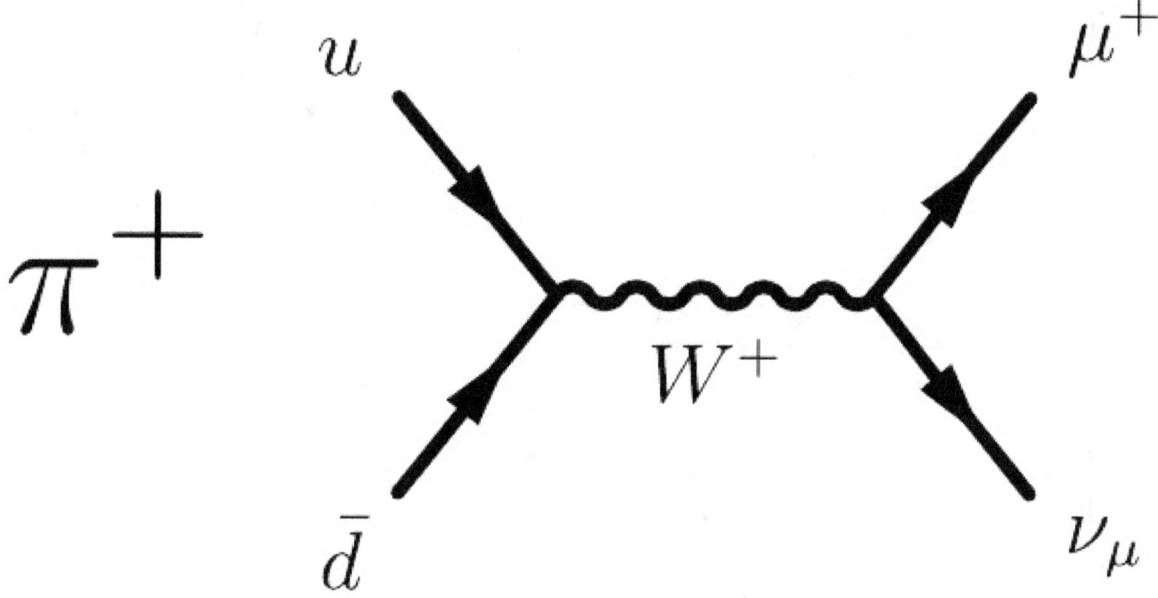

π+ decay through the weak interaction

In any given interaction, weak isospin is conserved: the sum of the weak isospin numbers of the particles entering the interaction equals the sum of the weak isospin numbers of the particles exiting that interaction. For example, a (left-handed) π+, with a weak isospin of 1 normally decays into a v
μ (+1/2) and a μ+ (as a right-handed antiparticle, +1/2).[13]

Following the development of the electroweak theory, another property, weak hypercharge, was developed. It is dependent on a particle's electrical charge and weak isospin, and is defined as:

$$Y_W = 2(Q - T_3)$$

where YW is the weak hypercharge of a given type of particle, Q is its electrical charge (in elementary charge units) and T_3 is its weak isospin. Whereas some particles have a weak isospin of zero, all particles, except gluons, have non-zero weak hypercharge. Weak hypercharge is the generator of the U(1) component of the electroweak gauge group.

19.3 Interaction types

There are two types of weak interaction (called *vertices*). The first type is called the "charged-current interaction" because it is mediated by particles that carry an electric charge (the W+ or W− bosons), and is responsible for the beta decay phenomenon. The second type is called the "neutral-current interaction" because it is mediated by a neutral particle, the Z boson.

19.3.1   Charged-current interaction

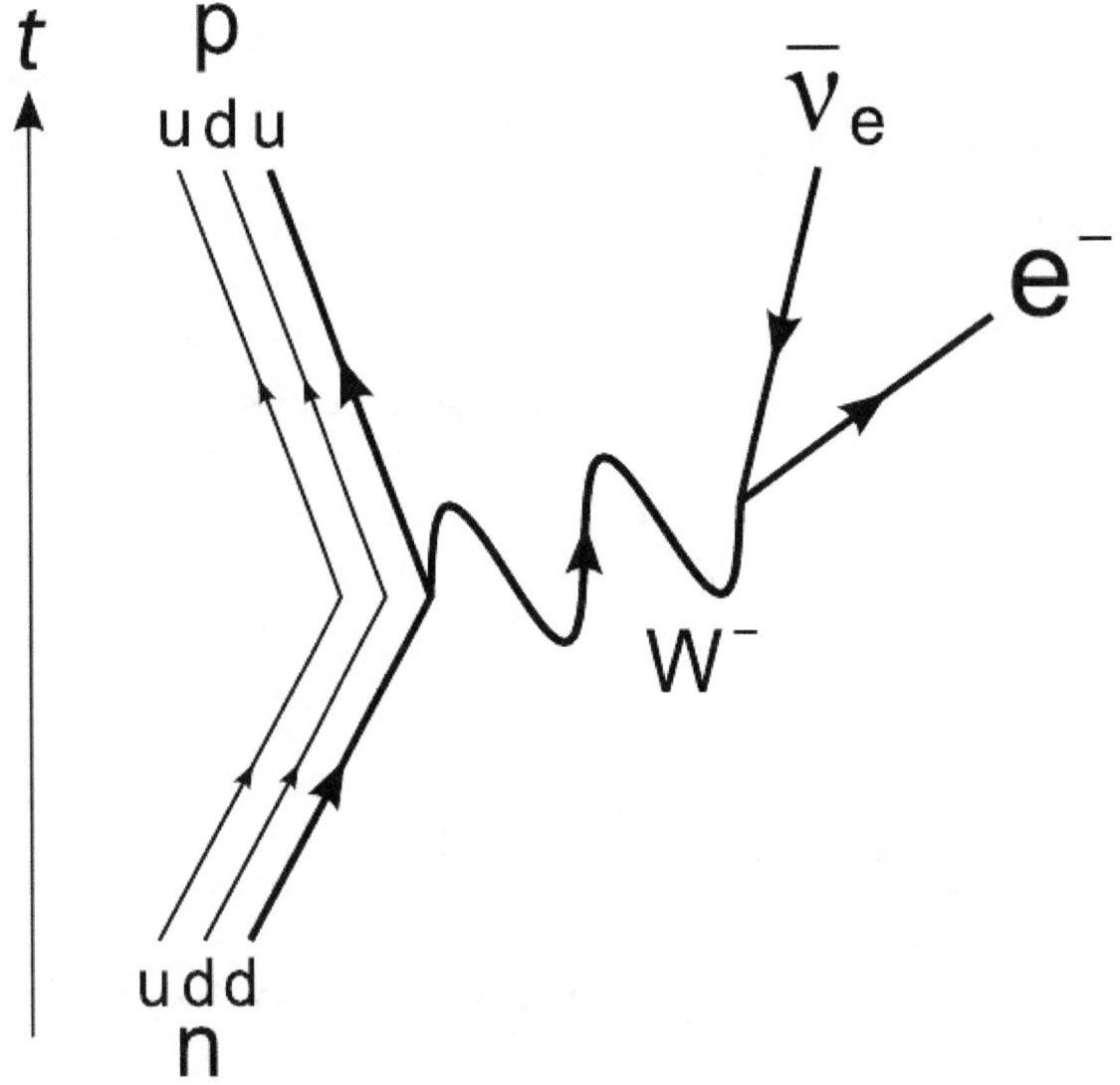

The Feynman diagram for beta-minus decay of a neutron into a proton, electron and electron anti-neutrino, via an intermediate heavy W− boson

In one type of charged current interaction, a charged lepton (such as an electron or a muon, having a charge of −1) can absorb a W+ boson (a particle with a charge of +1) and be thereby converted into a corresponding neutrino (with a charge of 0), where the type ("family") of neutrino (electron, muon or tau) is the same as the type of lepton in the interaction, for example:

$$\mu^- + W^+ \rightarrow \nu_\mu$$

Similarly, a down-type quark (d with a charge of $-\frac{1}{3}$) can be converted into an up-type quark (u, with a charge of $+\frac{2}{3}$), by emitting a W− boson or by absorbing a W+ boson. More precisely, the down-type quark becomes a quantum superposition of up-type quarks: that is to say, it has a possibility of becoming any one of the three up-type quarks, with the probabilities given in the CKM matrix tables. Conversely, an up-type quark can emit a W+ boson – or absorb a W− boson – and thereby be converted into a down-type quark, for example:

$$d \rightarrow u + W^-$$
$$d + W^+ \rightarrow u$$
$$c \rightarrow s + W^+$$
$$c + W^- \rightarrow s$$

The W boson is unstable so will rapidly decay, with a very short lifetime. For example:

$$W^- \rightarrow e^- + \bar{\nu}_e$$
$$W^+ \rightarrow e^+ + \nu_e$$

Decay of the W boson to other products can happen, with varying probabilities.[15]

In the so-called beta decay of a neutron (see picture, above), a down quark within the neutron emits a virtual W− boson and is thereby converted into an up quark, converting the neutron into a proton. Because of the energy involved in the process (i.e., the mass difference between the down quark and the up quark), the W− boson can only be converted into an electron and an electron-antineutrino.[16] At the quark level, the process can be represented as:

$$d \rightarrow u + e^- + \bar{\nu}_e$$

19.3.2   Neutral-current interaction

In neutral current interactions, a quark or a lepton (e.g., an electron or a muon) emits or absorbs a neutral Z boson. For example:

$$e^- \rightarrow e^- + Z^0$$

Like the W boson, the Z boson also decays rapidly,[15] for example:

$$Z^0 \rightarrow b + \bar{b}$$

19.4   Electroweak theory

Main article: Electroweak interaction

The Standard Model of particle physics describes the electromagnetic interaction and the weak interaction as two different aspects of a single electroweak interaction, the theory of which was developed around 1968 by Sheldon Glashow, Abdus Salam and Steven Weinberg. They were awarded the 1979 Nobel Prize in Physics for their work.[17] The Higgs mechanism provides an explanation for the presence of three massive gauge bosons (the three carriers of the weak interaction) and the massless photon of the electromagnetic interaction.[18]

According to the electroweak theory, at very high energies, the universe has four massless gauge boson fields similar to the photon and a complex scalar Higgs field doublet. However, at low energies, gauge symmetry is spontaneously broken down to the U(1) symmetry of electromagnetism (one of the Higgs fields acquires a vacuum expectation value). This symmetry breaking would produce three massless bosons, but they become integrated by three photon-like fields (through the Higgs mechanism) giving them mass. These three fields become the W+, W− and Z bosons of the weak interaction, while the fourth gauge field, which remains massless, is the photon of electromagnetism.[18]

This theory has made a number of predictions, including a prediction of the masses of the Z and W bosons before their discovery. On 4 July 2012, the CMS and the ATLAS experimental teams at the Large Hadron Collider independently announced that they had confirmed the formal discovery of a previously unknown boson of mass between 125–127 GeV/c^2, whose behaviour so far was "consistent with" a Higgs boson, while adding a cautious note that further data and analysis were needed before positively identifying the new boson as being a Higgs boson of some type. By 14 March 2013, the Higgs boson was tentatively confirmed to exist.[19]

19.5 Violation of symmetry

Left- and right-handed particles: p is the particle's momentum and S is its spin. Note the lack of reflective symmetry between the states.

The laws of nature were long thought to remain the same under mirror reflection, the reversal of one spatial axis. The results of an experiment viewed via a mirror were expected to be identical to the results of a mirror-reflected copy of the experimental apparatus. This so-called law of parity conservation was known to be respected by classical gravitation, electromagnetism and the strong interaction; it was assumed to be a universal law.[20] However, in the mid-1950s Chen Ning Yang and Tsung-Dao Lee suggested that the weak interaction might violate this law. Chien Shiung Wu and collaborators in 1957 discovered that the weak interaction violates parity, earning Yang and Lee the 1957 Nobel Prize in Physics.[21]

Although the weak interaction used to be described by Fermi's theory, the discovery of parity violation and renormalization theory suggested that a new approach was needed. In 1957, Robert Marshak and George Sudarshan and, somewhat later, Richard Feynman and Murray Gell-Mann proposed a **V−A** (vector minus axial vector or left-handed) Lagrangian for weak interactions. In this theory, the weak interaction acts only on left-handed particles (and right-handed antiparticles). Since the mirror reflection of a left-handed particle is right-handed, this explains the maximal violation of parity. Interestingly, the **V−A** theory was developed before the discovery of the Z boson, so it did not include the right-handed fields that enter in the neutral current interaction.

However, this theory allowed a compound symmetry **CP** to be conserved. **CP** combines parity **P** (switching left to right) with charge conjugation **C** (switching particles with antiparticles). Physicists were again surprised when in 1964, James

Cronin and Val Fitch provided clear evidence in kaon decays that CP symmetry could be broken too, winning them the 1980 Nobel Prize in Physics.[22] In 1973, Makoto Kobayashi and Toshihide Maskawa showed that CP violation in the weak interaction required more than two generations of particles,[23] effectively predicting the existence of a then unknown third generation. This discovery earned them half of the 2008 Nobel Prize in Physics.[24] Unlike parity violation, CP violation occurs in only a small number of instances, but remains widely held as an answer to the difference between the amount of matter and antimatter in the universe; it thus forms one of Andrei Sakharov's three conditions for baryogenesis.[25]

19.6 See also

- Weakless Universe – the postulate that weak interactions are not anthropically necessary

- Gravity

- Nuclear force

- Electromagnetism

19.7 References

19.7.1 Citations

[1] Griffiths, David (2009). *Introduction to Elementary Particles*. pp. 59–60. ISBN 978-3-527-40601-2.

[2] "The Nobel Prize in Physics 1979: Press Release". *NobelPrize.org*. Nobel Media. Retrieved 22 March 2011.

[3] Fermi, Enrico (1934). "Versuch einer Theorie der β-Strahlen. I". *Zeitschrift für Physik A* **88** (3–4): 161–177. Bibcode: . doi:10.1007/BF01351864.

[4] Wilson, Fred L. (December 1968). "Fermi's Theory of Beta Decay". *American Journal of Physics* **36** (12): 1150–1160. Bibcode:1968AmJPh..36.1150W. doi:10.1119/1.1974382.

[5] W.-M. Yao *et al.* (Particle Data Group) (2006). "Review of Particle Physics: Quarks" (PDF). *Journal of Physics G* **33**: 1–1232. arXiv:astro-ph/0601168. Bibcode:2006JPhG...33....1Y. doi:10.1088/0954-3899/33/1/001.

[6] Peter Watkins (1986). *Story of the W and Z*. Cambridge: Cambridge University Press. p. 70. ISBN 978-0-521-31875-4.

[7] "Coupling Constants for the Fundamental Forces". *HyperPhysics*. Georgia State University. Retrieved 2 March 2011.

[8] J. Christman (2001). "The Weak Interaction" (PDF). *Physnet*. Michigan State University.

[9] "Electroweak". *The Particle Adventure*. Particle Data Group. Retrieved 3 March 2011.

[10] Walter Greiner; Berndt Müller (2009). *Gauge Theory of Weak Interactions*. Springer. p. 2. ISBN 978-3-540-87842-1.

[11] Cottingham & Greenwood (1986, 2001), p.29

[12] Cottingham & Greenwood (1986, 2001), p.28

[13] Cottingham & Greenwood (1986, 2001), p.30

[14] Baez, John C.; Huerta, John (2009)."The Algebra of Grand Unified Theories".*Bull.Am.Math.Soc.***0904**: 483–552.arXiv6. Bibcode:2009arXiv0904.1556B. doi:10.1090/s0273-0979-10-01294-2. Retrieved 15 October 2013.

[15] K. Nakamura *et al.* (Particle Data Group) (2010). "Gauge and Higgs Bosons" (PDF). *Journal of Physics G* **37**. doi:10.1088/0954-3899/37/7a/075021.

[16] K. Nakamura *et al.* (Particle Data Group) (2010). "n" (PDF). *Journal of Physics G* **37**: 7. doi:10.1088/0954-3899/37/7a/075021.

[17] "The Nobel Prize in Physics 1979". *NobelPrize.org*. Nobel Media. Retrieved 26 February 2011.

[18] C. Amsler *et al.* (Particle Data Group) (2008). "Review of Particle Physics – Higgs Bosons: Theory and Searches" (PDF). *Physics Letters B* **667**: 1–6. Bibcode:2008PhLB..667....1P. doi:10.1016/j.physletb.2008.07.018.

[19] "New results indicate that new particle is a Higgs boson | CERN". Home.web.cern.ch. Retrieved 20 September 2013.

[20] Charles W. Carey (2006). "Lee, Tsung-Dao". *American scientists*. Facts on File Inc. p. 225. ISBN 9781438108070.

[21] "The Nobel Prize in Physics 1957". *NobelPrize.org*. Nobel Media. Retrieved 26 February 2011.

[22] "The Nobel Prize in Physics 1980". *NobelPrize.org*. Nobel Media. Retrieved 26 February 2011.

[23] M. Kobayashi, T. Maskawa (1973). "CP-Violation in the Renormalizable Theory of Weak Interaction". *Progress of Theoretical Physics* **49** (2): 652–657. Bibcode:1973PThPh..49..652K. doi:10.1143/PTP.49.652.

[24] "The Nobel Prize in Physics 1980". *NobelPrize.org*. Nobel Media. Retrieved 17 March 2011.

[25] Paul Langacker (2001) [1989]. "Cp Violation and Cosmology". In Cecilia Jarlskog. *CP violation*. London, River Edge: World Scientific Publishing Co. p. 552. ISBN 9789971505615.

19.7.2 General readers

- R. Oerter (2006). *The Theory of Almost Everything: The Standard Model, the Unsung Triumph of Modern Physics.* Plume. ISBN 978-0-13-236678-6.

- B.A. Schumm (2004). *Deep Down Things: The Breathtaking Beauty of Particle Physics.* Johns Hopkins University Press. ISBN 0-8018-7971-X.

19.7.3 Texts

- D.A. Bromley (2000). *Gauge Theory of Weak Interactions.* Springer. ISBN 3-540-67672-4.

- G.D. Coughlan, J.E. Dodd, B.M. Gripaios (2006). *The Ideas of Particle Physics: An Introduction for Scientists* (3rd ed.). Cambridge University Press. ISBN 978-0-521-67775-2.

- W. N. Cottingham; D. A. Greenwood (2001) [1986]. *An introduction to nuclear physics* (2nd ed.). Cambridge University Press. p. 30. ISBN 978-0-521-65733-4.

- D.J. Griffiths (1987). *Introduction to Elementary Particles.* John Wiley & Sons. ISBN 0-471-60386-4.

- G.L. Kane (1987). *Modern Elementary Particle Physics.* Perseus Books. ISBN 0-201-11749-5.

- D.H. Perkins (2000). *Introduction to High Energy Physics.* Cambridge University Press. ISBN 0-521-62196-8.

Chapter 20

Electromagnetism

Electromagnetism is a branch of physics which involves the study of the **electromagnetic force**, a type of physical interaction that occurs between electrically charged particles. The electromagnetic force usually shows electromagnetic fields, such as electric fields, magnetic fields, and light. The electromagnetic force is one of the four fundamental interactions in nature. The other three fundamental interactions are the strong interaction, the weak interaction, and gravitation.[1]

Lightning is an electrostatic discharge that travels between two charged regions.

The word *electromagnetism* is a compound form of two Greek terms, ἤλεκτρον, *ēlektron*, "amber", and μαγνῆτις λίθος *magnētis lithos*, which means "magnesian stone", a type of iron ore. The science of electromagnetic phenomena is defined in terms of the electromagnetic force, sometimes called the Lorentz force, which includes both electricity and magnetism as elements of one phenomenon.

The electromagnetic force plays a major role in determining the internal properties of most objects encountered in daily life. Ordinary matter takes its form as a result of intermolecular forces between individual molecules in matter. Electrons are bound by electromagnetic wave mechanics into orbitals around atomic nuclei to form atoms, which are the building blocks of molecules. This governs the processes involved in chemistry, which arise from interactions between the electrons of neighboring atoms, which are in turn determined by the interaction between electromagnetic force and the momentum of the electrons.

There are numerous mathematical descriptions of the electromagnetic field. In classical electrodynamics, electric fields are described as electric potential and electric current in Ohm's law, magnetic fields are associated with electromagnetic induction and magnetism, and Maxwell's equations describe how electric and magnetic fields are generated and altered by each other and by charges and currents.

The theoretical implications of electromagnetism, in particular the establishment of the speed of light based on properties of the "medium" of propagation (permeability and permittivity), led to the development of special relativity by Albert Einstein in 1905.

Although electromagnetism is considered one of the four fundamental forces, at high energy the weak force and electromagnetism are unified. In the history of the universe, during the quark epoch, the electroweak force split into the electromagnetic and weak forces.

20.1 History of the theory

See also: History of electromagnetic theory

Originally electricity and magnetism were thought of as two separate forces. This view changed, however, with the publication of James Clerk Maxwell's 1873 *A Treatise on Electricity and Magnetism* in which the interactions of positive and negative charges were shown to be regulated by one force. There are four main effects resulting from these interactions, all of which have been clearly demonstrated by experiments:

1. Electric charges attract or repel one another with a force inversely proportional to the square of the distance between them: unlike charges attract, like ones repel.

2. Magnetic poles (or states of polarization at individual points) attract or repel one another in a similar way and always come in pairs: every north pole is yoked to a south pole.

3. An electric current inside a wire creates a corresponding circular magnetic field outside the wire. Its direction (clockwise or counter-clockwise) depends on the direction of the current in the wire.

4. A current is induced in a loop of wire when it is moved towards or away from a magnetic field, or a magnet is moved towards or away from it, the direction of current depending on that of the movement.

While preparing for an evening lecture on 21 April 1820, Hans Christian Ørsted made a surprising observation. As he was setting up his materials, he noticed a compass needle deflected away from magnetic north when the electric current from the battery he was using was switched on and off. This deflection convinced him that magnetic fields radiate from all sides of a wire carrying an electric current, just as light and heat do, and that it confirmed a direct relationship between electricity and magnetism.

At the time of discovery, Ørsted did not suggest any satisfactory explanation of the phenomenon, nor did he try to represent the phenomenon in a mathematical framework. However, three months later he began more intensive investigations. Soon thereafter he published his findings, proving that an electric current produces a magnetic field as it flows through a wire. The CGS unit of magnetic induction (oersted) is named in honor of his contributions to the field of electromagnetism.

Hans Christian Ørsted.

His findings resulted in intensive research throughout the scientific community in electrodynamics. They influenced French physicist André-Marie Ampère's developments of a single mathematical form to represent the magnetic forces between current-carrying conductors. Ørsted's discovery also represented a major step toward a unified concept of energy.

This unification, which was observed by Michael Faraday, extended by James Clerk Maxwell, and partially reformulated by Oliver Heaviside and Heinrich Hertz, is one of the key accomplishments of 19th century mathematical physics. It had far-reaching consequences, one of which was the understanding of the nature of light. Unlike what was proposed in Electromagnetism, light and other electromagnetic waves are at the present seen as taking the form of quantized, self-propagating oscillatory electromagnetic field disturbances which have been called photons. Different frequencies of oscillation give rise to the different forms of electromagnetic radiation, from radio waves at the lowest frequencies, to visible light at intermediate frequencies, to gamma rays at the highest frequencies.

Ørsted was not the only person to examine the relationship between electricity and magnetism. In 1802, Gian Domenico Romagnosi, an Italian legal scholar, deflected a magnetic needle using electrostatic charges. Actually, no galvanic current existed in the setup and hence no electromagnetism was present. An account of the discovery was published in 1802 in an Italian newspaper, but it was largely overlooked by the contemporary scientific community.[2]

20.2   Fundamental forces

The electromagnetic force is one of the four known fundamental forces. The other fundamental forces are:

- the weak nuclear force, which binds to all known particles in the Standard Model, and causes certain forms of radioactive decay. (In particle physics though, the electroweak interaction is the unified description of two of the four known fundamental interactions of nature: electromagnetism and the weak interaction);

- the strong nuclear force, which binds quarks to form nucleons, and binds nucleons to form nuclei

- the gravitational force.

All other forces (e.g., friction) are derived from these four fundamental forces (including momentum which is carried by the movement of particles).

The electromagnetic force is the one responsible for practically all the phenomena one encounters in daily life above the nuclear scale, with the exception of gravity. Roughly speaking, all the forces involved in interactions between atoms can be explained by the electromagnetic force acting on the electrically charged atomic nuclei and electrons inside and around the atoms, together with how these particles carry momentum by their movement. This includes the forces we experience in "pushing" or "pulling" ordinary material objects, which come from the intermolecular forces between the individual molecules in our bodies and those in the objects. It also includes all forms of chemical phenomena.

A necessary part of understanding the intra-atomic to intermolecular forces is the effective force generated by the momentum of the electrons' movement, and that electrons move between interacting atoms, carrying momentum with them. As a collection of electrons becomes more confined, their minimum momentum necessarily increases due to the Pauli exclusion principle. The behaviour of matter at the molecular scale including its density is determined by the balance between the electromagnetic force and the force generated by the exchange of momentum carried by the electrons themselves.

20.3   Classical electrodynamics

Main article: Classical electrodynamics

The scientist William Gilbert proposed, in his *De Magnete* (1600), that electricity and magnetism, while both capable of causing attraction and repulsion of objects, were distinct effects. Mariners had noticed that lightning strikes had the ability to disturb a compass needle, but the link between lightning and electricity was not confirmed until Benjamin Franklin's proposed experiments in 1752. One of the first to discover and publish a link between man-made electric current and magnetism was Romagnosi, who in 1802 noticed that connecting a wire across a voltaic pile deflected a nearby compass needle. However, the effect did not become widely known until 1820, when Ørsted performed a similar experiment.[3] Ørsted's work influenced Ampère to produce a theory of electromagnetism that set the subject on a mathematical foundation.

A theory of electromagnetism, known as classical electromagnetism, was developed by various physicists over the course of the 19th century, culminating in the work of James Clerk Maxwell, who unified the preceding developments into a single theory and discovered the electromagnetic nature of light. In classical electromagnetism, the electromagnetic field obeys a set of equations known as Maxwell's equations, and the electromagnetic force is given by the Lorentz force law.

One of the peculiarities of classical electromagnetism is that it is difficult to reconcile with classical mechanics, but it is compatible with special relativity. According to Maxwell's equations, the speed of light in a vacuum is a universal constant, dependent only on the electrical permittivity and magnetic permeability of free space. This violates Galilean invariance, a long-standing cornerstone of classical mechanics. One way to reconcile the two theories (electromagnetism and classical mechanics) is to assume the existence of a luminiferous aether through which the light propagates. However, subsequent experimental efforts failed to detect the presence of the aether. After important contributions of Hendrik Lorentz and Henri Poincaré, in 1905, Albert Einstein solved the problem with the introduction of special relativity, which replaces classical kinematics with a new theory of kinematics that is compatible with classical electromagnetism. (For more information, see History of special relativity.)

In addition, relativity theory shows that in moving frames of reference a magnetic field transforms to a field with a nonzero electric component and vice versa; thus firmly showing that they are two sides of the same coin, and thus the term "electromagnetism". (For more information, see Classical electromagnetism and special relativity and Covariant formulation of classical electromagnetism.

20.4 Quantum mechanics

20.4.1 Photoelectric effect

Main article: Photoelectric effect

In another paper published in 1905, Albert Einstein undermined the very foundations of classical electromagnetism. In his theory of the photoelectric effect (for which he won the Nobel prize in physics) and inspired by the idea of Max Planck's "quanta", he posited that light could exist in discrete particle-like quantities as well, which later came to be known as photons. Einstein's theory of the photoelectric effect extended the insights that appeared in the solution of the ultraviolet catastrophe presented by Max Planck in 1900. In his work, Planck showed that hot objects emit electromagnetic radiation in discrete packets ("quanta"), which leads to a finite total energy emitted as black body radiation. Both of these results were in direct contradiction with the classical view of light as a continuous wave. Planck's and Einstein's theories were progenitors of quantum mechanics, which, when formulated in 1925, necessitated the invention of a quantum theory of electromagnetism. This theory, completed in the 1940s-1950s, is known as quantum electrodynamics (or "QED"), and, in situations where perturbation theory is applicable, is one of the most accurate theories known to physics.

20.4.2 Quantum electrodynamics

Main article: Quantum electrodynamics

All electromagnetic phenomena are underpinned by quantum mechanics, specifically by quantum electrodynamics (which includes classical electrodynamics as a limiting case) and this accounts for almost all physical phenomena observable to the unaided human senses, including light and other electromagnetic radiation, all of chemistry, most of mechanics (excepting gravitation), and, of course, magnetism and electricity.

20.4.3 Electroweak interaction

Main article: Electroweak interaction

The **electroweak interaction** is the unified description of two of the four known fundamental interactions of nature: electromagnetism and the weak interaction. Although these two forces appear very different at everyday low energies, the theory models them as two different aspects of the same force. Above the unification energy, on the order of 100 GeV, they would merge into a single **electroweak force**. Thus if the universe is hot enough (approximately 10^{15} K, a temperature exceeded until shortly after the Big Bang) then the electromagnetic force and weak force merge into a combined electroweak force. During the electroweak epoch, the electroweak force separated from the strong force. During the quark epoch, the electroweak force split into the electromagnetic and weak force.

20.5 Quantities and units

See also: List of physical quantities and List of electromagnetism equations

Electromagnetic units are part of a system of electrical units based primarily upon the magnetic properties of electric currents, the fundamental SI unit being the ampere. The units are:

- ampere (electric current)

- coulomb (electric charge)

- farad (capacitance)

- henry (inductance)

- ohm (resistance)

- tesla (magnetic flux density)

- volt (electric potential)

- watt (power)

- weber (magnetic flux)

In the electromagnetic cgs system, electric current is a fundamental quantity defined via Ampère's law and takes the permeability as a dimensionless quantity (relative permeability) whose value in a vacuum is unity. As a consequence, the square of the speed of light appears explicitly in some of the equations interrelating quantities in this system.

Formulas for physical laws of electromagnetism (such as Maxwell's equations) need to be adjusted depending on what system of units one uses. This is because there is no one-to-one correspondence between electromagnetic units in SI and those in CGS, as is the case for mechanical units. Furthermore, within CGS, there are several plausible choices of electromagnetic units, leading to different unit "sub-systems", including Gaussian, "ESU", "EMU", and Heaviside–Lorentz. Among these choices, Gaussian units are the most common today, and in fact the phrase "CGS units" is often used to refer specifically to CGS-Gaussian units.

20.6 See also

- Abraham–Lorentz force

- Aeromagnetic surveys

- Computational electromagnetics

- Double-slit experiment

- Electromagnet

- Electromagnetic induction

- Electromagnetic wave equation

- Electromechanics

- Magnetostatics

- Magnetoquasistatic field

- Optics

- Relativistic electromagnetism

- Wheeler–Feynman absorber theory

20.7   References

[1] Ravaioli, Fawwaz T. Ulaby, Eric Michielssen, Umberto (2010). *Fundamentals of applied electromagnetics* (6th ed.). Boston: Prentice Hall. p. 13. ISBN 978-0-13-213931-1.

[2] Martins, Roberto de Andrade. "Romagnosi and Volta's Pile: Early Difficulties in the Interpretation of Voltaic Electricity". In Fabio Bevilacqua and Lucio Fregonese (eds). *Nuova Voltiana: Studies on Volta and his Times* (PDF). vol. 3. Università degli Studi di Pavia. pp. 81–102. Retrieved 2010-12-02.

[3] Stern, Dr. David P.; Peredo, Mauricio (2001-11-25). "Magnetic Fields -- History". NASA Goddard Space Flight Center. Retrieved 2009-11-27.

[4] International Union of Pure and Applied Chemistry (1993). *Quantities, Units and Symbols in Physical Chemistry*, 2nd edition, Oxford: Blackwell Science. ISBN 0-632-03583-8. pp. 14–15. Electronic version.

20.8   Further reading

20.8.1   Web sources

- Nave, R. "Electricity and magnetism". *HyperPhysics*. Georgia State University. Retrieved 2013-11-12.

20.8.2   Lecture notes

- Littlejohn, Robert (Spring 2011). "Emission and absorption of radiation" (PDF). *Physics 221B: Quantum mechanics*. University of California Berkeley. Retrieved 2013-11-12.

- Littlejohn, Robert (Spring 2011). "The Classical Electromagnetic Field Hamiltonian" (PDF). *Physics 221B: Quantum mechanics*. University of California Berkeley. Retrieved 2013-11-12.

20.8.3   Textbooks

- G.A.G. Bennet (1974). *Electricity and Modern Physics* (2nd ed.). Edward Arnold (UK). ISBN 0-7131-2459-8.

- Dibner, Bern (2012). *Oersted and the discovery of electromagnetism*. Literary Licensing, LLC. ISBN 9781258335557.

- Durney, Carl H. and Johnson, Curtis C. (1969). *Introduction to modern electromagnetics*. McGraw-Hill. ISBN 0-07-018388-0.

- Feynman, Richard P. (1970). *The Feynman Lectures on Physics Vol II*. Addison Wesley Longman. ISBN 978-0-201-02115-8.

- Fleisch, Daniel (2008). *A Student's Guide to Maxwell's Equations*. Cambridge, UK: Cambridge University Press. ISBN 978-0-521-70147-1.

- I.S. Grant, W.R. Phillips, Manchester Physics (2008). *Electromagnetism* (2nd ed.). John Wiley & Sons. ISBN 978-0-471-92712-9.

- Griffiths, David J. (1998). *Introduction to Electrodynamics* (3rd ed.). Prentice Hall. ISBN 0-13-805326-X.

- Jackson, John D. (1998). *Classical Electrodynamics* (3rd ed.). Wiley. ISBN 0-471-30932-X.

- Moliton, André (2007). *Basic electromagnetism and materials*. *430 pages* (New York City: Springer-Verlag New York, LLC). ISBN 978-0-387-30284-3.

- Purcell, Edward M. (1985). *Electricity and Magnetism Berkeley Physics Course Volume 2 (2nd ed.)*. McGraw-Hill. ISBN 0-07-004908-4.

- Rao, Nannapaneni N. (1994). *Elements of engineering electromagnetics (4th ed.)*. Prentice Hall. ISBN 0-13-948746-8.

- Rothwell, Edward J.; Cloud, Michael J. (2001). *Electromagnetics*. CRC Press. ISBN 0-8493-1397-X.

- Tipler, Paul (1998). *Physics for Scientists and Engineers: Vol. 2: Light, Electricity and Magnetism* (4th ed.). W. H. Freeman. ISBN 1-57259-492-6.

- Wangsness, Roald K.; Cloud, Michael J. (1986). *Electromagnetic Fields (2nd Edition)*. Wiley. ISBN 0-471-81186-6.

20.8.4 General references

- A. Beiser (1987). *Concepts of Modern Physics* (4th ed.). McGraw-Hill (International). ISBN 0-07-100144-1.

- L.H. Greenberg (1978). *Physics with Modern Applications*. Holt-Saunders International W.B. Saunders and Co. ISBN 0-7216-4247-0.

- R.G. Lerner, G.L. Trigg (2005). *Encyclopaedia of Physics* (2nd ed.). VHC Publishers, Hans Warlimont, Springer. pp. 12–13. ISBN 978-0-07-025734-4.

- J.B. Marion, W.F. Hornyak (1984). *Principles of Physics*. Holt-Saunders International Saunders College. ISBN 4-8337-0195-2.

- H.J. Pain (1983). *The Physics of Vibrations and Waves* (3rd ed.). John Wiley & Sons,. ISBN 0-471-90182-2.

- C.B. Parker (1994). *McGraw Hill Encyclopaedia of Physics* (2nd ed.). McGraw Hill. ISBN 0-07-051400-3.

- R. Penrose (2007). *The Road to Reality*. Vintage books. ISBN 0-679-77631-1.

- P.A. Tipler, G. Mosca (2008). *Physics for Scientists and Engineers: With Modern Physics* (6th ed.). W.H. Freeman and Co. ISBN 9-781429-202657.

- P.M. Whelan, M.J. Hodgeson (1978). *Essential Principles of Physics* (2nd ed.). John Murray. ISBN 0-7195-3382-1.

20.9   External links

- Oppelt, Arnulf (2006-11-02). "magnetic field strength". Retrieved 2007-06-04.

- "magnetic field strength converter". Retrieved 2007-06-04.

- Electromagnetic Force - from Eric Weisstein's World of Physics

- Goudarzi, Sara (2006-08-15). "Ties That Bind Atoms Weaker Than Thought". *LiveScience.com*. Retrieved 2013-11-12.

- Quarked Electromagnetic force - A good introduction for kids

- The Deflection of a Magnetic Compass Needle by a Current in a Wire (video) on YouTube

- Electromagnetism abridged

André-Marie Ampère

Michael Faraday

James Clerk Maxwell

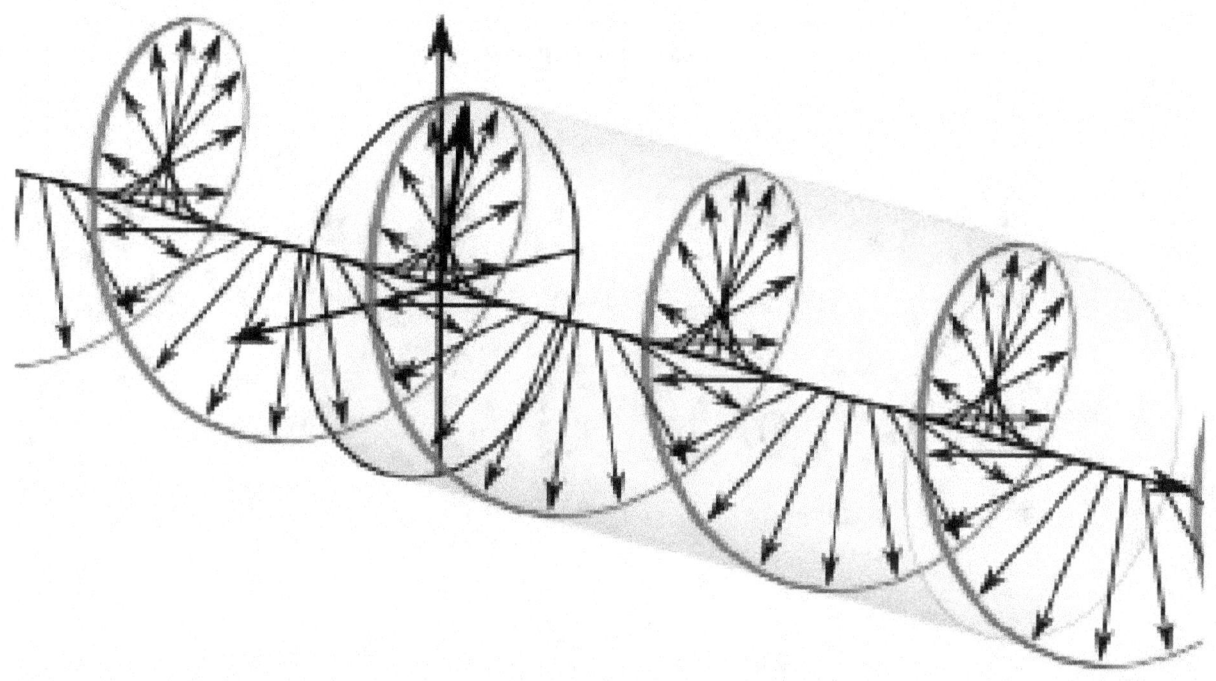

Representation of the electric field vector of a wave of circularly polarized electromagnetic radiation.

Chapter 21

Elementary particle

This article is about the physics concept. For the novel, see The Elementary Particles.
In particle physics, an **elementary particle** or **fundamental particle** is a particle whose substructure is unknown, thus

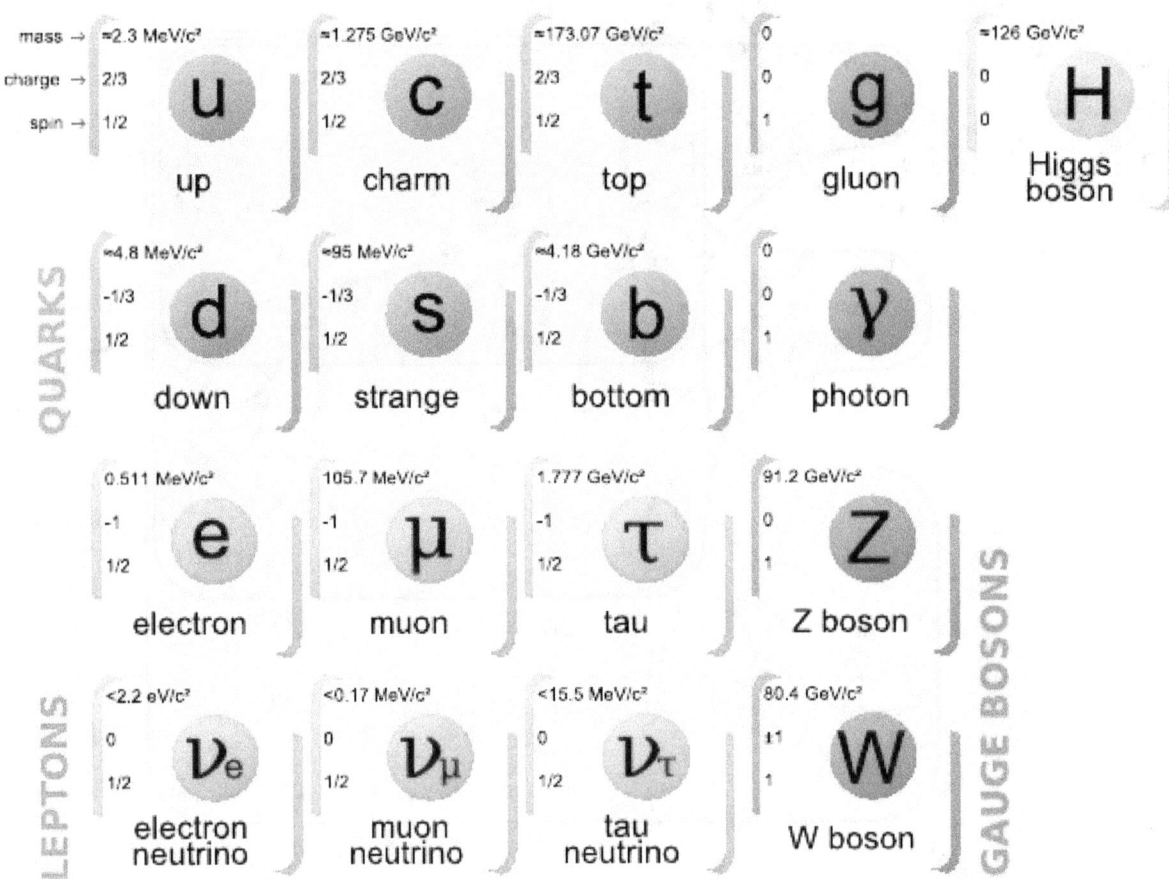

Elementary particles included in the Standard Model

it is unknown whether it is composed of other particles.[1] Known elementary particles include the fundamental fermions (quarks, leptons, antiquarks, and antileptons), which generally are "matter particles" and "antimatter particles", as well as the fundamental bosons (gauge bosons and Higgs boson), which generally are "force particles" that mediate interactions among fermions.[1] A particle containing two or more elementary particles is a *composite particle*.

Everyday matter is composed of atoms, once presumed to be matter's elementary particles—*atom* meaning "indivisible" in Greek—although the atom's existence remained controversial until about 1910, as some leading physicists regarded molecules as mathematical illusions, and matter as ultimately composed of energy.[1][2] Soon, subatomic constituents of the atom were identified. As the 1930s opened, the electron and the proton had been observed, along with the photon, the particle of electromagnetic radiation.[1] At that time, the recent advent of quantum mechanics was radically altering the conception of particles, as a single particle could seemingly span a field as would a wave, a paradox still eluding satisfactory explanation.[3][4][5]

Via quantum theory, protons and neutrons were found to contain quarks—up quarks and down quarks—now considered elementary particles.[1] And within a molecule, the electron's three degrees of freedom (charge, spin, orbital) can separate via wavefunction into three quasiparticles (holon, spinon, orbiton).[6] Yet a free electron—which, not orbiting an atomic nucleus, lacks orbital motion—appears unsplittable and remains regarded as an elementary particle.[6]

Around 1980, an elementary particle's status as indeed elementary—an *ultimate constituent* of substance—was mostly discarded for a more practical outlook,[1] embodied in particle physics' Standard Model, science's most experimentally successful theory.[5][7] Many elaborations upon and theories beyond the Standard Model, including the extremely popular supersymmetry, double the number of elementary particles by hypothesizing that each known particle associates with a "shadow" partner far more massive,[8][9] although all such superpartners remain undiscovered.[7][10] Meanwhile, an elementary boson mediating gravitation—the graviton—remains hypothetical.[1]

21.1   Overview

Main article: Standard Model
See also: Physics beyond the Standard Model
 All elementary particles are—depending on their *spin*—either bosons or fermions. These are differentiated via the spin-

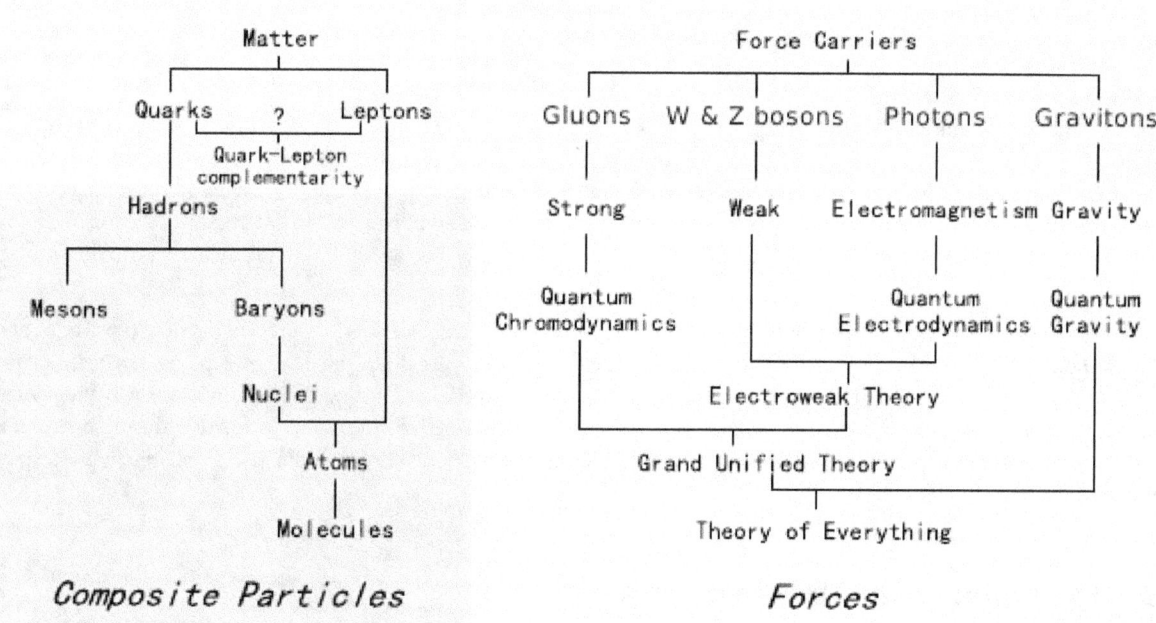

An overview of the various families of elementary and composite particles, and the theories describing their interactions

statistics theorem of quantum statistics. Particles of *half-integer* spin exhibit Fermi–Dirac statistics and are fermions.[1] Particles of *integer* spin, in other words full-integer, exhibit Bose–Einstein statistics and are bosons.[1]

Elementary fermions:

- Matter particles

  - Quarks:

    - up, down
    - charm, strange
    - top, bottom

  - Leptons:

    - electron, electron neutrino (a.k.a., "neutrino")
    - muon, muon neutrino
    - tau, tau neutrino

- Antimatter particles

  - Antiquarks

  - Antileptons

Elementary bosons:

- Force particles (gauge bosons):

  - photon
  - gluon (numbering eight)[1]
  - W^+, W^-, and Z^0 bosons
  - graviton (hypothetical)[1]

- Scalar boson

  - Higgs boson

A particle's mass is quantified in units of energy versus the electron's (electronvolts). Through conversion of energy into mass, any particle can be produced through collision of other particles at high energy,[1][11] although the output particle might not contain the input particles, for instance matter creation from colliding photons. Likewise, the composite fermions protons were collided at nearly light speed to produce a Higgs boson, which elementary boson is far more massive.[11] The most massive elementary particle, the top quark, rapidly decays, but apparently does not contain, lighter particles.

When probed at energies available in experiments, particles exhibit spherical sizes. In operating particle physics' Standard Model, elementary particles are usually represented for predictive utility as point particles, which, as zero-dimensional, lack spatial extension. Though extremely successful, the Standard Model is limited to the microcosm by its omission of gravitation, and has some parameters arbitrarily added but unexplained.[12] Seeking to resolve those shortcomings, string theory posits that elementary particles are ultimately composed of one-dimensional energy strings whose absolute minimum size is the Planck length.

21.2 Common elementary particles

Main article: cosmic abundance of elements

According to the current models of big bang nucleosynthesis, the primordial composition of visible matter of the universe should be about 75% hydrogen and 25% helium-4 (in mass). Neutrons are made up of one up and two down quark, while protons are made of two up and one down quark. Since the other common elementary particles (such as electrons, neutrinos, or weak bosons) are so light or so rare when compared to atomic nuclei, we can neglect their mass contribution to the observable universe's total mass. Therefore, one can conclude that most of the visible mass of the universe consists of protons and neutrons, which, like all baryons, in turn consist of up quarks and down quarks.

Some estimates imply that there are roughly 10^{80} baryons (almost entirely protons and neutrons) in the observable universe.[13][14][15]

The number of protons in the observable universe is called the Eddington number.

In terms of number of particles, some estimates imply that nearly all the matter, excluding dark matter, occurs in neutrinos, and that roughly 10^{86} elementary particles of matter exist in the visible universe, mostly neutrinos.[15] Other estimates imply that roughly 10^{97} elementary particles exist in the visible universe (not including dark matter), mostly photons, gravitons, and other massless force carriers.[15]

21.3   Standard Model

Main article: Standard Model

The Standard Model of particle physics contains 12 flavors of elementary fermions, plus their corresponding antiparticles, as well as elementary bosons that mediate the forces and the Higgs boson, which was reported on July 4, 2012, as having been likely detected by the two main experiments at the LHC (ATLAS and CMS). However, the Standard Model is widely considered to be a provisional theory rather than a truly fundamental one, since it is not known if it is compatible with Einstein's general relativity. There may be hypothetical elementary particles not described by the Standard Model, such as the graviton, the particle that would carry the gravitational force, and sparticles, supersymmetric partners of the ordinary particles.

21.3.1   Fundamental fermions

Main article: Fermion

The 12 fundamental fermionic flavours are divided into three generations of four particles each. Six of the particles are quarks. The remaining six are leptons, three of which are neutrinos, and the remaining three of which have an electric charge of -1: the electron and its two cousins, the muon and the tau.

Antiparticles

Main article: Antimatter

There are also 12 fundamental fermionic antiparticles that correspond to these 12 particles. For example, the antielectron (positron) $e+$ is the electron's antiparticle and has an electric charge of $+1$.

Quarks

Main article: Quark

Isolated quarks and antiquarks have never been detected, a fact explained by confinement. Every quark carries one of three color charges of the strong interaction; antiquarks similarly carry anticolor. Color-charged particles interact via gluon exchange in the same way that charged particles interact via photon exchange. However, gluons are themselves color-charged, resulting in an amplification of the strong force as color-charged particles are separated. Unlike the electromagnetic force, which diminishes as charged particles separate, color-charged particles feel increasing force.

However, color-charged particles may combine to form color neutral composite particles called hadrons. A quark may pair up with an antiquark: the quark has a color and the antiquark has the corresponding anticolor. The color and anticolor cancel out, forming a color neutral meson. Alternatively, three quarks can exist together, one quark being "red",

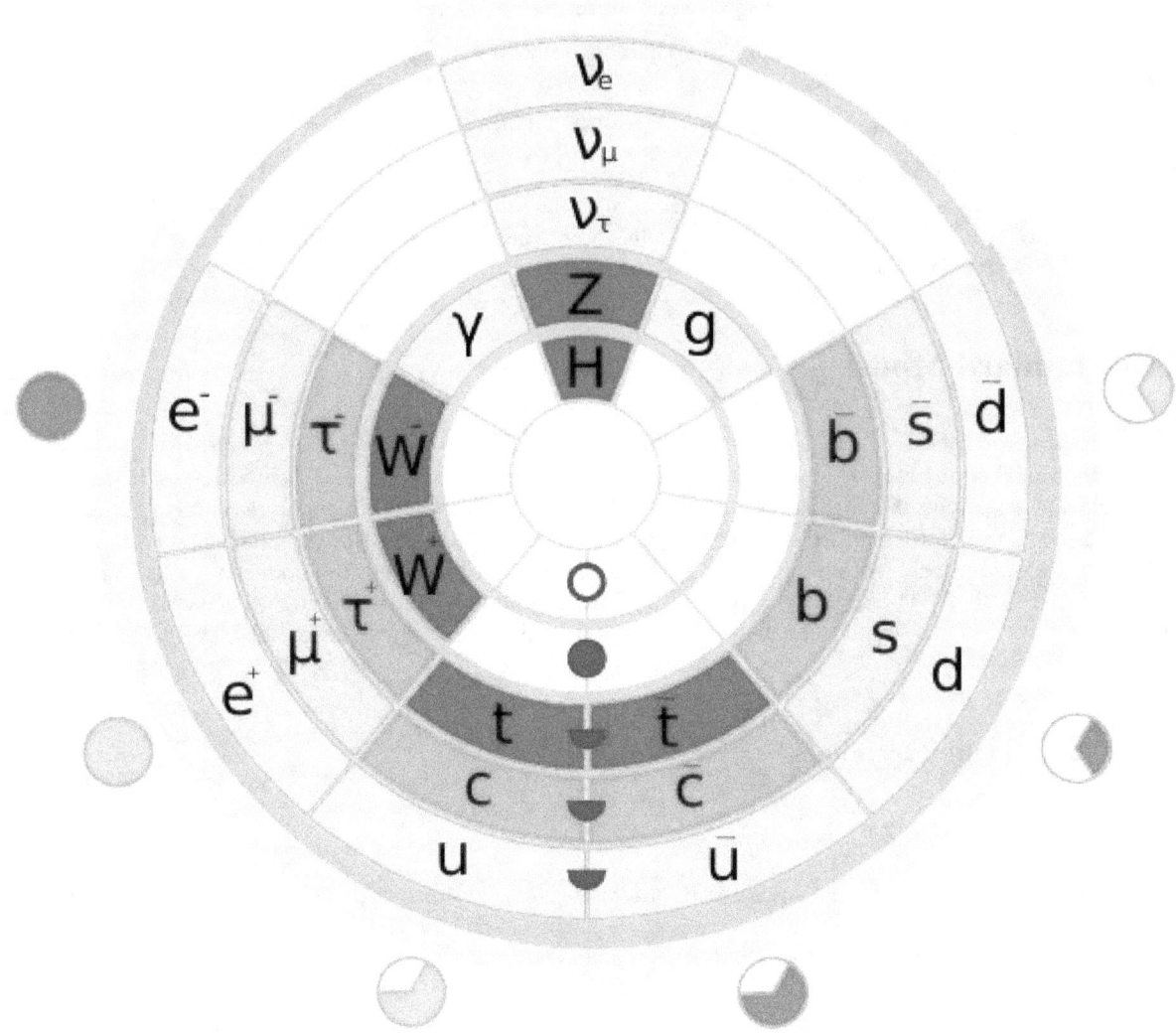

Graphic representation of the standard model. Spin, charge, mass and participation in different force interactions are shown. Click on the image to see the full description

another "blue", another "green". These three colored quarks together form a color-neutral baryon. Symmetrically, three antiquarks with the colors "antired", "antiblue" and "antigreen" can form a color-neutral antibaryon.

Quarks also carry fractional electric charges, but, since they are confined within hadrons whose charges are all integral, fractional charges have never been isolated. Note that quarks have electric charges of either +2/3 or −1/3, whereas antiquarks have corresponding electric charges of either −2/3 or +1/3.

Evidence for the existence of quarks comes from deep inelastic scattering: firing electrons at nuclei to determine the distribution of charge within nucleons (which are baryons). If the charge is uniform, the electric field around the proton should be uniform and the electron should scatter elastically. Low-energy electrons do scatter in this way, but, above a particular energy, the protons deflect some electrons through large angles. The recoiling electron has much less energy and a jet of particles is emitted. This inelastic scattering suggests that the charge in the proton is not uniform but split among smaller charged particles: quarks.

21.3.2   Fundamental bosons

Main article: Boson

In the Standard Model, vector (spin−1) bosons (gluons, photons, and the W and Z bosons) mediate forces, whereas the Higgs boson (spin-0) is responsible for the intrinsic mass of particles. Bosons differ from fermions in the fact that multiple bosons can occupy the same quantum state (Pauli exclusion principle). Also, bosons can be either elementary, like photons, or a combination, like mesons. The spin of bosons are integers instead of half integers.

Gluons

Main article: Gluon

Gluons mediate the strong interaction, which join quarks and thereby form hadrons, which are either baryons (three quarks) or mesons (one quark and one antiquark). Protons and neutrons are baryons, joined by gluons to form the atomic nucleus. Like quarks, gluons exhibit colour and anticolour—unrelated to the concept of visual color—sometimes in combinations, altogether eight variations of gluons.

Electroweak bosons

Main articles: W and Z bosons and Photon

There are three weak gauge bosons: W^+, W^-, and Z^0; these mediate the weak interaction. The W bosons are known for their mediation in nuclear decay. The W^- converts a neutron into a proton then decay into an electron and electron antineutrino pair. The Z^0 does not convert charge but rather changes momentum and is the only mechanism for elastically scattering neutrinos. The weak gauge bosons were discovered due to momentum change in electrons from neutrino-Z exchange. The massless photon mediates the electromagnetic interaction. These four gauge bosons form the electroweak interaction among elementary particles.

Higgs boson

Main article: Higgs boson

Although the weak and electromagnetic forces appear quite different to us at everyday energies, the two forces are theorized to unify as a single electroweak force at high energies. This prediction was clearly confirmed by measurements of cross-sections for high-energy electron-proton scattering at the HERA collider at DESY. The differences at low energies is a consequence of the high masses of the W and Z bosons, which in turn are a consequence of the Higgs mechanism. Through the process of spontaneous symmetry breaking, the Higgs selects a special direction in electroweak space that causes three electroweak particles to become very heavy (the weak bosons) and one to remain massless (the photon). On 4 July 2012, after many years of experimentally searching for evidence of its existence, the Higgs boson was announced to have been observed at CERN's Large Hadron Collider. Peter Higgs who first posited the existence of the Higgs boson was present at the announcement.[16] The Higgs boson is believed to have a mass of approximately 125 GeV.[17] The statistical significance of this discovery was reported as 5-sigma, which implies a certainty of roughly 99.99994%. In particle physics, this is the level of significance required to officially label experimental observations as a discovery. Research into the properties of the newly discovered particle continues.

Graviton

Main article: Graviton

The graviton is hypothesized to mediate gravitation, but remains undiscovered and yet is sometimes included in tables of elementary particles.[11] Its spin would be two—thus a boson—and it would lack charge or mass. Besides mediating an extremely feeble force, the graviton would have its own antiparticle and rapidly annihilate, rendering its detection extremely difficult even if it exists.

21.4 Beyond the Standard Model

Although experimental evidence overwhelmingly confirms the predictions derived from the Standard Model, some of its parameters were added arbitrarily, not determined by a particular explanation, which remain mysteries, for instance the hierarchy problem. Theories beyond the Standard Model attempt to resolve these shortcomings.

21.4.1 Grand unification

Main article: Grand Unified Theory

One extension of the Standard Model attempts to combine the electroweak interaction with the strong interaction into a single 'grand unified theory' (GUT). Such a force would be spontaneously broken into the three forces by a Higgs-like mechanism. The most dramatic prediction of grand unification is the existence of X and Y bosons, which cause proton decay. However, the non-observation of proton decay at the Super-Kamiokande neutrino observatory rules out the simplest GUTs, including SU(5) and SO(10).

21.4.2 Supersymmetry

Main article: Supersymmetry

Supersymmetry extends the Standard Model by adding another class of symmetries to the Lagrangian. These symmetries exchange fermionic particles with bosonic ones. Such a symmetry predicts the existence of supersymmetric particles, abbreviated as *sparticles*, which include the sleptons, squarks, neutralinos, and charginos. Each particle in the Standard Model would have a superpartner whose spin differs by 1/2 from the ordinary particle. Due to the breaking of supersymmetry, the sparticles are much heavier than their ordinary counterparts; they are so heavy that existing particle colliders would not be powerful enough to produce them. However, some physicists believe that sparticles will be detected by the Large Hadron Collider at CERN.

21.4.3 String theory

Main article: String theory

String theory is a model of physics where all "particles" that make up matter are composed of strings (measuring at the Planck length) that exist in an 11-dimensional (according to M-theory, the leading version) universe. These strings vibrate at different frequencies that determine mass, electric charge, color charge, and spin. A string can be open (a line) or closed in a loop (a one-dimensional sphere, like a circle). As a string moves through space it sweeps out something called a *world sheet*. String theory predicts 1- to 10-branes (a 1-brane being a string and a 10-brane being a 10-dimensional object) that prevent tears in the "fabric" of space using the uncertainty principle (E.g., the electron orbiting a hydrogen atom has the probability, albeit small, that it could be anywhere else in the universe at any given moment).

String theory proposes that our universe is merely a 4-brane, inside which exist the 3 space dimensions and the 1 time dimension that we observe. The remaining 6 theoretical dimensions either are very tiny and curled up (and too small to be macroscopically accessible) or simply do not/cannot exist in our universe (because they exist in a grander scheme called the "multiverse" outside our known universe).

Some predictions of the string theory include existence of extremely massive counterparts of ordinary particles due to vibrational excitations of the fundamental string and existence of a massless spin-2 particle behaving like the graviton.

21.4.4 Technicolor

Main article: Technicolor (physics)

Technicolor theories try to modify the Standard Model in a minimal way by introducing a new QCD-like interaction. This means one adds a new theory of so-called Techniquarks, interacting via so called Technigluons. The main idea is that the Higgs-Boson is not an elementary particle but a bound state of these objects.

21.4.5 Preon theory

Main article: Preon

According to preon theory there are one or more orders of particles more fundamental than those (or most of those) found in the Standard Model. The most fundamental of these are normally called preons, which is derived from "pre-quarks". In essence, preon theory tries to do for the Standard Model what the Standard Model did for the particle zoo that came before it. Most models assume that almost everything in the Standard Model can be explained in terms of three to half a dozen more fundamental particles and the rules that govern their interactions. Interest in preons has waned since the simplest models were experimentally ruled out in the 1980s.

21.4.6 Acceleron theory

Accelerons are the hypothetical subatomic particles that integrally link the newfound mass of the neutrino and to the dark energy conjectured to be accelerating the expansion of the universe.[18]

In theory, neutrinos are influenced by a new force resulting from their interactions with accelerons. Dark energy results as the universe tries to pull neutrinos apart.[18]

21.5   See also

- Asymptotic freedom

- List of particles

- Physical ontology

- Quantum field theory

- Quantum gravity

- Quantum triviality

- UV fixed point

21.6   Notes

[1] Sylvie Braibant; Giorgio Giacomelli; Maurizio Spurio (2012). *Particles and Fundamental Interactions: An Introduction to Particle Physics* (2nd ed.). Springer. pp. 1–3. ISBN 978-94-007-2463-1.

[2] Ronald Newburgh; Joseph Peidle; Wolfgang Rueckner (2006). "Einstein, Perrin, and the reality of atoms: 1905 revisited" (PDF). *American Journal of Physics*. **74** (6): 478–481. Bibcode:2006AmJPh..74..478N. doi:10.1119/1.2188962.

[3] Friedel Weinert (2004). *The Scientist as Philosopher: Philosophical Consequences of Great Scientific Discoveries*. Springer. p. 43. ISBN 978-3-540-20580-7.

[4] Friedel Weinert (2004). *The Scientist as Philosopher: Philosophical Consequences of Great Scientific Discoveries*. Springer. pp. 57–59. ISBN 978-3-540-20580-7.

[5] Meinard Kuhlmann (24 Jul 2013). "Physicists debate whether the world is made of particles or fields—or something else entirely". *Scientific American*.

[6] Zeeya Merali (18 Apr 2012). "Not-quite-so elementary, my dear electron: Fundamental particle 'splits' into quasiparticles, including the new 'orbiton'". *Nature*. doi:10.1038/nature.2012.10471.

[7] Ian O'Neill (24 Jul 2013). "LHC discovery maims supersymmetry, again". *Discovery News*. Retrieved 2013-08-28.

[8] Particle Data Group. "Unsolved mysteries—supersymmetry". *The Particle Adventure*. Berkeley Lab. Retrieved 2013-08-28.

[9] National Research Council (2006). *Revealing the Hidden Nature of Space and Time: Charting the Course for Elementary Particle Physics*. National Academies Press. p. 68. ISBN 978-0-309-66039-6.

[10] "CERN latest data shows no sign of supersymmetry—yet". *Phys.Org*. 25 Jul 2013. Retrieved 2013-08-28.

[11] Ryan Avent (19 Jul 2012). "The Q&A: Brian Greene—Life after the Higgs". *The Economist*. Retrieved 2013-08-28.

[12] Sylvie Braibant; Giorgio Giacomelli; Maurizio Spurio (2012). *Particles and Fundamental Interactions: An Introduction to Particle Physics* (2nd ed.). Springer. p. 384. ISBN 978-94-007-2463-1.

[13] Frank Heile. "Is the Total Number of Particles in the Universe Stable Over Long Periods of Time?". 2014.

[14] Jared Brooks. "Galaxies and Cosmology". 2014. p. 4. equation 16.

[15] Robert Munafo (24 Jul 2013). "Notable Properties of Specific Numbers". Retrieved 2013-08-28.

[16] Lizzy Davies (4 July 2014). "Higgs boson announcement live: CERN scientists discover subatomic particle". *The Guardian*. Retrieved 2012-07-06.

[17] Lucas Taylor (4 Jul 2014). "Observation of a new particle with a mass of 125 GeV". CMS. Retrieved 2012-07-06.

[18] "New theory links neutrino's slight mass to accelerating Universe expansion". *ScienceDaily*. 28 Jul 2004. Retrieved 2008-06-05.

21.7   Further reading

21.7.1   General readers

- Feynman, R.P. & Weinberg, S. (1987) *Elementary Particles and the Laws of Physics: The 1986 Dirac Memorial Lectures*. Cambridge Univ. Press.

- Ford, Kenneth W. (2005) *The Quantum World*. Harvard Univ. Press.

- Brian Greene (1999). *The Elegant Universe*. W.W.Norton & Company. ISBN 0-393-05858-1.

- John Gribbin (2000) *Q is for Quantum – An Encyclopedia of Particle Physics*. Simon & Schuster. ISBN 0-684-85578-X.

- Oerter, Robert (2006) *The Theory of Almost Everything: The Standard Model, the Unsung Triumph of Modern Physics*. Plume.

- Schumm, Bruce A. (2004) *Deep Down Things: The Breathtaking Beauty of Particle Physics*. Johns Hopkins University Press. ISBN 0-8018-7971-X.

- Martinus Veltman (2003). *Facts and Mysteries in Elementary Particle Physics*. World Scientific. ISBN 981-238-149-X.

- Frank Close (2004). *Particle Physics: A Very Short Introduction*. Oxford: Oxford University Press. ISBN 0-19-280434-0.

- Seiden, Abraham (2005). *Particle Physics – A Comprehensive Introduction*. Addison Wesley. ISBN 0-8053-8736-6.

21.7.2 Textbooks

- Bettini, Alessandro (2008) *Introduction to Elementary Particle Physics*. Cambridge Univ. Press. ISBN 978-0-521-88021-3

- Coughlan, G. D., J. E. Dodd, and B. M. Gripaios (2006) *The Ideas of Particle Physics: An Introduction for Scientists*, 3rd ed. Cambridge Univ. Press. An undergraduate text for those not majoring in physics.

- Griffiths, David J. (1987) *Introduction to Elementary Particles*. John Wiley & Sons. ISBN 0-471-60386-4.

- Kane, Gordon L. (1987). *Modern Elementary Particle Physics*. Perseus Books. ISBN 0-201-11749-5.

- Perkins, Donald H. (2000) *Introduction to High Energy Physics*, 4th ed. Cambridge Univ. Press.

21.8   External links

The most important address about the current experimental and theoretical knowledge about elementary particle physics is the Particle Data Group, where different international institutions collect all experimental data and give short reviews over the contemporary theoretical understanding.

- Particle Data Group

other pages are:

- Greene, Brian, "*Elementary particles*", The Elegant Universe, NOVA (PBS)

- particleadventure.org, a well-made introduction also for non physicists

- CERNCourier: Season of Higgs and melodrama

- Pentaquark information page

- Interactions.org, particle physics news

- Symmetry Magazine, a joint Fermilab/SLAC publication

- "Sized Matter: perception of the extreme unseen", Michigan University project for artistic visualisation of sub-atomic particles

- Elementary Particles made thinkable, an interactive visualisation allowing physical properties to be compared

Chapter 22

Planck energy

In physics, **Planck energy**, denoted by EP, is the unit of energy in the system of natural units known as Planck units.[1]

$$E_P = \sqrt{\frac{\hbar c^5}{G}} \approx 1.956 \times 10^9 \text{ J} \approx 1.22 \times 10^{28} \text{ eV} \approx 0.5433 \text{ MWh}$$

where c is the speed of light in a vacuum, ħ is the reduced Planck's constant, and G is the gravitational constant. EP is a *derived*, as opposed to *basic*, Planck unit.

An equivalent definition is:

$$E_P = \frac{\hbar}{t_P},$$

where tP is the Planck time.

Also:

$$E_P = m_P c^2,$$

where mP is the Planck mass.

The ultra-high-energy cosmic rays observed in 1991 had a measured energy of about 50 joules, equivalent to about 2.5×10^{-8} EP. Most Planck units are fantastically small and thus are unrelated to "macroscopic" phenomena (or fantastically large, as in the case of Planck density). One EP, on the other hand, is definitely macroscopic, approximately equaling the energy stored in an automobile gas tank (57.2 L of gasoline at 34.2 MJ/L of chemical energy).

Planck units are designed to normalize the physical constants, G and c, to 1. Hence given Planck units, the mass-energy equivalence $E = mc^2$ simplifies to $E = m$, so that the Planck energy and mass are numerically identical. In the equations of general relativity, G is often multiplied by 8π. Hence writings in particle physics and physical cosmology often normalize $8\pi G$ to 1. This normalization results in the **reduced Planck energy**, defined as:

$$\sqrt{\frac{\hbar c^5}{8\pi G}} \approx 0.390 \times 10^9 \text{ J} \approx 2.43 \times 10^{18} \text{ GeV}.$$

22.1 See also

- Max Planck

- Planck epoch

- Planck particle

- Planck units

- Quantum gravity

- Photon energy

22.2   References

[1] "Planck Energy". Cosmos, The SAO Encyclopedia, Swinburne University of Technology. Retrieved 18 September 2015.

Chapter 23

Supersymmetry

"SUSY" redirects here. For other uses, see Susy (disambiguation).
For the episode of the American TV series *Angel*, see Supersymmetry (Angel).

Supersymmetry (SUSY), a theory of particle physics, is a proposed type of spacetime symmetry that relates two basic classes of elementary particles: bosons, which have an integer-valued spin, and fermions, which have a half-integer spin.[1] Each particle from one group is associated with a particle from the other, known as its superpartner, the spin of which differs by a half-integer. In a theory with perfectly "unbroken" supersymmetry, each pair of superpartners would share the same mass and internal quantum numbers besides spin. For example, there would be a "selectron" (superpartner electron), a bosonic version of the electron with the same mass as the electron, that would be easy to find in a laboratory. Thus, since no superpartners have been observed, if supersymmetry exists it must be a spontaneously broken symmetry so that superpartners may differ in mass.[2][3] Spontaneously-broken supersymmetry could solve many mysterious problems in particle physics including the hierarchy problem. The simplest realization of spontaneously-broken supersymmetry, the so-called Minimal Supersymmetric Standard Model, is one of the best studied candidates for physics beyond the Standard Model.

There is only indirect evidence and motivation for the existence of supersymmetry. Direct confirmation would entail production of superpartners in collider experiments, such as the Large Hadron Collider (LHC). The first run of the LHC found no evidence for supersymmetry (all results were consistent with the Standard Model), and thus set limits on superpartner masses in supersymmetric theories. Whilst many remain enthusiastic about supersymmetry,[4] this first run at the LHC led some physicists to explore other ideas.[5] In any case, in 2015 the LHC resumed its search for supersymmetry and other new physics in its second run.

23.1 Motivations

There are numerous phenomenological motivations for supersymmetry close to the electroweak scale, as well as technical motivations for supersymmetry at any scale.

23.1.1 The hierarchy problem

Supersymmetry close to the electroweak scale ameliorates the hierarchy problem that afflicts the Standard Model. In the Standard Model, the electroweak scale receives enormous Planck-scale quantum corrections. The observed hierarchy between the electroweak scale and the Planck scale must be achieved with extraordinary fine tuning. In a supersymmetric theory, on the other hand, Planck-scale quantum corrections cancel between partners and superpartners (owing to a minus sign associated with fermionic loops). The hierarchy between the electroweak scale and the Planck scale is achieved in a natural manner, without miraculous fine-tuning.

23.1.2   Gauge coupling unification

The idea that the gauge symmetry groups unify at high-energy is called Grand unification theory. In the Standard Model, however, the weak, strong and electromagnetic couplings fail to unify at high energy. In a supersymmetry theory, the running of the gauge couplings are modified, and precise high-energy unification of the gauge couplings is achieved. The modified running also provides a natural mechanism for radiative electroweak symmetry breaking.

23.1.3   Dark matter

TeV-scale supersymmetry (augmented with a discrete symmetry) typically provides a candidate dark matter particle at a mass scale consistent with thermal relic abundance calculations.[6][7]

23.1.4   Other technical motivations

Supersymmetry is also motivated by solutions to several theoretical problems, for generally providing many desirable mathematical properties, and for ensuring sensible behavior at high energies. Supersymmetric quantum field theory is often much easier to analyze, as many more problems become exactly solvable. When supersymmetry is imposed as a *local* symmetry, Einstein's theory of general relativity is included automatically, and the result is said to be a theory of supergravity. It is also a necessary feature of the most popular candidate for a theory of everything, superstring theory.

Another theoretically appealing property of supersymmetry is that it offers the only "loophole" to the Coleman–Mandula theorem, which prohibits spacetime and internal symmetries from being combined in any nontrivial way, for quantum field theories like the Standard Model with very general assumptions. The Haag-Lopuszanski-Sohnius theorem demonstrates that supersymmetry is the only way spacetime and internal symmetries can be combined consistently.[8]

23.2   History

A supersymmetry relating mesons and baryons was first proposed, in the context of hadronic physics, by Hironari Miyazawa during 1966. This supersymmetry did not involve spacetime, that is, it concerned internal symmetry, and was broken badly. Miyazawa's work was largely ignored at the time.[9][10][11][12]

J. L. Gervais and B. Sakita (during 1971),[13] Yu. A. Golfand and E. P. Likhtman (also during 1971), and D.V. Volkov and V.P. Akulov (1972),[14] independently rediscovered supersymmetry in the context of quantum field theory, a radically new type of symmetry of spacetime and fundamental fields, which establishes a relationship between elementary particles of different quantum nature, bosons and fermions, and unifies spacetime and internal symmetries of microscopic phenomena. Supersymmetry with a consistent Lie-algebraic graded structure on which the Gervais–Sakita rediscovery was based directly first arose during 1971[15] in the context of an early version of string theory by Pierre Ramond, John H. Schwarz and André Neveu.

Finally, Julius Wess and Bruno Zumino (during 1974)[16] identified the characteristic renormalization features of four-dimensional supersymmetric field theories, which identified them as remarkable QFTs, and they and Abdus Salam and their fellow researchers introduced early particle physics applications. The mathematical structure of supersymmetry (Graded Lie superalgebras) has subsequently been applied successfully to other topics of physics, ranging from nuclear physics,[17][18] critical phenomena,[19] quantum mechanics to statistical physics. It remains a vital part of many proposed theories of physics.

The first realistic supersymmetric version of the Standard Model was proposed during 1977 by Pierre Fayet and is known as the Minimal Supersymmetric Standard Model or MSSM for short. It was proposed to solve, amongst other things, the hierarchy problem.

23.3   Applications

23.3.1   Extension of possible symmetry groups

One reason that physicists explored supersymmetry is because it offers an extension to the more familiar symmetries of quantum field theory. These symmetries are grouped into the Poincaré group and internal symmetries and the Coleman–Mandula theorem showed that under certain assumptions, the symmetries of the S-matrix must be a direct product of the Poincaré group with a compact internal symmetry group or if there is not any mass gap, the conformal group with a compact internal symmetry group. During 1971 Golfand and Likhtman were the first to show that the Poincaré algebra can be extended through introduction of four anticommuting spinor generators (in four dimensions), which later became known as supercharges. During 1975 the Haag-Lopuszanski-Sohnius theorem analyzed all possible superalgebras in the general form, including those with an extended number of the supergenerators and central charges. This extended super-Poincaré algebra paved the way for obtaining a very large and important class of supersymmetric field theories.

The supersymmetry algebra

Main article: Supersymmetry algebra

Traditional symmetries of physics are generated by objects that transform by the tensor representations of the Poincaré group and internal symmetries. Supersymmetries, however, are generated by objects that transform by the spinor representations. According to the spin-statistics theorem, bosonic fields commute while fermionic fields anticommute. Combining the two kinds of fields into a single algebra requires the introduction of a $\mathbf{Z}_2$-grading under which the bosons are the even elements and the fermions are the odd elements. Such an algebra is called a Lie superalgebra.

The simplest supersymmetric extension of the Poincaré algebra is the Super-Poincaré algebra. Expressed in terms of two Weyl spinors, has the following anti-commutation relation:

$$\{Q_\alpha, \bar{Q}\beta\} = 2(\sigma^\mu)_{\alpha\beta} P_\mu$$

and all other anti-commutation relations between the Qs and commutation relations between the Qs and Ps vanish. In the above expression $P_\mu = -i\partial_\mu$ are the generators of translation and σ^μ are the Pauli matrices.

There are representations of a Lie superalgebra that are analogous to representations of a Lie algebra. Each Lie algebra has an associated Lie group and a Lie superalgebra can sometimes be extended into representations of a Lie supergroup.

23.3.2   The Supersymmetric Standard Model

Main article: Minimal Supersymmetric Standard Model

Incorporating supersymmetry into the Standard Model requires doubling the number of particles since there is no way that any of the particles in the Standard Model can be superpartners of each other. With the addition of new particles, there are many possible new interactions. The simplest possible supersymmetric model consistent with the Standard Model is the Minimal Supersymmetric Standard Model (MSSM) which can include the necessary additional new particles that are able to be superpartners of those in the Standard Model.

One of the main motivations for SUSY comes from the quadratically divergent contributions to the Higgs mass squared. The quantum mechanical interactions of the Higgs boson causes a large renormalization of the Higgs mass and unless there is an accidental cancellation, the natural size of the Higgs mass is the greatest scale possible. This problem is known as the hierarchy problem. Supersymmetry reduces the size of the quantum corrections by having automatic cancellations between fermionic and bosonic Higgs interactions. If supersymmetry is restored at the weak scale, then the Higgs mass is related to supersymmetry breaking which can be induced from small non-perturbative effects explaining the vastly different scales in the weak interactions and gravitational interactions.

In many supersymmetric Standard Models there is a heavy stable particle (such as neutralino) which could serve as a weakly interacting massive particle (WIMP) dark matter candidate. The existence of a supersymmetric dark matter candidate is related closely to R-parity.

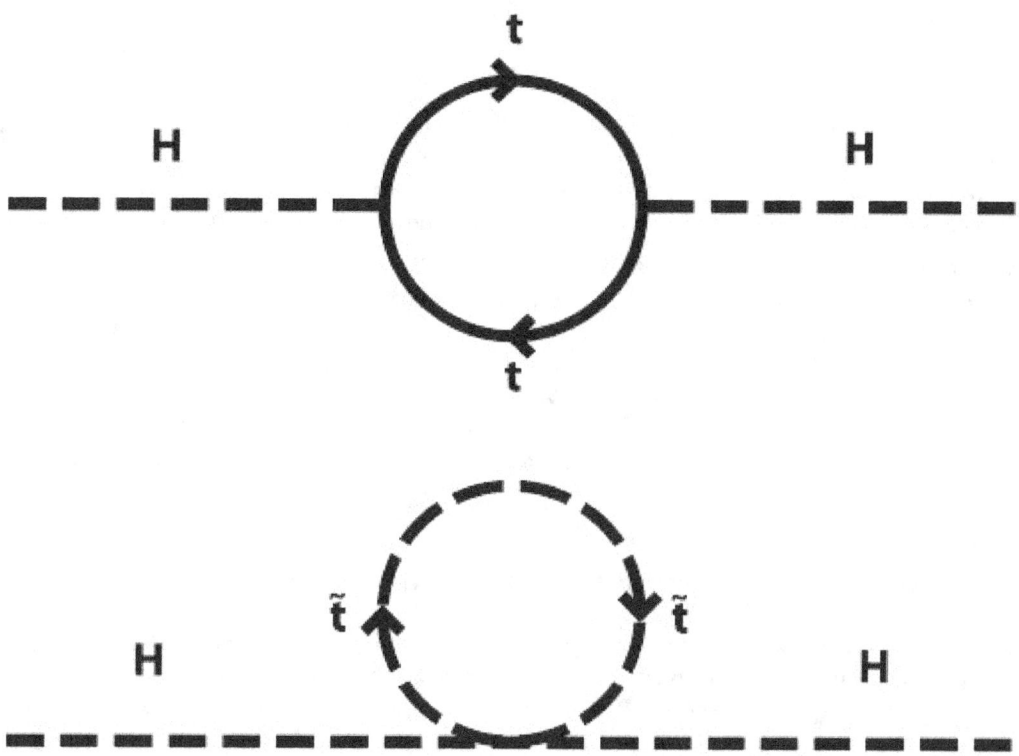

Cancellation of the Higgs boson quadratic mass renormalization between fermionic top quark loop and scalar stop squark tadpole Feynman diagrams in a supersymmetric extension of the Standard Model

The standard paradigm for incorporating supersymmetry into a realistic theory is to have the underlying dynamics of the theory be supersymmetric, but the ground state of the theory does not respect the symmetry and supersymmetry is broken spontaneously. The supersymmetry break can not be done permanently by the particles of the MSSM as they currently appear. This means that there is a new sector of the theory that is responsible for the breaking. The only constraint on this new sector is that it must break supersymmetry permanently and must give superparticles TeV scale masses. There are many models that can do this and most of their details do not matter. In order to parameterize the relevant features of supersymmetry breaking, arbitrary soft SUSY breaking terms are added to the theory which temporarily break SUSY explicitly but could never arise from a complete theory of supersymmetry breaking.

Gauge-coupling unification

Main article: Minimal Supersymmetric Standard Model § Gauge-coupling unification

One piece of evidence for supersymmetry existing is gauge coupling unification. The renormalization group evolution of the three gauge coupling constants of the Standard Model is somewhat sensitive to the present particle content of the theory. These coupling constants do not quite meet together at a common energy scale if we run the renormalization group using the Standard Model.[20] With the addition of minimal SUSY joint convergence of the coupling constants is projected at approximately 10^{16} GeV.[20]

23.3.3   Supersymmetric quantum mechanics

Main article: Supersymmetric quantum mechanics

Supersymmetric quantum mechanics adds the SUSY superalgebra to quantum mechanics as opposed to quantum field theory. Supersymmetric quantum mechanics often becomes relevant when studying the dynamics of supersymmetric solitons, and due to the simplified nature of having fields which are only functions of time (rather than space-time), a great deal of progress has been made in this subject and it is now studied in its own right.

SUSY quantum mechanics involves pairs of Hamiltonians which share a particular mathematical relationship, which are called *partner Hamiltonians*. (The potential energy terms which occur in the Hamiltonians are then known as *partner potentials*.) An introductory theorem shows that for every eigenstate of one Hamiltonian, its partner Hamiltonian has a corresponding eigenstate with the same energy. This fact can be exploited to deduce many properties of the eigenstate spectrum. It is analogous to the original description of SUSY, which referred to bosons and fermions. We can imagine a "bosonic Hamiltonian", whose eigenstates are the various bosons of our theory. The SUSY partner of this Hamiltonian would be "fermionic", and its eigenstates would be the theory's fermions. Each boson would have a fermionic partner of equal energy.

23.3.4   Supersymmetry: Applications to condensed matter physics

SUSY concepts have provided useful extensions to the WKB approximation. Additionally, SUSY has been applied to disorder averaged systems both quantum and non-quantum (through statistical mechanics), the Fokker-Planck equation being an example of a non-quantum theory. The 'supersymmetry' in all these systems arises from the fact that one is modelling one particle and as such the 'statistics' don't matter. The use of the supersymmetry method provides a mathematical rigorous alternative to the replica trick, but only in non-interacting systems, which attempts to address the so-called 'problem of the denominator' under disorder averaging. For more on the applications of supersymmetry in condensed matter physics see the book[21]

23.3.5   Supersymmetry in optics

Integrated optics was recently found[22] to provide a fertile ground on which certain ramifications of SUSY can be explored in readily-accessible laboratory settings. Making use of the analogous mathematical structure of the quantum-mechanical Schrödinger equation and the wave equation governing the evolution of light in one-dimensional settings, one may interpret the refractive index distribution of a structure as a potential landscape in which optical wave packets propagate. In this manner, a new class of functional optical structures with possible applications in phase matching, mode conversion[23] and space-division multiplexing becomes possible. SUSY transformations have been also proposed as a way to address inverse scattering problems in optics and as a one-dimensional transformation optics [24]

23.3.6   Mathematics

SUSY is also sometimes studied mathematically for its intrinsic properties. This is because it describes complex fields satisfying a property known as holomorphy, which allows holomorphic quantities to be exactly computed. This makes supersymmetric models useful "toy models" of more realistic theories. A prime example of this has been the demonstration of S-duality in four-dimensional gauge theories[25] that interchanges particles and monopoles.

The proof of the Atiyah-Singer index theorem is much simplified by the use of supersymmetric quantum mechanics.

23.4   General supersymmetry

Supersymmetry appears in many related contexts of theoretical physics. It is possible to have multiple supersymmetries and also have supersymmetric extra dimensions.

23.4.1   Extended supersymmetry

Main article: Extended supersymmetry

It is possible to have more than one kind of supersymmetry transformation. Theories with more than one supersymmetry transformation are known as extended supersymmetric theories. The more supersymmetry a theory has, the more constrained are the field content and interactions. Typically the number of copies of a supersymmetry is a power of 2, i.e. 1, 2, 4, 8. In four dimensions, a spinor has four degrees of freedom and thus the minimal number of supersymmetry generators is four in four dimensions and having eight copies of supersymmetry means that there are 32 supersymmetry generators.

The maximal number of supersymmetry generators possible is 32. Theories with more than 32 supersymmetry generators automatically have massless fields with spin greater than 2. It is not known how to make massless fields with spin greater than two interact, so the maximal number of supersymmetry generators considered is 32. This corresponds to an $N = 8$ supersymmetry theory. Theories with 32 supersymmetries automatically have a graviton.

For four dimensions there are the following theories, with the corresponding multiplets[26](CPT adds a copy, whenever they are not invariant under such symmetry)

- $N = 1$

Chiral multiplet: $(0,\frac{1}{2})$ Vector multiplet: $(\frac{1}{2},1)$ Gravitino multiplet: $(1,\frac{3}{2})$ Graviton multiplet: $(\frac{3}{2},2)$

- $N = 2$

hypermultiplet: $(-\frac{1}{2},0^2,\frac{1}{2})$ vector multiplet: $(0,\frac{1}{2}^2,1)$ supergravity multiplet: $(1,\frac{3}{2}^2,2)$

- $N = 4$

Vector multiplet: $(-1,-\frac{1}{2}^4,0^6,\frac{1}{2}^4,1)$ Supergravity multiplet: $(0,\frac{1}{2}^4,1^6,\frac{3}{2}^4,2)$

- $N = 8$

Supergravity multiplet: $(-2,-\frac{3}{2}^8,-1^{28},-\frac{1}{2}^{56},0^{70},\frac{1}{2}^{56},1^{28},\frac{3}{2}^8,2)$

23.4.2   Supersymmetry in alternate numbers of dimensions

It is possible to have supersymmetry in dimensions other than four. Because the properties of spinors change drastically between different dimensions, each dimension has its characteristic. In d dimensions, the size of spinors is approximately $2^{d/2}$ or $2^{(d-1)/2}$. Since the maximum number of supersymmetries is 32, the greatest number of dimensions in which a supersymmetric theory can exist is eleven.

23.5   Supersymmetry as a quantum group

Main article: Supersymmetry as a quantum group

Supersymmetry can be reinterpreted in the language of noncommutative geometry and quantum groups. In particular, it involves a mild form of noncommutativity, namely supercommutativity. See the main article for more details.

23.6   Supersymmetry in quantum gravity

Supersymmetry is part of a larger enterprise of theoretical physics to unify everything we know about the universe into a single consistent set of physical principles, known as the quest for a Theory of Everything (TOE). A significant part of this larger enterprise is the quest for a theory of quantum gravity, which would unify the classical theory of general relativity and the Standard Model, which explains the other three basic forces in physics (electromagnetism, the strong interaction, and the weak interaction), and provides a palette of fundamental particles upon which all four forces act. Two of the most active methods of forming a theory of quantum gravity are string theory and loop quantum gravity (LQG), although in theory, supersymmetry could be a component of other theories as well.

For string theory to be consistent, supersymmetry seems to be required at some level (although it may be a strongly broken symmetry). In particle theory, supersymmetry is recognized as a way to stabilize the hierarchy between the unification scale and the electroweak scale (or the Higgs boson mass), and can also provide a natural dark matter candidate. String theory also requires extra spatial dimensions which have to be compactified as in Kaluza–Klein theory.

Loop quantum gravity (LQG) predicts no additional spatial dimensions, nor anything else about particle physics. These theories can be formulated in three spatial dimensions and one dimension of time, although in some LQG theories dimensionality is an emergent property of the theory, rather than a fundamental assumption of the theory. Also, LQG is a theory of quantum gravity which does not require supersymmetry. Lee Smolin, one of the originators of LQG, has proposed that a loop quantum gravity theory incorporating either supersymmetry or extra dimensions, or both, be called "loop quantum gravity II".

If experimental evidence confirms supersymmetry in the form of supersymmetric particles such as the neutralino that is often believed to be the lightest superpartner, some people believe this would be a major boost to string theory. Since supersymmetry is a required component of string theory, any discovered supersymmetry would be consistent with string theory. If the Large Hadron Collider and other major particle physics experiments fail to detect supersymmetric partners or evidence of extra dimensions, many versions of string theory which had predicted certain low mass superpartners to existing particles may need to be significantly revised. The failure of experiments to discover either supersymmetric partners or extra spatial dimensions, as of 2013, has encouraged loop quantum gravity researchers.

23.7   Current status

Supersymmetric models are constrained by a variety of experiments, including measurements of low-energy observables – for example, the anomalous magnetic moment of the muon at Brookhaven; the WMAP dark matter density measurement and direct detection experiments – for example, XENON–100 and LUX; and by particle collider experiments, including B-physics, Higgs phenomenology and direct searches for superpartners (sparticles), at the Large Electron–Positron Collider, Tevatron and the LHC.

Historically, the tightest limits were from direct production at colliders. The first mass limits for squarks and gluinos were made at CERN by the UA1 experiment and the UA2 experiment at the Super Proton Synchrotron. LEP later set very strong limits.,[27] which in 2006 were extended by the D0 experiment at the Tevatron.[28][29] From 2003, WMAP's and Planck's dark matter density measurements have strongly constrained supersymmetry models, which, if they explain dark matter, have to be tuned to invoke a particular mechanism to sufficiently reduce the neutralino density.

Prior to the beginning of the LHC, in 2009 fits of available data to CMSSM and NUHM1 indicated that squarks and gluinos were most likely to have masses in the 500 to 800 GeV range, though values as high as 2.5 TeV were allowed with low probabilities. Neutralinos and sleptons were expected to be quite light, with the lightest neutralino and the lightest stau most likely to be found between 100 to 150 GeV.[30]

The first run of the LHC found no evidence for supersymmetry, and, as a result, surpassed existing experimental limits from the Large Electron–Positron Collider and Tevatron and partially excluded the aforementioned expected ranges.[31]

During 2011 and 2012, the LHC discovered a Higgs boson with a mass of about 125 GeV, and with couplings to fermions and bosons which are consistent with the Standard Model. The MSSM predicts that the mass of the lightest Higgs boson should not be much higher than the mass of the Z boson, and, in the absence of fine tuning (with the supersymmetry breaking scale on the order of 1 TeV), should not exceed 130 GeV. Furthermore, for values of the MSSM parameter *tan* β ≤ 3, it predicts a Higgs mass below 114 GeV over most of the parameter space.[32] This region of Higgs mass was excluded by LEP by 2000. The LHC result is somewhat problematic for the minimal supersymmetric model, as the value of 125 GeV is relatively large for the model and can only be achieved with large radiative loop corrections from top squarks, which many theorists consider to be "unnatural" (see naturalness and fine tuning).[33] On the other hand, the lightest Higgs boson in the MSSM is Standard Model-like, which is consistent with measurements of the Higgs boson couplings at the LHC.

In spite of the null searches and the heavy Higgs, a recent analysis of the constrained minimal supersymmetric Standard Model, the CMSSM, suggests that the model is still compatible with all present experimental constraints.[34][35] The preferred masses for squarks and gluinos is about 2 TeV. The resulting fine-tuning of the electroweak scale, however, is considered "unnatural" (see little hierarchy problem), and some theorists now favor extended supersymmetry models – for example, the NMSSM.

23.8   See also

- Supersymmetric gauge theory

- Wess–Zumino model

- Minimal Supersymmetric Standard Model

- Supersymmetry as a quantum group

- Quantum group

- Supercharge

- Superfield

- Supergeometry

- Supergravity

- Supergroup

- Superspace

23.9   References

[1] Haber, Howie. "SUPERSYMMETRY, PART I (THEORY)" (PDF). *Reviews, Tables and Plots*. Particle Data Group (PDG). Retrieved 8 July 2015.

[2] Martin, Stephen P. (1997). "A Supersymmetry Primer". arXiv:hep-ph/9709356.

[3] Dine, Michael (2007). *Supersymmetry and String Theory: Beyond the Standard Model*. p. 169.

[4] Ellis, John. "The Physics Landscape after the Higgs Discovery at the LHC". *arXiv*. Invited plenary talk at SILAFAE 2014. Retrieved 8 July 2015.

[5] Wolchover, Natalie (November 20, 2012). "Supersymmetry Fails Test, Forcing Physics to Seek New Ideas". *Quanta Magazine*.

[6] Jonathan Feng: Supersymmetric Dark Matter *(pdf)*, University of California, Irvine, 11 May 2007

[7] Torsten Bringmann: The WIMP "Miracle" *(pdf)* University of Hamburg

[8] R. Haag, J. T. Lopuszanski and M. Sohnius, "All Possible Generators Of Supersymmetries Of The S Matrix", Nucl. Phys. B 88 (1975) 257

[9] H. Miyazawa (1966). "Baryon Number Changing Currents".*Prog. Theor. Phys.***36**(6): 1266–1276.Bibcode:1966PThPh.M. doi:10.1143/PTP.36.1266.

[10]H. Miyazawa (1968). "Spinor Currents and Symmetries of Baryons and Mesons".*Phys. Rev.***170**(5): 1586–1590.Bibcode6M. doi:10.1103/PhysRev.170.1586.

[11] Michio Kaku, *Quantum Field Theory*, ISBN 0-19-509158-2, pg 663.

[12] Peter Freund, *Introduction to Supersymmetry*, ISBN 0-521-35675-X, pages 26-27, 138.

[13] Gervais, J. -L.; Sakita, B. (1971). "Field theory interpretation of supergauges in dual models". *Nuclear Physics B* **34** (2): 632. Bibcode:1971NuPhB..34..632G. doi:10.1016/0550-3213(71)90351-8.

[14] D.V. Volkov, V.P. Akulov, Pisma Zh.Eksp.Teor.Fiz. 16 (1972) 621; Phys.Lett. B46 (1973) 109; V.P. Akulov, D.V. Volkov, Teor.Mat.Fiz. 18 (1974) 39

[15] Ramond, P. (1971). "Dual Theory for Free Fermions". *Physical Review D* **3** (10): 2415. Bibcode:1971PhRvD...3.2415R. doi:10.1103/PhysRevD.3.2415.

[16]Wess, J.; Zumino, B. (1974). "Supergauge transformations in four dimensions" *Nuclear Physics B***70**: 39.Bibcode:1974NuPhW. doi:10.1016/0550-3213(74)90355-1.

[17] http://users.physik.fu-berlin.de/~{]kleinert/kleinert/?p=supersym suggested here

[18]Iachello, F. (1980). "Dynamical Supersymmetries in Nuclei".*Physical Review Letters***44**(12): 772.Bibcode:1980PhRvL..44.2I. doi:10.1103/PhysRevLett.44.772.

[19] Friedan, D.; Qiu, Z.; Shenker, S. (1984). "Conformal Invariance, Unitarity, and Critical Exponents in Two Dimensions". *Physical Review Letters* **52** (18): 1575. Bibcode:1984PhRvL..52.1575F. doi:10.1103/PhysRevLett.52.1575.

[20] Gordon L. Kane, *The Dawn of Physics Beyond the Standard Model*, Scientific American, June 2003, page 60 and *The frontiers of physics*, special edition, Vol 15, #3, page 8

[21] *Supersymmetry in Disorder and Chaos*, Konstantin Efetov, Cambridge university press, 1997.

[22] Miri, M.-A.; Heinrich, M.; El-Ganainy, R.; Christodoulides, D. N. (2013). "Superymmetric optical structures". *Physical Review Letters* (APS) **110** (23): 233902. arXiv:1304.6646. Bibcode:2013PhRvL.110w3902M. doi:10.1103/PhysRevLett.110.233902. PMID 25167493. Retrieved April 2014.

[23] Heinrich, M.; Miri, M.-A.; Stützer, S.; El-Ganainy, R.; Nolte, S.; Szameit, A.; Christodoulides, D. N. (2014). "Superymmetric mode converters". *Nature Communications* (NPG) **5**: 3698. arXiv:1401.5734. Bibcode:2014NatCo...5E3698H. doi:1 8.PMID24739256.Retrieved April 2014.

[24] Miri, M.-A.; Heinrich, Matthias; Christodoulides, D. N. (2014). "SUSY-inspired one-dimensional transformation optics". *Optica* (OSA) **1** (2): 89. arXiv:1408.0832. doi:10.1364/OPTICA.1.000089. Retrieved August 2014.

[25] Krasnitz, Michael (2002). *Correlation functions in supersymmetric gauge theories from supergravity fluctuafluctuations hHKtions* (PDF). Princeton University Department of Physics: Princeton University Department of Physics. p. 91.

[26] Polchinski,J. *String theory. Vol. 2: Superstring theory and beyond*, Appendix B

[27] LEPSUSYWG, ALEPH, DELPHI, L3 and OPAL experiments, charginos, large m0 LEPSUSYWG/01-03.1

[28] The D0-Collaboration (2009). "Search for associated production of charginos and neutralinos in the trilepton final state using 2.3 fb^{-1} of data". arXiv:0901.0646. Bibcode:2009PhLB..680...34D. doi:10.1016/j.physletb.2009.08.011.

[29] The D0 Collaboration (2006). "Search for squarks and gluinos in events with jets and missing transverse energy using 2.1 fb-1 of pp⁻ collision data at s=1.96 TeV". arXiv:0712.3805. Bibcode:2008PhLB..660..449D. doi:10.1016/j.physletb.2008.01.042.

[30] O. Buchmueller; et al. (2009). "Likelihood Functions for Supersymmetric Observables in Frequentist Analyses of the CMSSM and NUHM1". *The European Physical Journal C* **64** (3): 391–415. arXiv:0907.5568. Bibcode:2009EPJC...64..391B. doi: 009-1159-z.

[31] Roszkowski, Leszek; Sessolo, Enrico Maria; Williams, Andrew J. (11 August 2014). "What next for the CMSSM and the NUHM: improved prospects for superpartner and dark matter detection". *Journal of High Energy Physics* **2014** (8). doi:

[32] Marcela Carena and Howard E. Haber; Haber (1970). "Higgs Boson Theory and Phenomenology". *Progress in Particle and Nuclear Physics* **50**: 63. arXiv:hep-ph/0208209v3. Bibcode:2003PrPNP..50...63C. doi:10.1016/S0146-6410(02)00177-1.

[33] Patrick Draper; et al. (December 2011). "Implications of a 125 GeV Higgs for the MSSM and Low-Scale SUSY Breaking". *Physical Review D* **85** (9): 095007. arXiv:1112.3068. Bibcode:2012PhRvD..85i5007D. doi:10.1103/PhysRevD.85.095007.

[34] Bechtle, Philip. "How alive is constrained SUSY really?". *arXiv*. Retrieved 8 July 2015.

[35] Jan de Vries, Kees. "SUSY fits with full LHC Run I data". *arXiv*. Retrieved 8 July 2015.

23.10   Further reading

- Supersymmetry and Supergravity page in String Theory Wiki lists more books and reviews.

23.10.1   Theoretical introductions, free and online

- S. Martin (2011). "A Supersymmetry Primer". arXiv:hep-ph/9709356.

- Joseph D. Lykken (1996). "Introduction to Supersymmetry". arXiv:hep-th/9612114.

- Manuel Drees (1996). "An Introduction to Supersymmetry". arXiv:hep-ph/9611409.

- Adel Bilal (2001). "Introduction to Supersymmetry". arXiv:hep-th/0101055.

- An Introduction to Global Supersymmetry by Philip Arygres, 2001

23.10.2   Monographs

- Weak Scale Supersymmetry by Howard Baer and Xerxes Tata, 2006.

- Cooper, F.; Khare, A.; Sukhatme, U. (1995). "Supersymmetry and quantum mechanics". *Physics Reports* **251** (5–6): 267. doi:10.1016/0370-1573(94)00080-M. (arXiv:hep-th/9405029).

- Junker, G. (1996). "Supersymmetric Methods in Quantum and Statistical Physics". doi:10.1007/978-3-642-61194-0. ISBN 978-3-540-61591-0..

- Gordon L. Kane.*Supersymmetry: Unveiling the Ultimate Laws of Nature* Basic Books, New York (2001). ISBN 0-7382-0489-7.

- Gordon L. Kane and Shifman, M., eds. *The Supersymmetric World: The Beginnings of the Theory*, World Scientific, Singapore (2000). ISBN 981-02-4522-X.

- Weinberg, Steven, *The Quantum Theory of Fields, Volume 3: Supersymmetry*, Cambridge University Press, Cambridge, (1999). ISBN 0-521-66000-9.

- Wess, Julius, and Jonathan Bagger, *Supersymmetry and Supergravity*, Princeton University Press, Princeton, (1992). ISBN 0-691-02530-4.

- "Concise Encyclopedia of Supersymmetry". 2003. doi:10.1007/1-4020-4522-0. ISBN 978-1-4020-1338-6.

23.10.3   On experiments

- Bennett GW; Muon (g−2) Collaboration; Bousquet; Brown; Bunce; Carey; Cushman; Danby; Debevec; Deile; Deng; Dhawan; Druzhinin; Duong; Farley; Fedotovich; Gray; Grigoriev; Grosse-Perdekamp; Grossmann; Hare; Hertzog; Huang; Hughes; Iwasaki; Jungmann; Kawall; Khazin; Krienen; Kronkvist; et al. (2004). "Measurement of the negative muon anomalous magnetic moment to 0.7 ppm". *Physical Review Letters* **92** (16): 161802. arXiv:hep-ex/0401008. Bibcode:2004PhRvL..92p1802B. doi:10.1103/PhysRevLett.92.161802. PMID 15169217.

- Brookhaven National Laboratory (Jan. 8, 2004). *New g−2 measurement deviates further from Standard Model.* Press Release.

- Fermi National Accelerator Laboratory (Sept 25, 2006). *Fermilab's CDF scientists have discovered the quick-change behavior of the B-sub-s meson.* Press Release.

23.11   External links

- Supersymmetry (physics) at *Encyclopædia Britannica*

- What do current LHC results (mid-August 2011) imply about supersymmetry? Matt Strassler

- ATLAS Experiment Supersymmetry search documents

- CMS Experiment Supersymmetry search documents

- "Particle wobble shakes up supersymmetry", *Cosmos* magazine, September 2006

- LHC results put supersymmetry theory 'on the spot' BBC news 27/8/2011

- SUSY running out of hiding places BBC news 12/11/2012

- Supersymmetry in optics? "Skulls in the Stars" blog 22/08/2013

Chapter 24

Superstring theory

"Superstring" redirects here. For the converse relation of "substring", see Superstring (formal languages). For the bundle of firecrackers, see Superstring (fireworks).

Superstring theory is an attempt to explain all of the particles and fundamental forces of nature in one theory by modelling them as vibrations of tiny supersymmetric strings.

'Superstring theory' is a shorthand for **supersymmetric string theory** because unlike bosonic string theory, it is the version of string theory that incorporates fermions and supersymmetry.

Since the second superstring revolution the five superstring theories are regarded as different limits of a single theory tentatively called M-theory, or simply string theory.

24.1 Background

The deepest problem in theoretical physics is harmonizing the theory of general relativity, which describes gravitation and applies to large-scale structures (stars, galaxies, super clusters), with quantum mechanics, which describes the other three fundamental forces acting on the atomic scale.

The development of a quantum field theory of a force invariably results in infinite possibilities. Physicists have developed mathematical techniques (renormalization) to eliminate these infinities which work for three of the four fundamental forces—electromagnetic, strong nuclear and weak nuclear forces—but not for gravity. The development of a quantum theory of gravity must therefore come about by different means than those used for the other forces.[1]

According to the theory, the fundamental constituents of reality are strings of the Planck length (about 10^{-33} cm) which vibrate at resonant frequencies. Every string, in theory, has a unique resonance, or harmonic. Different harmonics determine different fundamental particles. The tension in a string is on the order of the Planck force (10^{44} newtons). The graviton (the proposed messenger particle of the gravitational force), for example, is predicted by the theory to be a string with wave amplitude zero.

24.1.1 Lack of experimental evidence

Superstring theory is based on supersymmetry. No supersymmetric particles have been discovered and recent research at LHC and Tevatron has excluded some of the ranges.[2][3][4][5] For instance, the mass constraint of the Minimal Supersymmetric Standard Model squarks has been up to 1.1 TeV, and gluinos up to 500 GeV.[6] No report on suggesting large extra dimensions has been delivered from LHC. There have been no principles so far to limit the number of vacua in the concept of a landscape of vacua.[7]

Some particle physicists became disappointed[8] by the lack of experimental verification of supersymmetry, and some have already discarded it; Jon Butterworth at the University College London said that we had no sign of supersymmetry, even in higher energy region, excluding the superpartners of the top quark up to a few TeV. Ben Allanach at the University of Cambridge states that if we do not discover any new particles in the next trial at the LHC, then we can say it is unlikely to discover supersymmetry at CERN in the foreseeable future.[8]

24.2 Extra dimensions

See also: Why does consistency require 10 dimensions?

Our physical space is observed to have only three large dimensions and—taken together with duration as the fourth dimension—a physical theory must take this into account. However, nothing prevents a theory from including more than 4 dimensions. In the case of string theory, consistency requires spacetime to have 10 (3+1+6) dimensions. The fact that we see only 3 dimensions of space can be explained by one of two mechanisms: either the extra dimensions are compactified on a very small scale, or else our world may live on a 3-dimensional submanifold corresponding to a brane, on which all known particles besides gravity would be restricted.

If the extra dimensions are compactified, then the extra six dimensions must be in the form of a Calabi–Yau manifold. Within the more complete framework of M-theory, they would have to take form of a G2 manifold. Calabi-Yaus are interesting mathematical spaces in their own right. A particular exact symmetry of string/M-theory called T-duality (which exchanges momentum modes for winding number and sends compact dimensions of radius R to radius 1/R),[9] has led to the discovery of equivalences between different Calabi-Yaus called Mirror Symmetry.

Superstring theory is not the first theory to propose extra spatial dimensions. It can be seen as building upon the Kaluza–Klein theory which proposed a 4+1-dimensional theory of gravity. When compactified on a circle, the gravity in the extra dimension precisely describes electromagnetism from the perspective of the 3 remaining large space dimensions. Thus the original Kaluza–Klein theory is a prototype for the unification of gauge and gravity interactions, at least at the classical level, however it is known to be insufficient to describe nature for a variety of reasons (missing weak and strong forces, lack of parity violation, etc.) A more complex compact geometry is needed to reproduce the known gauge forces. This is not all: In order to obtain a consistent, fundamental, quantum theory the upgrade to string theory is also necessary, not just the extra dimensions.

24.3 Number of superstring theories

Theoretical physicists were troubled by the existence of five separate string theories. A possible solution for this dilemma was suggested at the beginning of what is called the second superstring revolution in the 1990s, which suggests that the five string theories might be different limits of a single underlying theory, called M-theory. This remains a conjecture.[10]

The five consistent superstring theories are:

- The type I string has one supersymmetry in the ten-dimensional sense (16 supercharges). This theory is special in the sense that it is based on unoriented open and closed strings, while the rest are based on oriented closed strings.

- The type II string theories have two supersymmetries in the ten-dimensional sense (32 supercharges). There are actually two kinds of type II strings called type IIA and type IIB. They differ mainly in the fact that the IIA theory is non-chiral (parity conserving) while the IIB theory is chiral (parity violating).

- The heterotic string theories are based on a peculiar hybrid of a type I superstring and a bosonic string. There are two kinds of heterotic strings differing in their ten-dimensional gauge groups: the heterotic $E_8 \times E_8$ string and the heterotic SO(32) string. (The name heterotic SO(32) is slightly inaccurate since among the SO(32) Lie groups, string theory singles out a quotient $\text{Spin}(32)/Z_2$ that is not equivalent to SO(32).)

Chiral gauge theories can be inconsistent due to anomalies. This happens when certain one-loop Feynman diagrams cause a quantum mechanical breakdown of the gauge symmetry. The anomalies were canceled out via the Green–Schwarz mechanism.

Even though there are only five superstring theories, in order to make detailed predictions for real experiments, information is needed about exactly what physical configuration the theory is in. This considerably complicates efforts to test string theory because there is an astronomically high number – 10^{500} or more – of configurations that meet some of the basic requirements to be consistent with our world. Along with the extreme remoteness of the Planck scale, this is the other major reason it is hard to test superstring theory.

Another approach to the number of superstring theories refers to the mathematical structure called composition algebra. In the findings of abstract algebra there are just seven composition algebras over the field of real numbers. In 1990 physicists R. Foot and G.C. Joshi in Australia stated that "the seven classical superstring theories are in one-to-one correspondence to the seven composition algebras."[11]

24.4   Integrating general relativity and quantum mechanics

General relativity typically deals with situations involving large mass objects in fairly large regions of spacetime whereas quantum mechanics is generally reserved for scenarios at the atomic scale (small spacetime regions). The two are very rarely used together, and the most common case in which they are combined is in the study of black holes. Having "peak density", or the maximum amount of matter possible in a space, and very small area, the two must be used in synchrony in order to predict conditions in such places; yet, when used together, the equations fall apart, spitting out impossible answers, such as imaginary distances and less than one dimension.

The major problem with their congruence is that, at Planck scale (a fundamental small unit of length) lengths, general relativity predicts a smooth, flowing surface, while quantum mechanics predicts a random, warped surface, neither of which are anywhere near compatible. Superstring theory resolves this issue, replacing the classical idea of point particles with loops. These loops have an average diameter of the Planck length, with extremely small variances, which completely ignores the quantum mechanical predictions of Planck-scale length dimensional warping.

Singularities are avoided because the observed consequences of "Big Crunches" never reach zero size. In fact, should the universe begin a "big crunch" sort of process, string theory dictates that the universe could never be smaller than the size of a string, at which point it would actually begin expanding.

24.5   Mathematics

24.5.1   D-branes

D-branes are membrane-like objects in 10D string theory. They can be thought of as occurring as a result of a Kaluza–Klein compactification of 11D M-theory which contains membranes. Because compactification of a geometric theory produces extra vector fields the D-branes can be included in the action by adding an extra U(1) vector field to the string action.

$$\partial_z \to \partial_z + iA_z(z, \bar{z})$$

In **type I** open string theory, the ends of open strings are always attached to D-brane surfaces. A string theory with more gauge fields such as SU(2) gauge fields would then correspond to the compactification of some higher-dimensional theory above 11 dimensions which is not thought to be possible to date. Furthemore, the tachyons attached to the D-branes, show, the instability of those d-branes with respect to the annihilation. We will consider that tachyon total energy is (or reflects) the total energy of the D-branes.

24.5.2   Why five superstring theories?

For a 10 dimensional supersymmetric theory we are allowed a 32-component Majorana spinor. This can be decomposed into a pair of 16-component Majorana-Weyl (chiral) spinors. There are then various ways to construct an invariant depending on whether these two spinors have the same or opposite chiralities:

The heterotic superstrings come in two types SO(32) and $E_8 \times E_8$ as indicated above and the type I superstrings include open strings.

24.6 Beyond superstring theory

It is conceivable that the five superstring theories are approximated to a theory in higher dimensions possibly involving membranes. Because the action for this involves quartic terms and higher so is not Gaussian, the functional integrals are very difficult to solve and so this has confounded the top theoretical physicists. Edward Witten has popularised the concept of a theory in 11 dimensions M-theory involving membranes interpolating from the known symmetries of superstring theory. It may turn out that there exist membrane models or other non-membrane models in higher dimensions which may become acceptable when new unknown symmetries of nature are found, such as noncommutative geometry for example. It is thought, however, that 16 is probably the maximum since O(16) is a maximal subgroup of E8 the largest exceptional lie group and also is more than large enough to contain the Standard Model. Quartic integrals of the non-functional kind are easier to solve so there is hope for the future. This is the series solution which is always convergent when a is non-zero and negative:

$$\int_{-\infty}^{\infty} \exp(ax^4 + bx^3 + cx^2 + dx + f)\, dx = e^f \sum_{n,m,p=0}^{\infty} \frac{b^{4n}}{(4n)!} \frac{c^{2m}}{(2m)!} \frac{d^{4p}}{(4p)!} \frac{\Gamma(3n + m + p + \frac{1}{4})}{a^{3n+m+p+\frac{1}{4}}}$$

In the case of membranes the series would correspond to sums of various membrane interactions that are not seen in string theory.

24.6.1 Compactification

Investigating theories of higher dimensions often involves looking at the 10 dimensional superstring theory and interpreting some of the more obscure results in terms of compactified dimensions. For example, D-branes are seen as compactified membranes from 11D M-theory. Theories of higher dimensions such as 12D F-theory and beyond will produce other effects such as gauge terms higher than $U(1)$. The components of the extra vector fields (A) in the D-brane actions can be thought of as extra coordinates (X) in disguise. However, the *known* symmetries including supersymmetry currently restrict the spinors to have 32-components which limits the number of dimensions to 11 (or 12 if you include two time dimensions.) Some commentators (e.g. John Baez et al.) have speculated that the exceptional lie groups E_6, E_7 and E_8 having maximum orthogonal subgroups O(10), O(12) and O(16) may be related to theories in 10, 12 and 16 dimensions; 10 dimensions corresponding to string theory and the 12 and 16 dimensional theories being yet undiscovered but would be theories based on 3-branes and 7-branes respectively. However this is a minority view within the string community. Since E_7 is in some sense F_4 quaternified and E_8 is F_4 octonified, then the 12 and 16 dimensional theories, if they did exist, may involve the noncommutative geometry based on the quaternions and octonions respectively. From the above discussion, it can be seen that physicists have many ideas for extending superstring theory beyond the current 10 dimensional theory, but so far none have been successful.

24.6.2 Kac–Moody algebras

Since strings can have an infinite number of modes, the symmetry used to describe string theory is based on infinite dimensional Lie algebras. Some Kac–Moody algebras that have been considered as symmetries for M-theory have been E_{10} and E_{11} and their supersymmetric extensions.

24.7 See also

- AdS/CFT

- dS/CFT correspondence

- Grand unification theory

- Large Hadron Collider

- List of string theory topics

- Quantum gravity

- String field theory

24.8   Notes

[1] Polchinski, Joseph. *String Theory: Volume I*. Cambridge University Press, p. 4.

[2] Woit, Peter (February 22, 2011). "Implications of Initial LHC Searches for Supersymmetry".

[3] Cassel, S.; Ghilencea, D. M.; Kraml, S.; Lessa, A.; Ross, G. G. (2011). "Fine-tuning implications for complementary dark matter and LHC SUSY searches". *Journal of High Energy Physics* **2011** (5): 120. arXiv:1101.4664. Bibcode:2011JHEP...05..120C. doi:10.1007/JHEP05(2011)120.

[4] Falkowski, Adam (Jester) (February 16, 2011). "What LHC tells about SUSY". *resonaances.blogspot.com*. Archived from the original on March 22, 2014. Retrieved March 22, 2014.

[5] Tapper, Alex (24 March 2010). "Early SUSY searches at the LHC" (PDF). Imperial College London.

[6] CMS Collaboration (2011). "Search for Supersymmetry at the LHC in Events with Jets and Missing Transverse Energy". *Physical Review Letters* **107** (22): 221804. arXiv:1109.2352. Bibcode:2011PhRvL.107v1804C. doi:10.1103/PhysRevLett.107.224. PMID 22182023.

[7] Shifman, M. (2012). "Frontiers Beyond the Standard Model: Reflections and Impressionistic Portrait of the Conference". *Modern Physics Letters A* **27** (40): 1230043. Bibcode:2012MPLA...2730043S. doi:10.1142/S0217732312300431.

[8] Jha, Alok (August 6, 2013). "One year on from the Higgs boson find, has physics hit the buffers?". *The Guardian*. photograph: Harold Cunningham/Getty Images (London: GMG). ISSN 0261-3077. OCLC 60623878. Archived from the original on March 22, 2014. Retrieved March 22, 2014.

[9] Polchinski, Joseph. *String Theory: Volume I*. Cambridge University Press, p. 247.

[10] Polchinski, Joseph. *String Theory: Volume II*. Cambridge University Press, p. 198.

[11] Foot, R.; Joshi, G. C. (1990). "Nonstandard signature of spacetime, superstrings, and the split composition algebras". *Letters in Mathematical Physics* **19**: 65–71. Bibcode:1990LMaPh..19...65F. doi:10.1007/BF00402262.

24.9   References

- Kaku, Michio (1999). *Introduction to Superstring and M-Theory* (2nd ed.). New York, USA: Springer-Verlag.

- Shen, Sinyan (1982). *Introduction to Superfluidity* (2nd ed.). Beijing, China: Science Press.

- Greene, Brian (2000). *The Elegant Universe: Superstrings, Hidden Dimensions, and the Quest for the Ultimate Theory*. Random House Inc.

24.10   External links

- Wellcome Collection video on superstring theory

- The Official Superstring theory website: http://superstringtheory.com/index.html

Chapter 25

M-theory

For a more accessible and less technical introduction to this topic, see Introduction to M-theory.

M-theory is a theory in physics that unifies all consistent versions of superstring theory. The existence of such a theory was first conjectured by Edward Witten at a string theory conference at the University of Southern California in the spring of 1995. Witten's announcement initiated a flurry of research activity known as the second superstring revolution.

Prior to Witten's announcement, string theorists had identified five versions of superstring theory. Although these theories appeared at first to be very different, work by several physicists showed that the theories were related in intricate and nontrivial ways. In particular, physicists found that apparently distinct theories could be unified by mathematical transformations called S-duality and T-duality. Witten's conjecture was based in part on the existence of these dualities and in part on the relationship of the string theories to a field theory called eleven-dimensional supergravity.

Although a complete formulation of M-theory is not known, the theory should describe two- and five-dimensional objects called branes and should be approximated by eleven-dimensional supergravity at low energies. Modern attempts to formulate M-theory are typically based on matrix theory or the AdS/CFT correspondence. According to Witten, M should stand for "magic", "mystery", or "membrane" according to taste, and the true meaning of the title should be decided when a more fundamental formulation of the theory is known.[1]

Investigations of the mathematical structure of M-theory have spawned important theoretical results in physics and mathematics. More speculatively, M-theory may provide a framework for developing a unified theory of all of the fundamental forces of nature. Attempts to connect M-theory to experiment typically focus on compactifying its extra dimensions to construct candidate models of our four-dimensional world, although so far none have been verified to give rise to physics as observed at, for instance, the Large Hadron Collider.

25.1 Background

25.1.1 Quantum gravity and strings

Main articles: Quantum gravity and String theory

One of the deepest problems in modern physics is the problem of quantum gravity. The current understanding of gravity is based on Albert Einstein's general theory of relativity, which is formulated within the framework of classical physics. However, nongravitational forces are described within the framework of quantum mechanics, a radically different formalism for describing physical phenomena based on probability.[lower-alpha 1] A quantum theory of gravity is needed in order to reconcile general relativity with the principles of quantum mechanics,[lower-alpha 2] but difficulties arise when one attempts to apply the usual prescriptions of quantum theory to the force of gravity.[lower-alpha 3]

String theory is a theoretical framework that attempts to reconcile gravity and quantum mechanics. In string theory, the point-like particles of particle physics are replaced by one-dimensional objects called strings. String theory describes

The fundamental objects of string theory are open and closed strings.

how strings propagate through space and interact with each other. In a given version of string theory, there is only one kind of string, which may look like a small loop or segment of ordinary string, and it can vibrate in different ways. On distance scales larger than the string scale, a string will look just like an ordinary particle, with its mass, charge, and other properties determined by the vibrational state of the string. In this way, all of the different elementary particles may be viewed as vibrating strings. One of the vibrational states of a string gives rise to the graviton, a quantum mechanical particle that carries gravitational force.[lower-alpha 4]

There are several versions of string theory: type I, type IIA, type IIB, and two flavors of heterotic string theory ($SO(32)$ and $E_8 \times E_8$). The different theories allow different types of strings, and the particles that arise at low energies exhibit different symmetries. For example, the type I theory includes both open strings (which are segments with endpoints) and closed strings (which form closed loops), while types IIA and IIB include only closed strings.[2] Each of these five string theories arises as a special limiting case of M-theory. This theory, like its string theory predecessors, is an example of a quantum theory of gravity. It describes a force just like the familiar gravitational force subject to the rules of quantum mechanics.[3]

25.1.2   Number of dimensions

Main article: Compactification (physics)

 In everyday life, there are three familiar dimensions of space: height, width and depth. Einstein's general theory of relativity treats time as a dimension on par with the three spatial dimensions; in general relativity, space and time are not modeled as separate entities but are instead unified to a four-dimensional spacetime. In this framework, the phenomenon of gravity is viewed as a consequence of the geometry of spacetime.[4]

In spite of the fact that the universe is well described by four-dimensional spacetime, there are several reasons why physicists consider theories in other dimensions. In some cases, by modeling spacetime in a different number of dimensions, a theory becomes more mathematically tractable, and one can perform calculations and gain general insights more easily.[lower-alpha 5] There are also situations where theories in two or three spacetime dimensions are useful for describing phenomena in condensed matter physics.[5] Finally, there exist scenarios in which there could actually be more than four

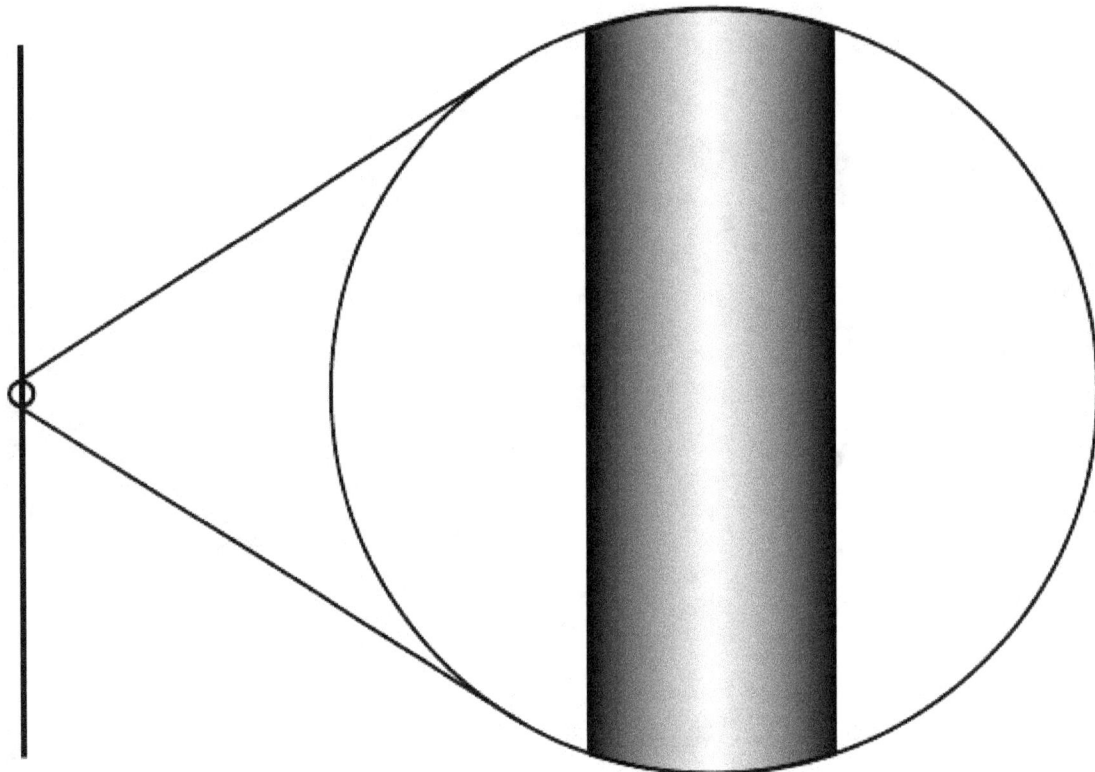

An example of compactification: At large distances, a two dimensional surface with one circular dimension looks one-dimensional.

dimensions of spacetime which have nonetheless managed to escape detection.[6]

One notable feature of string theory and M-theory is that these theories require extra dimensions of spacetime for their mathematical consistency. In string theory, spacetime is ten-dimensional, while in M-theory it is eleven-dimensional. In order to describe real physical phenomena using these theories, one must therefore imagine scenarios in which these extra dimensions would not be observed in experiments.[7]

Compactification is one way of modifying the number of dimensions in a physical theory.[lower-alpha 6] In compactification, some of the extra dimensions are assumed to "close up" on themselves to form circles.[8] In the limit where these curled up dimensions become very small, one obtains a theory in which spacetime has effectively a lower number of dimensions. A standard analogy for this is to consider a multidimensional object such as a garden hose. If the hose is viewed from a sufficient distance, it appears to have only one dimension, its length. However, as one approaches the hose, one discovers that it contains a second dimension, its circumference. Thus, an ant crawling on the surface of the hose would move in two dimensions.[lower-alpha 7]

25.1.3 Dualities

Main articles: S-duality and T-duality

Theories that arise as different limits of M-theory turn out to be related in highly nontrivial ways. One of the relationships that can exist between these different physical theories is called S-duality. This is a relationship which says that a collection of strongly interacting particles in one theory can, in some cases, be viewed as a collection of weakly interacting particles in a completely different theory. Roughly speaking, a collection of particles is said to be strongly interacting if they combine and decay often and weakly interacting if they do so infrequently. Type I string theory turns out to be equivalent by S-duality to the $SO(32)$ heterotic string theory. Similarly, type IIB string theory is related to itself in a nontrivial way by S-duality.[10]

Another relationship between different string theories is T-duality. Here one considers strings propagating around a circular extra dimension. T-duality states that a string propagating around a circle of radius R is equivalent to a string propagating around a circle of radius $1/R$ in the sense that all observable quantities in one description are identified with quantities in the dual description. For example, a string has momentum as it propagates around a circle, and it can also wind around the circle one or more times. The number of times the string winds around a circle is called the winding number. If a string has momentum p and winding number n in one description, it will have momentum n and winding number p in the dual description. For example, type IIA string theory is equivalent to type IIB string theory via T-duality, and the two versions of heterotic string theory are also related by T-duality.[10]

In general, the term *duality* refers to a situation where two seemingly different physical systems turn out to be equivalent in a nontrivial way. If two theories are related by a duality, it means that one theory can be transformed in some way so that it ends up looking just like the other theory. The two theories are then said to be *dual* to one another under the transformation. Put differently, the two theories are mathematically different descriptions of the same phenomena.[11]

25.1.4 Supersymmetry

Main article: Supersymmetry

Another important theoretical idea that plays a role in M-theory is supersymmetry. This is a mathematical relation that exists in certain physical theories between a class of particles called bosons and a class of particles called fermions. Roughly speaking, fermions are the constituents of matter, while bosons mediate interactions between particles. In theories with supersymmetry, each boson has a counterpart which is a fermion, and vice versa. When supersymmetry is imposed as a local symmetry, one automatically obtains a quantum mechanical theory that includes gravity. Such a theory is called a supergravity theory.[12]

A theory of strings that incorporates the idea of supersymmetry is called a superstring theory. There are several different versions of superstring theory which are all subsumed within the M-theory framework. At low energies, the superstring theories are approximated by supergravity in ten spacetime dimensions. Similarly, M-theory is approximated at low energies by supergravity in eleven dimensions.[3]

25.1.5 Branes

Main article: Brane

In string theory and related theories such as supergravity theories, a brane is a physical object that generalizes the notion of a point particle to higher dimensions. For example, a point particle can be viewed as a brane of dimension zero, while a string can be viewed as a brane of dimension one. It is also possible to consider higher-dimensional branes. In dimension p, these are called p-branes. Branes are dynamical objects which can propagate through spacetime according to the rules of quantum mechanics. They can have mass and other attributes such as charge. A p-brane sweeps out a $(p+1)$-dimensional volume in spacetime called its *worldvolume*. Physicists often study fields analogous to the electromagnetic field which live on the worldvolume of a brane. The word brane comes from the word "membrane" which refers to a two-dimensional brane.[13]

In string theory, the fundamental objects that give rise to elementary particles are the one-dimensional strings. Although the physical phenomena described by M-theory are still poorly understood, physicists know that the theory describes two- and five-dimensional branes. Much of the current research in M-theory attempts to better understand the properties of these branes.[lower-alpha 8]

25.2 History and development

25.2.1 Kaluza–Klein theory

Main article: Kaluza–Klein theory

In the early 20th century, physicists and mathematicians including Albert Einstein and Hermann Minkowski pioneered the use of four-dimensional geometry for describing the physical world.[14] These efforts culminated in the formulation of Einstein's general theory of relativity, which relates gravity to the geometry of four-dimensional spacetime.[15]

The success of general relativity led to efforts to apply higher dimensional geometry to explain other forces. In 1919, work by Theodor Kaluza showed that by passing to five-dimensional spacetime, one can unify gravity and electromagnetism into a single force.[15] This idea was improved by physicist Oskar Klein, who suggested that the additional dimension proposed by Kaluza could take the form of a circle with radius around 10^{-30} cm.[16]

The Kaluza–Klein theory and subsequent attempts by Einstein to develop unified field theory were never completely successful. In part this was because Kaluza–Klein theory predicted a particle that has never been shown to exist, and in part because it was unable to correctly predict the ratio of an electron's mass to its charge. In addition, these theories were being developed just as other physicists were beginning to discover quantum mechanics, which would ultimately prove successful in describing known forces such as electromagnetism, as well as new nuclear forces that were being discovered throughout the middle part of the century. Thus it would take almost fifty years for the idea of new dimensions to be taken seriously again.[17]

25.2.2 Early work on supergravity

Main article: Supergravity

New concepts and mathematical tools provided fresh insights into general relativity, giving rise to a period in the 1960s and 70s now known as the golden age of general relativity.[18] In the mid-1970s, physicists began studying higher-dimensional theories combining general relativity with supersymmetry, the so-called supergravity theories.[19]

General relativity does not place any limits on the possible dimensions of spacetime. Although the theory is typically formulated in four dimensions, one can write down the same equations for the gravitational field in any number of dimensions. Supergravity is more restrictive because it places an upper limit on the number of dimensions.[12] In 1978, work by Werner Nahm showed that the maximum spacetime dimension in which one can formulate a consistent supersymmetric theory is eleven.[20] In the same year, Eugene Cremmer, Bernard Julia, and Joel Scherk of the École Normale Supérieure showed that supergravity not only permits up to eleven dimensions but is in fact most elegant in this maximal number of dimensions.[21][22]

Initially, many physicists hoped that by compactifying eleven-dimensional supergravity, it might be possible to construct realistic models of our four-dimensional world. The hope was that such models would provide a unified description of the four fundamental forces of nature: electromagnetism, the strong and weak nuclear forces, and gravity. Interest in eleven-dimensional supergravity soon waned as various flaws in this scheme were discovered. One of the problems was that the laws of physics appear to distinguish between clockwise and counterclockwise, a phenomenon known as chirality. Edward Witten and others observed this chirality property cannot be readily derived by compactifying from eleven dimensions.[22]

In the first superstring revolution in 1984, many physicists turned to string theory as a unified theory of particle physics and quantum gravity. Unlike supergravity theory, string theory was able to accommodate the chirality of the standard model, and it provided a theory of gravity consistent with quantum effects.[22] Another feature of string theory that many physicists were drawn to in the 1980s and 1990s was its high degree of uniqueness. In ordinary particle theories, one can consider any collection of elementary particles whose classical behavior is described by an arbitrary Lagrangian. In string theory, the possibilities are much more constrained: by the 1990s, physicists had argued that there were only five consistent supersymmetric versions of the theory.[22]

25.2.3 Relationships between string theories

Although there were only a handful of consistent superstring theories, it remained a mystery why there was not just one consistent formulation.[22] However, as physicists began to examine string theory more closely, they realized that these theories are related in intricate and nontrivial ways.[23]

In the late 1970s, Claus Montonen and David Olive had conjectured a special property of certain physical theories.[24] A sharpened version of their conjecture concerns a theory called N=4 supersymmetric Yang–Mills theory, which describes particles similar to the quarks and gluons that make up atomic nuclei. The strength with which the particles of this theory interact is measured by a number called the coupling constant. The result of Montonen and Olive, now known as Montonen–Olive duality, states that N=4 supersymmetric Yang–Mills theory with coupling constant g is equivalent to the same theory with coupling constant 1/g. In other words, a system of strongly interacting particles (large coupling constant) has an equivalent description as a system of weakly interacting particles (small coupling constant) and vice versa.[25]

In the 1990s, several theorists generalized Montonen–Olive duality to the S-duality relationship, which connects different string theories. Ashoke Sen studied S-duality in the context of heterotic strings in four dimensions.[26][27] Chris Hull and Paul Townsend showed that type IIB string theory with a large coupling constant is equivalent via S-duality to the same theory with small coupling constant.[28] Theorists also found that different string theories may be related by T-duality. This duality implies that strings propagating on completely different spacetime geometries may be physically equivalent.[29]

25.2.4   Membranes and fivebranes

String theory extends ordinary particle physics by promoting zero-dimensional point particles to one-dimensional objects called strings. In the late 1980s, it was natural for theorists to attempt to formulate other extensions in which particles are replaced by two-dimensional supermembranes or by higher-dimensional objects called branes. Such objects had been considered as early as 1962 by Paul Dirac,[30] and they were reconsidered by a small but enthusiastic group of physicists in the 1980s.[22]

Supersymmetry severely restricts the possible number of dimensions of a brane. In 1987, Eric Bergshoeff, Ergin Sezgin, and Paul Townsend showed that eleven-dimensional supergravity includes two-dimensional branes.[31] Intuitively, these objects look like sheets or membranes propagating through the eleven-dimensional spacetime. Shortly after this discovery, Michael Duff, Paul Howe, Takeo Inami, and Kellogg Stelle considered a particular compactification of eleven-dimensional supergravity with one of the dimensions curled up into a circle.[32] In this setting, one can imagine the membrane wrapping around the circular dimension. If the radius of the circle is sufficiently small, then this membrane looks just like a string in ten-dimensional spacetime. In fact, Duff and his collaborators showed that this construction reproduces exactly the strings appearing in type IIA superstring theory.[25]

In 1990, Andrew Strominger published a similar result which suggested that strongly interacting strings in ten dimensions might have an equivalent description in terms of weakly interacting five-dimensional branes.[33] Initially, physicists were unable to prove this relationship for two important reasons. On the one hand, the Montonen–Olive duality was still unproven, and so Strominger's conjecture was even more tenuous. On the other hand, there were many technical issues related to the quantum properties of five-dimensional branes.[34] The first of these problems was solved in 1993 when Ashoke Sen established that certain physical theories require the existence of objects with both electric and magnetic charge which were predicted by the work of Montonen and Olive.[35]

In spite of this progress, the relationship between strings and five-dimensional branes remained conjectural because theorists were unable to quantize the branes. Starting in 1991, a team of researchers including Michael Duff, Ramzi Khuri, Jianxin Lu, and Ruben Minasian considered a special compactification of string theory in which four of the ten dimensions curl up. If one considers a five-dimensional brane wrapped around these extra dimensions, then the brane looks just like a one-dimensional string. In this way, the conjectured relationship between strings and branes was reduced to a relationship between strings and strings, and the latter could be tested using already established theoretical techniques.[29]

25.2.5   Second superstring revolution

Main article: Second superstring revolution

Speaking at the string theory conference at the University of Southern California in 1995, Edward Witten of the Institute for Advanced Study made the surprising suggestion that all five superstring theories were in fact just different limiting cases of a single theory in eleven spacetime dimensions. Witten's announcement drew together all of the previous results on S-

and T-duality and the appearance of two- and five-dimensional branes in string theory.[36] In the months following Witten's announcement, hundreds of new papers appeared on the Internet confirming that the new theory involved membranes in an important way.[37] Today this flurry of work is known as the second superstring revolution.[38]

One of the important developments following Witten's announcement was Witten's work in 1996 with string theorist Petr Hořava.[39][40] Witten and Hořava studied M-theory on a special spacetime geometry with two ten-dimensional boundary components. Their work shed light on the mathematical structure of M-theory and suggested possible ways of connecting M-theory to real world physics.[41]

25.2.6 Origin of the term

Initially, some physicists suggested that the new theory was a fundamental theory of membranes, but Witten was skeptical of the role of membranes in the theory. In a paper from 1996, Hořava and Witten wrote

> As it has been proposed that the eleven-dimensional theory is a supermembrane theory but there are some reasons to doubt that interpretation, we will non-committally call it the M-theory, leaving to the future the relation of M to membranes.[39]

In the absence of an understanding of the true meaning and structure of M-theory, Witten has suggested that the *M* should stand for "magic", "mystery", or "membrane" according to taste, and the true meaning of the title should be decided when a more fundamental formulation of the theory is known.[1]

25.3 Matrix theory

25.3.1 BFSS matrix model

Main article: Matrix theory (physics)

In mathematics, a matrix is a rectangular array of numbers or other data. In physics, a matrix model is a particular kind of physical theory whose mathematical formulation involves the notion of a matrix in an important way. A matrix model describes the behavior of a set of matrices within the framework of quantum mechanics.[42][43]

One important example of a matrix model is the BFSS matrix model proposed by Tom Banks, Willy Fischler, Stephen Shenker, and Leonard Susskind in 1997. This theory describes the behavior of a set of nine large matrices. In their original paper, these authors showed, among other things, that the low energy limit of this matrix model is described by eleven-dimensional supergravity. These calculations led them to propose that the BFSS matrix model is exactly equivalent to M-theory. The BFSS matrix model can therefore be used as a prototype for a correct formulation of M-theory and a tool for investigating the properties of M-theory in a relatively simple setting.[42]

25.3.2 Noncommutative geometry

Main articles: Noncommutative geometry and Noncommutative quantum field theory

In geometry, it is often useful to introduce coordinates. For example, in order to study the geometry of the Euclidean plane, one defines the coordinates x and y as the distances between any point in the plane and a pair of axes. In ordinary geometry, the coordinates of a point are numbers, so they can be multiplied, and the product of two coordinates does not depend on the order of multiplication. That is, $xy = yx$. This property of multiplication is known as the commutative law, and this relationship between geometry and the commutative algebra of coordinates is the starting point for much of modern geometry.[44]

Noncommutative geometry is a branch of mathematics that attempts to generalize this situation. Rather than working with ordinary numbers, one considers some similar objects, such as matrices, whose multiplication does not satisfy the commutative law (that is, objects for which xy is not necessarily equal to yx). One imagines that these noncommuting objects are coordinates on some more general notion of "space" and proves theorems about these generalized spaces by exploiting the analogy with ordinary geometry.[45]

In a paper from 1998, Alain Connes, Michael R. Douglas, and Albert Schwarz showed that some aspects of matrix models and M-theory are described by a noncommutative quantum field theory, a special kind of physical theory in which the coordinates on spacetime do not satisfy the commutativity property.[43] This established a link between matrix models and M-theory on the one hand, and noncommutative geometry on the other hand. It quickly led to the discovery of other important links between noncommutative geometry and various physical theories.[46][47]

25.4   AdS/CFT correspondence

25.4.1   Overview

Main article: AdS/CFT correspondence

The application of quantum mechanics to physical objects such as the electromagnetic field, which are extended in space and time, is known as quantum field theory.[lower-alpha 9] In particle physics, quantum field theories form the basis for our understanding of elementary particles, which are modeled as excitations in the fundamental fields. Quantum field theories are also used throughout condensed matter physics to model particle-like objects called quasiparticles.[lower-alpha 10]

One approach to formulating M-theory and studying its properties is provided by the anti-de Sitter/conformal field theory (AdS/CFT) correspondence. Proposed by Juan Maldacena in late 1997, the AdS/CFT correspondence is a theoretical result which implies that M-theory is in some cases equivalent to a quantum field theory.[48] In addition to providing insights into the mathematical structure of string and M-theory, the AdS/CFT correspondence has shed light on many aspects of quantum field theory in regimes where traditional calculational techniques are ineffective.[49]

In the AdS/CFT correspondence, the geometry of spacetime is described in terms of a certain vacuum solution of Einstein's equation called anti-de Sitter space.[50] In very elementary terms, anti-de Sitter space is a mathematical model of spacetime in which the notion of distance between points (the metric) is different from the notion of distance in ordinary Euclidean geometry. It is closely related to hyperbolic space, which can be viewed as a disk as illustrated on the left.[51] This image shows a tessellation of a disk by triangles and squares. One can define the distance between points of this disk in such a way that all the triangles and squares are the same size and the circular outer boundary is infinitely far from any point in the interior.[52]

Now imagine a stack of hyperbolic disks where each disk represents the state of the universe at a given time. The resulting geometric object is three-dimensional anti-de Sitter space.[51] It looks like a solid cylinder in which any cross section is a copy of the hyperbolic disk. Time runs along the vertical direction in this picture. The surface of this cylinder plays an important role in the AdS/CFT correspondence. As with the hyperbolic plane, anti-de Sitter space is curved in such a way that any point in the interior is actually infinitely far from this boundary surface.[52]

This construction describes a hypothetical universe with only two space dimensions and one time dimension, but it can be generalized to any number of dimensions. Indeed, hyperbolic space can have more than two dimensions and one can "stack up" copies of hyperbolic space to get higher-dimensional models of anti-de Sitter space.[51]

An important feature of anti-de Sitter space is its boundary (which looks like a cylinder in the case of three-dimensional anti-de Sitter space). One property of this boundary is that, within a small region on the surface around any given point, it looks just like Minkowski space, the model of spacetime used in nongravitational physics.[53] One can therefore consider an auxiliary theory in which "spacetime" is given by the boundary of anti-de Sitter space. This observation is the starting point for AdS/CFT correspondence, which states that the boundary of anti-de Sitter space can be regarded as the "spacetime" for a quantum field theory. The claim is that this quantum field theory is equivalent to the gravitational theory on the bulk anti-de Sitter space in the sense that there is a "dictionary" for translating entities and calculations in one theory into their counterparts in the other theory. For example, a single particle in the gravitational theory might correspond to some collection of particles in the boundary theory. In addition, the predictions in the two theories are quantitatively identical so that if two particles have a 40 percent chance of colliding in the gravitational theory, then the corresponding collections in the boundary theory would also have a 40 percent chance of colliding.[54]

25.4.2   6D (2,0) superconformal field theory

Main article: 6D (2,0) superconformal field theory

One particular realization of the AdS/CFT correspondence states that M-theory on the product space $AdS_7 \times S^4$ is equivalent to the so-called (2,0)-theory on the six-dimensional boundary.[48] Here "(2,0)" refers to the particular type of supersymmetry that appears in the theory. In this example, the spacetime of the gravitational theory is effectively seven-dimensional (hence the notation AdS_7), and there are four additional "compact" dimensions (encoded by the S^4 factor). In the real world, spacetime is four-dimensional, at least macroscopically, so this version of the correspondence does not provide a realistic model of gravity. Likewise, the dual theory is not a viable model of any real-world system since it describes a world with six spacetime dimensions.[lower-alpha 11]

Nevertheless, the (2,0)-theory has proven to be important for studying the general properties of quantum field theories. Indeed, this theory subsumes many mathematically interesting effective quantum field theories and points to new dualities relating these theories. For example, Luis Alday, Davide Gaiotto, and Yuji Tachikawa showed that by compactifying this theory on a surface, one obtains a four-dimensional quantum field theory, and there is a duality known as the AGT correspondence which relates the physics of this theory to certain physical concepts associated with the surface itself.[55] More recently, theorists have extended these ideas to study the theories obtained by compactifying down to three dimensions.[56]

In addition to its applications in quantum field theory, the (2,0)-theory has spawned important results in pure mathematics. For example, the existence of the (2,0)-theory was used by Witten to give a "physical" explanation for a conjectural relationship in mathematics called the geometric Langlands correspondence.[57] In subsequent work, Witten showed that the (2,0)-theory could be used to understand a concept in mathematics called Khovanov homology.[58] Developed by Mikhail Khovanov around 2000, Khovanov homology provides a tool in knot theory, the branch of mathematics that studies and classifies the different shapes of knots.[59] Another application of the (2,0)-theory in mathematics is the work of Davide Gaiotto, Greg Moore, and Andrew Neitzke, which used physical ideas to derive new results in hyperkähler geometry.[60]

25.4.3   ABJM superconformal field theory

Main article: ABJM superconformal field theory

Another realization of the AdS/CFT correspondence states that M-theory on $AdS_4 \times S^7$ is equivalent to a quantum field theory called the ABJM theory in three dimensions. In this version of the correspondence, seven of the dimensions of M-theory are curled up, leaving four non-compact dimensions. Since the spacetime of our universe is four-dimensional, this version of the correspondence provides a somewhat more realistic description of gravity.[61]

The ABJM theory appearing in this version of the correspondence is also interesting for a variety of reasons. Introduced by Aharony, Bergman, Jafferis, and Maldacena, it is closely related to another quantum field theory called Chern–Simons theory. The latter theory was popularized by Witten in the late 1980s because of its applications to knot theory.[62] In addition, the ABJM theory serves as a semi-realistic simplified model for solving problems that arise in condensed matter physics.[61]

25.5   Phenomenology

25.5.1   Overview

Main article: String phenomenology

In addition to being an idea of considerable theoretical interest, M-theory provides a framework for constructing models of real world physics that combine general relativity with the standard model of particle physics. Phenomenology is the branch of theoretical physics in which physicists construct realistic models of nature from more abstract theoretical ideas.

String phenomenology is the part of string theory that attempts to construct realistic models of particle physics based on string and M-theory.[63]

Typically, such models are based on the idea of compactification.[lower-alpha 12] Starting with the ten- or eleven-dimensional spacetime of string or M-theory, physicists postulate a shape for the extra dimensions. By choosing this shape appropriately, they can construct models roughly similar to the standard model of particle physics, together with additional undiscovered particles,[64] usually supersymmetric partners to analogues of known particles. One popular way of deriving realistic physics from string theory is to start with the heterotic theory in ten dimensions and assume that the six extra dimensions of spacetime are shaped like a six-dimensional Calabi–Yau manifold. This is a special kind of geometric object named after mathematicians Eugenio Calabi and Shing-Tung Yau.[65] Calabi–Yau manifolds offer many ways of extracting realistic physics from string theory. Other similar methods can be used to construct models with physics resembling to some extent that of our four-dimensional world based on M-theory.[66]

Partly because of theoretical and mathematical difficulties and partly because of the extremely high energies (beyond what is technologically possible for the foreseeable future) needed to test these theories experimentally, there is so far no experimental evidence that would unambiguously point to any of these models being a correct fundamental description of nature. This has led some in the community to criticize these approaches to unification and question the value of continued research on these problems.[67]

25.5.2   Compactification on G_2 manifolds

In one approach to M-theory phenomenology, theorists assume that the seven extra dimensions of M-theory are shaped like a G_2 manifold. This is a special kind of seven-dimensional shape constructed by mathematician Dominic Joyce of the University of Oxford.[68] These G_2 manifolds are still poorly understood mathematically, and this fact has made it difficult for physicists to fully develop this approach to phenomenology.[69]

For example, physicists and mathematicians often assume that space has a mathematical property called smoothness, but this property cannot be assumed in the case of a G_2 manifold if one wishes to recover the physics of our four-dimensional world. Another problem is that G_2 manifolds are not complex manifolds, so theorists are unable to use tools from the branch of mathematics known as complex analysis. Finally, there are many open questions about the existence, uniqueness, and other mathematical properties of G_2 manifolds, and mathematicians lack a systematic way of searching for these manifolds.[69]

25.5.3   Heterotic M-theory

Because of the difficulties with G_2 manifolds, most attempts to construct realistic theories of physics based on M-theory have taken a more indirect approach to compactifying eleven-dimensional spacetime. One approach, pioneered by Witten, Hořava, Burt Ovrut, and others, is known as heterotic M-theory. In this approach, one imagines that one of the eleven dimensions of M-theory is shaped like a circle. If this circle is very small, then the spacetime becomes effectively ten-dimensional. One then assumes that six of the ten dimensions form a Calabi–Yau manifold. If this Calabi–Yau manifold is also taken to be small, one is left with a theory in four-dimensions.[69]

Heterotic M-theory has been used to construct models of brane cosmology in which the observable universe is thought to exist on a brane in a higher dimensional ambient space. It has also spawned alternative theories of the early universe that do not rely on the theory of cosmic inflation.[69]

25.6   References

25.6.1   Notes

[1] For a standard introduction to quantum mechanics, see Griffiths 2004.

[2] The necessity of a quantum mechanical description of gravity follows from the fact that one cannot consistently couple a classical system to a quantum one. See Wald 1984, p. 382.

[3] From a technical point of view, the problem is that the theory one gets in this way is not renormalizable and therefore cannot be used to make meaningful physical predictions. See Zee 2010, p. 72 for a discussion of this issue.

[4] For an accessible introduction to string theory, see Greene 2000.

[5] For example, in the context of the AdS/CFT correspondence, theorists often formulate and study theories of gravity in unphysical numbers of spacetime dimensions.

[6] Dimensional reduction is another way of modifying the number of dimensions.

[7] This analogy is used for example in Greene 2000, p. 186.

[8] For example, see the subsections on the 6D (2,0) superconformal field theory and ABJM superconformal field theory.

[9] A standard text is Peskin and Schroeder 1995.

[10] For an introduction to the applications of quantum field theory to condensed matter physics, see Zee 2010.

[11] For a review of the (2,0)-theory, see Moore 2012.

[12] Brane world scenarios provide an alternative way of recovering real world physics from string theory. See Randall and Sundrum 1999.

25.6.2 Citations

[1] Duff 1996, sec. 1

[2] Zwiebach 2009, p. 324

[3] Becker, Becker, and Schwarz 2007, p. 12

[4] Wald 1984, p. 4

[5] Zee 2010, Parts V and VI

[6] Zwiebach 2009, p. 9

[7] Zwiebach 2009, p. 8

[8] Yau and Nadis 2010, Ch. 6

[9] Becker, Becker, and Schwarz 2007, pp. 339–347

[10] Becker, Becker, and Schwarz 2007

[11] Zwiebach 2009, p. 376

[12] Duff 1998, p. 64

[13] Moore 2005

[14] Yau and Nadis 2010, p. 9

[15] Yau and Nadis 2010, p. 10

[16] Yau and Nadis 2010, p. 12

[17] Yau and Nadis 2010, p. 13

[18] Wald 1984, p. 3

[19] van Nieuwenhuizen 1981

[20] Nahm 1978

[21] Cremmer, Julia, and Scherk 1978

[22] Duff 1998, p. 65

[23] Duff 1998

[24] Montonen and Olive 1977

[25] Duff 1998, p. 66

[26] Sen 1994a

[27] Sen 1994b

[28] Hull and Townsend 1995

[29] Duff 1998, p. 67

[30] Dirac 1962

[31] Bergshoeff, Sezgin, and Townsend 1987

[32] Duff et al. 1987

[33] Strominger 1990

[34] Duff 1998, pp 66–67

[35] Sen 1993

[36] Witten 1995

[37] Duff 1998, pp. 67–68

[38] Becker, Becker, and Schwarz 2007, p. 296

[39] Hořava and Witten 1996a

[40] Hořava and Witten 1996b

[41] Duff 1998, p. 68

[42] Banks et al. 1997

[43] Connes, Douglas, and Schwarz 1998

[44] Connes 1994, p. 1

[45] Connes 1994

[46] Nekrasov and Schwarz 1998

[47] Seiberg and Witten 1999

[48] Maldacena 1998

[49] Klebanov and Maldacena 2009

[50] Klebanov and Maldacena 2009, p. 28

[51] Maldacena 2005, p. 60

[52] Maldacena 2005, p. 61

[53] Zwiebach 2009, p. 552

[54] Maldacena 2005, pp. 61–62

[55] Alday, Gaiotto, and Tachikawa 2010

[56] Dimofte, Gaiotto, and Gukov 2010

[57] Witten 2009

[58] Witten 2012

[59] Khovanov 2000

[60] Gaiotto, Moore, and Neitzke 2013

[61] Aharony et al. 2008

[62] Witten 1989

[63] Dine 2000

[64] Candelas et al. 1985

[65] Yau and Nadis 2010, p. ix

[66] Yau and Nadis 2010, pp. 147–150

[67] Woit 2006

[68] Yau and Nadis 2010, p. 149

[69] Yau and Nadis 2010, p. 150

25.6.3   Bibliography

- Aharony, Ofer; Bergman, Oren; Jafferis, Daniel Louis; Maldacena, Juan (2008). "$N=6$ superconformal Chern-Simons-matter theories, M2-branes and their gravity duals". *Journal of High Energy Physics* **2008** (10): 091. arXiv:0806.1218. Bibcode:2008JHEP...10..091A. doi:10.1088/1126-6708/2008/10/091.

- Alday, Luis; Gaiotto, Davide; Tachikawa, Yuji (2010). "Liouville correlation functions from four-dimensional gauge theories". *Letters in Mathematical Physics* **91** (2): 167–197. arXiv:0906.3219. Bibcode:2010LMaPh..91..167A. doi:10.1007/s11005-010-0369-5.

- Banks, Tom; Fischler, Willy; Schenker, Stephen; Susskind, Leonard (1997). "M theory as a matrix model: A conjecture". *Physical Review D* **55** (8): 5112. arXiv:hep-th/9610043. Bibcode:1997PhRvD..55.5112B. doi:

- Becker, Katrin; Becker, Melanie; Schwarz, John (2007). *String theory and M-theory: A modern introduction.* Cambridge University Press. ISBN 978-0-521-86069-7.

- Bergshoeff, Eric; Sezgin, Ergin; Townsend, Paul (1987). "Supermembranes and eleven-dimensional supergravity". *Physics Letters B* **189** (1): 75–78. Bibcode:1987PhLB..189...75B. doi:10.1016/0370-2693(87)91272-X.

- Candelas, Philip; Horowitz, Gary; Strominger, Andrew; Witten, Edward (1985). "Vacuum configurations for superstrings". *Nuclear Physics B* **258**: 46–74. Bibcode:1985NuPhB.258...46C. doi:10.1016/0550-3213(85)90602-9.

- Connes, Alain (1994). *Noncommutative Geometry.* Academic Press. ISBN 978-0-12-185860-5.

- Connes, Alain; Douglas, Michael; Schwarz, Albert (1998). "Noncommutative geometry and matrix theory". *Journal of High Energy Physics.* 19981 (2): 003. arXiv:hep-th/9711162.Bibcode:1998JHEP...02..003C.doi:10.106-6708/1998/02/003.

- Cremmer, Eugene; Julia, Bernard; Scherk, Joel (1978). "Supergravity theory in eleven dimensions". *Physics Letters B* **76** (4): 409–412. Bibcode:1978PhLB...76..409C. doi:10.1016/0370-2693(78)90894-8.

- Dimofte, Tudor; Gaiotto, Davide; Gukov, Sergei (2010). "Gauge theories labelled by three-manifolds". *Communications in Mathematical Physics* **325** (2): 367–419. Bibcode:2014CMaPh.325..367D. doi:10.1007/s00220-013-1863-2.

- Dine, Michael (2000). "TASI Lectures on M Theory Phenomenology". arXiv:hep-th/0003175.

- Dirac, Paul (1962). "An extensible model of the electron". *Proceedings of the Royal Society of London*. A. Mathematical and Physical Sciences **268** (1332): 57–67. Bibcode:1962RSPSA.268...57D. doi:10.1098/rspa.1962.0124.

- Duff, Michael (1996). "M-theory (the theory formerly known as strings)". *International Journal of Modern Physics A* **11** (32): 6523–41. arXiv:hep-th/9608117. Bibcode:1996IJMPA..11.5623D. doi:10.1142/S0217751X96002583.

- Duff, Michael (1998). "The theory formerly known as strings".*Scientific American***278**(2): 64–9.doi:10.1038298-64.

- Duff, Michael; Howe, Paul; Inami, Takeo; Stelle, Kellogg (1987). "Superstrings in $D=10$ from supermembranes in $D=11$". *Nuclear Physics B* **191** (1): 70–74. Bibcode:1987PhLB..191...70D. doi:10.1016/0370-2693(87)91323-2.

- Gaiotto, Davide; Moore, Gregory; Neitzke, Andrew (2013). "Wall-crossing, Hitchin systems, and the WKB approximation". *Advances in Mathematics* **2341**: 239–403. arXiv:0907.3987. doi:10.1016/j.aim.2012.09.027.

- Greene, Brian (2000). *The Elegant Universe: Superstrings, Hidden Dimensions, and the Quest for the Ultimate Theory*. Random House. ISBN 978-0-9650888-0-0.

- Griffiths, David (2004). *Introduction to Quantum Mechanics*. Pearson Prentice Hall. ISBN 978-0-13-111892-8.

- Hořava, Petr; Witten, Edward (1996a). "Heterotic and Type I string dynamics from eleven dimensions". *Nuclear Physics B* **460** (3): 506–524. arXiv:hep-th/9510209. Bibcode:1996NuPhB.460..506H. doi:10.1016/0550-3213(95)00621-4.

- Hořava, Petr; Witten, Edward (1996b). "Eleven dimensional supergravity on a manifold with boundary". *Nuclear Physics B* **475** (1): 94–114. arXiv:hep-th/9603142. Bibcode:1996NuPhB.475...94H. doi:10.1016/0550-3213(96)00308-2.

- Hull, Chris; Townsend, Paul (1995). "Unity of superstring dualities". *Nuclear Physics B* **4381** (1): 109–137. arXiv:hep-th/9410167. Bibcode:1995NuPhB.438..109H. doi:10.1016/0550-3213(94)00559-W.

- Khovanov, Mikhail (2000). "A categorification of the Jones polynomial". *Duke Mathematical Journal* **1011** (3): 359–426. doi:10.1215/S0012-7094-00-10131-7.

- Klebanov, Igor; Maldacena, Juan (2009). "Solving Quantum Field Theories via Curved Spacetimes" (PDF). *Physics Today* **62**: 28. Bibcode:2009PhT....62a..28K. doi:10.1063/1.3074260. Retrieved May 2013.

- Maldacena, Juan (1998). "The Large N limit of superconformal field theories and supergravity". *Advances in Theoretical and Mathematical Physics* **2**: 231–252. arXiv:hep-th/9711200. Bibcode:1998AdTMP...2..231M. doi:10.1063/1.59653.

- Maldacena, Juan (2005)."The Illusion of Gravity"(PDF).*Scientific American***293**(5): 56–63.Bibcode:2005ScM. doi:10.1038/scientificamerican1105-56. PMID 16318027. Retrieved July 2013.

- Montonen, Claus; Olive, David (1977). "Magnetic monopoles as gauge particles?". *Physics Letters B* **72** (1): 117–120. Bibcode:1977PhLB...72..117M. doi:10.1016/0370-2693(77)90076-4.

- Moore, Gregory (2005). "What is ... a Brane?" (PDF). *Notices of the AMS* **52**: 214. Retrieved June 2013.

- Moore, Gregory (2012). "Lecture Notes for Felix Klein Lectures" (PDF). Retrieved 14 August 2013.

- Nahm, Walter (1978). "Supersymmetries and their representations".*Nuclear Physics B***135**(1): 1495..149N. doi:10.1016/0550-3213(78)90218-3.

- Nekrasov, Nikita; Schwarz, Albert (1998). "Instantons on noncommutative $\mathbf{R}^4$ and (2,0) superconformal six dimensional theory". *Communications in Mathematical Physics* **198** (3): 689–703. arXiv:hep-th/980208..689N. doi:10.1007/s002200050490.

- Peskin, Michael; Schroeder, Daniel (1995). *An Introduction to Quantum Field Theory*. Westview Press. ISBN 978-0-201-50397-5.

- Randall, Lisa; Sundrum, Raman (1999). "An alternative to compactification". *Physical Review Letters* **83** (23): 4690. arXiv:hep-th/9906064. Bibcode:1999PhRvL..83.4690R. doi:10.1103/PhysRevLett.83.4690.

- Seiberg, Nathan; Witten, Edward (1999). "String Theory and Noncommutative Geometry". *Journal of High Energy Physics* **1999** (9): 032. arXiv:hep-th/9908142.Bibcode:1999JHEP...09..032S.doi:10.1088/1126-6708/19992.

- Sen, Ashoke (1993). "Electric-magnetic duality in string theory". *Nuclear Physics B* **404** (1): 109–126. arXiv:hep-th/9207053. Bibcode:1993NuPhB.404..109S. doi:10.1016/0550-3213(93)90475-5.

- Sen, Ashoke (1994a). "Strong-weak coupling duality in four-dimensional string theory". *International Journal of Modern Physics A* **9** (21): 3707–3750. arXiv:hep-th/9402002.Bibcode:1994IJMPA...9.3707S.doi:10.1142/S497.

- Sen, Ashoke (1994b). "Dyon-monopole bound states, self-dual harmonic forms on the multi-monopole moduli space, and $SL(2,\mathbf{Z})$ invariance in string theory". *Physics Letters B* **329** (2): 217–221. arXiv:hep-th/9402032. Bibcode:1994PhLB..329..217S. doi:10.1016/0370-2693(94)90763-3.

- Strominger, Andrew (1990). "Heterotic solitons".*Nuclear Physics B***343**(1): 167–184.Bibcode:1990NuPhB.343S. doi:10.1016/0550-3213(90)90599-9.

- van Nieuwenhuizen, Peter (1981). "Supergravity". *Physics Reports* **68** (4): 189–398. Bibcode:1981PhR....68..189V. doi:10.1016/0370-1573(81)90157-5.

- Wald, Robert (1984). *General Relativity*. University of Chicago Press. ISBN 978-0-226-87033-5.

- Witten, Edward (1989). "Quantum Field Theory and the Jones Polynomial". *Communications in Mathematical Physics* **121** (3): 351–399. Bibcode:1989CMaPh.121..351W. doi:10.1007/BF01217730. MR 0990772.

- Witten, Edward (1995). "String theory dynamics in various dimensions". *Nuclear Physics B* **443** (1): 85–126. arXiv:hep-th/9503124. Bibcode:1995NuPhB.443...85W. doi:10.1016/0550-3213(95)00158-O.

- Witten, Edward (2009). "Geometric Langlands from six dimensions". arXiv:0905.2720 [hep-th].

- Witten, Edward (2012). "Fivebranes and knots". *Quantum Topology* **3** (1): 1–137. doi:10.4171/QT/26.

- Woit, Peter (2006). *Not Even Wrong: The Failure of String Theory and the Search for Unity in Physical Law*. Basic Books. p. 105. ISBN 0-465-09275-6.

- Yau, Shing-Tung; Nadis, Steve (2010). *The Shape of Inner Space: String Theory and the Geometry of the Universe's Hidden Dimensions*. Basic Books. ISBN 978-0-465-02023-2.

- Zee, Anthony (2010). *Quantum Field Theory in a Nutshell* (2nd ed.). Princeton University Press. ISBN 978-0-691-14034-6.

- Zwiebach, Barton (2009). *A First Course in String Theory*. Cambridge University Press. ISBN 978-0-521-88032-9.

25.7 External links

- The Elegant Universe—A three-hour miniseries with Brian Greene on the series *Nova* (original PBS broadcast dates: October 28, 8–10 p.m. and November 4, 8–9 p.m., 2003). Various images, texts, videos and animations explaining string theory and M-theory.

- Superstringtheory.com—The "Official String Theory Web Site", created by Patricia Schwarz. References on string theory and M-theory for the layperson and expert.

- Not Even Wrong—Peter Woit's blog on physics in general, and string theory in particular.

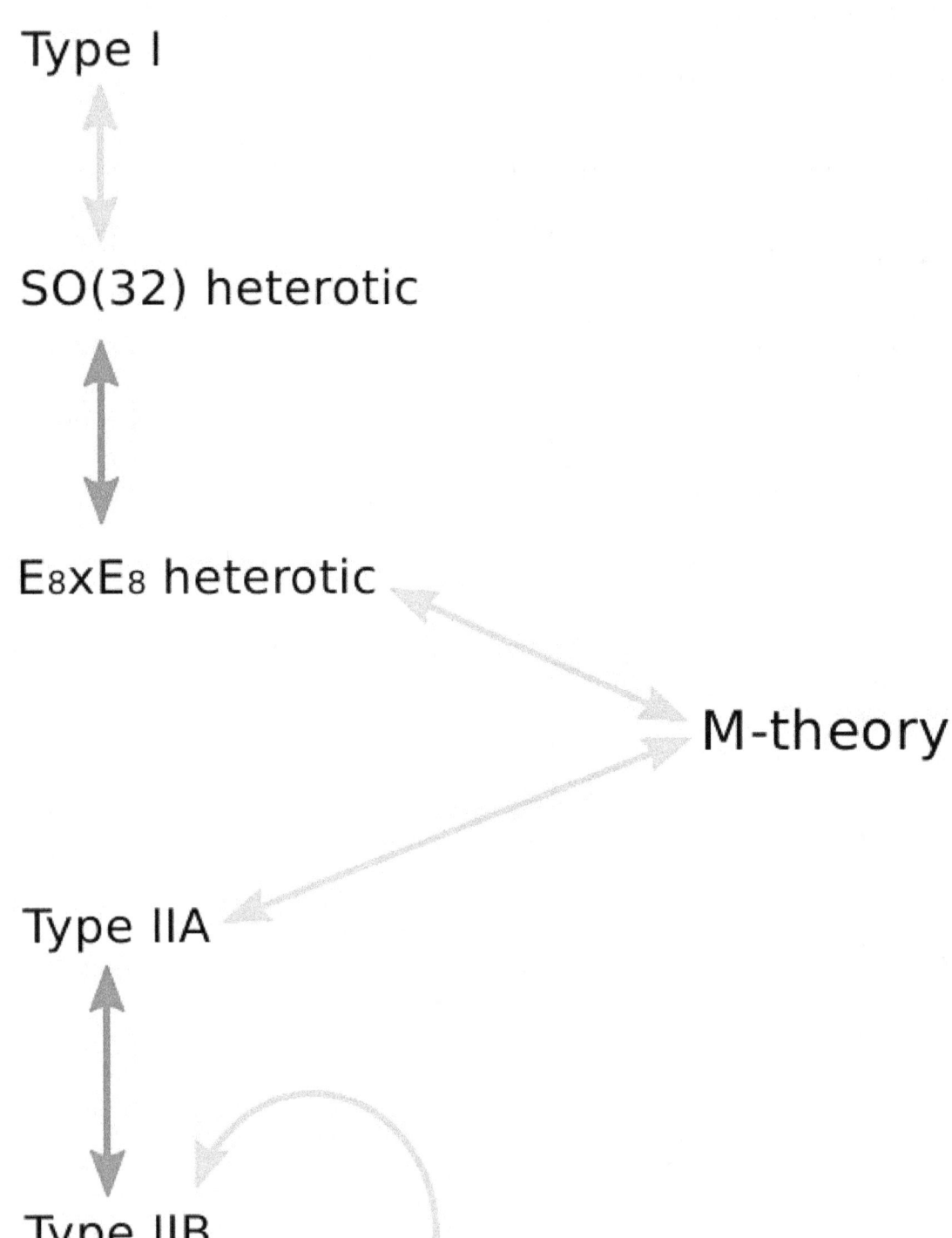

Type I

SO(32) heterotic

E₈xE₈ heterotic

M-theory

Type IIA

Type IIB

A diagram of string theory dualities. Yellow arrows indicate S-duality. Blue arrows indicate T-duality. These dualities may be combined to obtain equivalences of any of the five theories with M-theory.[91]

In the 1980s, Edward Witten contributed to the understanding of supergravity theories. In 1995, he introduced M-theory, sparking the second superstring revolution.

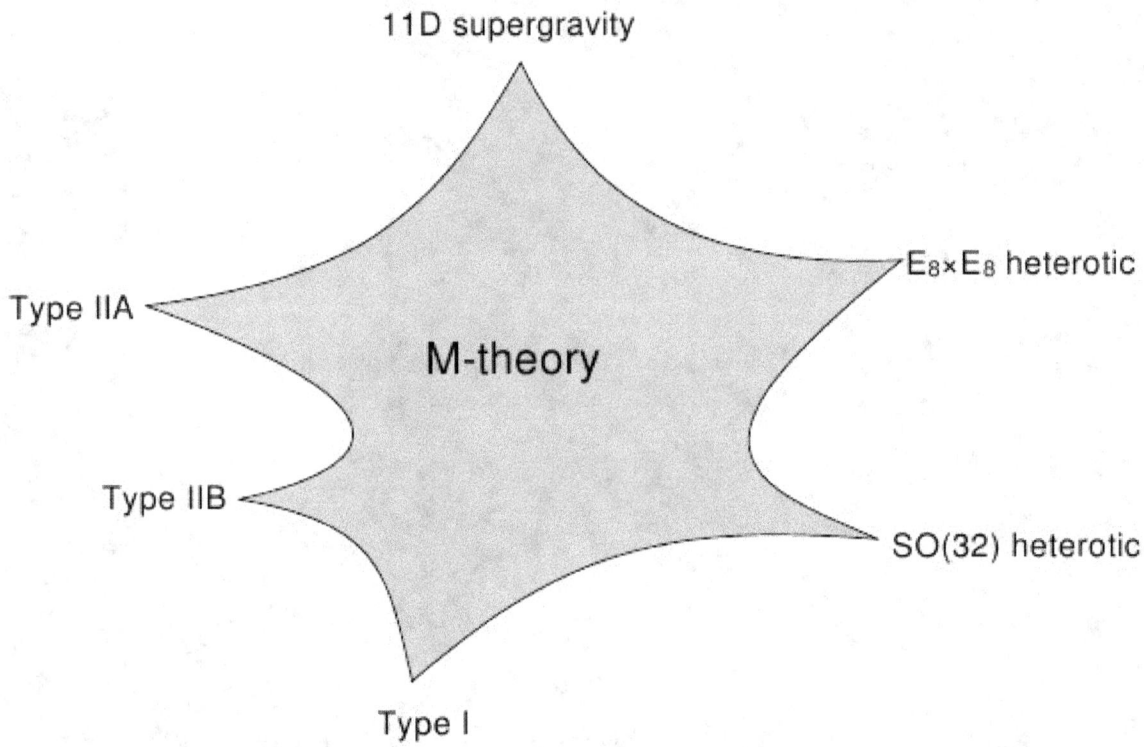

A schematic illustration of the relationship between M-theory, the five superstring theories, and eleven-dimensional supergravity. The shaded region represents a family of different physical scenarios that are possible in M-theory. In certain limiting cases corresponding to the cusps, it is natural to describe the physics using one of the six theories labeled there.

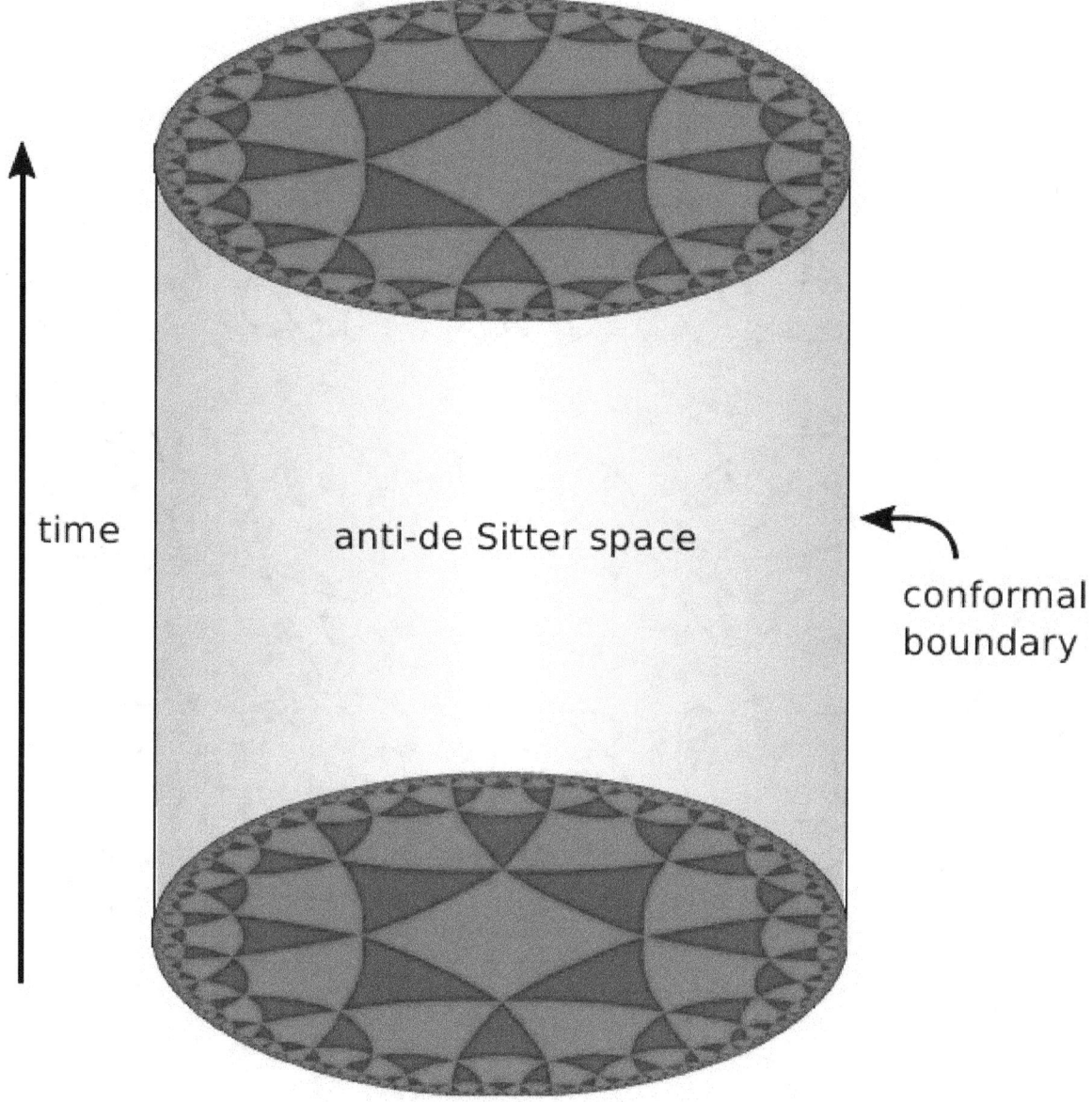

time

anti-de Sitter space

conformal
boundary

*Three-dimensional anti-de Sitter space is like a stack of hyperbolic disks, each one representing the state of the universe at a given time.
One can study theories of quantum gravity such as M-theory in the resulting spacetime.*

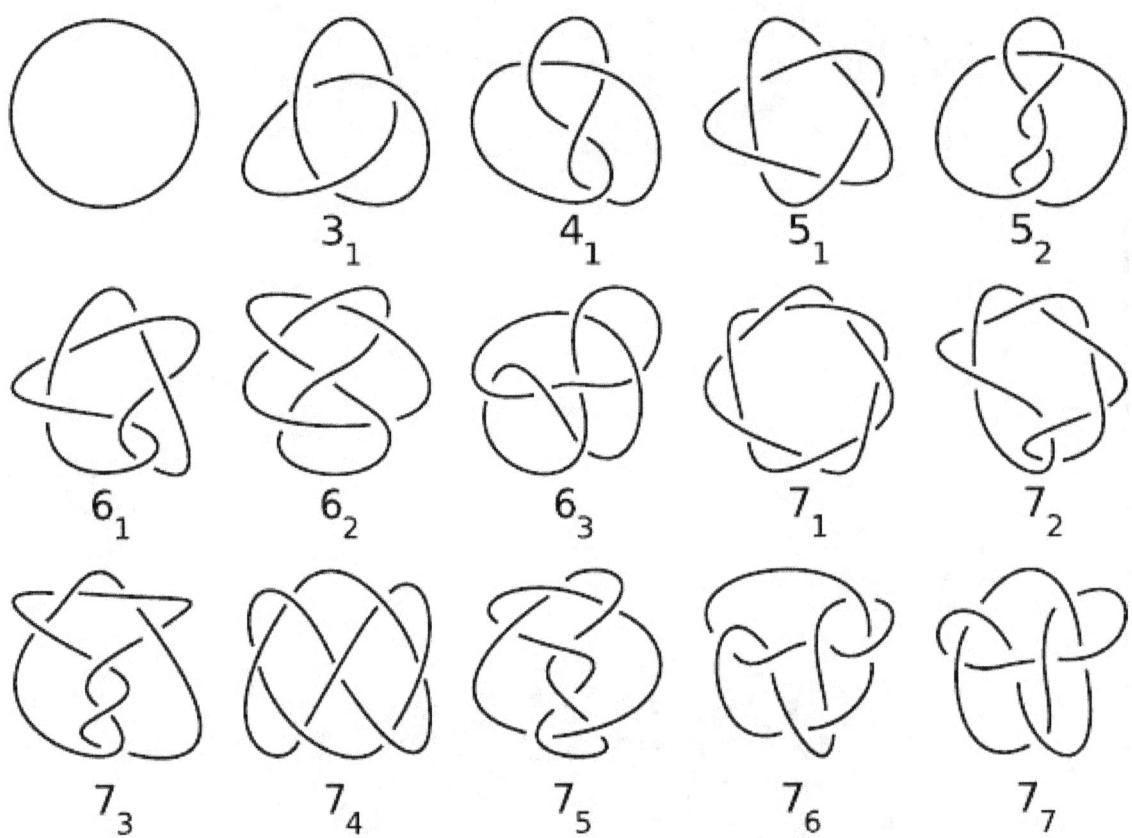

3_1 4_1 5_1 5_2

6_1 6_2 6_3 7_1 7_2

7_3 7_4 7_5 7_6 7_7

The six-dimensional (2,0)-theory has been used to understand results from the mathematical theory of knots.

Chapter 26

Hierarchy problem

In theoretical physics, the **hierarchy problem** is the large discrepancy between aspects of the weak force and gravity.[1] There is no scientific consensus on, for example, why the weak force is 10^{32} times stronger than gravity.

26.1 Technical definition

A hierarchy problem occurs when the fundamental value of some physical parameter, such as a coupling constant or a mass, in some Lagrangian is vastly different from its effective value, which is the value that gets measured in an experiment. This happens because the effective value is related to the fundamental value by a prescription known as renormalization, which applies quantum corrections to it. Typically the renormalized value of parameters are close to their fundamental values, but in some cases, it appears that there has been a delicate cancellation between the fundamental quantity and the quantum corrections. Hierarchy problems are related to fine-tuning problems and problems of naturalness.

Studying renormalization in hierarchy problems is difficult, because such quantum corrections are usually power-law divergent, which means that the shortest-distance physics are most important. Because we do not know the precise details of the shortest-distance theory of physics, we cannot even address how this delicate cancellation between two large terms occurs. Therefore, researchers are led to postulate new physical phenomena that resolve hierarchy problems without fine tuning.

26.2 The Higgs mass

In particle physics, the most important **hierarchy problem** is the question that asks why the weak force is 10^{32} times stronger than gravity. Both of these forces involve constants of nature, Fermi's constant for the weak force and Newton's constant for gravity. Furthermore if the Standard Model is used to calculate the quantum corrections to Fermi's constant, it appears that Fermi's constant is surprisingly large and is expected to be closer to Newton's constant, unless there is a delicate cancellation between the bare value of Fermi's constant and the quantum corrections to it.

More technically, the question is why the Higgs boson is so much lighter than the Planck mass (or the grand unification energy, or a heavy neutrino mass scale): one would expect that the large quantum contributions to the square of the Higgs boson mass would inevitably make the mass huge, comparable to the scale at which new physics appears, unless there is an incredible fine-tuning cancellation between the quadratic radiative corrections and the bare mass.

It should be remarked that the problem cannot even be formulated in the strict context of the Standard Model, for the Higgs mass cannot be calculated. In a sense, the problem amounts to the worry that a future theory of fundamental particles, in which the Higgs boson mass will be calculable, should not have excessive fine-tunings.

One proposed solution, popular amongst many physicists, is that one may solve the hierarchy problem via supersymmetry. Supersymmetry can explain how a tiny Higgs mass can be protected from quantum corrections. Supersymmetry removes

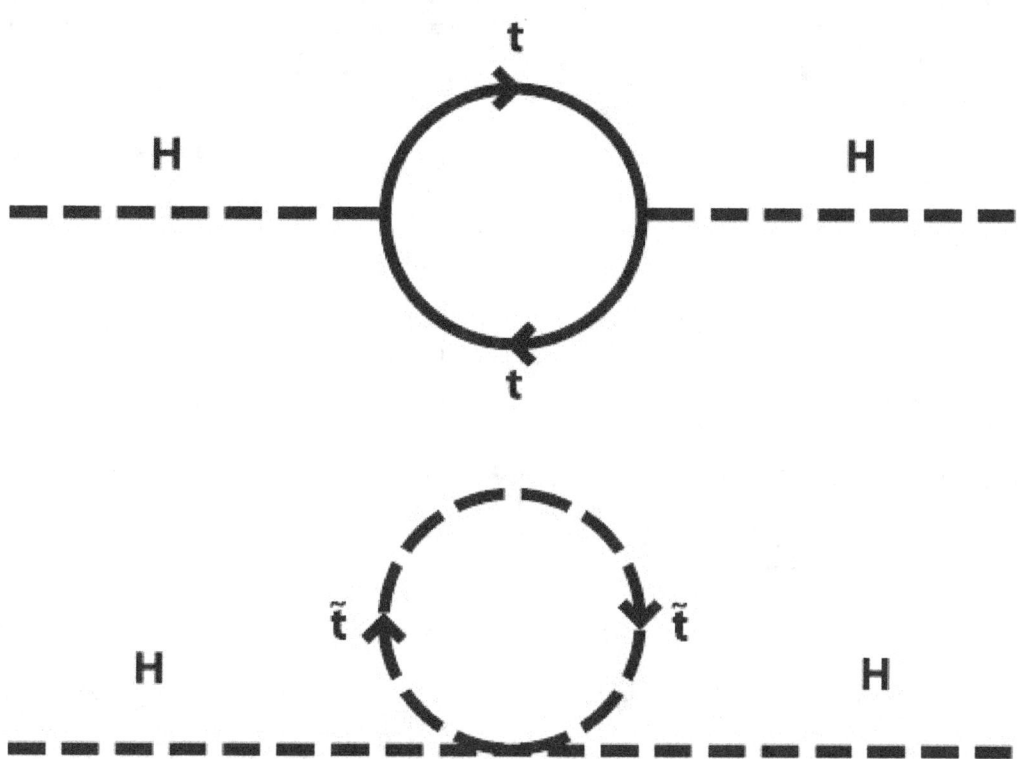

Cancellation of the Higgs boson quadratic mass renormalization between fermionic top quark loop and scalar stop squark tadpole Feynman diagrams in a supersymmetric extension of the Standard Model

the power-law divergences of the radiative corrections to the Higgs mass and solves the hierarchy problem as long as the supersymmetric particles are light enough to satisfy the Barbieri–Giudice criterion.[2] This still leaves open the mu problem, however. Currently the tenets of supersymmetry are being tested at the LHC, although no evidence has been found so far for supersymmetry.

26.3   Theoretical solutions

26.3.1   Supersymmetric solution

Each particle that couples to the Higgs field has a Yukawa coupling λ_f. The coupling with the Higgs field for fermions gives an interaction term $\mathcal{L}_{\text{Yukawa}} = -\lambda_f \bar{\psi} H \psi$, with ψ being the Dirac Field and H the Higgs Field. Also, the mass of a fermion is proportional to its Yukawa coupling, meaning that the Higgs boson will couple most to the most massive particle. This means that the most significant corrections to the Higgs mass will originate from the heaviest particles, most prominently the top quark. By applying the Feynman rules, one gets the quantum corrections to the Higgs mass squared from a fermion to be:

$$\Delta m_H^2 = -\frac{|\lambda_f|^2}{8\pi^2}[\Lambda_{\text{UV}}^2 + ...].$$

The Λ_{UV} is called the ultraviolet cutoff and is the scale up to which the Standard Model is valid. If we take this scale to be the Planck scale, then we have the quadratically diverging Lagrangian. However, suppose there existed two complex scalars (taken to be spin 0) such that:

$\lambda_S = |\lambda_f|^2$ (the couplings to the Higgs are exactly the same).

Then by the Feynman rules, the correction (from both scalars) is:

$$\Delta m_H^2 = 2 \times \frac{\lambda_S}{16\pi^2} [\Lambda_{UV}^2 + ...].$$

(Note that the contribution here is positive. This is because of the spin-statistics theorem, which means that fermions will have a negative contribution and bosons a positive contribution. This fact is exploited.)

This gives a total contribution to the Higgs mass to be zero if we include both the fermionic and bosonic particles. Supersymmetry is an extension of this that creates 'superpartners' for all Standard Model particles.

This section adapted from Stephen P. Martin's "A Supersymmetry Primer" on arXiv.[3]

26.3.2 Conformal solution

Without supersymmetry, a solution to the hierarchy problem has been proposed using just the Standard Model. The idea can be traced back to the fact that the term in the Higgs field that produces the uncontrolled quadratic correction upon renormalization is the quadratic one. If the Higgs field had no mass term, then no hierarchy problem arises. But by missing a quadratic term in the Higgs field, one must find a way to recover the breaking of electroweak symmetry through a non-null vacuum expectation value. This can be obtained using the Weinberg–Coleman mechanism with terms in the Higgs potential arising from quantum corrections. Mass obtained in this way is far too small with respect to what is seen in accelerator facilities and so a conformal Standard Model needs more than one Higgs particle. This proposal has been put forward in 2006 by Krzysztof Meissner and Hermann Nicolai[4] and is currently under scrutiny. But if no further excitation is observed beyond the one seen so far at LHC, this model would have to be abandoned.

26.3.3 Solution via extra dimensions

If we live in a 3+1 dimensional world, then we calculate the Gravitational Force via Gauss' law for gravity:

$$\mathbf{g}(\mathbf{r}) = -Gm\frac{\mathbf{e_r}}{r^2}$$

which is simply Newton's law of gravitation. Note that Newton's constant G can be rewritten in terms of the Planck mass.

$$\frac{1}{M_{Pl}^2}$$

If we extend this idea to δ extra dimensions, then we get:

$$\mathbf{g}(\mathbf{r}) = -m\frac{\mathbf{e_r}}{M_{Pl_{3+1+\delta}}^{2+\delta} r^{2+\delta}}$$

where $M_{Pl_{3+1+\delta}}$ is the 3+1+ δ dimensional Planck mass. However, we are assuming that these extra dimensions are the same size as the normal 3+1 dimensions. Let us say that the extra dimensions are of size $n \lll$ than normal dimensions. If we let $r \ll n$, then we get (2). However, if we let $r \gg n$, then we get our usual Newton's law. However, when $r \gg n$, the flux in the extra dimensions becomes a constant, because there is no extra room for gravitational flux to flow through. Thus the flux will be proportional to n^δ because this is the flux in the extra dimensions. The formula is:

$$\mathbf{g(r)} = -m \frac{\mathbf{e_r}}{M_{\text{Pl}_{3+1+\delta}}^{2+\delta} r^2 n^\delta}$$

$$-m \frac{\mathbf{e_r}}{M_{\text{Pl}}^2 r^2} = -m \frac{\mathbf{e_r}}{M_{\text{Pl}_{3+1+\delta}}^{2+\delta} r^2 n^\delta}$$

which gives:

$$\frac{1}{M_{\text{Pl}}^2 r^2} = \frac{1}{M_{\text{Pl}_{3+1+\delta}}^{2+\delta} r^2 n^\delta} \Rightarrow$$

$$M_{\text{Pl}}^2 = M_{\text{Pl}_{3+1+\delta}}^{2+\delta} n^\delta .$$

Thus the fundamental Planck mass (the extra dimensional one) could actually be small, meaning that gravity is actually strong, but this must be compensated by the number of the extra dimensions and their size. Physically, this means that gravity is weak because there is a loss of flux to the extra dimensions.

This section adapted from "Quantum Field Theory in a Nutshell" by A. Zee.[5]

Braneworld models

Main article: Brane cosmology

In 1998 Nima Arkani-Hamed, Savas Dimopoulos, and Gia Dvali proposed the **ADD model**, also known as the model with large extra dimensions, an alternative scenario to explain the weakness of gravity relative to the other forces.[6][7] This theory requires that the fields of the Standard Model are confined to a four-dimensional membrane, while gravity propagates in several additional spatial dimensions that are large compared to the Planck scale.[8]

In 1998/99 Merab Gogberashvili published on the arXiv (and subsequently in peer-reviewed journals) a number of articles where he showed that if the Universe is considered as a thin shell (a mathematical synonym for "brane") expanding in 5-dimensional space then it is possible to obtain one scale for particle theory corresponding to the 5-dimensional cosmological constant and Universe thickness, and thus to solve the hierarchy problem.[9][10][11] It was also shown that four-dimensionality of the Universe is the result of stability requirement since the extra component of the Einstein field equations giving the localized solution for matter fields coincides with one of the conditions of stability.

Subsequently, there were proposed the closely related Randall–Sundrum scenarios which offered their solution to the hierarchy problem.

Finite Groups

It has also been noted that the group order of the Baby Monster group is of the right order of magnitude, 4×10^{33}. It is known that the Monster Group is related to the symmetries of a particular bosonic string theory on the Leech lattice. However, there's no physical reason for why the size of the Monster Group or its subgroups should appear in the Lagrangian. Most physicists think this is merely a coincidence. Another coincidence is that in *reduced* Planck units, the Higgs mass is approximately $48.|M|^{-1/3} = 125.5$ GeV where |M| is the order of the Monster group. This suggests that the smallness of the Higgs mass may be due to a redundancy caused by a symmetry of the extra dimensions, which must be divided out. There are other groups that are also of the right order of magnitude for example $Weyl(E_8 \times E_8)$.

Extra dimensions

Until now, no experimental or observational evidence of extra dimensions has been officially reported. Analyses of results from the Large Hadron Collider severely constrain theories with large extra dimensions.[12] However, extra dimensions could explain why the gravity force is so weak, and why the expansion of the universe is faster than expected.[13]

26.4 The cosmological constant

In physical cosmology, current observations in favor of an accelerating universe imply the existence of a tiny, but nonzero cosmological constant. This is a hierarchy problem very similar to that of the Higgs boson mass problem, since the cosmological constant is also very sensitive to quantum corrections. It is complicated, however, by the necessary involvement of general relativity in the problem and may be a clue that we do not understand gravity on long distance scales (such as the size of the universe today). While quintessence has been proposed as an explanation of the acceleration of the Universe, it does not actually address the cosmological constant hierarchy problem in the technical sense of addressing the large quantum corrections. Supersymmetry does not address the cosmological constant problem, since supersymmetry cancels the $M^4 P_{lanck}$ contribution, but not the $M^2 P_{lanck}$ one (quadratically diverging).

26.5 See also

- CP violation

- Little hierarchy problem

- Quantum triviality

26.6 References

[1] http://profmattstrassler.com/articles-and-posts/particle-physics-basics/the-hierarchy-problem/

[2] R. Barbieri, G. F. Giudice (1988). "Upper Bounds on Supersymmetric Particle Masses".*Nucl. Phys.* **B306**: 63.Bibc..63B. doi:10.1016/0550-3213(88)90171-X.

[3] Stephen P. Martin, A Supersymmetry Primer

[4] K. Meissner, H. Nicolai (2006). "Conformal Symmetry and the Standard Model". *Physics Letters* **B648**: 312–317. arXiv:hep-th/0612165. Bibcode:2007PhLB..648..312M. doi:10.1016/j.physletb.2007.03.023.

[5] Zee, A. (2003). "Quantum field theory in a nutshell". Princeton University Press. Bibcode:2003qftn.book.....Z.

[6] N. Arkani-Hamed, S. Dimopoulos, G. Dvali (1998). "The Hierarchy problem and new dimensions at a millimeter". *Physics Letters* **B429**: 263–272. arXiv:hep-ph/9803315. Bibcode:1998PhLB..429..263A. doi:10.1016/S0370-2693(98)00466-3.

[7] N. Arkani-Hamed, S. Dimopoulos, G. Dvali (1999). "Phenomenology, astrophysics and cosmology of theories with submillimeter dimensions and TeV scale quantum gravity". *Physical Review* **D59**: 086004. arXiv:hep-ph/9807344.Bibcode:1999PhR4A. doi:10.1103/PhysRevD.59.086004.

[8] For a pedagogical introduction, see M. Shifman (2009). *Large Extra Dimensions: Becoming acquainted with an alternative paradigm*. Crossing the boundaries: Gauge dynamics at strong coupling. Singapore: World Scientific. arXiv:0907.3074.

[9] M. Gogberashvili, *Hierarchy problem in the shell universe model*, Arxiv:hep-ph/9812296.

[10] M. Gogberashvili, *Our world as an expanding shell*, Arxiv:hep-ph/9812365.

[11] M. Gogberashvili, *Four dimensionality in noncompact Kaluza-Klein model*, Arxiv:hep-ph/9904383.

[12] ATLAS Collaboration, "Search for Quantum Black Hole Production in High-Invariant-Mass Lepton+Jet Final States Using pp Collisions at s√=8 TeV and the ATLAS Detector" http://arxiv.org/abs/1311.2006

[13] CERN (20 January 2012). "Extra dimensions, gravitons, and tiny black holes".

Chapter 27

String theory landscape

The **string theory landscape** refers to the huge number of possible false vacua in string theory.[1] The large number of theoretically allowed configurations has prompted suggestions that certain physical mysteries, particularly relating to the fine-tuning of constants like the cosmological constant or the Higgs boson mass, may be explained not by a physical mechanism but by assuming that many different vacua are physically realized.[2] The *anthropic landscape* thus refers to the collection of those portions of the landscape that are suitable for supporting intelligent life, an application of the anthropic principle that selects a subset of the otherwise possible configurations.

In string theory the number of false vacua is thought to be somewhere between 10^{10} to 10^{100}.[1] The large number of possibilities arises from different choices of Calabi–Yau manifolds and different values of generalized magnetic fluxes over different homology cycles. If one assumes that there is no structure in the space of vacua, the problem of finding one with a sufficiently small cosmological constant is NP complete,[3] being a version of the subset sum problem.

27.1   Anthropic principle

Main article: Anthropic principle

The idea of the string theory landscape has been used to propose a concrete implementation of the anthropic principle, the idea that fundamental constants may have the values they have not for fundamental physical reasons, but rather because such values are necessary for life (and hence intelligent observers to measure the constants). In 1987, Steven Weinberg proposed that the observed value of the cosmological constant was so small because it is impossible for life to occur in a universe with a much larger cosmological constant.[4] In order to implement this idea in a concrete physical theory, it is necessary to postulate a multiverse in which fundamental physical parameters can take different values. This has been realized in the context of eternal inflation.

27.2   Bayesian probability

Main article: Bayesian probability

Some physicists, starting with Weinberg, have proposed that Bayesian probability can be used to compute probability distributions for fundamental physical parameters, where the probability $P(x)$ of observing some fundamental parameters x is given by,

$$P(x) = P_{\text{prior}}(x) \times P_{\text{selection}}(x),$$

where P_{prior} is the prior probability, from fundamental theory, of the parameters x and $P_{\text{selection}}$ is the anthropic selection function, determined by the number of "observers" that would occur in the universe with parameters x. These probabilistic arguments are the most controversial aspect of the landscape. Technical criticisms of these proposals have pointed out that:

- The function P_{prior} is completely unknown in string theory and may be impossible to define or interpret in any sensible probabilistic way.

- The function $P_{\text{selection}}$ is completely unknown, since so little is known about the origin of life. Simplified criteria (such as the number of galaxies) must be used as a proxy for the number of observers. Moreover, it may never be possible to compute it for parameters radically different from those of the observable universe.

(Interpreting probability in a context where it is only possible to draw one sample from a distribution is problematic in frequentist probability but not in Bayesian probability, which is not defined in terms of the frequency of repeated events.)

Various physicists have tried to address these objections, and the ideas remain extremely controversial both within and outside the string theory community. These ideas have been reviewed by Carroll.[5]

27.3 Simplified approaches

Tegmark *et al.* have recently considered these objections and proposed a simplified anthropic scenario for axion dark matter in which they argue that the first two of these problems do not apply.[6]

Vilenkin and collaborators have proposed a consistent way to define the probabilities for a given vacuum.[7]

A problem with many of the simplified approaches people have tried is that they "predict" a cosmological constant that is too large by a factor of 10–1000 (depending on one's assumptions) and hence suggest that the cosmic acceleration should be much more rapid than is observed.[8][9][10]

27.4 Criticism

Although few dispute the idea that string theory appears to have an unimaginably large number of metastable vacua, the existence - meaning and scientific relevance of the anthropic landscape - remain highly controversial. Prominent proponents of the idea include Andrei Linde, Sir Martin Rees and especially Leonard Susskind, who advocate it as a solution to the cosmological-constant problem. Opponents, such as David Gross, suggest that the idea is inherently unscientific, unfalsifiable or premature. A famous debate on the anthropic landscape of string theory is the Smolin–Susskind debate on the merits of the landscape.

The term "landscape" comes from evolutionary biology (see *Fitness landscape*) and was first applied to cosmology by Lee Smolin in his book.[11] It was first used in the context of string theory by Susskind.

There are several popular books about the anthropic principle in cosmology.[12] Two popular physics blogs are opposed to this use of the anthropic principle.[13]

27.5 See also

- Extra dimensions

- Compactification

27.6   References

[1] The most commonly quoted number is of the order 10^{500}. See M. Douglas, "The statistics of string / M theory vacua", *JHEP* **0305**, 46 (2003). arXiv:hep-th/0303194; S. Ashok and M. Douglas, "Counting flux vacua", *JHEP* **0401**, 060 (2004).

[2] L. Susskind, "The anthropic landscape of string theory", arXiv:hep-th/0302219.

[3] Frederik Denef; Douglas, Michael R. (2006). "Computational complexity of the landscape". *Annals of Physics* **322** (5): 1096–1142. arXiv:hep-th/0602072. Bibcode:2007AnPhy.322.1096D. doi:10.1016/j.aop.2006.07.013.

[4] S. Weinberg, "Anthropic bound on the cosmological constant", *Phys. Rev. Lett.* **59**, 2607 (1987).

[5] S. M. Carroll, "Is our universe natural?", arXiv:hep-th/0512148.

[6] M. Tegmark, A. Aguirre, M. Rees and F. Wilczek, "Dimensionless constants, cosmology and other dark matters", arXiv:astro-ph/0511774. F. Wilczek, "Enlightenment, knowledge, ignorance, temptation", arXiv:hep-ph/0512187. See also the discussion at .

[7] See, *e.g.* Alexander Vilenkin (2006). "A measure of the multiverse". *Journal of Physics A: Mathematical and Theoretical* **40** (25): 6777–6785. arXiv:hep-th/0609193. Bibcode:2007JPhA...40.6777V. doi:10.1088/1751-8113/40/25/S22.

[8] Abraham Loeb (2006). "An observational test for the anthropic origin of the cosmological constant". *JCAP* **0605**: 009. (subscription required (help)).

[9] Jaume Garriga & Alexander Vilenkin (2006). "Anthropic prediction for Lambda and the Q catastrophe". *Prog. Theor.Phys. Suppl.* **163**: 245–57. arXiv:hep-th/0508005. Bibcode:2006PThPS.163..245G. doi:10.1143/PTPS.163.245. (subscription required (help)).

[10] Delia Schwartz-Perlov & Alexander Vilenkin (2006). "Probabilities in the Bousso-Polchinski multiverse". *JCAP* **0606**: 010. (subscription required (help)).

[11] L. Smolin, "Did the universe evolve?", *Classical and Quantum Gravity* **9**, 173–191 (1992). L. Smolin, *The Life of the Cosmos* (Oxford, 1997)

[12] L. Susskind, *The cosmic landscape: string theory and the illusion of intelligent design* (Little, Brown, 2005). M. J. Rees, *Just six numbers: the deep forces that shape the universe* (Basic Books, 2001). R. Bousso and J. Polchinski, "The string theory landscape", *Sci. Am.* **291**, 60–69 (2004).

[13] Lubos Motl's blog criticized the anthropic principle and Peter Woit's blog frequently attacks the anthropic string landscape.

27.7   External links

- String theory landscape on arxiv.org

- On the computation of non-perturbative effective potentials in the string theory landscape, Mirjam Cvetič, Iñaki García-Etxebarria, James Halverson

Chapter 28

Loop quantum gravity

Loop quantum gravity (**LQG**) is a theory that attempts to describe the quantum properties of the universe and gravity. It is also a theory of quantum spacetime because, according to general relativity, gravity is a manifestation of the geometry of spacetime. LQG is an attempt to merge quantum mechanics and general relativity. The main output of the theory is a physical picture of space where space is granular. The granularity is a direct consequence of the quantization. It has the same nature as the granularity of the photons in the quantum theory of electromagnetism and the discrete levels of the energy of the atoms. Here, it is space itself that is discrete. In other words, there is a minimum distance possible to travel through it.

More precisely, space can be viewed as an extremely fine fabric or network "woven" of finite loops. These networks of loops are called spin networks. The evolution of a spin network over time is called a spin foam. The predicted size of this structure is the Planck length, which is approximately 10^{-35} meters. According to the theory, there is no meaning to distance at scales smaller than the Planck scale. Therefore, LQG predicts that not just matter, but space itself, has an atomic structure.

Today LQG is a vast area of research, developing in several directions, which involves about 30 research groups worldwide.[1] They all share the basic physical assumptions and the mathematical description of quantum space. The full development of the theory is being pursued in two directions: the more traditional canonical loop quantum gravity, and the newer covariant loop quantum gravity, more commonly called spin foam theory.

Research into the physical consequences of the theory is proceeding in several directions. Among these, the most well-developed is the application of LQG to cosmology, called loop quantum cosmology (LQC). LQC applies LQG ideas to the study of the early universe and the physics of the Big Bang. Its most spectacular consequence is that the evolution of the universe can be continued beyond the Big Bang. The Big Bang appears thus to be replaced by a sort of cosmic Big Bounce.

28.1   History

Main article: History of loop quantum gravity

In 1986, Abhay Ashtekar reformulated Einstein's general relativity in a language closer to that of the rest of fundamental physics. Shortly after, Ted Jacobson and Lee Smolin realized that the formal equation of quantum gravity, called the Wheeler–DeWitt equation, admitted solutions labelled by loops, when rewritten in the new Ashtekar variables, and Carlo Rovelli and Lee Smolin defined a nonperturbative and background-independent quantum theory of gravity in terms of these loop solutions. Jorge Pullin and Jerzy Lewandowski understood that the intersections of the loops are essential for the consistency of the theory, and the theory should be formulated in terms of intersecting loops, or graphs.

In 1994, Rovelli and Smolin showed that the quantum operators of the theory associated to area and volume have a discrete spectrum. That is, geometry is quantized. This result defines an explicit basis of states of quantum geometry, which turned out to be labelled by Roger Penrose's spin networks, which are graphs labelled by spins.

The canonical version of the dynamics was put on firm ground by Thomas Thiemann, who defined an anomaly-free Hamiltonian operator, showing the existence of a mathematically consistent background-independent theory. The covariant or spinfoam version of the dynamics developed during several decades, and crystallized in 2008, from the joint work of research groups in France, Canada, UK, Poland, and Germany, lead to the definition of a family of transition amplitudes, which in the classical limit can be shown to be related to a family of truncations of general relativity.[2] The finiteness of these amplitudes was proven in 2011.[3][4] It requires the existence of a positive cosmological constant, and this is consistent with observed acceleration in the expansion of the Universe.

28.2   General covariance and background independence

Main articles: General covariance, background-independent and diffeomorphism

In theoretical physics, general covariance is the invariance of the form of physical laws under arbitrary differentiable coordinate transformations. The essential idea is that coordinates are only artifices used in describing nature, and hence should play no role in the formulation of fundamental physical laws. A more significant requirement is the principle of general relativity that states that the laws of physics take the same form in all reference systems. This is a generalization of the principle of special relativity which states that the laws of physics take the same form in all inertial frames.

In mathematics, a diffeomorphism is an isomorphism in the category of smooth manifolds. It is an invertible function that maps one differentiable manifold to another, such that both the function and its inverse are smooth. These are the defining symmetry transformations of General Relativity since the theory is formulated only in terms of a differentiable manifold.

In general relativity, general covariance is intimately related to "diffeomorphism invariance". This symmetry is one of the defining features of the theory. However, it is a common misunderstanding that "diffeomorphism invariance" refers to the invariance of the physical predictions of a theory under arbitrary coordinate transformations; this is untrue and in fact every physical theory is invariant under coordinate transformations this way. Diffeomorphisms, as mathematicians define them, correspond to something much more radical; intuitively a way they can be envisaged is as simultaneously dragging all the physical fields (including the gravitational field) over the bare differentiable manifold while staying in the same coordinate system. Diffeomorphisms are the true symmetry transformations of general relativity, and come about from the assertion that the formulation of the theory is based on a bare differentiable manifold, but not on any prior geometry — the theory is background-independent (this is a profound shift, as all physical theories before general relativity had as part of their formulation a prior geometry). What is preserved under such transformations are the coincidences between the values the gravitational field take at such and such a "place" and the values the matter fields take there. From these relationships one can form a notion of matter being located with respect to the gravitational field, or vice versa. This is what Einstein discovered: that physical entities are located with respect to one another only and not with respect to the spacetime manifold. As Carlo Rovelli puts it: "No more fields on spacetime: just fields on fields.".[5] This is the true meaning of the saying "The stage disappears and becomes one of the actors"; space-time as a "container" over which physics takes place has no objective physical meaning and instead the gravitational interaction is represented as just one of the fields forming the world. This is known as the relationalist interpretation of space-time. The realization by Einstein that general relativity should be interpreted this way is the origin of his remark "Beyond my wildest expectations".

In LQG this aspect of general relativity is taken seriously and this symmetry is preserved by requiring that the physical states remain invariant under the generators of diffeomorphisms. The interpretation of this condition is well understood for purely spatial diffeomorphisms. However, the understanding of diffeomorphisms involving time (the Hamiltonian constraint) is more subtle because it is related to dynamics and the so-called "problem of time" in general relativity.[6] A generally accepted calculational framework to account for this constraint has yet to be found.[7][8] A plausible candidate for the quantum hamiltonian constraint is the operator introduced by Thiemann.[9]

LQG is formally background independent. The equations of LQG are not embedded in, or dependent on, space and time (except for its invariant topology). Instead, they are expected to give rise to space and time at distances which are large compared to the Planck length. The issue of background independence in LQG still has some unresolved subtleties. For example, some derivations require a fixed choice of the topology, while any consistent quantum theory of gravity should include topology change as a dynamical process.

28.3 Constraints and their Poisson bracket algebra

Main articles: Poisson bracket and Hamiltonian constraint

28.3.1 The constraints of classical canonical general relativity

Main article: Lie derivative

In the Hamiltonian formulation of ordinary classical mechanics the Poisson bracket is an important concept. A "canonical coordinate system" consists of canonical position and momentum variables that satisfy canonical Poisson-bracket relations,

$$\{q_i, p_j\} = \delta_{ij}$$

where the Poisson bracket is given by

$$\{f, g\} = \sum_{i=1}^{N} \left(\frac{\partial f}{\partial q_i} \frac{\partial g}{\partial p_i} - \frac{\partial f}{\partial p_i} \frac{\partial g}{\partial q_i} \right).$$

for arbitrary phase space functions $f(q_i, p_j)$ and $g(q_i, p_j)$. With the use of Poisson brackets, the Hamilton's equations can be rewritten as,

$$\dot{q}_i = \{q_i, H\},$$

$$\dot{p}_i = \{p_i, H\}.$$

These equations describe a "flow" or orbit in phase space generated by the Hamiltonian H. Given any phase space function $F(q, p)$, we have

$$\frac{d}{dt} F(q_i, p_i) = \{F, H\}.$$

Let us consider constrained systems, of which General relativity is an example. In a similar way the Poisson bracket between a constraint and the phase space variables generates a flow along an orbit in (the unconstrained) phase space generated by the constraint. There are three types of constraints in Ashtekar's reformulation of classical general relativity:

$SU(2)$ **Gauss gauge constraints**

The Gauss constraints

$$G_j(x) = 0.$$

This represents an infinite number of constraints one for each value of x. These come about from re-expressing General relativity as an $SU(2)$ Yang–Mills type gauge theory (Yang–Mills is a generalization of Maxwell's theory where the gauge field transforms as a vector under Gauss transformations, that is, the Gauge field is of the form $A_a^i(x)$ where i is an internal index. See Ashtekar variables). These infinite number of Gauss gauge constraints can be smeared with test fields with internal indices, $\lambda^j(x)$,

$$G(\lambda) = \int d^3x G_j(x) \lambda^j(x).$$

which we demand vanish for any such function. These smeared constraints defined with respect to a suitable space of smearing functions give an equivalent description to the original constraints.

In fact Ashtekar's formulation may be thought of as ordinary $SU(2)$ Yang–Mills theory together with the following special constraints, resulting from diffeomorphism invariance, and a Hamiltonian that vanishes. The dynamics of such a theory are thus very different from that of ordinary Yang–Mills theory.

Spatial diffeomorphisms constraints

The spatial diffeomorphism constraints

$$C_a(x) = 0$$

can be smeared by the so-called shift functions $\vec{N}(x)$ to give an equivalent set of smeared spatial diffeomorphism constraints,

$$C(\vec{N}) = \int d^3x C_a(x) N^a(x) .$$

These generate spatial diffeomorphisms along orbits defined by the shift function $N^a(x)$.

Hamiltonian constraints

The Hamiltonian

$$H(x) = 0$$

can be smeared by the so-called lapse functions $N(x)$ to give an equivalent set of smeared Hamiltonian constraints,

$$H(N) = \int d^3x H(x) N(x) .$$

These generate time diffeomorphisms along orbits defined by the lapse function $N(x)$.

In Ashtekar formulation the gauge field $A_a^i(x)$ is the configuration variable (the configuration variable being analogous to q in ordinary mechanics) and its conjugate momentum is the (densitized) triad (electrical field) $\hat{E}_i^a(x)$. The constraints are certain functions of these phase space variables.

We consider the action of the constraints on arbitrary phase space functions. An important notion here is the Lie derivative, $\mathcal{L}_V$, which is basically a derivative operation that infinitesimally "shifts" functions along some orbit with tangent vector V .

28.3.2   The Poisson bracket algebra

Of particular importance is the Poisson bracket algebra formed between the (smeared) constraints themselves as it completely determines the theory. In terms of the above smeared constraints the constraint algebra amongst the Gauss' law reads,

$$\{G(\lambda), G(\mu)\} = G([\lambda, \mu])$$

where $[\lambda, \mu]^k = \lambda_i \mu_j \epsilon^{ijk}$. And so we see that the Poisson bracket of two Gauss' law is equivalent to a single Gauss' law evaluated on the commutator of the smearings. The Poisson bracket amongst spatial diffeomorphisms constraints reads

$$\{C(\vec{N}), C(\vec{M})\} = C(\mathcal{L}_{\vec{N}} \vec{M})$$

and we see that its effect is to "shift the smearing". The reason for this is that the smearing functions are not functions of the canonical variables and so the spatial diffeomorphism does not generate diffeomorphims on them. They do however generate diffeomorphims on everything else. This is equivalent to leaving everything else fixed while shifting the smearing .The action of the spatial diffeomorphism on the Gauss law is

$$\{C(\vec{N}), G(\lambda)\} = G(\mathcal{L}_{\vec{N}} \lambda) .$$

again, it shifts the test field λ . The Gauss law has vanishing Poisson bracket with the Hamiltonian constraint. The spatial diffeomorphism constraint with a Hamiltonian gives a Hamiltonian with its smearing shifted,

$$\{C(\vec{N}), H(M)\} = H(\mathcal{L}_{\vec{N}} M) .$$

Finally, the poisson bracket of two Hamiltonians is a spatial diffeomorphism,

$$\{H(N), H(M)\} = C(K)$$

where K is some phase space function. That is, it is a sum over infinitesimal spatial diffeomorphisms constraints where the coefficients of proportionality are not constants but have non-trivial phase space dependence.

A (Poisson bracket) Lie algebra, with constraints C_I , is of the form

$$\{C_I, C_J\} = f_{IJ}^K C_K$$

where f_{IJ}^K are constants (the so-called structure constants). The above Poisson bracket algebra for General relativity does not form a true Lie algebra as we have structure functions rather than structure constants for the Poisson bracket between two Hamiltonians. This leads to difficulties.

28.3.3   Dirac observables

The constraints define a constraint surface in the original phase space. The gauge motions of the constraints apply to all phase space but have the feature that they leave the constraint surface where it is, and thus the orbit of a point in the hypersurface under gauge transformations will be an orbit entirely within it. Dirac observables are defined as phase space functions, O , that Poisson commute with all the constraints when the constraint equations are imposed,

$$\{G_j, O\}_{G_j = C_a = H = 0} = \{C_a, O\}_{G_j = C_a = H = 0} = \{H, O\}_{G_j = C_a = H = 0} = 0 \,,$$

that is, they are quantities defined on the constraint surface that are invariant under the gauge transformations of the theory.

Then, solving only the constraint $G_j = 0$ and determining the Dirac observables with respect to it leads us back to the ADM phase space with constraints H, C_a . The dynamics of general relativity is generated by the constraints, it can be shown that six Einstein equations describing time evolution (really a gauge transformation) can be obtained by calculating the Poisson brackets of the three-metric and its conjugate momentum with a linear combination of the spatial diffeomorphism and Hamiltonian constraint. The vanishing of the constraints, giving the physical phase space, are the four other Einstein equations.[10]

28.4   Quantization of the constraints – the equations of quantum general relativity

28.4.1   Pre-history and Ashtekar new variables

Main articles: Frame fields in general relativity, Ashtekar variables and Self-dual Palatini action

Many of the technical problems in canonical quantum gravity revolve around the constraints. Canonical general relativity was originally formulated in terms of metric variables, but there seemed to be insurmountable mathematical difficulties in promoting the constraints to quantum operators because of their highly non-linear dependence on the canonical variables. The equations were much simplified with the introduction of Ashtekars new variables. Ashtekar variables describe canonical general relativity in terms of a new pair canonical variables closer to that of gauge theories. The first step consists of using densitized triads $\tilde{E}_i^a$ (a triad E_i^a is simply three orthogonal vector fields labeled by $i = 1, 2, 3$ and the densitized triad is defined by $\tilde{E}_i^a = \sqrt{\det(q)} E_i^a$) to encode information about the spatial metric,

$$\det(q) q^{ab} = \tilde{E}_i^a \tilde{E}_j^b \delta^{ij} \,.$$

(where δ^{ij} is the flat space metric, and the above equation expresses that q^{ab} , when written in terms of the basis E_i^a , is locally flat). (Formulating general relativity with triads instead of metrics was not new.) The densitized triads are not unique, and in fact one can perform a local in space rotation with respect to the internal indices i . The canonically conjugate variable is related to the extrinsic curvature by $K_a^i = K_{ab} \tilde{E}^{ai} / \sqrt{\det(q)}$. But problems similar to using the metric formulation arise when one tries to quantize the theory. Ashtekar's new insight was to introduce a new configuration variable,

$$A_a^i = \Gamma_a^i - i K_a^i$$

that behaves as a complex SU(2) connection where Γ_a^i is related to the so-called spin connection via $\Gamma_a^i = \Gamma_{ajk} \epsilon^{jki}$. Here A_a^i is called the chiral spin connection. It defines a covariant derivative $\mathcal{D}_a$. It turns out that $\tilde{E}_i^a$ is the conjugate momentum of A_a^i , and together these form Ashtekar's new variables.

The expressions for the constraints in Ashtekar variables; the Gauss's law, the spatial diffeomorphism constraint and the (densitized) Hamiltonian constraint then read:

$$G^i = \mathcal{D}_a \tilde{E}^a_i = 0$$

$$C_a = \tilde{E}^b_i F^i_{ab} - A^i_a(\mathcal{D}_b \tilde{E}^b_i) = V_a - A^i_a G^i = 0,$$

$$\tilde{H} = \epsilon_{ijk}\tilde{E}^a_i \tilde{E}^b_j F^i_{ab} = 0$$

respectively, where F^i_{ab} is the field strength tensor of the connection A^i_a and where V_a is referred to as the vector constraint. The above-mentioned local in space rotational invariance is the original of the SU(2) gauge invariance here expressed by the Gauss law. Note that these constraints are polynomial in the fundamental variables, unlike as with the constraints in the metric formulation. This dramatic simplification seemed to open up the way to quantizing the constraints. (See the article Self-dual Palatini action for a derivation of Ashtekar's formulism).

With Ashtekar's new variables, given the configuration variable A^i_a, it is natural to consider wavefunctions $\Psi(A^i_a)$. This is the connection representation. It is analogous to ordinary quantum mechanics with configuration variable q and wavefunctions $\psi(q)$. The configuration variable gets promoted to a quantum operator via:

$$\hat{A}^i_a \Psi(A) = A^i_a \Psi(A),$$

(analogous to $\hat{q}\psi(q) = q\psi(q)$) and the triads are (functional) derivatives,

$$\hat{\tilde{E}}^a_i \Psi(A) = -i\frac{\delta \Psi(A)}{\delta A^a_i}.$$

(analogous to $\hat{p}\psi(q) = -i\hbar d\psi(q)/dq$). In passing over to the quantum theory the constraints become operators on a kinematic Hilbert space (the unconstrained SU(2) Yang–Mills Hilbert space). Note that different ordering of the A 's and $\tilde{E}$'s when replacing the $\tilde{E}$'s with derivatives give rise to different operators - the choice made is called the factor ordering and should be chosen via physical reasoning. Formally they read

$$\hat{G}_j|\psi\rangle = 0$$

$$\hat{C}_a|\psi\rangle = 0$$

$$\hat{\tilde{H}}|\psi\rangle = 0.$$

There are still problems in properly defining all these equations and solving them. For example the Hamiltonian constraint Ashtekar worked with was the densitized version instead of the original Hamiltonian, that is, he worked with $\tilde{H} = \sqrt{\det(q)}H$. There were serious difficulties in promoting this quantity to a quantum operator. Moreover, although Ashtekar variables had the virtue of simplifying the Hamiltonian, they are complex. When one quantizes the theory, it is difficult to ensure that one recovers real general relativity as opposed to complex general relativity.

28.4.2   Quantum constraints as the equations of quantum general relativity

We now move on to demonstrate an important aspect of the quantum constraints. We consider Gauss' law only. First we state the classical result that the Poisson bracket of the smeared Gauss' law $G(\lambda) = \int d^3x \lambda^j (D_a E^a)^j$ with the connections is

$$\{G(\lambda), A^i_a\} = \partial_a \lambda^i + g\epsilon^{ijk}A^j_a \lambda^k = (D_a\lambda)^i.$$

The quantum Gauss' law reads

$$\hat{G}_j\Psi(A) = -iD_a\frac{\delta\lambda\Psi[A]}{\delta A^j_a} = 0.$$

If one smears the quantum Gauss' law and study its action on the quantum state one finds that the action of the constraint on the quantum state is equivalent to shifting the argument of Ψ by an infinitesimal (in the sense of the parameter λ small) gauge transformation,

$$\left[1 + \int d^3x \lambda^j(x)\hat{G}_j\right]\Psi(A) = \Psi[A + D\lambda] = \Psi[A],$$

and the last identity comes from the fact that the constraint annihilates the state. So the constraint, as a quantum operator, is imposing the same symmetry that its vanishing imposed classically: it is telling us that the functions $\Psi[A]$ have to be gauge invariant functions of the connection. The same idea is true for the other constraints.

Therefore the two step process in the classical theory of solving the constraints $C_I = 0$ (equivalent to solving the admissibility conditions for the initial data) and looking for the gauge orbits (solving the 'evolution' equations) is replaced by a one step process in the quantum theory, namely looking for solutions Ψ of the quantum equations $\hat{C}_I \Psi = 0$. This is because it obviously solves the constraint at the quantum level and it simultaneously looks for states that are gauge invariant because $\hat{C}_I$ is the quantum generator of gauge transformations (gauge invariant functions are constant along the gauge orbits and thus characterize them).[11] Recall that, at the classical level, solving the admissibility conditions and evolution equations was equivalent to solving all of Einstein's field equations, this underlines the central role of the quantum constraint equations in canonical quantum gravity.

28.4.3   Introduction of the loop representation

Main articles: Holonomy, Wilson loop and Knot invariant

It was in particular the inability to have good control over the space of solutions to the Gauss' law and spacial diffeomorphism constraints that led Rovelli and Smolin to consider a new representation - the loop representation in gauge theories and quantum gravity.[12]

We need the notion of a holonomy. A holonomy is a measure of how much the initial and final values of a spinor or vector differ after parallel transport around a closed loop; it is denoted

$h_\gamma[A]$.

Knowledge of the holonomies is equivalent to knowledge of the connection, up to gauge equivalence. Holonomies can also be associated with an edge; under a Gauss Law these transform as

$$(h_e')_{\alpha\beta} = U_{\alpha\gamma}^{-1}(x)(h_e)_{\gamma\sigma}U_{\sigma\beta}(y) .$$

For a closed loop $x = y$ if we take the trace of this, that is, putting $\alpha = \beta$ and summing we obtain

$$(h_e')_{\alpha\alpha} = U_{\alpha\gamma}^{-1}(x)(h_e)_{\gamma\sigma}U_{\sigma\alpha}(x) = [U_{\sigma\alpha}(x)U_{\alpha\gamma}^{-1}(x)](h_e)_{\gamma\sigma} = \delta_{\sigma\gamma}(h_e)_{\gamma\sigma} = (h_e)_{\gamma\gamma}$$

or

$$\operatorname{Tr} h_\gamma' = \operatorname{Tr} h_\gamma. .$$

The trace of an holonomy around a closed loop is written

$W_\gamma[A]$

and is called a Wilson loop. Thus Wilson loops are gauge invariant. The explicit form of the Holonomy is

$$h_\gamma[A] = \mathcal{P} \exp\left\{ -\int_{\gamma_0}^{\gamma_1} ds \dot{\gamma}^a A_a^i(\gamma(s))T_i \right\}$$

where γ is the curve along which the holonomy is evaluated, and s is a parameter along the curve, $\mathcal{P}$ denotes path ordering meaning factors for smaller values of s appear to the left, and T_i are matrices that satisfy the SU(2) algebra

$$[T^i, T^j] = 2i\epsilon^{ijk}T^k .$$

The Pauli matrices satisfy the above relation. It turns out that there are infinitely many more examples of sets of matrices that satisfy these relations, where each set comprises $(N + 1) \times (N + 1)$ matrices with $N = 1, 2, 3, \ldots$, and where none of these can be thought to 'decompose' into two or more examples of lower dimension. They are called different irreducible representations of the SU(2) algebra. The most fundamental representation being the Pauli matrices. The holonomy is labelled by a half integer $N/2$ according to the irreducible representation used.

The use of Wilson loops explicitly solves the Gauss gauge constraint. To handle the spatial diffeomorphism constraint we need to go over to the loop representation. As Wilson loops form a basis we can formally expand any Gauss gauge invariant function as,

$$\Psi[A] = \sum_\gamma \Psi[\gamma]W_\gamma[A] .$$

This is called the loop transform. We can see the analogy with going to the momentum representation in quantum mechanics(see Position and momentum space). There one has a basis of states $\exp(ikx)$ labelled by a number k and one expands

$$\psi[x] = \int dk \psi(k) \exp(ikx) \, .$$

and works with the coefficients of the expansion $\psi(k)$.

The inverse loop transform is defined by

$$\Psi[\gamma] = \int [dA] \Psi[A] W_\gamma[A] \, .$$

This defines the loop representation. Given an operator $\hat{O}$ in the connection representation,

$$\Phi[A] = \hat{O} \Psi[A] \qquad Eq \ 1 \, ,$$

one should define the corresponding operator $\hat{O}'$ on $\Psi[\gamma]$ in the loop representation via,

$$\Phi[\gamma] = \hat{O}' \Psi[\gamma] \qquad Eq \ 2 \, ,$$

where $\Phi[\gamma]$ is defined by the usual inverse loop transform,

$$\Phi[\gamma] = \int [dA] \Phi[A] W_\gamma[A] \qquad Eq \ 3 \, .$$

A transformation formula giving the action of the operator $\hat{O}'$ on $\Psi[\gamma]$ in terms of the action of the operator $\hat{O}$ on $\Psi[A]$ is then obtained by equating the R.H.S. of $Eq \ 2$ with the R.H.S. of $Eq \ 3$ with $Eq \ 1$ substituted into $Eq \ 3$, namely

$$\hat{O}' \Psi[\gamma] = \int [dA] W_\gamma[A] \hat{O} \Psi[A] \, ,$$

or

$$\hat{O}' \Psi[\gamma] = \int [dA] (\hat{O}^\dagger W_\gamma[A]) \Psi[A] \, ,$$

where by $\hat{O}^\dagger$ we mean the operator $\hat{O}$ but with the reverse factor ordering (remember from simple quantum mechanics where the product of operators is reversed under conjugation). We evaluate the action of this operator on the Wilson loop as a calculation in the connection representation and rearranging the result as a manipulation purely in terms of loops (one should remember that when considering the action on the Wilson loop one should choose the operator one wishes to transform with the opposite factor ordering to the one chosen for its action on wavefunctions $\Psi[A]$). This gives the physical meaning of the operator $\hat{O}'$. For example if $\hat{O}^\dagger$ corresponded to a spatial diffeomorphism, then this can be thought of as keeping the connection field A of $W_\gamma[A]$ where it is while performing a spatial diffeomorphism on γ instead. Therefore the meaning of $\hat{O}'$ is a spatial diffeomorphism on γ , the argument of $\Psi[\gamma]$.

In the loop representation we can then solve the spatial diffeomorphism constraint by considering functions of loops $\Psi[\gamma]$ that are invariant under spatial diffeomorphisms of the loop γ . That is, we construct what mathematicians call knot invariants. This opened up an unexpected connection between knot theory and quantum gravity.

What about the Hamiltonian constraint? Let us go back to the connection representation. Any collection of non-intersecting Wilson loops satisfy Ashtekar's quantum Hamiltonian constraint. This can be seen from the following. With a particular ordering of terms and replacing $\tilde{E}_i^a$ by a derivative, the action of the quantum Hamiltonian constraint on a Wilson loop is

$$\hat{\tilde{H}}^\dagger W_\gamma[A] = -\epsilon_{ijk} \hat{F}_{ab}^k \frac{\delta}{\delta A_a^i} \frac{\delta}{\delta A_b^j} W_\gamma[A] \, .$$

When a derivative is taken it brings down the tangent vector, $\dot{\gamma}^a$, of the loop, γ . So we have something like

$$\hat{F}_{ab}^i \dot{\gamma}^a \dot{\gamma}^b \, .$$

However, as F_{ab}^i is anti-symmetric in the indices a and b this vanishes (this assumes that γ is not discontinuous anywhere and so the tangent vector is unique). Now let us go back to the loop representation.

We consider wavefunctions $\Psi[\gamma]$ that vanish if the loop has discontinuities and that are knot invariants. Such functions solve the Gauss law, the spatial diffeomorphism constraint and (formally) the Hamiltonian constraint. Thus we have identified an infinite set of exact (if only formal) solutions to all the equations of quantum general relativity![12] This generated a lot of interest in the approach and eventually led to LQG.

28.4.4   Geometric operators, the need for intersecting Wilson loops and spin network states

The easiest geometric quantity is the area. Let us choose coordinates so that the surface Σ is characterized by $x^3 = 0$. The area of small parallelogram of the surface Σ is the product of length of each side times $\sin\theta$ where θ is the angle between the sides. Say one edge is given by the vector $\vec{u}$ and the other by $\vec{v}$ then,

$$A = \|\vec{u}\|\|\vec{v}\|\sin\theta = \sqrt{\|\vec{u}\|^2\|\vec{v}\|^2(1-\cos^2\theta)} = \sqrt{\|\vec{u}\|^2\|\vec{v}\|^2 - (\vec{u}\cdot\vec{v})^2}$$

From this we get the area of the surface Σ to be given by

$$A_\Sigma = \int_\Sigma dx^1 dx^2 \sqrt{\det(q^{(2)})}$$

where $\det(q^{(2)}) = q_{11}q_{22} - q_{12}^2$ and is the determinant of the metric induced on Σ. This can be rewritten as

$$\det(q^{(2)}) = \frac{\epsilon^{3ab}\epsilon^{3cd}q_{ac}q_{bc}}{2} .$$

The standard formula for an inverse matrix is

$$q^{ab} = \frac{\epsilon^{acd}\epsilon^{bef}q_{ce}q_{df}}{3!\det(q)}$$

Note the similarity between this and the expression for $\det(q^{(2)})$. But in Ashtekar variables we have $\tilde{E}_i^a \tilde{E}^{bi} = \det(q)q^{ab}$. Therefore

$$A_\Sigma = \int_\Sigma dx^1 dx^2 \sqrt{\tilde{E}_i^3 \tilde{E}^{3i}} .$$

According to the rules of canonical quantization we should promote the triads $\tilde{E}_i^3$ to quantum operators,

$$\hat{\tilde{E}}_i^3 \sim \frac{\delta}{\delta A_3^i} .$$

It turns out that the area A_Σ can be promoted to a well defined quantum operator despite the fact that we are dealing with product of two functional derivatives and worse we have a square-root to contend with as well.[13] Putting $N = 2J$, we talk of being in the J-th representation. We note that $\sum_i T^i T^i = J(J+1)\mathbf{1}$. This quantity is important in the final formula for the area spectrum. We simply state the result below,

$$\hat{A}_\Sigma W_\gamma[A] = 8\pi\ell_{\text{Planck}}^2\beta\sum_I \sqrt{j_I(j_I+1)}W_\gamma[A]$$

where the sum is over all edges I of the Wilson loop that pierce the surface Σ.

The formula for the volume of a region R is given by

$$V = \int_R d^3x \sqrt{\det(q)} = \frac{1}{6}\int_R dx^3 \sqrt{\epsilon_{abc}\epsilon^{ijk}\tilde{E}_i^a \tilde{E}_j^b \tilde{E}_k^c} .$$

The quantization of the volume proceeds the same way as with the area. As we take the derivative, and each time we do so we bring down the tangent vector $\dot{\gamma}^a$, when the volume operator acts on non-intersecting Wilson loops the result vanishes. Quantum states with non-zero volume must therefore involve intersections. Given that the anti-symmetric summation is taken over in the formula for the volume we would need at least intersections with three non-coplanar lines. Actually it turns out that one needs at least four-valent vertices for the volume operator to be non-vanishing.

We now consider Wilson loops with intersections. We assume the real representation where the gauge group is $SU(2)$. Wilson loops are an over complete basis as there are identities relating different Wilson loops. These come about from the fact that Wilson loops are based on matrices (the holonomy) and these matrices satisfy identities. Given any two $SU(2)$ matrices $\mathbb{A}$ and $\mathbb{B}$ it is easy to check that,

$$\text{Tr}(\mathbb{A})\,\text{Tr}(\mathbb{B}) = \text{Tr}(\mathbb{AB}) + \text{Tr}(\mathbb{AB}^{-1}) .$$

This implies that given two loops γ and η that intersect, we will have,

$$W_\gamma[A]W_\eta[A] = W_{\gamma\circ\eta}[A] + W_{\gamma\circ\eta^{-1}}[A]$$

where by η^{-1} we mean the loop η traversed in the opposite direction and $\gamma\circ\eta$ means the loop obtained by going around the loop γ and then along η. See figure below. Given that the matrices are unitary one has that $W_\gamma[A] = W_{\gamma^{-1}}[A]$. Also given the cyclic property of the matrix traces (i.e. $Tr(\mathbb{AB}) = Tr(\mathbb{BA})$) one has that $W_{\gamma\circ\eta}[A] = W_{\eta\circ\gamma}[A]$. These identities can be combined with each other into further identities of increasing complexity adding more loops. These identities are the so-called Mandelstam identities. Spin networks certain are linear combinations of intersecting Wilson loops designed to address the over completeness introduced by the Mandelstam identities (for trivalent intersections they eliminate the over-compleness entirely) and actually constitute a basis for all gauge invariant functions.

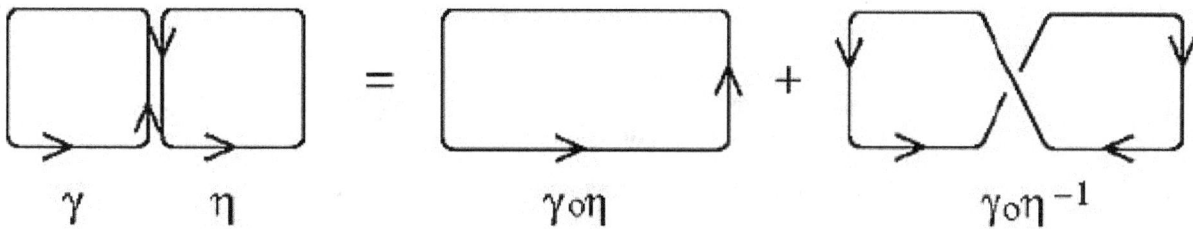

Graphical representation of the simplest non-trivial Mandestam identity relating different Wilson loops.

As mentioned above the holonomy tells you how to propagate test spin half particles. A spin network state assigns an amplitude to a set of spin half particles tracing out a path in space, merging and splitting. These are described by spin networks γ : the edges are labelled by spins together with 'intertwiners' at the vertices which are prescription for how to sum over different ways the spins are rerouted. The sum over rerouting are chosen as such to make the form of the intertwiner invariant under Gauss gauge transformations.

28.4.5   Real variables, modern analysis and LQG

Main article: Hamiltonian constraint of LQG

Let us go into more detail about the technical difficulties associated with using Ashtekar's variables:

With Ashtekar's variables one uses a complex connection and so the relevant gauge group as actually $SL(2, \mathbb{C})$ and not $SU(2)$. As $SL(2, \mathbb{C})$ is non-compact it creates serious problems for the rigorous construction of the necessary mathematical machinery. The group $SU(2)$ is on the other hand is compact and the relevant constructions needed have been developed.

As mentioned above, because Ashtekar's variables are complex it results in complex general relativity. To recover the real theory one has to impose what are known as the reality conditions. These require that the densitized triad be real and that the real part of the Ashtekar connection equals the compatible spin connection (the compatibility condition being $\nabla_a e_b^I = 0$) determined by the desitized triad. The expression for compatible connection Γ_a^i is rather complicated and as such non-polynomial formula enters through the back door.

Before we state the next difficulty we should give a definition; a tensor density of weight W transforms like an ordinary tensor, except that in additional the W th power of the Jacobian,

$$J = \left| \frac{\partial x^a}{\partial x'^b} \right|$$

appears as a factor, i.e.

$$T'^{a\ldots}_{b\ldots} = J^W \frac{\partial x'^a}{\partial x^c} \ldots \frac{\partial x^d}{\partial x'^b} T^{c\ldots}_{d\ldots} .$$

It turns out that it is impossible, on general grounds, to construct a UV-finite, diffeomorphism non-violating operator corresponding to $\sqrt{\det(q)}H$. The reason is that the rescaled Hamiltonian constraint is a scalar density of weight two while it can be shown that only scalar densities of weight one have a chance to result in a well defined operator. Thus, one is forced to work with the original unrescaled, density one-valued, Hamiltonian constraint. However, this is non-polynomial and the whole virtue of the complex variables is questioned. In fact, all the solutions constructed for Ashtekar's Hamiltonian constraint only vanished for finite regularization (physics), however, this violates spatial diffeomorphism invariance.

Without the implementation and solution of the Hamiltonian constraint no progress can be made and no reliable predictions are possible!

To overcome the first problem one works with the configuration variable

$$A_a^i = \Gamma_a^i + \beta K_a^i$$

where β is real (as pointed out by Barbero, who introduced real variables some time after Ashtekar's variables[14][15]).

The Guass law and the spatial diffeomorphism constraints are the same. In real Ashtekar variables the Hamiltonian is

$$H = \frac{\epsilon_{ijk} F_{ab}^k \tilde{E}_i^a \tilde{E}_j^b}{\sqrt{\det(q)}} + 2 \frac{\beta^2 + 1}{\beta^2} \frac{(\tilde{E}_i^a \tilde{E}_j^b - \tilde{E}_j^a \tilde{E}_i^b)}{\sqrt{\det(q)}} (A_a^i - \Gamma_a^i)(A_b^j - \Gamma_b^j) = H_E + H' \, .$$

The complicated relationship between Γ_a^i and the desitized triads causes serious problems upon quantization. It is with the choice $\beta = \pm i$ that the second more complicated term is made to vanish. However, as mentioned above Γ_a^i reappears in the reality conditions. Also we still have the problem of the $1/\sqrt{\det(q)}$ factor.

Thiemann was able to make it work for real β. First he could simplify the troublesome $1/\sqrt{\det(q)}$ by using the identity

$$\{A_c^k, V\} = \frac{\epsilon_{abc} \epsilon^{ijk} \tilde{E}_i^a \tilde{E}_j^b}{\sqrt{\det(q)}}$$

where V is the volume. The A_c^k and V can be promoted to well defined operators in the loop representation and the Poisson bracket is replaced by a commutator upon quantization; this takes care of the first term. It turns out that a similar trick can be used to treat the second term. One introduces the quantity

$$K = \int d^3x K_a^i \tilde{E}_i^a$$

and notes that

$$K_a^i = \{A_a^i, K\} \, .$$

We are then able to write

$$A_a^i - \Gamma_a^i = \beta K_a^i = \beta \{A_a^i, K\} \, .$$

The reason the quantity K is easier to work with at the time of quantization is that it can be written as

$$K = -\{V, \int d^3x H_E\}$$

where we have used that the integrated densitized trace of the extrinsic curvature, K, is the "time derivative of the volume".

In the long history of canonical quantum gravity formulating the Hamiltonian constraint as a quantum operator (Wheeler–DeWitt equation) in a mathematically rigorous manner has been a formidable problem. It was in the loop representation that a mathematically well defined Hamiltonian constraint was finally formulated in 1996.[9] We leave more details of its construction to the article Hamiltonian constraint of LQG. This together with the quantum versions of the Gauss law and spatial diffeomorphism constrains written in the loop representation are the central equations of LQG (modern canonical quantum General relativity).

Finding the states that are annihilated by these constraints (the physical states), and finding the corresponding physical inner product, and observables is the main goal of the technical side of LQG.

A very important aspect of the Hamiltonian operator is that it only acts at vertices (a consequence of this is that Thiemann's Hamiltonian operator, like Ashtekar's operator, annihilates non-intersecting loops except now it is not just formal and has rigorous mathematical meaning). More precisely, its action is non-zero on at least vertices of valence three and greater and results in a linear combination of new spin networks where the original graph has been modified by the addition of lines at each vertex together and a change in the labels of the adjacent links of the vertex.

28.4.6   Solving the quantum constraints

Main articles: spectrum, dual space and Rigged Hilbert space

We solve, at least approximately, all the quantum constraint equations and for the physical inner product to make physical predictions.

Before we move on to the constraints of LQG, lets us consider certain cases. We start with a kinematic Hilbert space $\mathcal{H}_{Kin}$ as so is equipped with an inner product—the kinematic inner product $\langle \phi, \psi \rangle_{Kin}$.

i) Say we have constraints $\hat{C}_I$ whose zero eigenvalues lie in their discrete spectrum. Solutions of the first constraint, $\hat{C}_1$, correspond to a subspace of the kinematic Hilbert space, $\mathcal{H}_1 \subset \mathcal{H}_{Kin}$. There will be a projection operator P_1 mapping

$\mathcal{H}_{\text{Kin}}$ onto $\mathcal{H}_1$. The kinematic inner product structure is easily employed to provide the inner product structure after solving this first constraint; the new inner product $\langle \phi, \psi \rangle_1$ is simply

$$\langle \phi, \psi \rangle_1 = \langle P\phi, P\psi \rangle_{\text{Kin}}$$

They are based on the same inner product and are states normalizable with respect to it.

ii) The zero point is not contained in the point spectrum of all the $\hat{C}_I$, there is then no non-trivial solution $\Psi \in \mathcal{H}_{\text{Kin}}$ to the system of quantum constraint equations $\hat{C}_I \Psi = 0$ for all I .

For example the zero eigenvalue of the operator

$$\hat{C} = \left(i\frac{d}{dx} - k \right)$$

on $L_2(\mathbb{R}, dx)$ lies in the continuous spectrum $\mathbb{R}$ but the formal "eigenstate" $\exp(-ikx)$ is not normalizable in the kinematic inner product,

$$\int_{-\infty}^{\infty} dx \psi^*(x)\psi(x) = \int_{-\infty}^{\infty} dx e^{ikx} e^{-ikx} = \int_{-\infty}^{\infty} dx = \infty$$

and so does not belong to the kinematic Hilbert space $\mathcal{H}_{\text{Kin}}$. In these cases we take a dense subset $\mathcal{S}$ of $\mathcal{H}_{\text{Kin}}$ (intuitively this means either any point in $\mathcal{S}$ is either in $\mathcal{H}_{\text{Kin}}$ or arbitrarily close to a point in $\mathcal{H}_{\text{Kin}}$) with very good convergence properties and consider its dual space $\mathcal{S}'$ (intuitively these map elements of $\mathcal{S}$ onto finite complex numbers in a linear manner), then $\mathcal{S} \subset \mathcal{H}_{\text{Kin}} \subset \mathcal{S}'$ (as $\mathcal{S}'$ contains distributional functions). The constraint operator is then implemented on this larger dual space, which contains distributional functions, under the adjoint action on the operator. One looks for solutions on this larger space. This comes at the price that the solutions must be given a new Hilbert space inner product with respect to which they are normalizable (see article on rigged Hilbert space). In this case we have a generalized projection operator on the new space of states. We cannot use the above formula for the new inner product as it diverges, instead the new inner product is given by the simply modification of the above,

$$\langle \phi, \psi \rangle_1 = \langle P\phi, \psi \rangle_{\text{Kin}}.$$

The generalized projector P is known as a rigging map.

Let us move to LQG, additional complications will arise from the fact the constraint algebra is not a Lie algebra due to the bracket between two Hamiltonian constraints.

The Gauss law is solved by the use of spin network states. They provide a basis for the Kinematic Hilbert space $\mathcal{H}_{\text{Kin}}$. The spatial diffeomorphism constraint has been solved. The induced inner product on $\mathcal{H}_{\text{Diff}}$ (we do not pursue the details) has a very simple description in terms of spin network states; given two spin networks s and s' , with associated spin network states ψ_s and $\psi_{s'}$, the inner product is 1 if s and s' are related to each other by a spatial diffeomorphism and zero otherwise.

The Hamiltonian constraint maps diffeomorphism invariant states onto non-diffeomorphism invaiant states as so does not preserve the diffeomorphism Hilbert space $\mathcal{H}_{\text{Diff}}$. This is an unavoidable consequence of the operator algebra, in particular the commutator:

$$[\hat{C}(\vec{N}), \hat{H}(M)] \propto \hat{H}(\mathcal{L}_{\vec{N}} M)$$

as can be seen by applying this to $\psi_s \in \mathcal{H}_{Diff}$,

$$(\vec{C}(\vec{N})\hat{H}(M) - \hat{H}(M)\vec{C}(\vec{N}))\psi_s \propto \hat{H}(\mathcal{L}_{\vec{N}} M)\psi_s$$

and using $\vec{C}(\vec{N})\psi_s = 0$ to obtain

$$\vec{C}(\vec{N})[\hat{H}(M)\psi_s] \propto \hat{H}(\mathcal{L}_{\vec{N}} M)\psi_s \neq 0$$

and so $\hat{H}(M)\psi_s$ is not in $\mathcal{H}_{Diff}$.

This means that you can't just solve the diffeomorphism constraint and then the Hamiltonian constraint. This problem can be circumvented by the introduction of the master constraint, with its trivial operator algebra, one is then able in principle to construct the physical inner product from $\mathcal{H}_{\text{Diff}}$.

28.5 Spin foams

Main articles: spin network, spin foam, BF model and Barrett–Crane model

In loop quantum gravity (LQG), a spin network represents a "quantum state" of the gravitational field on a 3-dimensional hypersurface. The set of all possible spin networks (or, more accurately, "s-knots" - that is, equivalence classes of spin networks under diffeomorphisms) is countable; it constitutes a basis of LQG Hilbert space.

In physics, a spin foam is a topological structure made out of two-dimensional faces that represents one of the configurations that must be summed to obtain a Feynman's path integral (functional integration) description of quantum gravity. It is closely related to loop quantum gravity.

28.5.1 Spin foam derived from the Hamiltonian constraint operator

The Hamiltonian constraint generates 'time' evolution. Solving the Hamiltonian constraint should tell us how quantum states evolve in 'time' from an initial spin network state to a final spin network state. One approach to solving the Hamiltonian constraint starts with what is called the Dirac delta function. This is a rather singular function of the real line, denoted $\delta(x)$, that is zero everywhere except at $x = 0$ but whose integral is finite and nonzero. It can be represented as a Fourier integral,

$$\delta(x) = \int e^{ikx} dk .$$

One can employ the idea of the delta function to impose the condition that the Hamiltonian constraint should vanish. It is obvious that

$$\prod_{x \in \Sigma} \delta(\hat{H}(x))$$

is non-zero only when $\hat{H}(x) = 0$ for all x in Σ. Using this we can 'project' out solutions to the Hamiltonian constraint. With analogy to the Fourier integral given above, this (generalized) projector can formally be written as

$$\int [dN] e^{i \int d^3 x N(x) \hat{H}(x)} .$$

Interestingly, this is formally spatially diffeomorphism-invariant. As such it can be applied at the spatially diffeomorphism-invariant level. Using this the physical inner product is formally given by

$$\left\langle \int [dN] e^{i \int d^3 x N(x) \hat{H}(x)} s_{int} s_{fin} \right\rangle_{Diff}$$

where s_{int} are the initial spin network and s_{fin} is the final spin network.

The exponential can be expanded

$$\left\langle \int [dN](1 + i \int d^3 x N(x) \hat{H}(x) + \frac{i^2}{2!} [\int d^3 x N(x) \hat{H}(x)][\int d^3 x' N(x') \hat{H}(x')] + \dots) s_{int}, s_{fin} \right\rangle_{Diff}$$

and each time a Hamiltonian operator acts it does so by adding a new edge at the vertex. The summation over different sequences of actions of $\hat{H}$ can be visualized as a summation over different histories of 'interaction vertices' in the 'time' evolution sending the initial spin network to the final spin network. This then naturally gives rise to the two-complex (a combinatorial set of faces that join along edges, which in turn join on vertices) underlying the spin foam description; we evolve forward an initial spin network sweeping out a surface, the action of the Hamiltonian constraint operator is to produce a new planar surface starting at the vertex. We are able to use the action of the Hamiltonian constraint on the vertex of a spin network state to associate an amplitude to each "interaction" (in analogy to Feynman diagrams). See figure below. This opens up a way of trying to directly link canonical LQG to a path integral description. Now just as a spin networks describe quantum space, each configuration contributing to these path integrals, or sums over history, describe 'quantum space-time'. Because of their resemblance to soap foams and the way they are labeled John Baez gave these 'quantum space-times' the name 'spin foams'.

There are however severe difficulties with this particular approach, for example the Hamiltonian operator is not self-adjoint, in fact it is not even a normal operator (i.e. the operator does not commute with its adjoint) and so the spectral theorem cannot be used to define the exponential in general. The most serious problem is that the $\hat{H}(x)$'s are not mutually

The action of the Hamiltonian constraint translated to the path integral or so-called spin foam description. A single node splits into three nodes, creating a spin foam vertex. $N(x_n)$ is the value of N at the vertex and H_{nop} are the matrix elements of the Hamiltonian constraint $\hat{H}$.

commuting, it can then be shown the formal quantity $\int [dN] e^{i \int d^3x N(x) \hat{H}(x)}$ cannot even define a (generalized) projector. The master constraint (see below) does not suffer from these problems and as such offers a way of connecting the canonical theory to the path integral formulation.

28.5.2 Spin foams from BF theory

It turns out there are alternative routes to formulating the path integral, however their connection to the Hamiltonian formalism is less clear. One way is to start with the BF theory. This is a simpler theory to general relativity. It has no local degrees of freedom and as such depends only on topological aspects of the fields. BF theory is what is known as a topological field theory. Surprisingly, it turns out that general relativity can be obtained from BF theory by imposing a constraint,[16] BF theory involves a field B_{ab}^{IJ} and if one chooses the field B to be the (anti-symmetric) product of two tetrads

$$B_{ab}^{IJ} = \tfrac{1}{2}(E_a^I E_b^J - E_b^I E_a^J)$$

(tetrads are like triads but in four spacetime dimensions), one recovers general relativity. The condition that the B field be given by the product of two tetrads is called the simplicity constraint. The spin foam dynamics of the topological field theory is well understood. Given the spin foam 'interaction' amplitudes for this simple theory, one then tries to implement the simplicity conditions to obtain a path integral for general relativity. The non-trivial task of constructing a spin foam model is then reduced to the question of how this simplicity constraint should be imposed in the quantum theory. The first attempt at this was the famous Barrett–Crane model.[17] However this model was shown to be problematic, for example there did not seem to be enough degrees of freedom to ensure the correct classical limit.[18] It has been argued that the simplicity constraint was imposed too strongly at the quantum level and should only be imposed in the sense of expectation values just as with the Lorenz gauge condition $\partial_\mu \hat{A}^\mu$ in the Gupta–Bleuler formalism of quantum electrodynamics. New models have now been put forward, sometimes motivated by imposing the simplicity conditions in a weaker sense.

Another difficulty here is that spin foams are defined on a discretization of spacetime. While this presents no problems for a topological field theory as it has no local degrees of freedom, it presents problems for GR. This is known as the problem triangularization dependence.

28.5.3 Modern formulation of spin foams

Just as imposing the classical simplicity constraint recovers general relativity from BF theory, one expects an appropriate quantum simplicity constraint will recover quantum gravity from quantum BF theory.

Much progress has been made with regard to this issue by Engle, Pereira, and Rovelli[19] and Freidal and Krasnov[20] in defining spin foam interaction amplitudes with much better behaviour.

An attempt to make contact between EPRL-FK spin foam and the canonical formulation of LQG has been made.[21]

28.5.4   Spin foam derived from the master constraint operator

See below.

28.6   The semi-classical limit

28.6.1   What is the semiclassical limit?

Main articles: Correspondence principle and classical limit

The **classical limit** or **correspondence limit** is the ability of a physical theory to approximate or "recover" classical mechanics when considered over special values of its parameters.[22] The classical limit is used with physical theories that predict non-classical behavior.

In physics, the **correspondence principle** states that the behavior of systems described by the theory of quantum mechanics (or by the old quantum theory) reproduces classical physics in the limit of large quantum numbers. In other words, it says that for large orbits and for large energies, quantum calculations must agree with classical calculations.[23]

The principle was formulated by Niels Bohr in 1920,[24] though he had previously made use of it as early as 1913 in developing his model of the atom.[25]

There are two basic requirements in establishing the semi-classical limit of any quantum theory:

i) reproduction of the Poisson brackets (of the diffeomorphism constraints in the case of general relativity). This is extremely important because, as noted above, the Poisson bracket algebra formed between the (smeared) constraints themselves completely determines the classical theory. This is analogous to establishing Ehrenfest's theorem;

ii) the specification of a complete set of classical observables whose corresponding operators (see complete set of commuting observables for the quantum mechanical definition of a complete set of observables) when acted on by appropriate semi-classical states reproduce the same classical variables with small quantum corrections (a subtle point is that states that are semi-classical for one class of observables may not be semi-classical for a different class of observables[26]).

This may be easily done, for example, in ordinary quantum mechanics for a particle but in general relativity this becomes a highly non-trivial problem as we will see below.

28.6.2   Why might LQG not have general relativity as its semiclassical limit?

Any candidate theory of quantum gravity must be able to reproduce Einstein's theory of general relativity as a classical limit of a quantum theory. This is not guaranteed because of a feature of quantum field theories which is that they have different sectors, these are analogous to the different phases that come about in the thermodynamical limit of statistical systems. Just as different phases are physically different, so are different sectors of a quantum field theory. It may turn out that LQG belongs to an unphysical sector - one in which you do not recover general relativity in the semi classical limit (in fact there might not be any physical sector at all).

Theorems establishing the uniqueness of the loop representation as defined by Ashtekar et al. (i.e. a certain concrete realization of a Hilbert space and associated operators reproducing the correct loop algebra - the realization that everybody was using) have been given by two groups (Lewandowski, Okolow, Sahlmann and Thiemann;[27] and Christian Fleischhack[28]). Before this result was established it was not known whether there could be other examples of Hilbert spaces with operators invoking the same loop algebra, other realizations, not equivalent to the one that had been used so far. These uniqueness theorems imply no others exist and so if LQG does not have the correct semiclassical limit then this would mean the end of the loop representation of quantum gravity altogether.

28.6.3   Difficulties checking the semiclassical limit of LQG

There are difficulties in trying to establish LQG gives Einstein's theory of general relativity in the semi classical limit. There are a number of particular difficulties in establishing the semi-classical limit

1. There is no operator corresponding to infinitesimal spacial diffeomorphisms (it is not surprising that the theory has no generator of infinitesimal spatial `translations' as it predicts spatial geometry has a discrete nature, compare to the situation in condensed matter). Instead it must be approximated by finite spatial diffeomorphisms and so the Poisson bracket structure of the classical theory is not exactly reproduced. This problem can be circumvented with the introduction of the so-called master constraint (see below)[29]

2. There is the problem of reconciling the discrete combinatorial nature of the quantum states with the continuous nature of the fields of the classical theory.

3. There are serious difficulties arising from the structure of the Poisson brackets involving the spatial diffeomorphism and Hamiltonian constraints. In particular, the algebra of (smeared) Hamiltonian constraints does not close, it is proportional to a sum over infinitesimal spatial diffeomorphisms (which, as we have just noted, does not exist in the quantum theory) where the coefficients of proportionality are not constants but have non-trivial phase space dependence - as such it does not form a Lie algebra. However, the situation is much improved by the introduction of the master constraint.[29]

4. The semi-classical machinery developed so far is only appropriate to non-graph-changing operators, however, Thiemann's Hamiltonian constraint is a graph-changing operator - the new graph it generates has degrees of freedom upon which the coherent state does not depend and so their quantum fluctuations are not suppressed. There is also the restriction, so far, that these coherent states are only defined at the Kinematic level, and now one has to lift them to the level of $\mathcal{H}_{Diff}$ and $\mathcal{H}_{Phys}$. It can be shown that Thiemann's Hamiltonian constraint is required to be graph changing in order to resolve problem 3 in some sense. The master constraint algebra however is trivial and so the requirement that it be graph changing can be lifted and indeed non-graph changing master constraint operators have been defined.

5. Formulating observables for classical general relativity is a formidable problem by itself because of its non-linear nature and space-time diffeomorphism invariance. In fact a systematic approximation scheme to calculate observables has only been recently developed.[30][31]

Difficulties in trying to examine the semi classical limit of the theory should not be confused with it having the wrong semi classical limit.

28.6.4   Progress in demonstrating LQG has the correct semiclassical limit

Much details here to be written up...

Concerning issue number 2 above one can consider so-called weave states. Ordinary measurements of geometric quantities are macroscopic, and planckian discreteness is smoothed out. The fabric of a T-shirt is analogous. At a distance it is a smooth curved two-dimensional surface. But a closer inspection we see that it is actually composed of thousands of one-dimensional linked threads. The image of space given in LQG is similar, consider a very large spin network formed by a very large number of nodes and links, each of Planck scale. But probed at a macroscopic scale, it appears as a three-dimensional continuous metric geometry.

As far as the editor knows problem 4 of having semi-classical machinery for non-graph changing operators is as the moment still out of reach.

To make contact with familiar low energy physics it is mandatory to have to develop approximation schemes both for the physical inner product and for Dirac observables.

The spin foam models have been intensively studied can be viewed as avenues toward approximation schemes for the physical inner product.

Markopoulou et al. adopted the idea of noiseless subsystems in an attempt to solve the problem of the low energy limit in background independent quantum gravity theories[32][33][34] The idea has even led to the intriguing possibility of matter of the standard model being identified with emergent degrees of freedom from some versions of LQG (see section below: *LQG and related research programs*).

As Wightman emphasized in the 1950s, in Minkowski QFTs the $n-$ point functions

$$W(x_1, \ldots, x_n) = \langle 0|\phi(x_n) \ldots \phi(x_1)|0 \rangle,$$

completely determine the theory. In particular, one can calculate the scattering amplitudes from these quantities. As explained below in the section on the *Background independent scattering amplitudes*, in the background-independent context, the $n-$ point functions refer to a state and in gravity that state can naturally encode information about a specific geometry which can then appear in the expressions of these quantities. To leading order LQG calculations have been shown to agree in an appropriate sense with the $n-$ point functions calculated in the effective low energy quantum general relativity.

28.7 Improved dynamics and the master constraint

Main articles: Hamiltonian (quantum mechanics), Hamiltonian constraint of LQG and Friedrichs extension

28.7.1 The master constraint

Thiemann's master constraint should not be confused with the master equation which has to do with random processes. The Master Constraint Programme for Loop Quantum Gravity (LQG) was proposed as a classically equivalent way to impose the infinite number of Hamiltonian constraint equations

$$H(x) = 0$$

(x being a continuous index) in terms of a single master constraint,

$$M = \int d^3x \frac{|H(x)|^2}{\sqrt{\det(q(x))}}.$$

which involves the square of the constraints in question. Note that $H(x)$ were infinitely many whereas the master constraint is only one. It is clear that if M vanishes then so do the infinitely many $H(x)$'s. Conversely, if all the $H(x)$'s vanish then so does M, therefore they are equivalent. The master constraint M involves an appropriate averaging over all space and so is invariant under spatial diffeomorphisms (it is invariant under spatial "shifts" as it is a summation over all such spatial "shifts" of a quantity that transforms as a scalar). Hence its Poisson bracket with the (smeared) spacial diffeomorphism constraint, $C(\vec{N})$, is simple:

$$\{M, C(\vec{N})\} = 0.$$

(it is $su(2)$ invariant as well). Also, obviously as any quantity Poisson commutes with itself, and the master constraint being a single constraint, it satisfies

$$\{M, M\} = 0.$$

We also have the usual algebra between spatial diffeomorphisms. This represents a dramatic simplification of the Poisson bracket structure, and raises new hope in understanding the dynamics and establishing the semi-classical limit.[35]

An initial objection to the use of the master constraint was that on first sight it did not seem to encode information about the observables; because the Mater constraint is quadratic in the constraint, when you compute its Poisson bracket with any quantity, the result is proportional to the constraint, therefore it always vanishes when the constraints are imposed and as such does not select out particular phase space functions. However, it was realized that the condition

$$\{\{M, O\}, O\}_{M=0} = 0$$

is equivalent to O being a Dirac observable. So the master constraint does capture information about the observables. Because of its significance this is known as the Master equation.[35]

That the master constraint Poisson algebra is an honest Lie algebra opens up the possibility of using a certain method, known as group averaging, in order to construct solutions of the infinite number of Hamiltonian constraints, a physical inner product thereon and Dirac observables via what is known as refined algebraic quantization RAQ[36]

28.7.2   The quantum master constraint

Define the quantum master constraint (regularisation issues aside) as

$$\hat{M} := \int d^3x \left(\widehat{\frac{H}{\det(q(x))^{1/4}}} \right)^\dagger (x) \left(\widehat{\frac{H}{\det(q(x))^{1/4}}} \right) (x) \ .$$

Obviously,

$$\left(\widehat{\frac{H}{\det(q(x))^{1/4}}} \right) (x) \Psi = 0$$

for all x implies $\hat{M}\Psi = 0$. Conversely, if $\hat{M}\Psi = 0$ then

$$0 = < \Psi, \hat{M}\Psi > = \int d^3x \left\| \left(\widehat{\frac{H}{\det(q(x))^{1/4}}} \right) (x) \Psi \right\|^2 \qquad Eq\ 4$$

implies

$$\left(\widehat{\frac{H}{\det(q(x))^{1/4}}} \right) (x) \Psi = 0 \ .$$

What is done first is, we are able to compute the matrix elements of the would-be operator $\hat{M}$, that is, we compute the quadratic form Q_M . It turns out that as Q_M is a graph changing, diffeomorphism invariant quadratic form it cannot exist on the kinematic Hilbert space H_{Kin} , and must be defined on H_{Diff} . The fact that the master constraint operator $\hat{M}$ is densely defined on H_{Diff} , it is obvious that $\hat{M}$ is a positive and symmetric operator in H_{Diff} . Therefore, the quadratic form Q_M associated with $\hat{M}$ is closable. The closure of Q_M is the quadratic form of a unique self-adjoint operator $\overline{M}$, called the Friedrichs extension of $\hat{M}$. We relabel $\overline{M}$ as $\hat{M}$ for simplicity. (Note that the presence of an inner product, viz Eq 4, means there are no superfluous solutions i.e. there are no Ψ such that $\left(\widehat{\frac{H}{\det(q(x))^{1/4}}} \right) (x) \Psi \neq 0$ but for which $\hat{M}\Psi = 0$).

It is also possible to construct a quadratic form Q_{M_E} for what is called the extended master constraint (discussed below) on H_{Kin} which also involves the weighted integral of the square of the spatial diffeomorphism constraint (this is possible because Q_{M_E} is not graph changing).

The spectrum of the master constraint may not contain zero due to normal or factor ordering effects which are finite but similar in nature to the infinite vacuum energies of background-dependent quantum field theories. In this case it turns out to be physically correct to replace $\hat{M}$ with $\hat{M}' := \hat{M} - min(spec(\hat{M}))\hat{1}$ provided that the "normal ordering constant" vanishes in the classical limit, that is, $\lim_{h \to 0} min(spec(\hat{M})) = 0$, so that $\hat{M}'$ is a valid quantisation of M .

28.7.3   Testing the master constraint

The constraints in their primitive form are rather singular, this was the reason for integrating them over test functions to obtain smeared constraints. However, it would appear that the equation for the master constraint, given above, is even more singular involving the product of two primitive constraints (although integrated over space). Squaring the constraint is dangerous as it could lead to worsened ultraviolent behaviour of the corresponding operator and hence the master constraint programme must be approached with due care.

In doing so the master constraint programme has been satisfactorily tested in a number of model systems with non-trivial constraint algebras, free and interacting field theories.[37][38][39][40][41] The master constraint for LQG was established as a genuine positive self-adjoint operator and the physical Hilbert space of LQG was shown to be non-empty,[42] an obvious consistency test LQG must pass to be a viable theory of quantum General relativity.

28.7.4   Applications of the master constraint

The master constraint has been employed in attempts to approximate the physical inner product and define more rigorous path integrals.[43][44][45][46]

The Consistent Discretizations approach to LQG,[47][48] is an application of the master constraint program to construct the physical Hilbert space of the canonical theory.

28.7.5   Spin foam from the master constraint

It turns out that the master constraint is easily generalized to incorporate the other constraints. It is then referred to as the extended master constraint, denoted M_E. We can define the extended master constraint which imposes both the Hamiltonian constraint and spatial diffeomorphism constraint as a single operator,

$$M_E = \int_\Sigma d^3x \frac{H(x)^2 - q^{ab}V_a(x)V_b(x)}{\sqrt{det(q)}} \ .$$

Setting this single constraint to zero is equivalent to $H(x) = 0$ and $V_a(x) = 0$ for all x in Σ. This constraint implements the spatial diffeomorphism and Hamiltonian constraint at the same time on the Kinematic Hilbert space. The physical inner product is then defined as

$$\langle \phi, \psi \rangle_{\text{Phys}} = \lim_{T\to\infty} \left\langle \phi, \int_{-T}^{T} dt e^{it\hat{M}_E} \psi \right\rangle$$

(as $\delta(\hat{M}_E) = \lim_{T\to\infty} \int_{-T}^{T} dt e^{it\hat{M}_E}$). A spin foam representation of this expression is obtained by splitting the t-parameter in discrete steps and writing

$$e^{it\hat{M}_E} = \lim_{n\to\infty} [e^{it\hat{M}_E/n}]^n = \lim_{n\to\infty} [1 + it\hat{M}_E/n]^n.$$

The spin foam description then follows from the application of $[1 + it\hat{M}_E/n]$ on a spin network resulting in a linear combination of new spin networks whose graph and labels have been modified. Obviously an approximation is made by truncating the value of n to some finite integer. An advantage of the extended master constraint is that we are working at the kinematic level and so far it is only here we have access semi-classical coherent states. Moreover, one can find none graph changing versions of this master constraint operator, which are the only type of operators appropriate for these coherent states.

28.7.6   Algebraic quantum gravity

The master constraint programme has evolved into a fully combinatorial treatment of gravity known as Algebraic Quantum Gravity (AQG).[49] The non-graph changing master constraint operator is adapted in the framework of algebraic quantum gravity. While AQG is inspired by LQG, it differs drastically from it because in AQG there is fundamentally no topology or differential structure - it is background independent in a more generalized sense and could possibly have something to say about topology change. In this new formulation of quantum gravity AQG semiclassical states always control the fluctuations of all present degrees of freedom. This makes the AQG semiclassical analysis superior over that of LQG, and progress has been made in establishing it has the correct semiclassical limit and providing contact with familiar low energy physics.[50][51] See Thiemann's book for details.

28.8   Physical applications of LQG

28.8.1   Black hole entropy

Main articles: Black hole thermodynamics, Isolated horizon and Immirzi parameter

The Immirzi parameter (also known as the Barbero-Immirzi parameter) is a numerical coefficient appearing in loop quantum gravity. It may take real or imaginary values.

An artist depiction of two black holes merging, a process in which the laws of thermodynamics are upheld.

Black hole thermodynamics is the area of study that seeks to reconcile the laws of thermodynamics with the existence of black hole event horizons. The no hair conjecture of general relativity states that a black hole is characterized only by its mass, its charge, and its angular momentum; hence, it has no entropy. It appears, then, that one can violate the second law of thermodynamics by dropping an object with nonzero entropy into a black hole.[52] Work by Stephen Hawking and Jacob Bekenstein showed that one can preserve the second law of thermodynamics by assigning to each black hole a *black-hole entropy*

$$S_{\mathrm{BH}} = \frac{k_{\mathrm{B}} A}{4 \ell_{\mathrm{P}}^2},$$

where A is the area of the hole's event horizon, k_{B} is the Boltzmann constant, and $\ell_{\mathrm{P}} = \sqrt{G\hbar/c^3}$ is the Planck length.[53] The fact that the black hole entropy is also the maximal entropy that can be obtained by the Bekenstein bound (wherein the Bekenstein bound becomes an equality) was the main observation that led to the holographic principle.[52]

An oversight in the application of the no-hair theorem is the assumption that the relevant degrees of freedom accounting for the entropy of the black hole must be classical in nature; what if they were purely quantum mechanical instead and had non-zero entropy? Actually, this is what is realized in the LQG derivation of black hole entropy, and can be seen as a consequence of its background-independence – the classical black hole spacetime comes about from the semi-classical limit of the quantum state of the gravitational field, but there are many quantum states that have the same semiclassical limit. Specifically, in LQG[54] it is possible to associate a quantum geometrical interpretation to the microstates: These are the quantum geometries of the horizon which are consistent with the area, A , of the black hole and the topology of the horizon (i.e. spherical). LQG offers a geometric explanation of the finiteness of the entropy and of the proportionality of the area of the horizon.[55][56] These calculations have been generalized to rotating black holes.[57]

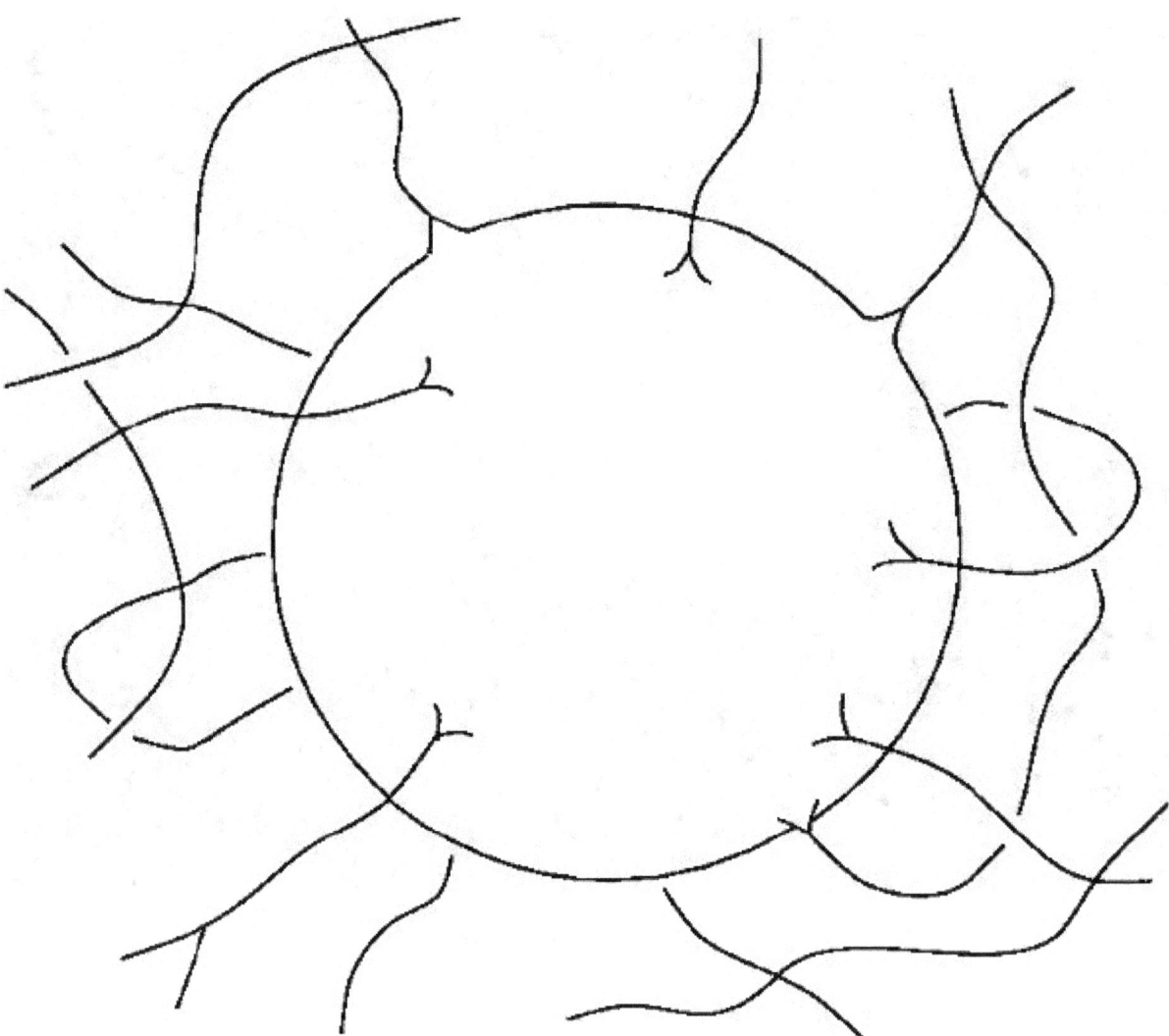

Representation of quantum geometries of the horizon. Polymer excitations in the bulk puncture the horizon, endowing it with quantized area. Intrinsically the horizon is flat except at punctures where it acquires a quantized deficit angle or quantized amount of curvature. These deficit angles add up to 4π .

It is possible to derive, from the covariant formulation of full quantum theory (Spinfoam) the correct relation between energy and area (1st law), the Unruh temperature and the distribution that yields Hawking entropy.[58] The calculation makes use of the notion of dynamical horizon and is done for non-extremal black holes.

A recent success of the theory in this direction is the computation of the entropy of all non singular black holes directly from theory and independent of Immirzi parameter.[59] The result is the expected formula $S = A/4$, where S is the entropy and A the area of the black hole, derived by Bekenstein and Hawking on heuristic grounds. This is the only known derivation of this formula from a fundamental theory, for the case of generic non singular black holes. Older attempts at this calculation had difficulties. The problem was that although Loop quantum gravity predicted that the entropy of a black hole is proportional to the area of the event horizon, the result depended on a crucial free parameter in the theory, the above-mentioned Immirzi parameter. However, there is no known computation of the Immirzi parameter, so it had to be fixed by demanding agreement with Bekenstein and Hawking's calculation of the black hole entropy.

28.8.2   Loop quantum cosmology

Main articles: loop quantum cosmology, Big bounce and inflation (cosmology)

The popular and technical literature makes extensive references to LQG-related topic of loop quantum cosmology. LQC was mainly developed by Martin Bojowald, it was popularized Loop quantum cosmology in *Scientific American* for predicting a Big Bounce prior to the Big Bang. Loop quantum cosmology (LQC) is a symmetry-reduced model of classical general relativity quantized using methods that mimic those of loop quantum gravity (LQG) that predicts a "quantum bridge" between contracting and expanding cosmological branches.

Achievements of LQC have been the resolution of the big bang singularity, the prediction of a Big Bounce, and a natural mechanism for inflation (cosmology).

LQC models share features of LQG and so is a useful toy model. However, the results obtained are subject to the usual restriction that a truncated classical theory, then quantized, might not display the true behaviour of the full theory due to artificial suppression of degrees of freedom that might have large quantum fluctuations in the full theory. It has been argued that singularity avoidance in LQC are by mechanisms only available in these restrictive models and that singularity avoidance in the full theory can still be obtained but by a more subtle feature of LQG.[60][61]

28.8.3   Loop quantum gravity phenomenology

Quantum gravity effects are notoriously difficult to measure because the Planck length is so incredibly small. However recently physicists have started to consider the possibility of measuring quantum gravity effects, mostly from astrophysical observations and gravitational wave detectors.The energy of those fluctuations at scales this small cause space-perturbations which are visible at higher scales.

28.8.4   Background independent scattering amplitudes

Loop quantum gravity is formulated in a background-independent language. No spacetime is assumed a priori, but rather it is built up by the states of theory themselves - however scattering amplitudes are derived from n -point functions (Correlation function (quantum field theory)) and these, formulated in conventional quantum field theory, are functions of points of a background space-time. The relation between the background-independent formalism and the conventional formalism of quantum field theory on a given spacetime is far from obvious, and it is far from obvious how to recover low-energy quantities from the full background-independent theory. One would like to derive the n -point functions of the theory from the background-independent formalism, in order to compare them with the standard perturbative expansion of quantum general relativity and therefore check that loop quantum gravity yields the correct low-energy limit.

A strategy for addressing this problem has been suggested;[62] the idea is to study the boundary amplitude, namely a path integral over a finite space-time region, seen as a function of the boundary value of the field.[63] In conventional quantum field theory, this boundary amplitude is well–defined[64][65] and codes the physical information of the theory; it does so in quantum gravity as well, but in a fully background–independent manner.[66] A generally covariant definition of n -point functions can then be based on the idea that the distance between physical points –arguments of the n -point function is determined by the state of the gravitational field on the boundary of the spacetime region considered.

Progress has been made in calculating background independent scattering amplitudes this way with the use of spin foams. This is a way to extract physical information from the theory. Claims to have reproduced the correct behaviour for graviton scattering amplitudes and to have recovered classical gravity have been made. "We have calculated Newton's law starting from a world with no space and no time." - Carlo Rovelli.

28.9   Gravitons, string theory, supersymmetry, extra dimensions in LQG

Main articles: graviton, string theory, supersymmetry, Kaluza–Klein theory and supergravity

Some quantum theories of gravity posit a spin-2 quantum field that is quantized, giving rise to gravitons. In string theory one generally starts with quantized excitations on top of a classically fixed background. This theory is thus described as background dependent. Particles like photons as well as changes in the spacetime geometry (gravitons) are both described as excitations on the string worldsheet. The background dependence of string theory can have important physical consequences, such as determining the number of quark generations. In contrast, loop quantum gravity, like general relativity, is manifestly background independent, eliminating the background required in string theory. Loop quantum gravity, like string theory, also aims to overcome the nonrenormalizable divergences of quantum field theories.

LQG never introduces a background and excitations living on this background, so LQG does not use gravitons as building blocks. Instead one expects that one may recover a kind of semiclassical limit or weak field limit where something like "gravitons" will show up again. In contrast, gravitons play a key role in string theory where they are among the first (massless) level of excitations of a superstring.

LQG differs from string theory in that it is formulated in 3 and 4 dimensions and without supersymmetry or Kaluza-Klein extra dimensions, while the latter requires both to be true. There is no experimental evidence to date that confirms string theory's predictions of supersymmetry and Kaluza–Klein extra dimensions. In a 2003 paper A dialog on quantum gravity,[67] Carlo Rovelli regards the fact LQG is formulated in 4 dimensions and without supersymmetry as a strength of the theory as it represents the most parsimonious explanation, consistent with current experimental results, over its rival string/M-theory. Proponents of string theory will often point to the fact that, among other things, it demonstrably reproduces the established theories of general relativity and quantum field theory in the appropriate limits, which Loop Quantum Gravity has struggled to do. In that sense string theory's connection to established physics may be considered more reliable and less speculative, at the mathematical level. Peter Woit in Not Even Wrong and Lee Smolin in The Trouble with Physics regard string/M-theory to be in conflict with current known experimental results.

Since LQG has been formulated in 4 dimensions (with and without supersymmetry), and M-theory requires supersymmetry and 11 dimensions, a direct comparison between the two has not been possible. It is possible to extend mainstream LQG formalism to higher-dimensional supergravity, general relativity with supersymmetry and Kaluza–Klein extra dimensions should experimental evidence establish their existence. It would therefore be desirable to have higher-dimensional Supergravity loop quantizations at one's disposal in order to compare these approaches. In fact a series of recent papers have been published attempting just this.[68][69][70][71][72][73][74][75] Most recently, Thiemann (and alumni) have made progress toward calculating black hole entropy for supergravity in higher dimensions. It will be interesting to compare these results to the corresponding super string calculations.[76][77]

As of April 2013 LHC has failed to find evidence of supersymmetry or Kaluza–Klein extra dimensions, which has encouraged LQG researchers. Shaposhnikov in his paper "Is there a new physics between electroweak and Planck scales?" has proposed the neutrino minimal standard model,[78] which claims the most parsimonious theory is a standard model extended with neutrinos, plus gravity, and that extra dimensions, GUT physics, and supersymmetry, string/M-theory physics are unrealized in nature, and that any theory of quantum gravity must be four dimensional, like loop quantum gravity.

28.10 LQG and related research programs

Main articles: noncommutative geometry, twistor theory, entropic gravity, Sundance Bilson-Thompson, Asymptotic safety in quantum gravity, Causal dynamical triangulation, group field theory and consistent discretizations

Several research groups have attempted to combine LQG with other research programs: Johannes Aastrup, Jesper M. Grimstrup et al. research combines noncommutative geometry with loop quantum gravity,[79] Laurent Freidel, Simone Speziale, et al., spinors and twistor theory with loop quantum gravity,[80] and Lee Smolin et al. with Verlinde entropic gravity and loop gravity.[81] Stephon Alexander, Antonino Marciano and Lee Smolin have attempted to explain the origins of weak force chirality in terms of Ashketar's variables, which describe gravity as chiral,[82] and LQG with Yang–Mills theory fields[83] in four dimensions. Sundance Bilson-Thompson, Hackett et al.,[84][85] has attempted to introduce standard model via LQG"s degrees of freedom as an emergent property (by employing the idea noiseless subsystems a useful notion introduced in more general situation for constrained systems by Fotini Markopoulou-Kalamara et al.[86]) LQG has also drawn philosophical comparisons with causal dynamical triangulation[87] and asymptotically safe gravity,[88] and the spinfoam with group field theory and AdS/CFT correspondence.[89] Smolin and Wen have suggested combining LQG with String-net liquid, tensors, and Smolin and Fotini Markopoulou-Kalamara Quantum Graphity. There is the consistent discretizations approach. In addition to what has already mentioned above, Pullin and Gambini provide a framework to connect the path integral and canonical approaches to quantum gravity. They may help reconcile the spin foam and canonical loop representation approaches. Recent research by Chris Duston and Matilde Marcolli introduces topology change via topspin networks.[90]

28.11 Problems and comparisons with alternative approaches

Main article: List of unsolved problems in physics

Some of the major unsolved problems in physics are theoretical, meaning that existing theories seem incapable of explaining a certain observed phenomenon or experimental result. The others are experimental, meaning that there is a difficulty in creating an experiment to test a proposed theory or investigate a phenomenon in greater detail.

Can quantum mechanics and general relativity be realized as a fully consistent theory (perhaps as a quantum field theory)? Is spacetime fundamentally continuous or discrete? Would a consistent theory involve a force mediated by a hypothetical graviton, or be a product of a discrete structure of spacetime itself (as in loop quantum gravity)? Are there deviations from the predictions of general relativity at very small or very large scales or in other extreme circumstances that flow from a quantum gravity theory?

The theory of LQG is one possible solution to the problem of quantum gravity, as is string theory. There are substantial differences however. For example, string theory also addresses unification, the understanding of all known forces and particles as manifestations of a single entity, by postulating extra dimensions and so-far unobserved additional particles and symmetries. Contrary to this, LQG is based only on quantum theory and general relativity and its scope is limited to understanding the quantum aspects of the gravitational interaction. On the other hand, the consequences of LQG are radical, because they fundamentally change the nature of space and time and provide a tentative but detailed physical and mathematical picture of quantum spacetime.

Presently, no semiclassical limit recovering general relativity has been shown to exist. This means it remains unproven that LQG's description of spacetime at the Planck scale has the right continuum limit (described by general relativity with possible quantum corrections). Specifically, the dynamics of the theory is encoded in the Hamiltonian constraint, but there is no candidate Hamiltonian.[91] Other technical problems include finding off-shell closure of the constraint algebra and physical inner product vector space, coupling to matter fields of Quantum field theory, fate of the renormalization of the graviton in perturbation theory that lead to ultraviolet divergence beyond 2-loops (see One-loop Feynman diagram in Feynman diagram).[91]

While there has been a recent proposal relating to observation of naked singularities,[92] and doubly special relativity as a part of a program called loop quantum cosmology, there is no experimental observation for which loop quantum gravity makes a prediction not made by the Standard Model or general relativity (a problem that plagues all current theories of quantum gravity). Because of the above-mentioned lack of a semiclassical limit, LQG has not yet even reproduced the predictions made by general relativity.

An alternative criticism is that general relativity may be an effective field theory, and therefore quantization ignores the fundamental degrees of freedom.

28.12 See also

28.13 Notes

[1] Rovelli, Carlo (August 2008). "Loop Quantum Gravity" (PDF). CERN. Retrieved 14 September 2014.

[2] Rovelli, C. (2011). "Zakopane lectures on loop gravity". arXiv:1102.3660 [gr-qc].

[3] Muxin, H. (2011). "Cosmological constant in loop quantum gravity vertex amplitude". *Physical Review D* **84** (6): 064010. arXiv:1105.2212. Bibcode:2011PhRvD..84f4010H. doi:10.1103/PhysRevD.84.064010.

[4] Fairbairn, W. J.; Meusburger, C. (2011). "q-Deformation of Lorentzian spin foam models". arXiv:1112.2511 [gr-qc].

[5] Rovelli, C. (2004). *Quantum Gravity*. Cambridge Monographs on Mathematical Physics. p. 71. ISBN 978-0-521-83733-0.

[6] Kauffman, S.; Smolin, L. (7 April 1997). "A Possible Solution For The Problem Of Time In Quantum Cosmology". *Edge.org*. Retrieved 2014-08-20.

[7] Smolin, L. (2006). "The Case for Background Independence". In Rickles, D.; French, S.; Saatsi, J. T. *The Structural Foundations of Quantum Gravity*. Clarendon Press. pp. 196*ff*. arXiv:hep-th/0507235. ISBN 978-0-19-926969-3.

[8] Rovelli, C. (2004). *Quantum Gravity*. Cambridge Monographs on Mathematical Physics. p. 13ff. ISBN 978-0-521-83733-0.

[9] Thiemann, T. (1996). "Anomaly-free formulation of non-perturbative, four-dimensional Lorentzian quantum gravity". *Physics Letters B* **380**: 257–264. arXiv:gr-qc/9606088. Bibcode:1996PhLB..380..257T. doi:10.1016/0370-2693(96)00532-1.

[10] Baez, J.; de Muniain, J. P. (1994). *Gauge Fields, Knots and Quantum Gravity*. Series on Knots and Everything. Vol. 4. World Scientific. Part III, chapter 4. ISBN 978-981-02-1729-7.

[11] Thiemann, T. (2003). "Lectures on Loop Quantum Gravity". *Lecture Notes in Physics* **631**: 41–135. arXiv:gr-qc/0210094. Bibcode:2003LNP...631...41T. doi:10.1007/978-3-540-45230-0_3.

[12] Rovelli, C.; Smolin, L.(1988). "Knot Theory and Quantum Gravity". *Physical Review Letters* **61**(10): 1155–1958.Bib55R. doi:10.1103/PhysRevLett.61.1155.

[13] Gambini, R.; Pullin, J. (2011). *A First Course in Loop Quantum Gravity*. Oxford University Press. Section 8.2. ISBN 978-0-19-959075-9.

[14] Fernando, J.; Barbero, G.). "Reality Conditions and Ashtekar Variables: A Different Perspective". *Physical Review D* **51**: 5498–5506. arXiv:gr-qc/9410013. Bibcode:1995PhRvD..51.5498B. doi:10.1103/PhysRevD.51.5498.

[15] Fernando, J.; Barbero, G. (1995). "Real Ashtekar Variables for Lorentzian Signature Space-times". *Physical Review D* **51**: 5507–5520. arXiv:gr-qc/9410014. Bibcode:1995PhRvD..51.5507B. doi:10.1103/PhysRevD.51.5507.

[16] Bojowald, M.; Alejandro, P. "Spin Foam Quantization and Anomalies". arXiv:gr-qc/0303026 [gr-qc].

[17] Barrett, J.; Crane, L. (2000). "A Lorentzian signature model for quantum general relativity". *Classical and Quantum Gravity* **17**: 3101–3118. arXiv:gr-qc/9904025. Bibcode:2000CQGra..17.3101B. doi:10.1088/0264-9381/17/16/302..

[18] Rovelli, C.; Alesci, E. (2007). "The complete LQG propagator I. Difficulties with the Barrett–Crane vertex". *Physical Review D* **76**: 104012. arXiv:hep-th/0703074. Bibcode:2007PhRvD..76b4012B. doi:10.1103/PhysRevD.76.024012.

[19] Engle, J.; Pereira, R.; Rovelli, C. (2009). "Loop-Quantum-Gravity Vertex Amplitude". *Physical Review Letters* **99**: 161301. arXiv:0705.2388. Bibcode:2007PhRvL..99p1301E. doi:10.1103/physrevlett.99.161301.

[20] Freidal, L.; Krasnov, K. (2008). "A new spin foam model for 4D gravity". *Classical and Quantum Gravity* **25**: 125018. arXiv:0708.1595. Bibcode:2008CQGra..25l5018F. doi:10.1088/0264-9381/25/12/125018.

[21] Alesci, E.; Thiemann, T.; Zipfel, A. (2011). "Linking covariant and canonical LQG: new solutions to the Euclidean Scalar Constraint". arXiv:1109.1290.

[22] Bohm, D. (1989). *Quantum Theory*. Dover Publications. ISBN 978-0-486-65969-5.

[23] Tipler, P.; Llewellyn, R. (2008). *Modern Physics* (5th ed.). W. H. Freeman and Co. pp. 160–161. ISBN 978-0-7167-7550-8.

[24] Bohr, N. (1920). "Über die Serienspektra der Element". *Zeitschrift für Physik* **2** (5): 423–478. Bibcode:1920ZPhy....2..423B. doi:10.1007/BF01329978. (English translation in Bohr 1976, pp. 241–282)

[25] Jammer, M. (1989). *The Conceptual Development of Quantum Mechanics* (2nd ed.). Tomash Publishers. Section 3.2. ISBN 978-0-88318-617-6.

[26] Ashtekar, A.; Bombelli, L.; Corichi, A. (2005). "Semiclassical States for Constrained Systems". *Physical Review D* **72**: 025008. arXiv:hep-ph/0504114. Bibcode:2005PhRvD..72a5008C. doi:10.1103/PhysRevD.72.015008.

[27] Lewandowski, J.; Okołów, A.; Sahlmann, H.; Thiemann, T. (2005). "Uniqueness of Diffeomorphism Invariant States on Holonomy-Flux Algebras". *Communications in Mathematical Physics* **267**: 703–733. arXiv:gr-qc/0504147.Bibcode:03L. doi:10.1007/s00220-006-0100-7.

[28] Fleischhack, C. (2006). "Irreducibility of the Weyl algebra in loop quantum gravity". *Physical Review Letters* **97**: 061302. Bibcode:2006PhRvL..97f1302F. doi:10.1103/physrevlett.97.061302.

[29] Thiemann, T. (2008). *Modern Canonical General Relativity*. Cambridge Monographs on Mathematical Physics. Cambridge University Press. Section 10.6. ISBN 978-0-521-74187-3.

[30] "Partial and Complete Observables for Hamiltonian Constrained Systems". *General Relativity and Gravitation* **39**: 1891–1927. 2007. arXiv:gr-qc/0411013. Bibcode:2007GReGr..39.1891D. doi:10.1007/s10714-007-0495-2.

[31] "Partial and Complete Observables for Canonical General Relativity". *Classical and Quantum Gravity* **23**: 6155–6184. arXiv:gr-qc/0507106. Bibcode:2006CQGra..23.6155D. doi:10.1088/0264-9381/23/22/006.

[32] Dreyer, O.; Markopoulou, f.; Smolin, L. (2006). "Symmetry and entropy of black hole horizons". *Nuclear Physics B* **774**: 1–13. arXiv:hep-th/0409056. Bibcode:2006NuPhB.744....1D. doi:10.1016/j.nuclphysb.2006.02.045.

[33] Kribs, D. W.; Markopoulou, F. "Geometry from quantum particles". arXiv:gr-qc/0510052.

[34] Markopoulou, F.; Poulin, D. "Noiseless subsystems and the low energy limit of spin foam models" (unpublished).

[35] *The Phoenix Project: Master Constraint Programme for Loop Quantum Gravity*, Class.Quant.Grav.23:2211-2248,2006 or http://fr.arxiv.org/pdf/gr-qc/0305080

[36] *Modern Canonical Quantum General Relativity* by Thomas Thiemann

[37] *Testing the Master Constraint Programme for Loop Quantum Gravity I. General Framework*, Bianca Dittrich, Thomas Thiemann. Class.Quant.Grav. 23 (2006) 1025-1066.

[38] *Testing the Master Constraint Programme for Loop Quantum Gravity II. Finite Dimensional Systems*, Bianca Dittrich, Thomas Thiemann, Class.Quant.Grav. 23 (2006) 1067-1088.

[39] *Testing the Master Constraint Programme for Loop Quantum Gravity III. SL(2,R) Models*, Bianca Dittrich, Thomas Thiemann, Class.Quant.Grav. 23 (2006) 1089-1120.

[40] *Testing the Master Constraint Programme for Loop Quantum Gravity IV. Free Field Theories*, Bianca Dittrich, Thomas Thiemann, Class.Quant.Grav. 23 (2006) 1121-1142.

[41] *Testing the Master Constraint Programme for Loop Quantum Gravity V. Interacting Field Theories*, Bianca Dittrich, Thomas Thiemann, Class.Quant.Grav. 23 (2006) 1143-1162.

[42] *Quantum Spin Dynamics VIII. The Master Constraint*, Thomas Thiemann, Class.Quant.Grav. 23 (2006) 2249-2266.

[43] *Approximating the physical inner product of Loop Quantum Cosmology*, Benjamin Bahr, Thomas Thiemann, Class.Quant.09-2138,2007.

[44] *On the Relation between Operator Constraint --, Master Constraint --, Reduced Phase Space --, and Path Integral Quantisation*, Muxin Han, Thomas Thiemann, Class.Quant.Grav.27:225019,2010.

[45] *On the Relation between Rigging Inner Product and Master Constraint Direct Integral Decomposition*, Muxin Han, Thomas Thiemann, J.Math.Phys.51:092501,2010.

[46] *A Path-integral for the Master Constraint of Loop Quantum Gravity*, Muxin Han, Class.Quant.Grav.27:215009,2010

[47] *Emergent diffeomorphism invariance in a discrete loop quantum gravity model*, Rodolfo Gambini, Jorge Pullin, Class.Quant.2009

[48] Section 10.2.2 *A First Course in Loop quantum Gravity*, Rodolfo Gambinni, Jorge Pullin, Oxford University Press, first published 2011.

[49] *Algebraic Quantum Gravity (AQG) I. Conceptual Setup*, K. Giesel, T. Thiemann, Class.Quant.Grav.24:2465-2498,2007.

[50] *Algebraic Quantum Gravity (AQG) II. Semiclassical Analysis*, K. Giesel, T. Thiemann, Class.Quant.Grav.24:2499-2564,2007.

[51] *Algebraic Quantum Gravity (AQG) III. Semiclassical Perturbation Theory*, K. Giesel, T. Thiemann, Class.Quant.Grav.24:2565-2588,2007.

[52] Bousso, Raphael (2002). "The Holographic Principle". *Reviews of Modern Physics* **74** (3): 825–874. arXiv:hep-th/0203101. Bibcode:2002RvMP...74..825B. doi:10.1103/RevModPhys.74.825.

[53] Majumdar, Parthasarathi (1998). "Black Hole Entropy and Quantum Gravity" **73**. p. 147. arXiv:gr-qc/9807045.Bi47M.

[54] See List of loop quantum gravity researchers

[55] Rovelli, Carlo (1996). "Black Hole Entropy from Loop Quantum Gravity". *Physical Review Letters* **77** (16): 3288–3291. arXiv:gr-qc/9603063. Bibcode:1996PhRvL..77.3288R. doi:10.1103/PhysRevLett.77.3288.

[56] Ashtekar, Abhay; Baez, John; Corichi, Alejandro; Krasnov, Kirill (1998). "Quantum Geometry and Black Hole Entropy". *Physical Review Letters* **80** (5): 904–907. arXiv:gr-qc/9710007. Bibcode:1998PhRvL..80..904A. doi:10.1103/PhysRevLett.80.904.

[57] *Quantum horizons and black hole entropy: Inclusion of distortion and rotation*, Abhay Ashtekar, Jonathan Engle, Chris Van Den Broeck, Class.Quant.Grav.22:L27-L34, 2005.

[58] Bianchi, Eugenio (2012). "Entropy of Non-Extremal Black Holes from Loop Gravity". arXiv:1204.5122.

[59] http://inspirehep.net/record/940357?ln=en. http://inspirehep.net/record/1111991.

[60] *On (Cosmological) Singularity Avoidance in Loop Quantum Gravity*, Johannes Brunnemann, Thomas Thiemann, Class.Quant.Grav. 23 (2006) 1395-1428.

[61] *Unboundedness of Triad -- Like Operators in Loop Quantum Gravity*, Johannes Brunnemann, Thomas Thiemann, Class.Quant.Gr. 23 (2006) 1429-1484.

[62] L. Modesto, C. Rovelli:*Particle scattering in loop quantum gravity*, Phys Rev Lett 95 (2005) 191301

[63] R Oeckl, *A 'general boundary' formulation for quantum mechanics and quantum gravity*, Phys Lett B575 (2003) 318-324 : *Schrodinger's cat and the clock: lessons for quantum gravity*, Class Quant Grav 20 (2003) 5371-5380l

[64] F. Conrady, C. Rovelli *Generalized Schrodinger equation in Euclidean field theory"*, Int J Mod Phys A 19, (2004) 1-32.

[65] L Doplicher, *Generalized Tomonaga-Schwinger equation from the Hadamard formula*, Phys Rev D70 (2004) 064037

[66] F. Conrady, L. Doplicher, R. Oeckl, C. Rovelli, M. Testa. *Minkowski vacuum in background independent quantum gravity*, Phys Rev D69 (2004) 064019.

[67] http://arxiv.org/abs/arXiv:hep-th/0310077

[68] *New Variables for Classical and Quantum Gravity in all Dimensions I. Hamiltonian Analysis*, Norbert Bodendorfer, Thomas Thiemann, Andreas Thurn, Class. Quantum Grav. 30 (2013) 045001

[69] *New Variables for Classical and Quantum Gravity in all Dimensions II. Lagrangian Analysis*, Norbert Bodendorfer, Thomas Thiemann, Andreas Thurn, Quantum Grav. 30 (2013) 045002

[70] *New Variables for Classical and Quantum Gravity in all Dimensions III. Quantum Theory*, Norbert Bodendorfer, Thomas Thiemann, Andreas Thurn, Class. Quantum Grav. 30 (2013) 045003

[71] *New Variables for Classical and Quantum Gravity in all Dimensions IV. Matter Coupling*, Norbert Bodendorfer, Thomas Thiemann, Andreas Thurn, Class. Quantum Grav. 30 (2013) 045004

[72] *On the Implementation of the Canonical Quantum Simplicity Constraint*, Norbert Bodendorfer, Thomas Thiemann, Andreas Thurn, Class. Quantum Grav. 30 (2013) 045005

[73] *Towards Loop Quantum Supergravity (LQSG) I. Rarita-Schwinger Sector*, Norbert Bodendorfer, Thomas Thiemann, Andreas Thurn, Class. Quantum Grav. 30 (2013) 045006

[74] *Towards Loop Quantum Supergravity (LQSG) II. p-Form Sector*, Norbert Bodendorfer, Thomas Thiemann, Andreas Thurn, Class. Quantum Grav. 30 (2013) 045007

[75] *Towards Loop Quantum Supergravity (LQSG)*, Norbert Bodendorfer, Thomas Thiemann, Andreas Thurn, Phys. Lett. B 711: 205-211 (2012)

[76] *New Variables for Classical and Quantum Gravity in all Dimensions V. Isolated Horizon Boundary Degrees of Freedom*, Norbert Bodendorfer, Thomas Thiemann, Andreas Thurn, http://uk.arxiv.org/pdf/1304.2679.

[77] *Black hole entropy from loop quantum gravity in higher dimensions*, Norbert Bodendorfer http://uk.arxiv.org/pdf/1307.5029

[78] http://arxiv.org/abs/0708.3550

[79] http://arxiv.org/abs/1203.6164

[80] http://arxiv.org/abs/1006.0199

[81] http://arxiv.org/abs/1001.3668

[82] http://arxiv.org/abs/1212.5246

[83] http://arxiv.org/abs/1105.3480

[84] *Quantum gravity and the standard model*, Sundance O. Bilson-Thompson, Fotini Markopoulou, Lee Smolin, Class.Quant.G-3994,2007.

[85] For a precise review and outlook of this research see: *Emergent Braided Matter of Quantum Geometry*, Sundance Bilson-Thompson, Jonathan Hackett, Louis Kauffman, Yidun Wan, SIGMA 8 (2012), 014, 43 pages.

[86] *Constrained Mechanics and Noiseless Subsystems*, Tomasz Konopka, Fotini Markopoulou, arXiv:gr-qc/0601028.

[87] http://www.perimeterinstitute.ca/people/renate-loll

[88] wwnpqft.inln.cnrs.fr/pdf/Bianchi.pdf

[89] http://arxiv.org/abs/0804.0632

[90] http://arxiv.org/abs/1308.2934

[91] Nicolai, Hermann; Peeters, Kasper; Zamaklar, Marija (2005). "Loop quantum gravity: an outside view". *Classical and Quantum Gravity* **22** (19): R193–R247. arXiv:hep-th/0501114. Bibcode:2005CQGra..22R.193N. doi:10.1088/0264-9381/22/19/R01.

[92] Goswami; Joshi, Pankaj S.; Singh, Parampreet; et al. (2006). "Quantum evaporation of a naked singularity". *Physical Review Letters* **96** (3): 31302. arXiv:gr-qc/0506129. Bibcode:2006PhRvL..96c1302G. doi:10.1103/PhysRevLett.96.031302.

28.14   References

- Topical Reviews

  - Rovelli, Carlo (2011). "Zakopane lectures on loop gravity". arXiv:1102.3660.

  - Rovelli, Carlo (1998). "Loop Quantum Gravity". *Living Reviews in Relativity* **1**. Retrieved 2008-03-13.

  - Thiemann, Thomas (2003). "Lectures on Loop Quantum Gravity". *Lectures Notes in Physics*. Lecture Notes in Physics **631**: 41–135. arXiv:gr-qc/0210094. Bibcode:2003LNP...631...41T. doi:10.1007/978-3-540-45230-0_3. ISBN 978-3-540-40810-9.

  - Ashtekar, Abhay; Lewandowski, Jerzy (2004). "Background Independent Quantum Gravity: A Status Report". *Classical and Quantum Gravity* **21** (15): R53–R152. arXiv:gr-qc/0404018. Bibcode:2004CQGra..2. doi:10.1088/0264-9381/21/15/R01.

  - Carlo Rovelli and Marcus Gaul, *Loop Quantum Gravity and the Meaning of Diffeomorphism Invariance*, e-print available as gr-qc/9910079.

  - Lee Smolin, *The case for background independence*, e-print available as hep-th/0507235.

  - Alejandro Corichi, *Loop Quantum Geometry: A primer*, e-print available as .

  - Alejandro Perez, *Introduction to loop quantum gravity and spin foams*, e-print available as .

  - Hermann Nicolai and Kasper Peeters *Loop and spin foam quantum gravity: A Brief guide for beginners.*, e-print available as .

- Popular books:

    - Lee Smolin, *Three Roads to Quantum Gravity*

    - Carlo Rovelli, *Che cos'è il tempo? Che cos'è lo spazio?*, Di Renzo Editore, Roma, 2004. French translation: *Qu'est ce que le temps? Qu'est ce que l'espace?*, Bernard Gilson ed, Brussel, 2006. English translation: *What is Time? What is space?*, Di Renzo Editore, Roma, 2006.

    - Julian Barbour, *The End of Time: The Next Revolution in Our Understanding of the Universe*

    - Musser, George (2008). "The Complete Idiot's Guide to String Theory". *The Physics Teacher* (Indianapolis: Alpha) **47** (2): 368. Bibcode:2009PhTea..47Q.128H. doi:10.1119/1.3072469. ISBN 978-1-59257-702-6. – Focuses on string theory but has an extended discussion of loop gravity as well.

- Magazine articles:

    - Lee Smolin, "Atoms of Space and Time", *Scientific American*, January 2004

    - Martin Bojowald, "Following the Bouncing Universe", *Scientific American*, October 2008

- Easier introductory, expository or critical works:

    - Abhay Ashtekar, *Gravity and the quantum*, e-print available as gr-qc/0410054 (2004)

    - John C. Baez and Javier Perez de Muniain, *Gauge Fields, Knots and Quantum Gravity*, World Scientific (1994)

    - Carlo Rovelli, *A Dialog on Quantum Gravity*, e-print available as hep-th/0310077 (2003)

    - Rodolfo Gambini and Jorge Pullin, *A First Course in Loop Quantum Gravity*, Oxford (2011)

    - Carlo Rovelli and Francesca Vidotto, *Covariant Loop Quantum Gravity*, Cambridge (2014); draft available online

- More advanced introductory/expository works:

    - Carlo Rovelli, *Quantum Gravity*, Cambridge University Press (2004); draft available online

    - Thomas Thiemann, *Introduction to modern canonical quantum general relativity*, e-print available as gr-qc/0110034

    - Thomas Thiemann, *Introduction to Modern Canonical Quantum General Relativity*, Cambridge University Press (2007)

    - Abhay Ashtekar, *New Perspectives in Canonical Gravity*, Bibliopolis (1988).

    - Abhay Ashtekar, *Lectures on Non-Perturbative Canonical Gravity*, World Scientific (1991)

    - Rodolfo Gambini and Jorge Pullin, *Loops, Knots, Gauge Theories and Quantum Gravity*, Cambridge University Press (1996)

    - Hermann Nicolai, Kasper Peeters, Marija Zamaklar, *Loop quantum gravity: an outside view*, e-print available as hep-th/0501114

    - H. Nicolai and K. Peeters, *Loop and Spin Foam Quantum Gravity: A Brief Guide for Beginners*, e-print available as hep-th/0601129

    - T. Thiemann The LQG – String: Loop Quantum Gravity Quantization of String Theory (2004)

- Conference proceedings:

    - John C. Baez (ed.), *Knots and Quantum Gravity*

- Fundamental research papers:

    - Ashtekar, Abhay (1986). "New variables for classical and quantum gravity". *Physical Review Letters* **57** (18): 2244–2247. Bibcode:1986PhRvL..57.2244A. doi:10.1103/PhysRevLett.57.2244. PMID 10033673

    - Ashtekar, Abhay (1987). "New Hamiltonian formulation of general relativity". *Physical Review D* **36** (6): 1587–1602. Bibcode:1987PhRvD..36.1587A. doi:10.1103/PhysRevD.36.1587

- Roger Penrose, *Angular momentum: an approach to combinatorial space-time* in *Quantum Theory and Beyond*, ed. Ted Bastin, Cambridge University Press, 1971

- Rovelli, Carlo; Smolin, Lee (1988). "Knot theory and quantum gravity". *Physical Review Letters* **61** (10): 1155–1158. Bibcode:1988PhRvL..61.1155R. doi:10.1103/PhysRevLett.61.1155.

- Rovelli, Carlo; Smolin, Lee (1990). "Loop space representation of quantum general relativity". *Nuclear Physics* **B331**: 80–152.

- Carlo Rovelli and Lee Smolin, *Discreteness of area and volume in quantum gravity*. Nucl. Phys., **B442** (1995) 593-622, e-print available as gr-qc/9411005

- Kuchař, Karel (1973). "Canonical Quantization of Gravity". In Israel, Werner. *Relativity, Astrophysics and Cosmology*. D. Reidel. pp. 237–288. ISBN 90-277-0369-8.

- Thiemann, Thomas (2006). "Loop Quantum Gravity: An Inside View". *Approaches to Fundamental Physics*. Lecture Notes in Physics **721**: 185–263. arXiv:hep-th/0608210. Bibcode:2007LNP...721..185T. doi:10.1007/978-3-540-71117-9_10. ISBN 978-3-540-71115-5.

28.15   External links

- "Loop Quantum Gravity" by Carlo Rovelli Physics World, November 2003

- Quantum Foam and Loop Quantum Gravity

- Abhay Ashtekar: Semi-Popular Articles . Some excellent popular articles suitable for beginners about space, time, GR, and LQG.

- Loop Quantum Gravity: Lee Smolin.

- Loop Quantum Gravity on arxiv.org

- A list of LQG references catered to fresh graduates

- Loop Quantum Gravity Lectures Online by Lee Smolin

- Spin networks, spin foams and loop quantum gravity

- Wired magazine, News: *Moving Beyond String Theory*

- April 2006 Scientific American Special Issue, *A Matter of Time*, has Lee Smolin LQG Article *Atoms of Space and Time*

- September 2006, The Economist, article *Looping the loop*

- Gamma-ray Large Area Space Telescope: http://glast.gsfc.nasa.gov/

- Zeno meets modern science. Article from Acta Physica Polonica B by Z.K. Silagadze.

- Did pre-big bang universe leave its mark on the sky? - According to a model based on "loop quantum gravity" theory, a parent universe that existed before ours may have left an imprint (*New Scientist*, 10 April 2008)

Chapter 29

Gauge theory

For a more accessible and less technical introduction to this topic, see Introduction to gauge theory.

In physics, a **gauge theory** is a type of field theory in which the Lagrangian is invariant under a continuous group of local transformations.

The term *gauge* refers to redundant degrees of freedom in the Lagrangian. The transformations between possible gauges, called *gauge transformations*, form a Lie group—referred to as the *symmetry group* or the *gauge group* of the theory. Associated with any Lie group is the Lie algebra of group generators. For each group generator there necessarily arises a corresponding vector field called the *gauge field*. Gauge fields are included in the Lagrangian to ensure its invariance under the local group transformations (called *gauge invariance*). When such a theory is quantized, the quanta of the gauge fields are called *gauge bosons*. If the symmetry group is non-commutative, the gauge theory is referred to as *non-abelian*, the usual example being the Yang–Mills theory.

Many powerful theories in physics are described by Lagrangians that are invariant under some symmetry transformation groups. When they are invariant under a transformation identically performed at *every* point in the space in which the physical processes occur, they are said to have a global symmetry. The requirement of local symmetry, the cornerstone of gauge theories, is a stricter constraint. In fact, a global symmetry is just a local symmetry whose group's parameters are fixed in space-time.

Gauge theories are important as the successful field theories explaining the dynamics of elementary particles. Quantum electrodynamics is an abelian gauge theory with the symmetry group $U(1)$ and has one gauge field, the electromagnetic four-potential, with the photon being the gauge boson. The Standard Model is a non-abelian gauge theory with the symmetry group $U(1) \times SU(2) \times SU(3)$ and has a total of twelve gauge bosons: the photon, three weak bosons and eight gluons.

Gauge theories are also important in explaining gravitation in the theory of general relativity. Its case is somewhat unique in that the gauge field is a tensor, the Lanczos tensor. Theories of quantum gravity, beginning with gauge gravitation theory, also postulate the existence of a gauge boson known as the graviton. Gauge symmetries can be viewed as analogues of the principle of general covariance of general relativity in which the coordinate system can be chosen freely under arbitrary diffeomorphisms of spacetime. Both gauge invariance and diffeomorphism invariance reflect a redundancy in the description of the system. An alternative theory of gravitation, gauge theory gravity, replaces the principle of general covariance with a true gauge principle with new gauge fields.

Historically, these ideas were first stated in the context of classical electromagnetism and later in general relativity. However, the modern importance of gauge symmetries appeared first in the relativistic quantum mechanics of electrons – quantum electrodynamics, elaborated on below. Today, gauge theories are useful in condensed matter, nuclear and high energy physics among other subfields.

29.1   History and importance

The earliest field theory having a gauge symmetry was Maxwell's formulation of electrodynamics in 1864. The importance of this symmetry remained unnoticed in the earliest formulations. Similarly unnoticed, Hilbert had derived the Einstein field equations by postulating the invariance of the action under a general coordinate transformation. Later Hermann Weyl, in an attempt to unify general relativity and electromagnetism, conjectured that *Eichinvarianz* or invariance under the change of scale (or "gauge") might also be a local symmetry of general relativity. After the development of quantum mechanics, Weyl, Vladimir Fock and Fritz London modified gauge by replacing the scale factor with a complex quantity and turned the scale transformation into a change of phase, which is a U(1) gauge symmetry. This explained the electromagnetic field effect on the wave function of a charged quantum mechanical particle. This was the first widely recognised gauge theory, popularised by Pauli in the 1940s.[1]

In 1954, attempting to resolve some of the great confusion in elementary particle physics, Chen Ning Yang and Robert Mills introduced **non-abelian gauge theories** as models to understand the strong interaction holding together nucleons in atomic nuclei. (Ronald Shaw, working under Abdus Salam, independently introduced the same notion in his doctoral thesis.) Generalizing the gauge invariance of electromagnetism, they attempted to construct a theory based on the action of the (non-abelian) SU(2) symmetry group on the isospin doublet of protons and neutrons. This is similar to the action of the U(1) group on the spinor fields of quantum electrodynamics. In particle physics the emphasis was on using **quantized gauge theories**.

This idea later found application in the quantum field theory of the weak force, and its unification with electromagnetism in the electroweak theory. Gauge theories became even more attractive when it was realized that non-abelian gauge theories reproduced a feature called asymptotic freedom. Asymptotic freedom was believed to be an important characteristic of strong interactions. This motivated searching for a strong force gauge theory. This theory, now known as quantum chromodynamics, is a gauge theory with the action of the SU(3) group on the color triplet of quarks. The Standard Model unifies the description of electromagnetism, weak interactions and strong interactions in the language of gauge theory.

In the 1970s, Sir Michael Atiyah began studying the mathematics of solutions to the classical Yang–Mills equations. In 1983, Atiyah's student Simon Donaldson built on this work to show that the differentiable classification of smooth 4-manifolds is very different from their classification up to homeomorphism. Michael Freedman used Donaldson's work to exhibit exotic $\mathbf{R}^4$s, that is, exotic differentiable structures on Euclidean 4-dimensional space. This led to an increasing interest in gauge theory for its own sake, independent of its successes in fundamental physics. In 1994, Edward Witten and Nathan Seiberg invented gauge-theoretic techniques based on supersymmetry that enabled the calculation of certain topological invariants (the Seiberg–Witten invariants). These contributions to mathematics from gauge theory have led to a renewed interest in this area.

The importance of gauge theories in physics is exemplified in the tremendous success of the mathematical formalism in providing a unified framework to describe the quantum field theories of electromagnetism, the weak force and the strong force. This theory, known as the Standard Model, accurately describes experimental predictions regarding three of the four fundamental forces of nature, and is a gauge theory with the gauge group SU(3) × SU(2) × U(1). Modern theories like string theory, as well as general relativity, are, in one way or another, gauge theories.

See Pickering[2] for more about the history of gauge and quantum field theories.

29.2   Description

29.2.1   Global and local symmetries

In physics, the mathematical description of any physical situation usually contains excess degrees of freedom; the same physical situation is equally well described by many equivalent mathematical configurations. For instance, in Newtonian dynamics, if two configurations are related by a Galilean transformation (an inertial change of reference frame) they represent the same physical situation. These transformations form a group of "symmetries" of the theory, and a physical situation corresponds not to an individual mathematical configuration but to a class of configurations related to one another by this symmetry group.

This idea can be generalized to include local as well as global symmetries, analogous to much more abstract "changes of coordinates" in a situation where there is no preferred "inertial" coordinate system that covers the entire physical system. A gauge theory is a mathematical model that has symmetries of this kind, together with a set of techniques for making physical predictions consistent with the symmetries of the model.

29.2.2   Example of global symmetry

When a quantity occurring in the mathematical configuration is not just a number but has some geometrical significance, such as a velocity or an axis of rotation, its representation as numbers arranged in a vector or matrix is also changed by a coordinate transformation. For instance, if one description of a pattern of fluid flow states that the fluid velocity in the neighborhood of $(x=1, y=0)$ is 1 m/s in the positive x direction, then a description of the same situation in which the coordinate system has been rotated clockwise by 90 degrees states that the fluid velocity in the neighborhood of $(x=0, y=1)$ is 1 m/s in the positive y direction. The coordinate transformation has affected both the coordinate system used to identify the *location* of the measurement and the basis in which its *value* is expressed. As long as this transformation is performed globally (affecting the coordinate basis in the same way at every point), the effect on values that represent the *rate of change* of some quantity along some path in space and time as it passes through point P is the same as the effect on values that are truly local to P.

29.2.3   Use of fiber bundles to describe local symmetries

In order to adequately describe physical situations in more complex theories, it is often necessary to introduce a "coordinate basis" for some of the objects of the theory that do not have this simple relationship to the coordinates used to label points in space and time. (In mathematical terms, the theory involves a fiber bundle in which the fiber at each point of the base space consists of possible coordinate bases for use when describing the values of objects at that point.) In order to spell out a mathematical configuration, one must choose a particular coordinate basis at each point (a *local section* of the fiber bundle) and express the values of the objects of the theory (usually "fields" in the physicist's sense) using this basis. Two such mathematical configurations are equivalent (describe the same physical situation) if they are related by a transformation of this abstract coordinate basis (a change of local section, or *gauge transformation*).

In most gauge theories, the set of possible transformations of the abstract gauge basis at an individual point in space and time is a finite-dimensional Lie group. The simplest such group is U(1), which appears in the modern formulation of quantum electrodynamics (QED) via its use of complex numbers. QED is generally regarded as the first, and simplest, physical gauge theory. The set of possible gauge transformations of the entire configuration of a given gauge theory also forms a group, the *gauge group* of the theory. An element of the gauge group can be parameterized by a smoothly varying function from the points of spacetime to the (finite-dimensional) Lie group, such that the value of the function and its derivatives at each point represents the action of the gauge transformation on the fiber over that point.

A gauge transformation with constant parameter at every point in space and time is analogous to a rigid rotation of the geometric coordinate system; it represents a global symmetry of the gauge representation. As in the case of a rigid rotation, this gauge transformation affects expressions that represent the rate of change along a path of some gauge-dependent quantity in the same way as those that represent a truly local quantity. A gauge transformation whose parameter is *not* a constant function is referred to as a local symmetry; its effect on expressions that involve a derivative is qualitatively different from that on expressions that don't. (This is analogous to a non-inertial change of reference frame, which can produce a Coriolis effect.)

29.2.4   Gauge fields

The "gauge covariant" version of a gauge theory accounts for this effect by introducing a gauge field (in mathematical language, an Ehresmann connection) and formulating all rates of change in terms of the covariant derivative with respect to this connection. The gauge field becomes an essential part of the description of a mathematical configuration. A configuration in which the gauge field can be eliminated by a gauge transformation has the property that its field strength (in mathematical language, its curvature) is zero everywhere; a gauge theory is *not* limited to these configurations. In other words, the distinguishing characteristic of a gauge theory is that the gauge field does not merely compensate for a poor choice of coordinate system; there is generally no gauge transformation that makes the gauge field vanish.

When analyzing the dynamics of a gauge theory, the gauge field must be treated as a dynamical variable, similarly to other objects in the description of a physical situation. In addition to its interaction with other objects via the covariant derivative, the gauge field typically contributes energy in the form of a "self-energy" term. One can obtain the equations for the gauge theory by:

- starting from a naïve ansatz without the gauge field (in which the derivatives appear in a "bare" form);

- listing those global symmetries of the theory that can be characterized by a continuous parameter (generally an abstract equivalent of a rotation angle);

- computing the correction terms that result from allowing the symmetry parameter to vary from place to place; and

- reinterpreting these correction terms as couplings to one or more gauge fields, and giving these fields appropriate self-energy terms and dynamical behavior.

This is the sense in which a gauge theory "extends" a global symmetry to a local symmetry, and closely resembles the historical development of the gauge theory of gravity known as general relativity.

29.2.5   Physical experiments

Gauge theories are used to model the results of physical experiments, essentially by:

- limiting the universe of possible configurations to those consistent with the information used to set up the experiment, and then

- computing the probability distribution of the possible outcomes that the experiment is designed to measure.

The mathematical descriptions of the "setup information" and the "possible measurement outcomes" (loosely speaking, the "boundary conditions" of the experiment) are generally not expressible without reference to a particular coordinate system, including a choice of gauge. (If nothing else, one assumes that the experiment has been adequately isolated from "external" influence, which is itself a gauge-dependent statement.) Mishandling gauge dependence in boundary conditions is a frequent source of anomalies in gauge theory calculations, and gauge theories can be broadly classified by their approaches to anomaly avoidance.

29.2.6   Continuum theories

The two gauge theories mentioned above (continuum electrodynamics and general relativity) are examples of continuum field theories. The techniques of calculation in a continuum theory implicitly assume that:

- given a completely fixed choice of gauge, the boundary conditions of an individual configuration can in principle be completely described;

- given a completely fixed gauge and a complete set of boundary conditions, the principle of least action determines a unique mathematical configuration (and therefore a unique physical situation) consistent with these bounds;

- the likelihood of possible measurement outcomes can be determined by:

    - establishing a probability distribution over all physical situations determined by boundary conditions that are consistent with the setup information,

    - establishing a probability distribution of measurement outcomes for each possible physical situation, and

- convolving these two probability distributions to get a distribution of possible measurement outcomes consistent with the setup information; and

- fixing the gauge introduces no anomalies in the calculation, due either to gauge dependence in describing partial information about boundary conditions or to incompleteness of the theory.

These assumptions are close enough to be valid across a wide range of energy scales and experimental conditions, to allow these theories to make accurate predictions about almost all of the phenomena encountered in daily life, from light, heat, and electricity to eclipses and spaceflight. They fail only at the smallest and largest scales (due to omissions in the theories themselves) and when the mathematical techniques themselves break down (most notably in the case of turbulence and other chaotic phenomena).

29.2.7 Quantum field theories

Other than these classical continuum field theories, the most widely known gauge theories are quantum field theories, including quantum electrodynamics and the Standard Model of elementary particle physics. The starting point of a quantum field theory is much like that of its continuum analog: a gauge-covariant action integral that characterizes "allowable" physical situations according to the principle of least action. However, continuum and quantum theories differ significantly in how they handle the excess degrees of freedom represented by gauge transformations. Continuum theories, and most pedagogical treatments of the simplest quantum field theories, use a gauge fixing prescription to reduce the orbit of mathematical configurations that represent a given physical situation to a smaller orbit related by a smaller gauge group (the global symmetry group, or perhaps even the trivial group).

More sophisticated quantum field theories, in particular those that involve a non-abelian gauge group, break the gauge symmetry within the techniques of perturbation theory by introducing additional fields (the Faddeev–Popov ghosts) and counterterms motivated by anomaly cancellation, in an approach known as BRST quantization. While these concerns are in one sense highly technical, they are also closely related to the nature of measurement, the limits on knowledge of a physical situation, and the interactions between incompletely specified experimental conditions and incompletely understood physical theory . The mathematical techniques that have been developed in order to make gauge theories tractable have found many other applications, from solid-state physics and crystallography to low-dimensional topology.

29.3 Classical gauge theory

29.3.1 Classical electromagnetism

Historically, the first example of gauge symmetry discovered was classical electromagnetism. In electrostatics, one can either discuss the electric field, **E**, or its corresponding electric potential, V. Knowledge of one makes it possible to find the other, except that potentials differing by a constant, $V \to V + C$, correspond to the same electric field. This is because the electric field relates to *changes* in the potential from one point in space to another, and the constant C would cancel out when subtracting to find the change in potential. In terms of vector calculus, the electric field is the gradient of the potential, $\mathbf{E} = -\nabla V$. Generalizing from static electricity to electromagnetism, we have a second potential, the vector potential **A**, with

$$\mathbf{E} = -\nabla V - \frac{\partial \mathbf{A}}{\partial t}$$
$$\mathbf{B} = \nabla \times \mathbf{A}$$

The general gauge transformations now become not just $V \to V + C$ but

$$\mathbf{A} \to \mathbf{A} + \nabla f$$
$$V \to V - \frac{\partial f}{\partial t}$$

where f is any function that depends on position and time. The fields remain the same under the gauge transformation, and therefore Maxwell's equations are still satisfied. That is, Maxwell's equations have a gauge symmetry.

29.3.2   An example: Scalar O(n) gauge theory

The remainder of this section requires some familiarity with classical or quantum field theory, and the use of Lagrangians.

Definitions in this section: gauge group, gauge field, interaction Lagrangian, gauge boson.

The following illustrates how local gauge invariance can be "motivated" heuristically starting from global symmetry properties, and how it leads to an interaction between originally non-interacting fields.

Consider a set of n non-interacting real scalar fields, with equal masses m. This system is described by an action that is the sum of the (usual) action for each scalar field φ_i

$$S = \int \mathrm{d}^4 x \sum_{i=1}^{n} \left[\frac{1}{2} \partial_\mu \varphi_i \partial^\mu \varphi_i - \frac{1}{2} m^2 \varphi_i^2 \right]$$

The Lagrangian (density) can be compactly written as

$$\mathcal{L} = \frac{1}{2} (\partial_\mu \Phi)^T \partial^\mu \Phi - \frac{1}{2} m^2 \Phi^T \Phi$$

by introducing a vector of fields

$$\Phi = (\varphi_1, \varphi_2, \ldots, \varphi_n)^T$$

The term ∂_μ is Einstein notation for the partial derivative of Φ in each of the four dimensions.

It is now transparent that the Lagrangian is invariant under the transformation

$$\Phi \mapsto \Phi' = G\Phi$$

whenever G is a *constant* matrix belonging to the n-by-n orthogonal group O(n). This is seen to preserve the Lagrangian, since the derivative of Φ transforms identically to Φ and both quantities appear inside dot products in the Lagrangian (orthogonal transformations preserve the dot product).

$$(\partial_\mu \Phi) \mapsto (\partial_\mu \Phi)' = G \partial_\mu \Phi$$

This characterizes the *global* symmetry of this particular Lagrangian, and the symmetry group is often called the **gauge group**; the mathematical term is **structure group**, especially in the theory of G-structures. Incidentally, Noether's theorem implies that invariance under this group of transformations leads to the conservation of the *currents*

$$J_\mu^a = i \partial_\mu \Phi^T T^a \Phi$$

where the T^a matrices are generators of the SO(n) group. There is one conserved current for every generator.

Now, demanding that this Lagrangian should have *local* O(n)-invariance requires that the G matrices (which were earlier constant) should be allowed to become functions of the space-time coordinates x.

In this case, the G matrices do not "pass through" the derivatives, when $G = G(x)$.

$$\partial_\mu (G\Phi) \neq G(\partial_\mu \Phi)$$

The failure of the derivative to commute with "G" introduces an additional term (in keeping with the product rule), which spoils the invariance of the Lagrangian. In order to rectify this we define a new derivative operator such that the derivative of Φ again transforms identically with Φ

$$(D_\mu \Phi)' = GD_\mu \Phi$$

This new "derivative" is called a (gauge) covariant derivative and takes the form

$$D_\mu = \partial_\mu + ig A_\mu$$

Where g is called the coupling constant; a quantity defining the strength of an interaction. After a simple calculation we can see that the **gauge field** $A(x)$ must transform as follows

$$A'_\mu = GA_\mu G^{-1} + \frac{i}{g}(\partial_\mu G)G^{-1}$$

The gauge field is an element of the Lie algebra, and can therefore be expanded as

$$A_\mu = \sum_a A^a_\mu T^a$$

There are therefore as many gauge fields as there are generators of the Lie algebra.

Finally, we now have a *locally gauge invariant* Lagrangian

$$\mathcal{L}_{\text{loc}} = \frac{1}{2}(D_\mu \Phi)^T D^\mu \Phi - \frac{1}{2}m^2 \Phi^T \Phi$$

Pauli uses the term *gauge transformation of the first type* to mean the transformation of Φ, while the compensating transformation in A is called a *gauge transformation of the second type*.

The difference between this Lagrangian and the original *globally gauge-invariant* Lagrangian is seen to be the **interaction Lagrangian**

$$\mathcal{L}_{\text{int}} = i\frac{g}{2}\Phi^T A^T_\mu \partial^\mu \Phi + i\frac{g}{2}(\partial_\mu \Phi)^T A^\mu \Phi - \frac{g^2}{2}(A_\mu \Phi)^T A^\mu \Phi$$

This term introduces interactions between the n scalar fields just as a consequence of the demand for local gauge invariance. However, to make this interaction physical and not completely arbitrary, the mediator $A(x)$ needs to propagate in space. That is dealt with in the next section by adding yet another term, $\mathcal{L}_{\text{gf}}$, to the Lagrangian. In the quantized version of the obtained classical field theory, the quanta of the gauge field $A(x)$ are called gauge bosons. The interpretation of the interaction Lagrangian in quantum field theory is of scalar bosons interacting by the exchange of these gauge bosons.

29.3.3 The Yang–Mills Lagrangian for the gauge field

Main article: Yang–Mills theory

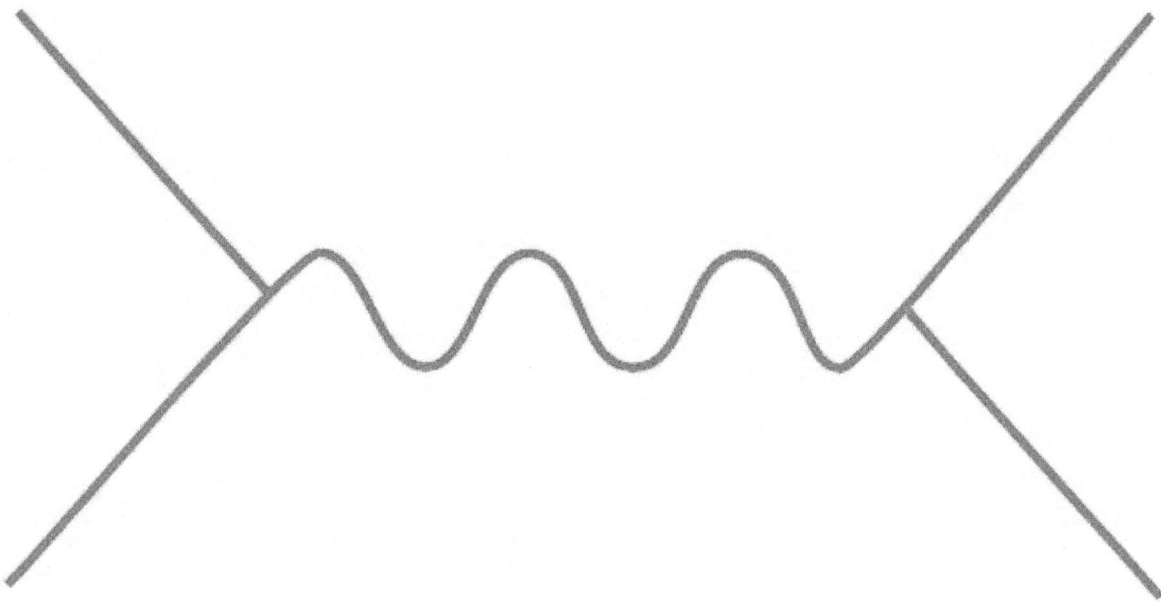

Feynman diagram of scalar bosons interacting via a gauge boson

The picture of a classical gauge theory developed in the previous section is almost complete, except for the fact that to define the covariant derivatives D, one needs to know the value of the gauge field $A(x)$ at all space-time points. Instead of manually specifying the values of this field, it can be given as the solution to a field equation. Further requiring that the Lagrangian that generates this field equation is locally gauge invariant as well, one possible form for the gauge field Lagrangian is (conventionally) written as

$$\mathcal{L}_{\mathrm{gf}} = -\frac{1}{2}\,\mathrm{Tr}(F^{\mu\nu}F_{\mu\nu})$$

with

$$F_{\mu\nu} = \frac{1}{ig}[D_\mu, D_\nu]$$

and the trace being taken over the vector space of the fields. This is called the **Yang–Mills action**. Other gauge invariant actions also exist (e.g., nonlinear electrodynamics, Born–Infeld action, Chern–Simons model, theta term, etc.).

Note that in this Lagrangian term there is no field whose transformation counterweighs the one of A. Invariance of this term under gauge transformations is a particular case of *a priori* classical (geometrical) symmetry. This symmetry must be restricted in order to perform quantization, the procedure being denominated gauge fixing, but even after restriction, gauge transformations may be possible.[3]

The complete Lagrangian for the gauge theory is now

$$\mathcal{L} = \mathcal{L}_{\mathrm{loc}} + \mathcal{L}_{\mathrm{gf}} = \mathcal{L}_{\mathrm{global}} + \mathcal{L}_{\mathrm{int}} + \mathcal{L}_{\mathrm{gf}}$$

29.3.4 An example: Electrodynamics

As a simple application of the formalism developed in the previous sections, consider the case of electrodynamics, with only the electron field. The bare-bones action that generates the electron field's Dirac equation is

$$S = \int \bar{\psi}(i\hbar c\, \gamma^\mu \partial_\mu - mc^2)\psi \, \mathrm{d}^4 x$$

The global symmetry for this system is

$$\psi \mapsto e^{i\theta}\psi$$

The gauge group here is U(1), just rotations of the phase angle of the field, with the particular rotation determined by the constant θ.

"Localising" this symmetry implies the replacement of θ by $\theta(x)$. An appropriate covariant derivative is then

$$D_\mu = \partial_\mu - i\frac{e}{\hbar} A_\mu$$

Identifying the "charge" e (not to be confused with the mathematical constant e in the symmetry description) with the usual electric charge (this is the origin of the usage of the term in gauge theories), and the gauge field $A(x)$ with the four-vector potential of electromagnetic field results in an interaction Lagrangian

$$\mathcal{L}_{\text{int}} = \frac{e}{\hbar} \bar{\psi}(x)\gamma^\mu \psi(x) A_\mu(x) = J^\mu(x) A_\mu(x)$$

where $J^\mu(x)$ is the usual four vector electric current density. The gauge principle is therefore seen to naturally introduce the so-called minimal coupling of the electromagnetic field to the electron field.

Adding a Lagrangian for the gauge field $A_\mu(x)$ in terms of the field strength tensor exactly as in electrodynamics, one obtains the Lagrangian used as the starting point in quantum electrodynamics.

$$\mathcal{L}_{\text{QED}} = \bar{\psi}(i\hbar c\, \gamma^\mu D_\mu - mc^2)\psi - \frac{1}{4\mu_0} F_{\mu\nu} F^{\mu\nu}$$

See also: Dirac equation, Maxwell's equations, Quantum electrodynamics

29.4 Mathematical formalism

Gauge theories are usually discussed in the language of differential geometry. Mathematically, a *gauge* is just a choice of a (local) section of some principal bundle. A **gauge transformation** is just a transformation between two such sections.

Although gauge theory is dominated by the study of connections (primarily because it's mainly studied by high-energy physicists), the idea of a connection is not central to gauge theory in general. In fact, a result in general gauge theory shows that affine representations (i.e., affine modules) of the gauge transformations can be classified as sections of a jet bundle satisfying certain properties. There are representations that transform covariantly pointwise (called by physicists gauge transformations of the first kind), representations that transform as a connection form (called by physicists gauge transformations of the second kind, an affine representation)—and other more general representations, such as the B field in BF theory. There are more general nonlinear representations (realizations), but these are extremely complicated. Still, nonlinear sigma models transform nonlinearly, so there are applications.

If there is a principal bundle P whose base space is space or spacetime and structure group is a Lie group, then the sections of P form a principal homogeneous space of the group of gauge transformations.

Connections (gauge connection) define this principal bundle, yielding a covariant derivative ∇ in each associated vector bundle. If a local frame is chosen (a local basis of sections), then this covariant derivative is represented by the connection form A, a Lie algebra-valued 1-form, which is called the **gauge potential** in physics. This is evidently not an intrinsic but a frame-dependent quantity. The curvature form F, a Lie algebra-valued 2-form that is an intrinsic quantity, is constructed from a connection form by

$$\mathbf{F} = d\mathbf{A} + \mathbf{A} \wedge \mathbf{A}$$

where d stands for the exterior derivative and $\wedge$ stands for the wedge product. ($\mathbf{A}$ is an element of the vector space spanned by the generators T^a, and so the components of $\mathbf{A}$ do not commute with one another. Hence the wedge product $\mathbf{A} \wedge \mathbf{A}$ does not vanish.)

Infinitesimal gauge transformations form a Lie algebra, which is characterized by a smooth Lie-algebra-valued scalar, ε. Under such an infinitesimal gauge transformation,

$$\delta_\varepsilon \mathbf{A} = [\varepsilon, \mathbf{A}] - d\varepsilon$$

where $[\cdot, \cdot]$ is the Lie bracket.

One nice thing is that if $\delta_\varepsilon X = \varepsilon X$, then $\delta_\varepsilon DX = \varepsilon DX$ where D is the covariant derivative

$$DX \stackrel{\text{def}}{=} dX + \mathbf{A}X$$

Also, $\delta_\varepsilon \mathbf{F} = \varepsilon \mathbf{F}$, which means $\mathbf{F}$ transforms covariantly.

Not all gauge transformations can be generated by infinitesimal gauge transformations in general. An example is when the base manifold is a compact manifold without boundary such that the homotopy class of mappings from that manifold to the Lie group is nontrivial. See instanton for an example.

The *Yang–Mills action* is now given by

$$\frac{1}{4g^2} \int \mathrm{Tr}[*F \wedge F]$$

where * stands for the Hodge dual and the integral is defined as in differential geometry.

A quantity which is **gauge-invariant** (i.e., invariant under gauge transformations) is the Wilson loop, which is defined over any closed path, γ, as follows:

$$\chi^{(\rho)} \left(\mathcal{P} \left\{ e^{\oint_\gamma A} \right\} \right)$$

where χ is the character of a complex representation ρ and $\mathcal{P}$ represents the path-ordered operator.

29.5 Quantization of gauge theories

Main article: Quantum gauge theory

Gauge theories may be quantized by specialization of methods which are applicable to any quantum field theory. However, because of the subtleties imposed by the gauge constraints (see section on Mathematical formalism, above) there are many technical problems to be solved which do not arise in other field theories. At the same time, the richer structure of gauge theories allows simplification of some computations: for example Ward identities connect different renormalization constants.

29.5.1 Methods and aims

The first gauge theory quantized was quantum electrodynamics (QED). The first methods developed for this involved gauge fixing and then applying canonical quantization. The Gupta–Bleuler method was also developed to handle this problem. Non-abelian gauge theories are now handled by a variety of means. Methods for quantization are covered in the article on quantization.

The main point to quantization is to be able to compute quantum amplitudes for various processes allowed by the theory. Technically, they reduce to the computations of certain correlation functions in the vacuum state. This involves a renormalization of the theory.

When the running coupling of the theory is small enough, then all required quantities may be computed in perturbation theory. Quantization schemes intended to simplify such computations (such as canonical quantization) may be called **perturbative quantization schemes**. At present some of these methods lead to the most precise experimental tests of gauge theories.

However, in most gauge theories, there are many interesting questions which are non-perturbative. Quantization schemes suited to these problems (such as lattice gauge theory) may be called **non-perturbative quantization schemes**. Precise computations in such schemes often require supercomputing, and are therefore less well-developed currently than other schemes.

29.5.2 Anomalies

Some of the symmetries of the classical theory are then seen not to hold in the quantum theory; a phenomenon called an **anomaly**. Among the most well known are:

- The scale anomaly, which gives rise to a *running coupling constant*. In QED this gives rise to the phenomenon of the Landau pole. In Quantum Chromodynamics (QCD) this leads to asymptotic freedom.

- The chiral anomaly in either chiral or vector field theories with fermions. This has close connection with topology through the notion of instantons. In QCD this anomaly causes the decay of a pion to two photons.

- The gauge anomaly, which must cancel in any consistent physical theory. In the electroweak theory this cancellation requires an equal number of quarks and leptons.

29.6 Pure gauge

A pure gauge is the set of field configurations obtained by a gauge transformation on the null-field configuration, i.e., a gauge-transform of zero. So it is a particular "gauge orbit" in the field configuration's space.

Thus, in the abelian case, where $A_\mu(x) \to A'_\mu(x) = A_\mu(x) + \partial_\mu f(x)$, the pure gauge is just the set of field configurations $A'_\mu(x) = \partial_\mu f(x)$ for all $f(x)$.

29.7 See also

29.8 References

[1] Wolfgang Pauli (1941) "Relativistic Field Theories of Elementary Particles," *Rev. Mod. Phys.* **13**: 203–32.

[2] Pickering, A. (1984). *Constructing Quarks*. University of Chicago Press. ISBN 0-226-66799-5.

[3] Sakurai, *Advanced Quantum Mechanics*, sect 1–4

29.9 Bibliography

General readers

- Schumm, Bruce (2004) *Deep Down Things*. Johns Hopkins University Press. Esp. chpt. 8. A serious attempt by a physicist to explain gauge theory and the Standard Model with little formal mathematics.

Texts

- Bromley, D.A. (2000). *Gauge Theory of Weak Interactions*. Springer. ISBN 3-540-67672-4.

- Cheng, T.-P.; Li, L.-F. (1983). *Gauge Theory of Elementary Particle Physics*. Oxford University Press. ISBN 0-19-851961-3.

- Frampton, P. (2008). *Gauge Field Theories* (3rd ed.). Wiley-VCH.

- Kane, G.L. (1987). *Modern Elementary Particle Physics*. Perseus Books. ISBN 0-201-11749-5.

Articles

- Becchi, C. (1997). "Introduction to Gauge Theories". p. 5211. arXiv:hep-ph/9705211. Bibcode:1997hep.p1B.

- Gross, D. (1992). "Gauge theory – Past, Present and Future" (PDF). Retrieved 2009-04-23.

- Jackson, J.D. (2002). "From Lorenz to Coulomb and other explicit gauge transformations". *Am.J.Phys* **70** (9): 917–928. arXiv:physics/0204034. Bibcode:2002AmJPh..70..917J. doi:10.1119/1.1491265.

- Svetlichny, George (1999). "Preparation for Gauge Theory". p. 2027. arXiv:math-ph/9902027. Bibcode:17S.

29.10 External links

- Hazewinkel, Michiel, ed. (2001), "Gauge transformation", *Encyclopedia of Mathematics*, Springer, ISBN 978-1-55608-010-4

- Yang–Mills equations on DispersiveWiki

- Gauge theories on Scholarpedia

Chapter 30

Dimensionless physical constant

In physics, a **dimensionless physical constant**, sometimes called **fundamental physical constant**, is a physical constant that is dimensionless – having no units attached, having a numerical value that is the same under all possible systems of units. A common example is the fine-structure constant α, with approximate value $7.29735257 \times 10^{-3}$.[1]

The term *fundamental physical constant* has also been used to refer to universal but dimensioned physical constants such as the speed of light c, vacuum permittivity ε_0, Planck's constant h, or the gravitational constant G.[2] However, the numerical values of these constants are not fundamental, since they are dependent on the units used to express them. Increasingly, physicists reserve the use of the term *fundamental physical constant* for dimensionless physical constants that cannot be derived from any other source.

30.1 Introduction

The numerical values of dimensional physical constants depend on the units used to express these physical constants. As such it is possible to define a basis set of units so that selected dimensional physical constants are normalized to 1 solely because of the choice of units. The basis set may consist of time, length, mass, charge, and temperature, or an equivalent set. A choice of units is called a system of units.

For example, the International System of Units (SI) is a system of units solely defined as convenient. The numerical values of dimensional physical constants have no natural significance. The Planck units define a system of natural units so that the numerical values of the vacuum speed of light, the universal gravitational constant, and the constants of Planck, Coulomb, and Boltzmann are unity. Because, merely from the choice of units, these five dimensional physical constants disappear from equations of physical laws, they are considered not fundamental in an operationally distinguishable sense.[3][4]

In contrast, the numerical values of dimensionless physical constants are independent of the units used. These constants cannot be eliminated by any choice of a system of units. Such constants include:

- α, the fine structure constant, the coupling constant for the electromagnetic interaction ($\approx 1/137.036$). Also the square of the electron charge, expressed in Planck units, which defines the scale of charge of elementary particles with charge.

- μ or β, the proton-to-electron mass ratio, the rest mass of the proton divided by that of the electron (≈ 1836.15). More generally, the ratio of the rest masses of any pair of elementary particles.

- α_s, the coupling constant for the strong force (≈ 1)

- αG, the gravitational coupling constant ($\approx 10^{-45}$) which is the square of the electron mass, expressed in Planck units. This defines the scale of the masses of elementary particles and has also been used to express the relative strength of gravitation.

Unlike mathematical constants, the values of the dimensionless fundamental physical constants cannot be calculated; they are determined only by physical measurement. This is one of the unsolved problems of physics.

One of the dimensionless fundamental constants is the fine structure constant:

$$\alpha = \frac{e^2}{\hbar c \, 4\pi\varepsilon_0} \approx \frac{1}{137.03599908},$$

where e is the elementary charge, $\hbar$ is the reduced Planck's constant, c is the speed of light in a vacuum, and ε_0 is the permittivity of free space. The fine structure constant is fixed to the strength of the electromagnetic force. At low energies, $\alpha \approx 1/137$, whereas at the scale of the Z boson, about 90 GeV, one measures $\alpha \approx 1/127$. There is no accepted theory explaining the value of α; Richard Feynman elaborates:

> There is a most profound and beautiful question associated with the observed coupling constant, e – the amplitude for a real electron to emit or absorb a real photon. It is a simple number that has been experimentally determined to be close to 0.08542455. (My physicist friends won't recognize this number, because they like to remember it as the inverse of its square: about 137.03597 with an uncertainty of about 2 in the last decimal place. It has been a mystery ever since it was discovered more than fifty years ago, and all good theoretical physicists put this number up on their wall and worry about it.) Immediately you would like to know where this number for a coupling comes from: is it related to pi or perhaps to the base of natural logarithms? Nobody knows. It's one of the greatest damn mysteries of physics: a magic number that comes to us with no understanding by man. You might say the "hand of God" wrote that number, and "we don't know how He pushed his pencil." We know what kind of a dance to do experimentally to measure this number very accurately, but we don't know what kind of dance to do on the computer to make this number come out, without putting it in secretly!
> — Richard Feynman, Richard P. Feynman (1985). *QED: The Strange Theory of Light and Matter*. Princeton University Press. p. 129. ISBN 0-691-08388-6.

The analog of the fine structure constant for gravitation is the gravitational coupling constant. This constant requires the arbitrary choice of a pair of objects having mass. The electron and proton are natural choices because they are stable, and their properties are well measured and well understood. If αG is calculated from two protons, its value is $\approx 10^{-38}$.

The list of dimensionless physical constants increases in length whenever experiments measure new relationships between physical phenomena. The list of fundamental dimensionless constants, however, decreases when advances in physics show how some previously known constant can be computed in terms of others. A long-sought goal of theoretical physics is to find first principles from which all of the fundamental dimensionless constants can be calculated and compared to the measured values. A successful "Theory of Everything" would allow such a calculation, but so far, this goal has remained elusive.

30.2   Constants in the standard model and in cosmology

The original standard model of particle physics from the 1970s contained 19 fundamental dimensionless constants describing the masses of the particles and the strengths of the electroweak and strong forces. In the 1990s, neutrinos were discovered to have nonzero mass, and a quantity called the vacuum angle was found to be indistinguishable from zero.

The complete standard model requires 25 fundamental dimensionless constants (Baez, 2011). At present, their numerical values are not understood in terms of any widely accepted theory and are determined only from measurement. These 25 constants are:

- the fine structure constant;

- the strong coupling constant;

- fifteen masses of the fundamental particles (relative to the Planck mass Mp=1.22089(6)×10^{19} GeV/c^2), namely:

  - six quarks

  - six leptons

  - the Higgs boson

  - the W boson

  - the Z boson

- four parameters of the CKM matrix, describing how quarks oscillate between different forms;

- four parameters of the Pontecorvo–Maki–Nakagawa–Sakata matrix, which does the same thing for neutrinos.

One constant is required for cosmology:

- the cosmological constant (in terms of Planck units) of Einstein's equations for general relativity, having a value of approximately 10^{-122}.

Thus, currently there are 26 known fundamental dimensionless physical constants. However, this number may not be the final one. For example, if neutrinos turn out to be Majorana fermions, the Maki-Nakagawa-Sakata matrix has two additional parameters. Secondly, if dark matter is discovered, or if the description of dark energy requires more than the cosmological constant, further fundamental constants will be needed.

30.3 Well-known subsets

Certain dimensionless constants are discussed more frequently than others.

30.3.1 Barrow and Tipler

Barrow and Tipler (1986) anchor their broad-ranging discussion of astrophysics, cosmology, quantum physics, teleology, and the anthropic principle in the fine structure constant, the proton-to-electron mass ratio (which they, along with Barrow (2002), call β), and the coupling constants for the strong force and gravitation.

30.3.2 Martin Rees's Six Numbers

Martin Rees, in his book *Just Six Numbers*, mulls over the following six dimensionless constants, whose values he deems fundamental to present-day physical theory and the known structure of the universe:

- $N \approx 10^{36}$: the ratio of the fine structure constant (the dimensionless coupling constant for electromagnetism) to the gravitational coupling constant, the latter defined using two protons. In Barrow and Tipler (1986) and elsewhere in Wikipedia, this ratio is denoted $\alpha/\alpha G$. N governs the relative importance of gravity and electrostatic attraction/repulsion in explaining the properties of baryonic matter;[5]

- $\varepsilon \approx 0.007$: The fraction of the mass of four protons that is released as energy when fused into a helium nucleus. ε governs the energy output of stars, and is determined by the coupling constant for the strong force;[6]

- $\Omega \approx 0.3$: the ratio of the actual density of the universe to the critical (minimum) density required for the universe to eventually collapse under its gravity. Ω determines the ultimate fate of the universe. If $\Omega>1$, the universe will experience a Big Crunch. If $\Omega < 1$, the universe will expand forever;[5]

- $\lambda \approx 0.7$: The ratio of the energy density of the universe, due to the cosmological constant, to the critical density of the universe. Others denote this ratio by Ω_Λ ;[7]

- $Q \approx 10^{-5}$: The energy required to break up and disperse an instance of the largest known structures in the universe, namely a galactic cluster or supercluster, expressed as a fraction of the energy equivalent to the rest mass m of that structure, namely mc^2;[8]

- $D = 3$: the number of macroscopic spatial dimensions.

N and ε govern the fundamental interactions of physics. The other constants (D excepted) govern the size, age, and expansion of the universe. These five constants must be estimated empirically. D, on the other hand, is necessarily a nonzero natural number and cannot be measured. Hence most physicists would not deem it a dimensionless physical constant of the sort discussed in this entry.

Any plausible fundamental physical theory must be consistent with these six constants, and must either derive their values from the mathematics of the theory, or accept their values as empirical.

30.4   Calculation attempts

No formulae for the fundamental physical constants are known to this day.

The mathematician Simon Plouffe has made an extensive search of computer databases of mathematical formulae, seeking formulae for the mass ratios of the fundamental particles.

One example of numerology is by the astrophysicist Arthur Eddington. He set out alleged mathematical reasons why the reciprocal of the fine structure constant had to be exactly 136. When its value was discovered to be closer to 137, he changed his argument to match that value. Experiments have since shown that Eddington was wrong; to six significant digits, the reciprocal of the fine-structure constant is 137.036.

An empirical relation between the masses of the electron, muon and tau has been discovered by physicist Yoshio Koide, but this formula remains unexplained.

30.5   See also

- Cabibbo–Kobayashi–Maskawa matrix (Cabibbo angle)

- Coupling constant

- Dimensionless quantity

- Fine-structure constant

- Gravitational coupling constant

- Neutrino oscillation

- Physical cosmology

- Standard Model

- Weinberg angle

- Fine-tuned Universe

- Koide formula

30.6 References

[1] "CODATA Value: fine-structure constant". *The NIST Reference on Constants, Units, and Uncertainty*. US National Institute of Standards and Technology. June 2015. Retrieved 2015-09-25. External link in |work= (help)

[2] http://physics.nist.gov/cuu/Constants/ NIST

[3] Michael Duff (2002) "Comment on time-variation of fundamental constants"

[4] Michael Duff, O. Okun and Gabriele Veneziano (2002) "Trialogue on the number of fundamental constants." *Journal of High Energy Physics* **3**: 023.

[5] Rees, M. (2000), p. .

[6] Rees, M. (2000), p. 53.

[7] Rees, M. (2000), p. 110.

[8] Rees, M. (2000), p. 118.

30.7 Bibliography

- Martin Rees, 1999. *Just Six Numbers: The Deep Forces that Shape the Universe*. London: Weidenfeld & Nicolson. ISBN 0-7538-1022-0

30.8 External articles

General

- John D. Barrow, 2002. *The Constants of Nature; From Alpha to Omega – The Numbers that Encode the Deepest Secrets of the Universe*. Pantheon Books. ISBN 0-375-42221-8.

- Barrow, John D.; Tipler, Frank J. (1988). *The Anthropic Cosmological Principle*. Oxford University Press. ISBN 978-0-19-282147-8. LCCN 87028148.

- Michio Kaku, 1994. *Hyperspace: A Scientific Odyssey Through Parallel Universes, Time Warps, and the Tenth Dimension*. Oxford University Press.

- Fundamental Physical Constants from NIST

- Values of fundamental constants. CODATA, 2002.

- John Baez, 2002, "How Many Fundamental Constants Are There?"

- Plouffe, Simon, 2004, "A search for a mathematical expression for mass ratios using a large database."

Articles on variance of the fundamental constants

- John Bahcall, Charles Steinhardt, and David Schlegel, 2004, "Does the fine-structure constant vary with cosmological epoch?" *Astrophys. J.* **600**: 520.

- John D. Barrow and Webb, J. K., "Inconstant Constants – Do the inner workings of nature change with time?" *Scientific American* (June 2005).

- Michael Duff, 2002 "Comment on time-variation of fundamental constants."

- Marion, H., et al. 2003, "A search for variations of fundamental constants using atomic fountain clocks," *Phys.* *90*: 150801.

- Martins, J.A.P. et al., 2004, "WMAP constraints on varying α and the promise of reionization," *Phys.Lett.* *B585*: 29–34.

- Olive, K.A., et al., 2002, "Constraints on the variations of the fundamental couplings," *Phys.Rev.* *D66*: 045022.

- Uzan, J-P, 2003, "The fundamental constants and their variation: observational status and theoretical motivations," *Rev.Mod.Phys.* *75*: 403.

- Webb, J.K. et al., 2001, "Further evidence for cosmological evolution of the fine-structure constant," *Phys. Rev. Lett.* *87*: 091301.

30.9 Text and image sources, contributors, and licenses

30.9.1 Text

- **Theory of everything** *Source:* https://en.wikipedia.org/wiki/Theory_of_everything?oldid=683741317 *Contributors:* AxelBoldt, Paul Drye, CYD, The Anome, Eclecticology, Toby Bartels, Roadrunner, Zippy, Stevertigo, Lorenzarius, Michael Hardy, Rojelague, Nixdorf, Takuya-Murata, Karada, Skysmith, Kosebamse, CesarB, Anders Feder, Angela, Julesd, Salsa Shark, Ugen64, Poor Yorick, Evercat, Schneelocke, Feedmecereal, Timwi, Dcoetzee, Dysprosia, Jitse Niesen, Wik, Jakenelson, Omegatron, Raul654, Nnh, Kevin M C Harkess, UninvitedCompany, Fredrik, Altenmann, Nurg, Naddy, Gandalf61, Mirv, Academic Challenger, Rursus, Blainster, Caknuck, Wereon, Diberri, Pengo, Tobias Bergemann, Hooloovoo, Ancheta Wis, Dbenbenn, Mporter, Jabra, Ferkelparade, Bfinn, Xerxes314, Curps, Alison, FeloniousMonk, McGravin, Behnam, Gzornenplatz, JRR Trollkien, Steuard, Andycjp, Sonjaaa, Antandrus, Kim54, Tomruen, Lumidek, Gseshoyru, WpZurp, TJSwoboda, Zondor, JimJast, Discospinster, Rich Farmbrough, H0riz0n, Pjacobi, Vsmith, Pluke, Autiger, Mal-enwiki, Pavel Vozenilek, Floorsheim, El C, Lycurgus, Sourcecode, Oldsoul, PhilHibbs, Sietse Snel, Jpgordon, Atraxani, Smalljim, Slicky, LostLeviathan, Matpitka, Juesch, Danski14, Alansohn, Gary, DariuszT, ShardPhoenix, Kocio, Pion, Hdeasy, Bart133, Schaefer, BanyanTree, ClockworkSoul, Tycho, Suruena, Count Iblis, DV8 2XL, Gene Nygaard, Euphrosyne, Squidwina, Ott, Siafu, Roylee, Woohookitty, Mindmatrix, RHaworth, TigerShark, Savantnavas, Mr-Darcy, Mpatel, GregorB, Athletec64, Christopher Thomas, Aarghdvaark, Ashmoo, BD2412, Drbogdan, Rjwilmsi, Kinu, Strait, Lordsatri, Dennis Estenson II, HappyCamper, LjL, Bubba73, The wub, Yamamoto Ichiro, JohnDBuell, FayssalF, ColinJF, Wragge, Windchaser, Musical Linguist, Mindloss, RexNL, Gurch, Pete.Hurd, Lmatt, Diza, Zayani, Spencerk, Chobot, Sharkface217, DVdm, Hmonroe, Bgwhite, Ptah-enwiki, Ugha, Wavelength, Hillman, StuffOfInterest, Phantomsteve, Arado, John Smith's, Zigamorph, SpuriousQ, Jobe457, Stephenb, Cambridge-BayWeather, Rsrikanth05, Vibritannia, Neilbeach, Salsb, Big Brother 1984, Anomalocaris, NawlinWiki, Joncolvin, ErkDemon, Trovatore, ETTan, Schrei, THB, Syrthiss, Wknight94, Richardcavell, FF2010, CWenger, Kevin, Caco de vidro, Katieh5584, Banus, Sbyrnes321, Narkstraws, SmackBot, R.E. Freak, Kurochka, DuoDeathscyther 02, Bayardo, McGeddon, Delldot, Kintetsubuffalo, Portillo, Rmosler2100, Bluebot, Jjalexand, 7777777s, Silly rabbit, George Church, Colonies Chris, A. B., Calc rulz, Nicknitro71, Zsinj, TallyJoe, John Hyams, Jamse, Scott3, Jefffire, Serenity-Fr, Bilgrau, Avb, Rrburke, Addshore, DrL, Mr.LMNOP, Rassisi, Spanyard, Byelf2007, Nishkid64, Giovanni33, Soap, Cronholm144, Loadmaster, Stupid Corn, Benjaminlobato, FredrickS, SirFozzie, Waggers, Alexander Gieg, Gcavep, Abel Cavaşi, Newone, Courcelles, Tubezone, Esn, Dave Runger, Valoem, JRSpriggs, Kurtan-enwiki, 0-8, Duduong, Friendly Neighbour, CRGreathouse, Geremia, Tkoeppe, Ken Gallager, DepartedUser2, Cydebot, Vanished user 2340rujowierfj08234irjwfw4, Ninguém, Steel, Peterdjones, Hebrides, David edwards, Michael C Price, Raoul NK, Wortzman, Ulnevets, Konradek, Mojo Hand, Raymond Feilner, Headbomb, Marek69, Inve40, Twejr, Duncan McB, KrakatoaKatie, Luna Santin, Gdo01, Byrgenwulf, Myanw, Knotwork, Len Raymond, JAnDbot, Barek, MER-C, Txomin, Inks.LWC, Matthew Fennell, Instinct, MoralMajority, Promking, Bongwarrior, VoABot II, JamesBWatson, JBKramer, DAGwyn, Theroadislong, Lenschulwitz, 28421u2232nfenfcenc, Peatbog, Allstarecho, Fang 23, Spellmaster, Philg88, Peter J Schoen, Denis tarasov, MartinBot, R'n'B, JCarlos, J.delanoy, Pharaoh of the Wizards, Maurice Carbonaro, LordAnubisBOT, Pyrospirit, AntiSpamBot, NewEnglandYankee, DadaNeem, Cometstyles, WJBscribe, Foofighter20x, Econofire, Squids and Chips, Germanium, Reelrt, ChaosCon343, Danwills, RingtailedFox, Jeff G., TXiKiBoT, Nxavar, Rei-bot, Vishal144, IllaZilla, Pouya sh, Corvus cornix, Michael H 34, Martin451, Cheffoxx, Betanon, BotKung, Everything counts, Popopp, MrMelonhead, Stephenmolesey, James McBride, Deanlsinclair, Pageman-enwiki, Monty845, Logan, Kpa4941, PaddyLeahy, Dogah, SieBot, Tiddly Tom, Robdunst, Wing gundam, Gammanon, Bentogoa, Likebox, Tiptoety, SteakN-Shake, Momo san, Freeman501, BartekChom, Monkeyspangler, Lightmouse, Anakin101, Divinestuff, Carbogen, Ayleuss, Soporaeternus, ArepoEn, ClueBot, LAX, Cliff, Ian the Aussie, Monomath1, Boing! said Zebedee, Heldbacktheband, LonelyBeacon, Neverquick, Excirial, WikiZorro, Tamaratrouts, Wndl42, Brews ohare, PhySusie, Morel, Mastertek, Mikaey, 7, Crowsnest, Thinking Stone, TimothyRias, Pat-Dunphey, JKeck, XLinkBot, Bvssvni, Ougner, Truthnlove, YeAaMsLtA, Thatguyflint, Tayste, Balungifrancis, Addbot, Proofreader77, Some jerk on the Internet, Uruk2008, DOI bot, Couchie, Johnchang6868, Discrepancy, Mjamja, Bobtron5000, Fluffernutter, KaityJoe, MrOllie, Favonian, Barak Sh, F Notebook, Tide rolls, Scientryst, WikiDreamer Bot, Meisam, Blah28948, Yobot, Finiter, Ptbotgourou, Ezequiels.90, Jgmoxness, Amble, Mirandamir, RDemelo, AnomieBOT, ^musaz, Girl Scout cookie, 9258fahsflkh917fas, Theunify, Anxfisa, Kanat Abildinov, Materialscientist, Citation bot, Subhajit Ganguly, Fleaman5000, Amareto2, Addihockey10, Smk65536, Mlpearc, GrouchoBot, Rwmeo, Omnipaedista, Shirik, RibotBOT, Fa.alt3r3g0, Fsdjfsdfk, Chaseroads, 敏敏, FrescoBot, Paine Ellsworth, Ribashka, Steven Avraham Rosten, PhysicsExplorer, Ottokar-enwiki, Tank hasmukh Khimjibhai, Tank theorist of everything, Hasmukh Khimjibhai Tank, DivineAlpha, Citation bot 1, Gil987, Three887, Tom.Reding, A8UDI, NarSakSasLee, Casimir9999, AndrewGrieder, Aknochel, IVAN3MAN, SchreyP, Noel Edward, Natwatchmaker, Weedwhacker128, Suffusion of Yellow, Koozedine, RjwilmsiBot, Specal ops, Afteread, DASHBot, Golumbo, EmausBot, Ikerus, Katherine, Dewritech, RA0808, K6ka, Zero939, Thecheesykid, Hhhippo, CanonLawJunkie, Traxs7, Arbnos, SporkBot, DanielBurnstein, FinalRapture, Aatu Koskensilta, Staszek Lem, Sridattadev, M00se1989, Wiggles007, Andrushkkutza, Maschen, Vedoder, Donner60, GIAN PHIL, Davidaedwards, WHF Christie, Terra Novus, Matevz91, Isocliff, Sanno89, Cgt, Will Beback Auto, ClueBot NG, Stein Sivertsen, ClaudeDes, Lord God Almighty, Hindustanilanguage, Helpful Pixie Bot, Nightingale.zj, B21O303V3941W42371, Bibcode Bot, Wiki13, Akashankitjain, Neutral current, Aranea Mortem, Stimulieconomy, Steven.w.kowalski, MathewTownsend, Flyerbri, GroupT, Megajakeroo, La marts boys, Zofo, LightandDark2000, Joseph1404, Nickhwee, Davidyevgeny, Kingcircle, Vith Nix, Illuusio, Davidyevgenyroven, QuantumNico, Vladimir Leonov, Friek555, HesterShaw, Sol1, Phaedrx, Jwratner1, Jmassion, HeymynamesJon, Bigfootrobert, Elitousson, Mdsheraj, Kdmeaney, JaconaFrere, Somecdnguy4, Monkbot, LollyBear12, StacyPoyPie, Mujii loving, Mayojohns, Gronk Oz, Yoyosami, Hakan tomaşoğlu, Pfpguy, Cirksena, Svm sudhan, 39Debangshu, Quantalogos, KasparBot, Christos Theopoulos, Patrickmantonio, Quackriot and Anonymous: 519

- **Theory of everything (philosophy)** *Source:* https://en.wikipedia.org/wiki/Theory_of_everything_(philosophy)?oldid=672050122 *Contributors:* RK, Goethean, Blainster, Alison, Bastique, Tim Smith, Tabor, M Alan Kazlev, Calréfa Wéná, BD2412, Windchaser, Kmorozov, Spencerk, Fogelmatrix, Trovatore, Bobet, Bluebot, Klichka, Cybercobra, Lapaz, Alexander Gieg, Xinyu, Whyvern, Gregbard, Peterdjones, AntiVandalBot, Byrgenwulf, PhilKnight, SlamDiego, Wilberg, Athaenara, Mikael Häggström, Enix150, Squids and Chips, Truthnlove, Muzekal Mike, Ronhjones, OlEnglish, Yobot, Materialscientist, Citation bot, Omnipaedista, DrilBot, Thecheesykid, Donner60, ClueBot NG, Snotbot, Stimulieconomy, Cawhee, Guymillion, Uqbarianer and Anonymous: 22

- **Chronology of the universe** *Source:* https://en.wikipedia.org/wiki/Chronology_of_the_universe?oldid=683100637 *Contributors:* AxelBoldt, Bryan Derksen, The Anome, Roadrunner, Heron, Modemac, Edward, Patrick, Boud, PhilipMW, Earth, Looxix-enwiki, Mkweise, Ahoerstemeier, Jebba, Darkwind, EdH, Timwi, A1r, Reddi, Doradus, Zoicon5, Maximus Rex, Topbanana, Lord Emsworth, TravelingDude, Chrisjj,

Flockmeal, BenRG, Robbot, Astronautics–enwiki, Jotomicron, Gandalf61, Raeky, Tobias Bergemann, Ancheta Wis, Giftlite, Lethe, Bkonrad, BrendanRyan, Eequor, Edcolins, OldakQuill, Coldacid, Opera hat, Piotrus, CSTAR, Tothebarricades.tk, Icairns, JDoolin, Lumidek, Darksun, Njh@bandsman.co.uk, D6, FT2, Pjacobi, ArnoldReinhold, Dave souza, Dbachmann, El C, Art LaPella, Drhex, CDN99, Bobo192, Army1987, Evolauxia, Maurreen, I9Q79oL78KiL0QTFHgyc, La goutte de pluie, Holdek, Pearle, Demi, *Kat*, RJFJR, Geraldshields11, Vuo, Gene Nygaard, Mindmatrix, FeanorStar7, Jeff3000, Duncan.france, Colin Watson, Joke137, Alan Canon, Christopher Thomas, Richard-Weiss, Surnôle, Zalasur, Drbogdan, Rjwilmsi, Zbxgscqf, Captain Disdain, Mike Peel, SeanMack, Krash, Yamamoto Ichiro, Nihiltres, Harmil, Who, SouthernNights, Pathoschild, Mathrick, Masnevets, Wjfox2005, YurikBot, Spacepotato, Mushin, Vuvar1, Jimp, TheDoober, Ericorbit, Stephenb, NawlinWiki, Siddiqui, Chrisbrl88, Raven4x4x, Pudist, Dbfirs, Bota47, Superiority, Leptictidium, Lenis, Enormousdude, Aremisasling, Arthur Rubin, Tevildo, GraemeL, Peyna, CWenger, ScoutNZ, Kevin, Geoffrey.landis, Albert ip, RG2, Hirudo, Paul Erik, GrinBot–enwiki, Serendipodous, KnightRider–enwiki, SmackBot, RDBury, MrDemeanour, Mehranwahid, Ashill, WildElf, AnOddName, David G Brault, Gilliam, BirdValiant, Saros136, JRSP, Bluebot, Stevenwagner, MrDrBob, Mergatroidal, Colonies Chris, Scwlong, Vladislav, Hve, Chlewbot, Andy120290, Jgoulden, Huon, Emre D., John D. Croft, Pwjb, Yevgeny Kats, CIS, SashatoBot, Lambiam, Swatjester, Writtenonsand, Count Caspian, Ckatz, Ex nihil, JHunterJ, Stwalkerster, SQGibbon, Hypnosifl, JeffW, Vanished user, Pegasus1138, Chetvorno, Fairyhairycarpetfluff, CWY2190, MaxEnt, Ramitmahajan, LouisBB, Gogo Dodo, Huysman, Michael C Price, Abtract, Thijs!bot, Fournax, Quarkchord, Sopranos-mob781, Headbomb, Najro, Marek69, James086, Second Quantization, Peter Gulutzan, AntiVandalBot, Obiwankenobi, Prolog, DarkAudit, Jdfekete, Shlomi Hillel, JAnDbot, MER-C, Xeno, Rothorpe, Ophion, Dward Fardhard, VoABot II, Coredumperror, Catgut, Cgingold, Tenacious221, DinoBot, MartinBot, Anaxial, CommonsDelinker, Leyo, J.delanoy, Love Krittaya, ARTE, Clemm, Madhava 1947, STBotD, Fuenfundachtzig, GrahamHardy, Idioma-bot, Sheliak, Black Kite, VolkovBot, TreasuryTag, Jmrowland, Cosmic Latte, Andrius.v, Clarince63, Seraphim, UnitedStatesian, Madhero88, Hanjabba, Richwil, James McBride, Hibou21, MCTales, Alemaeonid, NHRHS2010, Thw1309, Weezymagic, Llehctimj, Momo san, Oxymoron83, Globaleducator, Steven Crossin, Sean.hoyland, Namiluko, Denisarona, Escape Orbit, Martarius, ClueBot, Snigbrook, Taffboyz, Gawaxay, Wanderer57, Wysprgr2005, Cp111, Nondescripts, TheOldJacobite, CounterVandalismBot, Thomas Kist, ChandlerMapBot, Rprpr, DragonBot, Excirial, Gnome de plume, CrazyChemGuy, Eeekster, Scog, Soccerguy7735, Panos84, Subash.chandran007, DumZiBoT, XLinkBot, Tarlneustaedter, Scottman162, Jmacwiki, Aunt Entropy, CosmologyProfessor, Maldek, Addbot, Some jerk on the Internet, Medich1985, Spiderbo13, Ersik, E.pajer, Markkujn, 84user, F Notebook, Tide rolls, 봄봄, Whatismetric, Yinweichen, Yobot, TaBOT-zerem, THEN WHO WAS PHONE?, Robert Treat, AnomieBOT, Materialscientist, Citation bot, LilHelpa, Obersachsebot, HenryCorp, Wikikone, ProtectionTaggingBot, Omnipaedista, The Wiki Octopus, Smallman12q, Kikuyu3, Sae1962, Oashi, Citation bot 1, Dogaru Florin, Pinethicket, Tom.Reding, Naturehead, Seryo93, Saayiit, Extra999, Canuck100, JLincoln, Orca54, DeltaAce, Michaelbernal, Carlborden, DexDor, Tesseract2, EmausBot, GoingBatty, Rowan casey, Solomonfromfinland, Hhhippo, John Cline, Cogiati, Susfele, Medeis, Thedropsoffire, Mikeyfresh1556, Makecat, Wagino 20100516, Kirothereaper, RockMagnetist, Pandeist, AUN4, ClueBot NG, Astrocog, Jack Greenmaven, Gilderien, Piast93, Bped1985, James childs, Snotbot, Frietjes, Tinyplanetmusic, Cpluhar, Finchy0, Apopp8333, Bibcode Bot, Lowercase sigmabot, Rarelight, Cadiomals, Harizotoh9, Andi2011, Duxwing, Amphibio, ShizlGzngar, Justincheng12345-bot, Soulbust, Avneref, Wjs64, Mixert, 069952497a, SomeFreakOnTheInternet, AlbertoEspancho, IvanderClarent, Jwratner1, Kogge, SpeedEvil, Concord hioz, Monkbot, Sofia Koutsouveli, Wikipedian 2, Officer Sid X, Julietdeltalima, OmoiEgaite, Tetra quark, Xytreyum, Isambard Kingdom, Therewasnoothernameguys1, KSFT, Esadri21 and Anonymous: 347

- **Theoretical physics** *Source:* https://en.wikipedia.org/wiki/Theoretical_physics?oldid=666209419 *Contributors:* Zundark, The Anome, Ed Poor, XJaM, SimonP, D, Michael Hardy, Lexor, TakuyaMurata, Karada, Ellywa, Reddi, Silvonen, Tpbradbury, Grendelkhan, Bevo, Rbellin, SD6-Agent, Robbot, Moink, Mor–enwiki, Ancheta Wis, Giftlite, Everyking, Edcolins, Antandrus, Vina, Karol Langner, Boojum, Lumidek, Eep², Brianjd, D6, Jayjg, Stevenmattern, Discospinster, Khh3rd, CDN99, Bobo192, Ntmatter, I9Q79oL78KiL0QTFHgyc, Kjkolb, Pearle, Phils, Ranveig, Gary, Yhr, BryanD, Kocio, Knowledge Seeker, Jheald, Dirac1933, H2g2bob, Oleg Alexandrov, The Wordsmith, Jok2000, Waldir, Sjakkalle, Matt Deres, FlaBot, Margosbot–enwiki, Introvert, Chobot, Roboto de Ajvol, YurikBot, Wavelength, Crazytales, Member, NawlinWiki, Mipadi, Ragesoss, Rbarreira, Bota47, Googl, Tzustrategy, 21655, Deville, Jules.LT, Msuzen, Reyk, MaNeMeBasat, Jmeden2000, SmackBot, Unyoyega, David Shear, KocjoBot–enwiki, Jagged 85, HalfShadow, Ohnoitsjamie, Doublestein, Chris the speller, Silly rabbit, SchfiftyThree, Rdt–enwiki, Yttire, Voyajer, Thisisbossi, Stevemidgley, Cybercobra, Coolbho3000, Steve Pucci, Sadi Carnot, Andrei Stroe, Derek farn, SS2005, Cronholm144, Loodog, Pat Payne, Matt489, Sreekanthv, Arne Saknussemm–enwiki, Mike Doughney, Theoryinpractice–enwiki, Raystorm, Emote, Xod, Nnp, Myasuda, Alton, Brian Josephson, Gregbard, Equendil, AndrewHowse, Meno25, Gogo Dodo, ST47, DumbBOT, Epbr123, Aravindashwin, Impreziv, Headbomb, Mathforms05, AntiVandalBot, Seaphoto, Autocracy, Salgueiro–enwiki, Vivek Vyas, Canadian-Bacon, Husond, Kigali1, Ericoides, Adjwilley, Yill577, Clementvidal, VoABot II, P64, Kuyabribri, Jpod2, Etale, Schnieder, Joshua Davis, Neonblak, MartinBot, Ktsordia, Simaalusmani, Hans Dunkelberg, Maurice Carbonaro, Hodja Nasreddin, Moonaysl, DarkFalls, AntiSpamBot, NewEnglandYankee, Fountains of Bryn Mawr, Cmichael, Sailor1889, Idioma-bot, VolkovBot, Jeff G., TXiKiBoT, BotKung, Gnf1, SiliconSlick.I.Shmoove, Rknasc, Fieldworld, Newbyguesses, SieBot, Rockstone35, DBishop1984, JerrySteal, MaynardClark, The-G-Unit-Boss, BenoniBot–enwiki, JL-Bot, LoveMed, Not Ross Almighty, Martarius, ClueBot, Darthveda, Mild Bill Hiccup, Lordtzer0, Addisonstrack, SamuelTheGhost, Atzatzatz, Djr32, IMNTU, Micheal54, Ducarmont, Versus22, Crowsnest, Vanished user uih38riiw4hjlsd, TimothyRias, Dulcet man, Tyler Bronfstein, MidwestGeek, Daniel andersson, Truthnlove, T.M.M. Dowd, Addbot, Yoenit, AdrianW.Elder, CanadianLinuxUser, MrOllie, Michaelwuzthere, Acecrack, Nate-sama, Tassedethe, Censusgray, DubaiTerminator, Lightbot, Totorotroll, Zorrobot, Kein Einstein, Yndurain, Legobot, Luckas-bot, Yobot, Ptbotgourou, Uugedsaz, Tamtamar, AnomieBOT, Götz, JackieBot, Piano non troppo, Hades1228, Materialscientist, Soldarat, Bci2, LovesMacs, Xqbot, Tasudrty, Drilnoth, Acebulf, Bjorn.kallen, Francine Rogers, Wizardist, QMarion II, Strapuff, Nokkenbuer, GT5162, FrescoBot, Ironboy11, Снифт, Ottokar–enwiki, Ammarsakaji1967, Citation bot 1, Ractogon, Jeff Nixon, Matt1011025a, Calmer Waters, RedBot, TobeBot, TBG1997, Mathewsyriac, EmausBot, Dr Aaij, DrGordonJFreeman, Tdodds, SammyG1234, AvicAWB, Falxix, Mrmekz, Jakabob, Maschen, Anita5192, SusikMkr, Jacobean2213, BG19bot, Ymblanter, Acmedogs, Gqu, Cadiomals, Altaïr, Will Gladstone, Physicsch, Ema–or, MahdiBot, DanielMurdoch, Camyoung54, Nigellwh, Noyster, Aristarchus11, KasparBot and Anonymous: 290

- **List of unsolved problems in physics** *Source:* https://en.wikipedia.org/wiki/List_of_unsolved_problems_in_physics?oldid=682963757 *Contributors:* CYD, Zundark, The Anome, Ed Poor, XJaM, Roadrunner, SimonP, Hephaestos, Twilsonb, Stevertigo, Boud, PhilipMW, Ken Arromdee, Gabbe, Ixfd64, Yann, William M. Connolley, Julesd, Glenn, RadRafe, Bogdangiusca, Schneelocke, Feedmecereal, Timwi, Reddi, Mjklin, Lord Kenneth, Tpbradbury, Fvincent, Nnh, BenRG, Noeckel, Shantavira, Paranoid, Scott McNay, Lowellian, Gandalf61, Merovingian, Alan Liefting, Enochlau, Ancheta Wis, DavidCary, Dratman, Jason Quinn, Tddpirate, Andycjp, Nova77, Geni, LiDaobing, ConradPino, Pcarbonn, Antandrus, Beland, Ot, Vina, Karol Langner, Icairns, Tzarius, Deglr6328, Grstain, JTN, Urvabara, Discospinster, StephanKetz, Deep-

WhatamIdoing, Eldumpo, Allstarecho, User A1, Mollwollfumble, Chris G, Archen-enwiki, Thompson.matthew, STBot, Mermaid from the Baltic Sea, Shentino, Mschel, CommonsDelinker, Pbroks13, J.delanoy, DrKiernan, R. Baley, Numbo3, Leafsfan85, Lantonov, M C Y 1008, Mathlabster, Zedmelon, Aboutmovies, C quest000, Tcisco, Marrilpet, Aatomic1, Potatoswatter, Kolja21, Lseixas, Rémih, Caracalocelot, DemonicInfluence, Sheliak, Deor, Part Deux, JohnBlackburne, Philip Trueman, TXiKiBoT, Coder Dan, GimmeBot, Gombo, Hqb, Rei-bot, IPSOS, Qxz, T dotfing, Molinogi, Fizzackerly, JhsBot, Leafyplant, Geometry guy, Ilyushka88, Thebigbendizzle, SwordSmurf, Andy Dingley, Gabrielsleitao, Lamro, Antixt, Vector Potential, James-Chin, Arcfrk, Ccheese4, StevenJohnston, Katzmik, YohanN7, Dnarby, SieBot, Tiddly Tom, Work permit, Yintan, RadicalOne, Wizzard2k, SteakNShake, Arbor to SJ, Babareddeer, JSpung, Phil Bridger, Wmpearl, Oxymoron83, Henry Delforn (old), Csloomis, Thehotelambush, Lightmouse, BrightRoundCircle, OpTioNiGhT, The-G-Unit-Boss, Emgg, AWeishaupt, Divinestuff, Coldcreation, Adam Cuerden, Duae Quartunciae, Heptarchy of teh Anglo-Saxons, baby, Randomblue, TFCforever, Danthewhale, Martarius, Sfan00 IMG, ClueBot, The Thing That Should Not Be, Rjd0060, Metaprimer, Wwheaton, Der Golem, JTBX, TheAmigo42, CounterVandalismBot, Viran, Blanchardb, Rotational, Agge1000, Itzguru, Tanketz, CohesionBot, Eeekster, Stealth500, Brews ohare, NuclearWarfare, PhySusie, SockPuppetForTomruen, SchreiberBike, Another Believer, RubenGarciaHernandez, AC+79 3888, MasterOfHisOwnDomain, He6kd, TimothyRias, Lazyrussian, PseudoOne, Skarebo, NellieBly, JinJian, Truthnlove, Everydayidiot, Tayste, Balungifrancis, Addbot, Mortense, Some jerk on the Internet, Fizzycyst, DOI bot, Mistyocean3, Metagraph, Stariki, Fluffernutter, Schmoolik, MrOllie, Download, EconoPhysicist, Delaszk, Favonian, LinkFA-Bot, Tuition, Tassedethe, Nnedass, Tide rolls, Lightbot, Knutls, Luckas-bot, Ptbotgourou, Legobot II, Julia W, Trickyboarder93, Superamoeba, AnomieBOT, Kristen Eriksen, Giordano.ferdinandi, Jim1138, Jo3sampl, Materialscientist, Wandering Courier, The High Fin Sperm Whale, Citation bot, Xqbot, Stlwebs, Sionus, Amareto2, Unigfjkl, Nickkid5, Stsang, GrouchoBot, Collin21594, RibotBOT, Rucko123, GhalyBot, Acannas, LucienBOT, Paine Ellsworth, Lagelspeil, Steve Quinn, Knowandgive, Pokyrek, Citation bot 1, Citation bot 4, Electrozity8, Pinethicket, LittleWink, Jonesey95, A412, Tom.Reding, Yougeeaw, Barras, Jauhienij, Meier99, Citator, Comet Tuttle, Hughston, Defender of torch, Duoduoduo, Aribashka, libbmm, Diannaa, Earthandmoon, Tbhotch, Brambleclawx, Marie Poise, RjwilmsiBot, Aznhero3793, Ripchip Bot, EmausBot, WikitanvirBot, Immunize, Zhaskey, Fly by Night, DuKu, GoingBatty, Jmencisom, Slightsmile, Hhhippo, JSquish, ZéroBot, Cogiati, Stanford96, Empty Buffer, Sanford123456, H3llBot, Quondum, REkaxkjdsc, Monterey Bay, Mr little irish, TonyMath, Brandmeister, Maschen, Puffin, Carmichael, Newstv11, RockMagnetist, Sona11235, WizardofCalculus, Milk Coffee, Whoop whoop pull up, Mjbmrbot, Helpsome, ClueBot NG, Manubot, Hagenfeldt, This lousy T-shirt, SusikMkr, Ggonzalm, Jj1236, Mgvongoeden, Snotbot, Widr, Jamester234, Pluma, Ginger.spice14, Bibcode Bot, Jeraphine Gryphon, Lowercase sigmabot, Quarkgluonsoup, Bolatbek, Marsambe, Amp71, Mark Arsten, Lovepool1220, Marsambe1, Benzband, ENG.F.Younis, 123matt123, DeviantFrog, IrishDevil2, F=q(E+v^B), Egbertus2, Harizotoh9, Doctor Lipschitz, Snow Blizzard, Zoldyick, Roozitaa, BattyBot, Reed07, Vanobamo, JoshuSasori, Stigmatella aurantiaca, Cyberbot II, Abhay ravi, ChrisGualtieri, Maestro814, Deathlasersonline, Plokijnu, Billyshiverstick, Read Blooded, Theeditor6079, Flyer1997, Dexbot, Suffian Akhtar, Kryomaxim, Twhitguy14, CuriousMind01, J0437-4715, Jamesx12345, Among Men, WorldWideJuan, Devinray1991, 1888software, EvergreenFir, Enchantedscience, Mohamed F. El-Hewie, Vai ra'a toa Taina, NeapleBerlina, Jwratner1, Gigantmozg, Ginsuloft, SirKesuma, Anrnusna, JaconaFrere, Osamabin7, Juenni32, Filedelinkerbot, SantiLak, Aryabhatt 21, Willbh15, S11027158, Cris Cyborg, PeterShawhan, Evgeniy E., Sweeeeeeeed, Tetra quark, Praveece, JuanLT2045, Jf2839, GeneralizationsAreBad, KasparBot, Lemonberry622, Pizzaman62, Dgray101 and Anonymous: 700

- **Quantum field theory** *Source:* https://en.wikipedia.org/wiki/Quantum_field_theory?oldid=682800089 *Contributors:* AxelBoldt, CYD, Mav, The Anome, XJaM, Roadrunner, Stevertigo, Michael Hardy, Tim Starling, IZAK, TakuyaMurata, SebastianHelm, Looxix-enwiki, Ahoerstemeier, Cyp, Glenn, Rotem Dan, Stupidmoron, Charles Matthews, Timwi, Jitse Niesen, Kbk, Rudminjd, Wik, Phys, Bevo, BenRG, Northgrove, Robbot, Bkalafut, Gandalf61, Rursus, Fuelbottle, Tobias Bergemann, Ancheta Wis, Giftlite, Lethe, Dratman, Alison, St3vo, Mboverload, DefLog-enwiki, ConradPino, Amarvc, Pearbonn, Karol Langner, APH, AmarChandra, D6, CALR, Urvabara, Discospinster, Guanabot, Igorivanov-enwiki, Masudr, Pjacobi, Vsmith, Nvj, MuDavid, Bender235, Pt, El C, Shanes, Sietse Snel, Physicistjedi, KarlHallowell, PWilkinson, Helix84, Thialfi, Varuna, Gcbirzan, Docboat, Count Iblis, Egg, Mpatel, Marudubshinki, Graham87, Opie, Vanderdecken, Rjwilmsi, MarSch, Earin, R.e.b., RE, Strobilomyces, Arnero, Itinerant1, Alfred Centauri, Srleffler, Chobot, UkPaolo, Wavelength, Bambaiah, Hairy Dude, RussBot, TimNelson, Archelon, CambridgeBayWeather, SCZenz, Odddmonster, E2mb0t-enwiki, Semperf, Tetracube, Garion96, Erik J, Robert L. Banus, RG2, SmackBot, Stephan Schneider, Tom Lougheed, Melchoir, KocjoBot-enwiki, Mcld, Dauto, Chris the speller, Complexica, Threepounds, RuudVisser, QFT, Jmnbatista, Cybercobra, Rebooted, Victor Eremita, DJIndica, Lambiam, Mgiganteus1, Zarniwoot, Jim.belk, Stwalkerster, SirFozzie, Hu12, Dan Gluck, Iridescent, Joseph Solis in Australia, Albertod4, Van helsing, BeenAroundAWhile, Witten Is God, Cydebot, Jamie Lokier, Meno25, Michael C Price, The 80s chick, Mendicus-enwiki, AstroPig7, Msebast-enwiki, Mbell, Headbomb, Nick Number, Mentifisto, AntiVandalBot, Bt414, Bananan-enwiki, Martin Kostner, Moltrix, Kasimann, Kromatol, Puksik, Lerman, LLHolm, RogueNinja, Tlabshier, JEH, Nikolas Karalis, Størkk, JAnDbot, Igodard, Four Dog Night, N shaji, Bongwarrior, Andrea Allais, Soulbot, Etale, Maliz, Custos0, HEL, J.delanoy, Acalamari, Jeepday, Policron, Bickavnger, Juliancolton, Skou, Telecomtom, GrahamHardy, Sheliak, Cuzkatzimhut, VolkovBot, Bktennis2006, Marksr, HowardFrampton, The Original Wildbear, Dj thegreat, Markisgreen, TBond, Lejarrag, Moose-32, Raphtee, Sue Rangell, Neparis, Drschawrz, YohanN7, SieBot, TCO, Yintan, Likebox, Paolo.dL, Tugjob, Henry Delforn (old), Jecht (Final Fantasy X), OKBot, StewartMH, ClueBot, EoGuy, Wwheaton, The Wild West guy, Shvav-enwiki, Bob108, Brews ohare, Thingg, Count Truthstein, XLinkBot, PSimeon, SilvonenBot, Truthnlove, HexaChord, Addbot, ConCompS, Pinkgoanna, Leapold-enwiki, Dmhowarth26, Glane23, Hanish.polavarapu, Lightbot, Scientryst, R.ductor, Ettrig, Yndurain, Legobot, Luckas-bot, Yobot, Ht686rg90, Niout, Tamtamar, AnomieBOT, Ciphers, Palpher, IRP, Gjsreejith, Materialscientist, Citation bot, Bci2, ArthurBot, Northryde, LilHelpa, Caracolillo, Amareto2, MIRROR, Professor J Lawrence, Plasmon1248, Omnipaedista, RibotBOT, Spellage, JayJay, FrescoBot, Kenneth Dawson, D'ohBot, Knowandgive, N4tur4le, Hyqeom, Newt Winkler, Hickorybark, Lotje, Dinamik-bot, LilyKitty, Fortesque666, Reaper Eternal, Minimac, Marie Poise, Yaush, Dylan1946, EmausBot, Racerx11, GoingBatty, Carbosi, Thecheesykid, ZéroBot, Cogiati, Jjspinorfield1, Suslindisambiguator, Quondum, Maschen, Zueignung, Davidaedwards, RockMagnetist, Lom Konkreta, ClueBot NG, Gilderien, Iloveandrea, Vacation9, Heyheyheyhohoho, Fortune432, The ubik, Zak.estrada, Widr, Helpful Pixie Bot, Evanescent7, Ykentluo, Martin.uecker, Walterpfeifer, Pfeiferwalter, Klilidiplomus, W.D., CarrieVS, Khazar2, Momo1381, Dexbot, Cerabot-enwiki, Garuda0001, AHusain314, Thepalerider2012, A.entropy, Mark viking, Faizan, Aj7s6, संजीव कुमार, Lemnaminor, BerFinelli, Axel.P.Hedstrom, Kelongstocking, Mutley1989, I art a troler, Liquidityinsta, Prokaryotes, DemonThuum, Dingdong2680, Asherkirschbaum, Monkbot, Gjbayes, Thedarkcheese, BradNorton1979, UareNumber6, Teelaskeletor, YeOldeGentleman, Mret81, KasparBot and Anonymous: 297

- **Standard Model** *Source:* https://en.wikipedia.org/wiki/Standard_Model?oldid=679087631 *Contributors:* AxelBoldt, Derek Ross, CYD, Bryan Derksen, The Anome, Ed Poor, Andre Engels, Roadrunner, David spector, Isis-enwiki, Youandme, Ram-Man, Stevertigo, Edward, Patrick, Boud, Michael Hardy, SebastianHelm, Looxix-enwiki, Julesd, Glenn, AugPi, Mxn, Raven in Orbit, Reddi, Phr, Tpbradbury, Populus, Haoherb428, Phys, Floydian, Bevo, Pierre Boreal, AnonMoos, BenRG, Jeffq, Dmytro, Drxenocide, Robbot, Nurg, Securiger, Texture, Roscoe x, Fuelbottle, Mattflaschen, Tobias Bergemann, Alan Liefting, Ancheta Wis, Giftlite, Dbenbenn, Harp, Herbee, Monedula, LeYaYa, Xerxes314, Dratman, Alison, JeffBobFrank, Dmmaus, Pharotic, Brockert, Bodhitha, Andyjp, Sonjaaa, HorsePunchKid, APH, Icairns, AmarChandra, Gscshoyru, Kate, Arivero, FT2, Rama, Vsmith, David Schaich, Xezbeth, D-Notice, Dfan, Bender235, Pt, El C, Laurascudder, Shanes, Drhex, Fogger-enwiki, Brim, Rbj, Jeodesic, Jumbuck, Alansohn, Gary, ChristopherWillis, Guy Harris, Axl, Sligocki, Kocio, Stillnotelf, Alinor, Wtmitchell, Egg, TenOfAllTrades, H2g2bob, Killing Vector, Linas, Mindmatrix, Benbest, Dodiad, Mpatel, Faethon, TPickup, Faethon34, Palica, Dysepsion, Faethon36, Qwertyca, Drbogdan, Rjwilmsi, Zbxgscqf, Macumba, Strangethingintheland, Dstudent, R.e.b., Bubba73, Drrngrvy, Agasicles, FlaBot, Naraht, Agasides, DannyWilde, Dave1g, Itinerant1, Gparker, Jrtayloriv, Goudzovski, Chobot, Bgwhite, FrankTobia, YurikBot, Bambaiah, Ohwilleke, VoxMoose, Bhny, JabberWok, Bovineone, Krbabu, SCZenz, JulesH, Davemck, Lomn, E2mb0t-enwiki, Dna-webmaster, Jrf, Dv82matt, Tetracube, Hirak 99, Arthur Rubin, Netrapt, JLaTondre, Caco de vidro, RG2, GrinBot-enwiki, That Guy, From That Show!, Hal peridol, SmackBot, YellowMonkey, Tom Lougheed, Melchoir, Bazza 7, KocjoBot-enwiki, Jagged 85, Thunderboltz, Setanta747 (locked), Skizzik, Dauto, Chris the speller, Bluebot, TimBentley, Sirex98, Silly rabbit, Complexica, Metacomet, DHN-bot-enwiki, MovGP0, QFT, Kittybrewster, Addshore, Jmnbatista, Cybercobra, Jgwacker, BullRangifer, Soarhead77, Daniel.Cardenas, Yevgeny Kats, Byelf2007, TriTertButoxy, Craig Bolon, Ajnosek, Ekjon Lok, Bjankuloski06, Tarcieri, Waggers, JarahE, Michaelbusch, Lottamiata, Newone, Twas Now, IanOfNorwich, Srain, Patrickwooldridge, J Milburn, Mosaffa, Gatortpk, Vessels42, Geremia, Van helsing, Harrigan, Phatom87, Cydebot, David edwards, Verdy p, Michael C Price, Xantharius, Crum375, JamesAM, Thijs!bot, Epbr123, Headbomb, Phy1729, Stannered, Tariqhada, Seaphoto, Orionus, Voyaging, Gnixon, Jbaranao, Jrw@pobox.com, Len Raymond, Narssarssuaq, Bakken, CattleGirl, Davidoaf, Vanished user ty12kl89jq10, Lvwarren, Taborgate, Leyo, HEL, J.delanoy, Hans Dunkelberg, Stephanwehner, Wbellido, Aoosten, Jacksonwalters, The Transliterator, DadaNeem, Student7, Joshmt, WJBscribe, Jozwolf, Hexane2000, BernardZ, Awren, Sheliak, Physicist brazuca, Schucker, Goop Goop, Fences and windows, Dextrose, Mcewan, Swamy g, TXiKiBoT, Sharikkamur, Thrawn562, Voorlandt, Escalona, Setreset, PDFbot, Pleroma, UnitedStatesian, Piyush Sriva, Kaeser, Billinghurst, Francis Flinch, Moose-32, Ptrslv72, David Barnard, SieBot, ShiftFn, Robdunst, Jim E. Black, SheepNotGoats, Gerakibot, Nozzer42, Mr swordfish, Wing gundam, Bamkin, Likebox, Arthur Smart, HungarianBarbarian, Commutator, KathrynLybarger, Iomesus, C0nanPayne, Crazz bug 5, ClueBot, Superwj5, Wwheaton, Garyzx, SuperHamster, Elsweyn, Maldmac, DragonBot, Djr32, Diagramma Della Verita, Nymf, Eeekster, Brews ohare, NuclearWarfare, PhySusie, Ordovico, Mastertek, DumZiBoT, BodhisattvaBot, Guarracino, Mitch Ames, Truthnlove, Stephen Poppitt, Tayste, Addbot, Deepmath, Eric Drexler, DWHalliday, Mjamja, Leszek Jańczuk, NjardarBot, Mwoldin, Bassbonerocks, Barak Sh, AgadaUrbanit, Lightbot, Smeagol 17, Abjiklam, Ve744, Luckas-bot, Yobot, Orion11M87, AnomieBOT, JackieBot, Icalanise, Citation bot, ArthurBot, Northryde, LilHelpa, Xqbot, Sionus, Professor J Lawrence, Tomwsulcer, Edsegal, GrouchoBot, Trongphu, QMarion II, Ernsts, A. di M., Bytbox, FrescoBot, Paine Ellsworth, Aliotra, Steve Quinn, Citation bot 1, Rameshngbot, MJ94, RedBot, MastiBot, Aknochel, Sijothankam, Puzl bustr, Beta Orionis, Physics therapist, Bj norge, Innotata, Jesse V., RjwilmsiBot, Mathewsyriac, Afteread, EmausBot, Bookalign, WikitanvirBot, Wilhelm-physiker, Bdijkstra, DerNeedle, Kenmint, Dbraize, Tanner Swett, HeptishHotik, بدلار دمنشين, Suslindisambiguator, Quondum, Webbeh, UniversumExNihilo, Vanished user fijw983kjaslkekfhj45, Maschen, RockMagnetist, Stormymountain, Zετα ζ, Whoop whoop pull up, Isocliff, ClueBot NG, Smtchahal, Snotbot, Tonypak, O.Koslowski, CharleyQuinton, Dsperlich, Theopolisme, ZakMarksbury, Helpful Pixie Bot, Bibcode Bot, BG19bot, Tirebiter78, AvocatoBot, Lukys-enwiki, Stapletongrey, Ownedroad9, Chip123456, ChrisGualtieri, Khazar2, Billyfesh399, Rhlozier, JYBot, Dexbot, Doom636, Rongended, Cerabot-enwiki, CuriousMind01, Cjean42, Jayanta mallick, Joeinwiki, Kowtje, JPaestpreornJeolhlna, Eyesnore, Euan Richard, Nigstomper, Particle physicist, Prokaryotes, Jernahthern, Ginsuloft, Dimension10, JNrgbKLM, Krabaey, IcodesterS, FelixRosch, Monkbot, Delbert7, BradNorton1979, Lathamboyle, Tetra quark, KasparBot, Buckbill10 and Anonymous: 357

- **Physics beyond the Standard Model** *Source:* https://en.wikipedia.org/wiki/Physics_beyond_the_Standard_Model?oldid=681614974 *Contributors:* David spector, Ewen, Michael Hardy, Andrewman327, IceKarma, Donarreiskoffer, Nurg, Rursus, David Gerard, Alison, David Schaich, RJHall, El C, Kwamikagami, I9Q79oL78KiL0QTFHgyc, Jeodesic, 4v4l0n42, Alinor, Count Iblis, Rjwilmsi, Strait, Eyu100, HappyCamper, Lmatt, BradBeattie, Ohwilleke, Bhny, SCZenz, CecilWard, Karl Andrews, Nlu, Dna-webmaster, Pawyilee, 2over0, Caco de vidro, Jaysbro, SmackBot, Mdj, Nickst, Chris the speller, Bluebot, Scwlong, QFT, Pepsidrinka, Jgwacker, Yevgeny Kats, Doug Bell, John, Dspitzle, RandomCritic, JarahE, Kurtan-enwiki, Headbomb, Peter Gulutzan, N shaji, Lenny Kaufman, VoABot II, Email4mobile, Maliz, R'n'B, HEL, Natsirtguy, Rod57, Tarotcards, DadaNeem, Goop Goop, Fences and windows, Michael H 34, Venny85, Lamro, Wing gundam, Beast of traal, Bhuna71, Mild Bill Hiccup, Djr32, Excirial, RCalabraro, Brews ohare, Mastertek, TimothyRias, Truthnlove, Addbot, Luckas-bot, Zhitelew, Yobot, AnomieBOT, Citation bot, LilHelpa, Smk65536, Stevebow, Omnipaedista, Seeleschneider, A. di M., Kenneth Dawson, Steve Quinn, Citation bot 2, Aturen, Tom.Reding, ErgSlider, Physics therapist, Gistmass, Bj norge, Vstarsky, Serketan, ZéroBot, Galaktiker, Arbnos, Suslindisambiguator, Wiggles007, Maschen, Smtchahal, ClaudeDes, Braincricket, Widr, Helpful Pixie Bot, Mike9110, DryRun, Bibcode Bot, BG19bot, Brainssturm, Qtom.masters, ThePeriodicTable123, M0532062613, Andyhowlett, Cinaro, I am One of Many, Kowtje, CtrlAltBackspace, 22merlin, Monkbot, Delbert7, Tetra quark, TQuentin, Christos Theopoulos, MauiPhoenix and Anonymous: 74

- **Grand Unified Theory** *Source:* https://en.wikipedia.org/wiki/Grand_Unified_Theory?oldid=683790061 *Contributors:* AxelBoldt, Lee Daniel Crocker, Mav, AstroNomer-enwiki, XJaM, Heron, Michael Hardy, Zocky, CesarB, Looxix-enwiki, Emperorbma, Dysprosia, Phys, Omegatron, Northgrove, Robbot, Securiger, Lowellian, Meelar, Caknuck, Giftlite, Jmnbpt, Herbee, Fropuff, Xerxes314, Golbez, Ary29, Sam Hocevar, Lumidek, IcycleMort, M1sslontomars2k4, Mike Rosoft, Discospinster, 4pq1injbok, Pjacobi, Silence, JustinWick, Bobo192, Smalljim, John Vandenberg, Apyule, Foobaz, I9Q79oL78KiL0QTFHgyc, Jeodesic, Physicistjedi, Jérôme, Alansohn, Krischik, Sligocki, Mac Davis, GeorgeStepanek, RJFJR, Lee-Anne, DV8 2XL, Falcorian, Simetrical, Linas, Mindmatrix, FeanorStar7, GregorB, Ruziklan, Mekong Bluesman, Ashmoo, Rachel1, Rjwilmsi, Strait, Drmgrvy, FlaBot, Margosbot-enwiki, DannyWilde, Rune.welsh, BradBeattie, Snailwalker, Phoenix2-enwiki, Guanxi, DVdm, YurikBot, Ugha, Hairy Dude, Michael Slone, Gaius Cornelius, CambridgeBayWeather, NawlinWiki, Wiki alf, Joel7687, JocK, Zwobot, IceCreamAntisocial, Ms2ger, Noclip, Moogsi, CWenger, Smurrayinchester, Curpsbot-unicodify, Caco de vidro, Jaysbro, Sbyrnes321, SmackBot, Tom Lougheed, Eskimbot, Dauto, Silly rabbit, DHN-bot-enwiki, Colonies Chris, Scwlong, QFT, Addshore, Wen D House, Dreadstar, Pwjb, Gbinal, Thorsen, Vina-iwbot-enwiki, ArglebargleIV, Rory096, Ben Jos, Mr. Lefty, Ckatz, SirFozzie, Quaeler, Baderyp, Richwhite10, Cydebot, Peripitus, Michael C Price, Tawkerbot4, Thijs!bot, Headbomb, J.christianson, Luna Santin, Alphachimpbot, JAnDbot, Satarsa-enwiki, Homy, Mbarbier, Durianking, Danmctaggart, Maliz, JCarlos, AstroHurricane001, Adavidb, Bogey97, Qatter, Jeepday, Econofire, Lseixas, Jaffar33, Eismc2, Alphanon, Praveen pillay, KabbalistPhysicist, PaddyLeahy, Hemadh, Will Scot 55, Datpol, Moffitma, ClueBot, DFRussia, James edmiston, Ordinaterr, Djr32, PixelBot, Weysheehai, Sun Creator, Subdolous, Dekisugi, Louis925, AnonyScientist, TimothyRias, SilvonenBot, Bywater100, Truthnlove, Balungifrancis, Addbot, Micromaster, Favonian, Mohitsridhar, 84user, OlEnglish, WikiDan61, Amirobot, AnomieBOT, Girl Scout cookie, Theunify, Karanmohan, Materialscientist, Citation bot, Obersachsebot,

Under22Entreprenuer, Dale Ritter, Senouf, Ernsts, FrescoBot, Paine Ellsworth, Steven Avraham Rosten, Ironboy11, Thamntamil, Slawomir Biały, GreenRoot, Ysyoon, John85, Gil987, Stupidsimple, Tom.Reding, Casimir9999, RobinK, Aknochel, Grandunifier, RjwilmsiBot, Afteread, EmausBot, Slawekb, Arbnos, I. Kensington, ClueBot NG, ClaudeDes, Helpful Pixie Bot, Bibcode Bot, Bernard Rementilla, Kkumer, Wer900, Dilaton, Hilander316, Ryanr666, Davidyevgeny, Cjean42, Franzl aus tirol, Sagnac, GabeIglesia, Lmboyer04, Ovidiu cupsa, Jwratner1, Gilitejman1, Soumilm, Tetra quark, BuzzBloom, JD Wilcox, KasparBot and Anonymous: 131

• **Quantum gravity** *Source:*https://en.wikipedia.org/wiki/Quantum_gravity?oldid=683803465 *Contributors:*AstroNomer−enwiki, Matusz, Miguel Roadrunner, Stevertigo, Ubiquity, Bobby D. Bryant, Mcarling, NuclearWinner, Anders Feder, Susurrus, Coren, Charles Matthews, Timwi, Reddi, Tpbradbury, Phys, Bevo, Raul654, BenRG, Frazzydee, Jeffq, Sdedeo, Rholton, Wereon, Ilya (usurped), Seth Ilys, Ancheta Wis, Giftlite, Herbee, Fropuff, Endlessnameless, Malyctenar, Jason Quinn, Finn-Zoltan, YapaTi−enwiki, Lumidek, Marcus2, Joyous!, TJSwoboda, Vitaleyes, Davidclifford, JimJast, Guanabot, FT2, Masudr, Pjacobi, Pie4all88, David Schaich, Bender235, Clement Cherlin, El C, PhilHibbs, Army1987, Apyule, VBGFscJUn3, PWilkinson, Daniel Arteaga−enwiki, Keenan Pepper, Cjthellama, DonJStevens, Velella, Dabbler, Tycho, Cal 1234, RJFJR, Count Iblis, ThomasWinwood, Anarchimede, Scarykitty, Woohookitty, Igny, ToddFincannon, Mpatel, GregorB, Joke137, Christopher Thomas, Marudubshinki, Graham87, Yurik, Kroggz, Rjwilmsi, Eoghanacht, Jrasowsky, JHMM13, Smithfarm, Ems57fcva, FayssalF, Itinerant1, Lmatt, Chobot, Hmonroe, YurikBot, Hillman, ErkDemon, JocK, SCZenz, Roy Brumback, Bota47, Zunaid, JonathanD, 2over0, Arthur Rubin, Modify, LeonardoRob0t, Caco de vidro, RG2, KasugaHuang, Resolute, SmackBot, Samdutton, Vald, Eskimbot, Hbackman, Onebravemonkey, Chris the speller, Ben.c.roberts, Cthuljew, Silly rabbit, Complexica, Colonies Chris, QFT, Soosed, Theanphibian, Shushruth, Ck lostsword, Yevgeny Kats, DJIndica, Lambiam, Vampus, Vincenzo.romano, Jaganath, JorisvS, RoboDick−enwiki, IronGargoyle, Dicklyon, SirFozzie, Treyp, Twunchy, Piccor, Kurtan−enwiki, Harold f, CalebNoble, Duduong, Paulmlieberman, TVC 15, UncleBubba, TAz69x, Sam Staton, ST47, B. Patrick O'Leary, Epbr123, Koeplinger, Klasovsky, Markus Pössel, Keraunos, Headbomb, Marek69, MichaelMaggs, Tim Shuba, MER-C, ParadiZio, Clementvidal, Perlygatekeeper, VoABot II, Alvatros−enwiki, Bdalevin, SHCarter, Jpod2, DAGwyn, Nucleophilic, LorenzoB, Rickard Vogelberg, DancingPenguin, Rettetast, Victor Blacus, AstroHurricane001, Yonidebot, Acalamari, Mstuomel, Fullmetal2887, NewEnglandYankee, DorganBot, CardinalDan, Idioma-bot, Sheliak, VolkovBot, Pleasantville, Seattle Skier, AlnoktaBOT, TXiKiBoT, Dllahr, Rdekleer, Saibod, Cyberchip, Wikiwikimoore, Carlorovelli, LoreMiles, StevenJohnston, SieBot, LeadSongDog, Bentogoa, Coldcreation, ReluctantPhilosopher, StaticG, GarbagEcol, ClueBot, The Thing That Should Not Be, EoGuy, Polyamorph, Andwor9, Notburnt, Tms9, Alexbot, Resoru, Eeekster, Tamaratrouts, Brews ohare, SchreiberBike, Askahrc, BOTarate, Lambtron, DumZiBoT, XLinkBot, Rror, Facts707, SilvonenBot, Theonlydavewilliams, Mhsb, Truthnlove, Ttimespan, Trifonov−enwiki, Addbot, Mortense, Grayfell, Eric Drexler, Gravitophoton, DOI bot, AkhtaBot, CanadianLinuxUser, Frosty726, LaaknorBot, Delaszk, Tassedethe, Tide rolls, Taketa, Titan1129, Krano, Luckas-bot, Yobot, WikiDan61, Pigetrational, Wireader, Allowgolf−enwiki, Wiki Roxor, Jim1138, IRP, Sz-iwbot, Quantity, Materialscientist, Citation bot, ArthurBot, LilHelpa, Amareto2, Ekwos, KrisBogdanov, Rolfguthmann, StealthCopyEditor, 图图, Dan6hell66, Rabsmith, Hep thinker, Paine Ellsworth, DrArthurRubinPHD, Lagelspeil, Nunc aut numquam, Vacuunaut, Van Speijk, Knowandgive, Craig Pemberton, Udifuchs, Citation bot 2, Citation bot 1, Citation bot 4, Jonesey95, Hirvenkürpa, Tom.Reding, Pmokeefe, Casimir9999, Dac04, Dude1818, Valeriy Pischenko, Follyland, TrueTeargem, N0814444, Earthandmoon, Korepin, DARTH SIDIOUS 2, Musictime4me, RjwilmsiBot, EmausBot, Francophile124, Octaazacubane, Fotoni, Slightsmile, Garfield Salazar, Hhhippo, JSquish, John Cline, Fæ, LostAlone, Brazmyth, Throwmeaway, Arbnos, Ebrambot, Kusername, DanielBurnstein, TonyMath, I. Kensington, Maschen, Donner60, Parusaro, Apratim07, Terra Novus, Isocliff, Googledin!, ClueBot NG, SpikeTorontoRCP, Science writer, Preon, Raidr, Jhmmok, 336, Widr, Helpful Pixie Bot, Bibcode Bot, Bardsley Rides a Segway, Apelikedawg, FiveColourMap, Trevayne08, Mr.viktor.stepanov, Brainssturm, BattyBot, Jimw338, Ryanr666, Kryomaxim, Garuda0001, Saehry, TwoTwoHello, Sanathdevalapurkar, Andyhowlett, GabeIglesia, Sanathlab, Roiwallace, Spencer.mccormick, Spencerfjase, MrShlongNo1, Marc D. Garrett, D00d00ballz, Gigantmozg, Susan.grayeff, Polytope24, Frinthruit, Anrnusna, Dfyytj, Monkbot, Umut Alihan Dikel, Amortias, Klj1234, Pfpguy, KasparBot and Anonymous: 294

• **String theory** *Source:* https://en.wikipedia.org/wiki/String_theory?oldid=683652719 *Contributors:* AxelBoldt, Sodium, Mav, Bryan Derksen, Zundark, The Anome, Tarquin, Taw, Eean, Malcolm Farmer, Hephaestos, Olivier, Drseudo, Stevertigo, Spiff−enwiki, Edward, PhilipMW, Michael Hardy, Bewildebeast, Dante Alighieri, Gabbe, Graue, Tgeorgescu, Mcarling, CesarB, Looxix−enwiki, Ahoerstemeier, Theresa knott, Suisui, Angela, Den fjättrade ankan−enwiki, Jdforrester, Julesd, Salsa Shark, Schneelocke, Charles Matthews, Timwi, Bemoeial, Jitse Niesen, 4lex, Greenrd, ErikStewart, Furrykef, Saltine, Phys, Omegatron, Bevo, Topbanana, Trent, Nufy8, Robbot, Craig Stuntz, Fredrik, Chris 73, R3m0t, COGDEN, Mirv, Wjhonson, Sverdrup, Academic Challenger, DHN, Hadal, Khlo, ElBenevolente, HaeB, Tobias Bergemann, Giftlite, DocWatson42, Christopher Parham, Awolf002, Mporter, Amorim Parga, Mikez, Harp, Kim Bruning, Tom harrison, Ferkelparade, Leffyman, Fropuff, No Guru, Anville, Moyogo, Curps, Pashute, Nomad−enwiki, Mboverload, Solipsist, SWAdair, DemonThing, Wmahan, Btphelps, MSTCrow, Decoy, Chowbok, Gadfium, Steuard, Pgan002, Quadell, Carandol−enwiki, Antandrus, Beland, JoJan, Khaosworks, Tothebarricades.tk, Thincat, Tomruen, Shidobu, Icairns, Lumidek, NoPetrol, Avihu, Fanghong−enwiki, Trevor MacInnis, Lacrimosus, Zro, Mike Rosoft, D6, Urvabara, Felix Wan, Jkl, Discospinster, ElTyrant, Rich Farmbrough, Rhobite, Pjacobi, Alien life form, Vapour, Silence, Kzzl, LindsayH, Mani1, Pavel Vozenilek, Paul August, Bender235, Kjoonlee, Mashford, Kelvinc, Perlman10s, Panu−enwiki, Brian0918, Dpotter, Livajo, El C, Laurascudder, Shanes, Zegoma beach, RoyBoy, Causa sui, Bobo192, Directorstratton, Janna Isabot, Smalljim, John Vandenberg, Flxmghvgvk, I9Q79oL78KiL0QTFHgyc, Physicistjedi, Bongoo, 4v4l0n42, Merope, Geschichte, Linuxlad, Phils, Merenta, Alansohn, Gary, JYolkowski, Enirac Sum, Ryanmcdaniel, Arthena, Borisblue, Rd232, Plumbago, Axl, R Calvete, Lightdarkness, Kocio, Bart133, Wtmitchell, Isaac, Tycho, Cal 1234, Fadereu, CloudNine, Seiurinæ, Computerjoe, Kusma, DV8 2XL, Pwqn, Gene Nygaard, Ringbang, Ceyockey, Falcorian, Bobrayner, Joriki, Mel Etitis, Linas, BillC, Jacobolus, HFarmer, Before My Ken, Netdragon, MONGO, GeorgeOrr, Mpatel, Bbatsell, GregorB, 图图图图图, Joke137, Christopher Thomas, Dysepsion, GSlicer, Jan.bannister, Graham87, Magister Mathematicae, Hillbrand, BD2412, Elvey, Galwhaa, Raymond Hill, JIP, RxS, Athelwulf, Edison, Sjakkalle, Rjwilmsi, Xgamer4, Jake Wartenberg, Arabani, MarSch, TheRingess, Jmcc150, Aero66, Crazynas, Juan Marquez, R.e.b., Bubba73, DoubleBlue, Zelos, AlisonW, Asafavi, Lionelbrits, Conorific, Zunz, Mathbot, Crazycomputers, RexNL, Gurch, Algri, TeaDrinker, Zifnabxar, XAXISx, Erik4, Phoenix2−enwiki, Antimatter15, Ggb667, Chobot, Visor, DVdm, Mhking, VolatileChemical, Bgwhite, Algebraist, Ben Tibbetts, YurikBot, Ugha, Wavelength, Borgx, NuclearFusion−enwiki, Angus Lepper, Hairy Dude, Jimp, Hillman, Cyferx, Wolfmankurd, Pip2andahalf, RussBot, Moronoman, Crazytales, Pippo2001, Bhny, Pigman, SpuriousQ, Branman515, Stephenb, Gaius Cornelius, Eleassar, Bovineone, Cheesus, Shanel, NawlinWiki, Tong−enwiki, Mike18xx, SCZenz, Cleared as filed, Bdiah, Pym98, SColombo, Haemo, FF2010, Closedmouth, Reyk, Brina700, Chris Brennan, Vicarious, Brianlucas, Geoffrey.landis, Hitchhiker89, Spliffy, Pred, ArielGold, Roy Fulton, Ilmari Karonen, Katieh5584, Pentasyllabic, Lunch, DVD R W, WikiFew, That Guy, From That Show!, Street Scholar, AndrewWTaylor, QSquared, Sardanaphalus, Vanka5, MacsBug, Hvitlys, SmackBot, Kurochka, Zazaban, Tom Lougheed, Prodego, KnowledgeOfSelf, Hydrogen Iodide, Melchoir, Vald, Skrewtape, Atomota, Canthusus, GaeusOctavius,

Cool3, Andyvn22, Skizzik, RobertM525, Dauto, Bluebot, SSJ 5, Keegan, Aidan Croft, Thumperward, Oli Filth, Silly rabbit, Timneu22, SchfiftyThree, Moshe Constantine Hassan Al-Silverburg, Complexica, Rediahs, RayAYang, Aero77, Adamstevenson, Ikiroid, Epastore, Baronnet, Ned Scott, Sbharris, Colonies Chris, Konstable, Sct72, Scwlong, Can't sleep, clown will eat me, Timothy Clemans, Onorem, Neilanderson, EvelinaB, TKD, KerathFreeman, Addshore, UU, The tooth, Pepsidrinka, Somebody2292, --=The Doctor=--, Fuhghettaboutit, Cybercobra, Irish Souffle, Nakon, Jdlambert, James McNally, MichaelBillington, Lostart, Insineratehymn, Drphilharmonic, SpiderJon, DMacks, Ihatetoregister, Where, Michael IFA, Yevgeny Kats, Vasiliy Faronov, Byelf2007, Angela26, Visium, Rory096, Zymurgy, Harryboyles, Mdl53711, T-dot, Titus III, Ergative rlt, MagnaMopus, UberCryxic, Vgy7ujm, Linnell, Mgiganteus1, Nonsuch, IronGargoyle, Ckatz, DoItAgain, AstroGod, Kirbytime, Jimbo Mahoney, FredrickS, Invisifan, Ryulong, Ryanjunk, MathStuf, Mike Doughney, Norm mit, Hindol, Dan Gluck, Huntscorpio, Iridescent, K. Sunoco, You? Me? Us?, CzarB, Rabinzkaman, JoeBot, Lottamiata, Tony Fox, Vrkaul, Torrazzo, Gil Gamesh, Areldyb, Courcelles, Tawkerbot2, Gebrah, Shamvil, DKqwerty, Lbr123, Harold f, Heqs, Devourer09, Duduong, Sarvagnya, Dewayne76, JForget, Cg-realms, InvisibleK, CRGreathouse, CmdrObot, Earthlyreason, Van helsing, Olaf Davis, CBM, Rawling, Jibal, Witten Is God, Nunquam Dormio, Giko, KnightLago, Thubsch, Leujohn, SlashDot, TheTito, Karenjc, Myasuda, Emarv, Cydebot, Gmusser, Gogo Dodo, Jkokavec, Kahananite, Quajafrie, Michael C Price, Doug Weller, DumbBOT, Narayanese, AlphaNumeric, SRoughsedge, Vanished User jdksfajlasd, Woland37, Zalgo, Daniel Olsen, UberScienceNerd, Bkazaz, DJBullfish, Thijs!bot, Epbr123, Rwmnau, Babemachine, Pimpin101, Mbell, O, Faigl.ladislav, Ucanlookitup, Andyjsmith, Headbomb, Tcturner2002, Marek69, Brahmajnani, Arthurcprado-enwiki, Y.t., D3gtrd, Babemonkey, Dark dude, Duncan McB, EdJohnston, MichaelMaggs, Ancientanubis, Natalie Erin, Hempfel, Jomoal99, Mmortal03, Mentifisto, Geekdom04, AntiVandalBot, Luna Santin, Seaphoto, Ed270791, Opelio, Doc Tropics, David136a, NithinBekal, Dotdotdotdash, Helicoptor, Poshzombie, MontanNito, Dylan Lake, Maximilian77, Shlomi Hillel, Db63376, SamIAmNot, Knotwork, Res2216firestar, Superior IQ Genius, MER-C, Andonic, Sitethief, 100110100, TallulahBelle, Nestamachine, Savant13, Daynightrader, Goldenglove, Charibdis, Acroterion, Ophion, Aigisthos, Editmyhandman, Aruben537, Magioladitis, WolfmanSF, Bongwarrior, VoABot II, Yandman, JamesBWatson, بابيام, Qutt, Jespinos, Kevinmon, Aka042, Froid, DAGwyn, Catgut, Panser Born, Ensign beedrill, Perspectival, JJ Harrison, Dirac66, Justanother, Aziz1005, Cpl Syx, ChazBeckett, Teardrop onthefire, WLU, Stephen shenker, Robin S, SkepticVK, Joshua Davis, Mkroh, B9 hummingbird hovering, S3000, Hdt83, MartinBot, FlieGerFaUstMe262, Ytomem, Shimwell, Arjun01, KrishSundaresan, Anaxial, Jay Litman, Alexcalamaro, Andrej.westermann, Smokizzy, LedgendGamer, Cyrus Andiron, Peteryoung144, Tgeairn, Artaxiad, HEL, AlphaEta, J.delanoy, AstroHurricane001, Maurice Carbonaro, Yonidebot, Morris729, M C Y 1008, 69gangsta420, It Is Me Here, Shawn in Montreal, Janus Shadowsong, Bailo26, Fredsie, Madagaskar07, Duchesserin, AntiSpamBot, CHIAGEHYANG, Chiswick Chap, Watsup1313, Belovedfreak, HaloInverse, NewEnglandYankee, Scott1329m, Thesis4Eva, Policron, Jrcla2, WJBscribe, Rnricklefs, Jamesofur, Eyelidlessness, Jonnyk aus, Kvdveer, JavierMC, Izno, Xiahou, CardinalDan, Sheliak, HamatoKameko, Malik Shabazz, Concertmusic, JohnBlackburne, Justin-Hagstrom, Fences and windows, Wooba doob, Philip Trueman, DoorsAjar, HowardFrampton, TXiKiBoT, Zidonuke, Red Act, Kriak, Calwiki, Technopat, Hqb, Andrius.v, Anonymous Dissident, Crohnie, AlysTarr, Qxz, Vanished user ikijeirw34iuaeolaseriffic, Impunv, Seraphim, Martin451, Don4of4, ABigGreenHippo, Huperphuff, LeaveSleaves, Kaenneth, StringyGuy, Maxim, Erth64net, Meters, Lamro, Rickstaduhar, Enviroboy, Turgan, Anna512, PhysPhD, Northfox, NPguy, Matthew Sanders, Luke Walkerson, Newbyguesses, MissMJ, SieBot, Escher26, J.A.Ireland, BA (IHPST), 4wajzkd02, Robdunst, Dreamafter, Pallab1234, Dbelange, MTHarden, Lemonflash, Kylemew, Yintan, GlassCobra, Wpegden, Likebox, Flyer22, Exert, ProGeek314, Arbor to SJ, Babawhitemoose, Caidh, Dhatfield, Audree, Oxymoron83, Pretty Green, Weaselstomp, Manway, Alex.muller, Taco Manipulator, Tschach, Manheat84, Anchor Link Bot, Mikebernstein, ImperialismGo, Nergaal, Ionfield, Ayleuss, Sh4w20r, Naturespace, Martarius, Phyte, ClueBot, The Thing That Should Not Be, String4d, Illusion96, Polyamorph, Mpd1989, Alexdeburca18, Wiggl3sLimited, Excirial, Kjramesh, Jusdafax, Resoru, WikiZorro, Eeekster, Verum-enwiki, Tamaratrouts, Gtstricky, Humanino, Brews ohare, NuclearWarfare, Cenarium, Arjayay, Razorflame, Scoobey, BOTarate, Sideswiper, Thingg, Capudo, BVBede, Versus22, Introductory adverb clause, MelonBot, SoxBot III, Egmontaz, Notpayingthepsychiatrist, DumZiBoT, BahTab, TimothyRias, Aj00200, Reaperfromhell, Dunkaroo207, XLinkBot, AlexGWU, Impshum, Saeed.Veradi, Little Mountain 5, Guy392, David424, Truthnlove, Qweeveen, Tayste, Addbot, Steven66s, Denali134, Elemented9, Varrey280303, Eric Drexler, Some jerk on the Internet, Fizzycyst, Uruk2008, DOI bot, Jojhutton, AngryBacon, Captain-tucker, Auspex1729, Kongr43gpen, Fgnievinski, Rhetoric Of A Sophist, Ronhjones, CanadianLinuxUser, Cst17, Download, Glane23, Bassbonerocks, Chzz, Favonian, Kronix35, LinkFA-Bot, Udugunit, Aktsu, Tassedethe, Numbo3-bot, Anpecota, Tide rolls, HerpesVirus, SDJ, OlEnglish, Scourge of God, Davidmedlar, Couldbenoway66, Yobot, Maxdamantus, Terrisknickers, Kartano, TaBOT-zerem, Julia W, Unique and proud of it, FireMouseHQ, Terrifictriffid, ArchonMagnus, CinchBug, Synchronism, AnomieBOT, Cleeseheb, 1exec1, Charlesvi, Bigdaddy4x4, Gitman4, Jim1138, IRP, Mintrick, Drweetmola, Ornamentalone, M00npirate, Gautam10, Csigabi, PoliPsy, Materialscientist, 90 Auto, Citation bot, Teleprinter Sleuth, Vuerqex, Twri, Frankenpuppy, Fuzzy Bob Saget, DirlBot, Georgepowell2008, Heidisql, Cureden, Ekwos, Capricorn42, Gensanders, NFD9001, Anna Frodesiak, Tomwsulcer, A23649, Pra1998, Coretheapple, Ruy Pugliesi, Jagbag2, Vandalism destroyer, Ab1, Omnipaedista, Bandit5005, Shirik, RibotBOT, Waleswatcher, Saalstin, Amaury, Aaron35510, Caz34, Doulos Christos, Sewblon, Born Gay, Capricorn24, SchnitzelMannGreek, A. di M., SpacePyjamas, Kierkkadon, A.amitkumar, Dougofborg, StringLove, Nobelprizewinner, Astiburg, FrescoBot, Fortdj33, Paine Ellsworth, Goodbye Galaxy, HJ Mitchell, Steve Quinn, Vhann, Kwiki, Xhaoz, Citation bot 1, Batong, Gil987, Pinethicket, I dream of horses, Tallboyhoops1991, Three887, Steveo27five, RedBot, Sardinita, Serols, Vhsatheeshkumar, Swisstiger, DeletionUK, Reconsider the static, IVAN3MAN, Remingtonhill1, Orenburg1, Coltonhs, Willy Weazley, Smamaret, Bethovenn, Dinamik-bot, Dc987, Oswaldo Zapata, Egemont, Syebo, Alaithiran, Reaper Eternal, Seahorseruler, Ybungalobill, Quaker phil, Specs112, Dr. Aakash Patel, Tbhotch, StormbringerUK, Minimac, Mathgenius3141592, Keegscee, Omgwaffels, Mick le pick, Solancel, Aznhero3793, Dwielark, Afteread, Enauspeaker, EmausBot, MaooaM, Immunize, Az29, Milkocookie, Faolin42, Fotoni, RA0808, RenamedUser01302013, 8digits, Yukiseaside, Slightsmile, Tommy2010, Winner 42, Wikipelli, JonezyKiDx, Joe Gazz84, ZéroBot, Timeitsways, John Cline, Cogiati, Quaqa, Chrispaps2413, Nasulikid, Vollrath2323, Benjamin1414141414141414, Arbnos, Green Lane, A930913, Bamyers99, Azeraphale, H3llBot, Encyclopadia, Danga1988, Ollainen, PoisonGM, Wayne Slam, OnePt618, Knome335, L Kensington, Lulzprotuns, Kranix, Rpcappello, Maschen, Vastly-enwiki, Donner60, CatFiggy, CountMacula, Orange Suede Sofa, Etov, M1k3 101, Bill william compton, Wakabaloola, TERBAFAN, Nickslspride34, NeuralLotus, Isocliff, Brechbill123, Xanchester, ClueBot NG, Martti Muukkonen, KagakuKyouju, Jeff Song, This lousy T-shirt, Satellizer, Name Omitted, Marcdean123, Wiki incorp, Frietjes, O.Koslowski, Alexdamaino9, Dream of Nyx, Blackhall616, Widr, Sashhere, WikiPuppies, Stu181, T00g00d96, Pluma, Storm.sarup, Helpful Pixie Bot, Manzeet, Waffleboy36, HMSSolent, Mikeshelton1, Bibcode Bot, 2001:db8, Phillip.phillipson, Hoaxinator, Lowercase sigmabot, Thor cherubim, Mrshabam, Nishch, Flowerhat15, AvocatoBot, Housegeek224, MahRanch, Benzband, Altaïr, Benhenchdickthomas, Shreyakstring, Sweaty maori sphincter, DaFalk, Dsabo74, Ratanmaitra, MM4EVAH, Steven.w.kowalski, Minsbot, JGallardo2600, Dylanlatham, Myfriendganesha, OCCullens, Likeaboss189, Sean271293, LinusE8, BattyBot, Several Pending, Aldrich2122, CommanderMoka, The Illusive Man, ChrisGualtieri, KoalamaN2, Trevorkid45, Catsloveit07, Alex Modzz, Rustyjamsen, Goh ryangoh, Dexbot, Exolius, Hilander316, Alman1234321, SuperCalzer,

LightandDark2000, MeekMelange, BQND, Cdarrai1, Kephir, TheMonkeyboy524, Michael Anon, Mattfat8, Lugia2453, Anruy, Rachel weld, Jamesx12345, AHusain314, BossEditors, Hillbillyholiday, Joeinwiki, Mattninja, Theshadow444, Asaa82, Jakemarz197, Kzhang1025, Epicgenius, Spongbob456789, ⁊, TestMaster, Ianreisterariola, GrapperJ, Makeitnasty, Moemajdi, I am One of Many, NualaIvy, BAZINGASS, St3fanPC, Eyesnore, Isaac grozd, Jordanissexyaf1999, Baruch6525, Mosbruckerej, Ihatedirac2k13, Jonamithy121314, 123physicsquantum, Jt198, RaphaelQS, HeyJude70, AParker628, DimReg, A.k.blaze1, Joshuk, Zenibus, Nianoobasik, Ihelpapplen, Gamo To Apoel, SacredLabyrinth, Ginsuloft, Vampre1122, Dimension10, Howard Wolowitz, AddWittyNameHere, Polytope24, Elysion, Tutun12$, Longerboats5, SimonWombat8, Konveyor Belt, Vtank54, Micheal545, Hck24, Caliae19, Hexafish, Simpick, TheRealTheKoi, Bballbro62, Monkbot, ArmyPath, TheQ Editor, Jtsmith098, Joshmiller1, Hanseer360, XXvPIEvXx, Dbennett 24, Ghikpenos, Nick65633, Saundra03, Thehippothatknows, Sewwgers, Teelaskeletor, Cirksena, Balockaye1234, PloppyDoo, Yesufu29, Lumpy2k14, Podayeruma, Abstract92, Sbenfiel, Monkman2k4, Swegwegdgfyetkfoffkkfkfkv, John95541234, Poopman224, ScrapIronIV, Tetra quark, GeneralizationsAreBad, Shivansh2014n, KasparBot, SHUCKYLUCKY, Fabiotheoto, FartGoblin, Joca potato, Joshcool246, Theoretical Physisist4444 and Anonymous: 1546

- **Planck length** *Source:* https://en.wikipedia.org/wiki/Planck_length?oldid=681938635 *Contributors:* AxelBoldt, Mav, Uriyan, Roadrunner, GrahamN, Bdesham, Michael Hardy, Dominus, Dcljr, Alf, Dying, Dysprosia, Doradus, Nnh, BenRG, Digizen, Blainster, DHN, Netjeff, Mattflaschen, Giftlite, JamesMLane, Python eggs, Pne, Mckaysalisbury, YapaTi~enwiki, Utcursch, Sonjaaa, Superborsuk, JimWae, Icairns, Tdent, Yuriz, Chris Howard, Sysy, Hidaspal, Luqui, Florian Blaschke, Alistair1978, Bender235, Pt, El C, Rbj, Obradovic Goran, Wendell, Jhertel, DanielLC, Keenan Pepper, Axl, Sligocki, PAR, Dominic, Oleg Alexandrov, GVOLTT, StradivariusTV, GregorB, Christopher Thomas, Radiant!, JIP, Zbxgscqf, FlaBot, John Baez, DevastatorIIC, QuicksilverJohny, LeCire~enwiki, DVdm, Jimp, RussBot, Amakuha, Elizabeyth, NorsemanII, JonathanD, Ayeomans, Gtdp, Neoliten, TLSuda, Groyolo, KasugaHuang, SmackBot, Gold333, DLH, Skizzik, Amux, Bluebot, Sewlong, Voyajer, Mr.Z-man, Valenciano, Sadi Carnot, Tesseran, Lambiam, Petr Kopač, Adj08, Korean alpha for knowledge, JH-man, Loadmaster, JHunterJ, Slakr, Ashaver, Krispos42, Achoo5000, Roland.barrat, Zureks, Zginder, Quibik, Agony, Nonagonal Spider, Greg L, Oreo Priest, Defaultdotxbe, Tyco.skinner, Spartaz, Blaine Steinert, 100110100, Magioladitis, Nakkisormi, Harelx, Mtiffany71, Tanvirzaman, Stuver, Meam5555, Brian.ellis, BJ Axel, Peter Chastain, Melamed katz, Maurice Carbonaro, Foober, Grshiplett, Sigmundur, Rotemdanzig, VolkovBot, Thurth, TXiKiBoT, UnitedStatesian, Cojones893, Newbyguesses, Likebox, Ellohir, Fhburton, Danthewhale, ClueBot, Cygnis insignis, Unitfreak, Brews ohare, NuclearWarfare, BOTarate, Faramarz.M, PL290, Truthnlove, Addbot, Luckas-bot, Yobot, Crispmuncher, Linktex, AnomieBOT, DemocraticLuntz, The High Fin Sperm Whale, Xqbot, Brufydsy, Br77rino, Srich32977, Champlax, حامد میرزاحسینی, Brycee, Sewblon, Constructive editor, FrescoBot, Jonesey95, Tom.Reding, RobinK, DixonDBot, Fama Clamosa, Javierito92, Is it Protagoras?, Autumnalmonk, EmausBot, John of Reading, Bookalign, GoingBatty, Ihowardsr1, Medeis, Quondum, Rocketrod1960, ClueBot NG, ClaudeDes, Widr, StanS, Helpful Pixie Bot, CrannySmith, Bibcode Bot, BG19bot, Chess, Mynameisnoted, Mark Arsten, Phil.gagner, Klilidiplomus, Penguinstorm300, Pinkie Pie, Timothy Gu, Dtm1234, 069952497a, Jiawhein, ReconditeRodent, Zane6324, JustBerry, Ginsuloft, FDMS4, Crabbyhamster, Coreyemotela, Tobywheeer, Tobyepic, Monkbot, Alexander Klimets, Martijnvans, Gamebuster19901, Dtrumpnairobanbirthcertificate, Nertuop and Anonymous: 205

- **Unified field theory** *Source:* https://en.wikipedia.org/wiki/Unified_field_theory?oldid=683555666 *Contributors:* M~enwiki, Michael Hardy, Ahoerstemeier, William M. Connolley, Charles Matthews, Reddi, Dysprosia, Pakaran, SJRubenstein, Josh Cherry, Academic Challenger, Rursus, Dbenbenn, JamesMLane, Xerxes314, StargateX1, Gzornenplatz, Karol Langner, Gscshoyru, Vitaleyes, Svdb, Caroline Thompson, Brianhe, H0riz0n, Cfailde, DPFJr, Pjacobi, Paul August, Dmr2, ESkog, Lentando~enwiki, BenjBot, Oldsoul, Etimbo, Noren, Bobo192, Ablathanalba, Mordemur, John Vandenberg, I9Q79oL78KiL0QTFHgyc, Nsaa, Hdeasy, Count Iblis, Falcorian, Blaze Labs Research, Simetrical, Linas, Mindmatrix, Decrease789, Savantnavas, Mpatel, Knuckles, Marudubshinki, BD2412, Nightscream, Macumba, R.e.b., Protez, Srleffler, Chobot, Moocha, GangofOne, YurikBot, RadioFan2 (usurped), Rsrikanth05, NawlinWiki, Trovatore, Syrthiss, Steve G~enwiki, KasugaHuang, SmackBot, McGeddon, Gilliam, Chris the speller, Bluebot, Jjalexand, Crazy8s, Jeysaba, Silly rabbit, Timneu22, Redattore, Colonies Chris, QFT, InnocentMind, Xyzzyplugh, Jgwacker, Corby, RolandR, Marcus Brute, Sadi Carnot, Vina-iwbot~enwiki, Lambiam, Nishkid64, ArglebargleIV, Titus III, John, Gobonobo, Kevlarmry, Ckatz, Slakr, Rainwarrior, Trounce, Twas Now, Bridg, Courcelles, Shedsan, Prof.Maque, JForget, Will314159, Friendly Neighbour, NickW557, Gregbard, Peripitus, Gogo Dodo, Michael C Price, Qwyrxian, Roger Anderton, Headbomb, Marek69, Twcjr, KrakatoaKatie, AntiVandalBot, Emeraldcityserendipity, Tim Shuba, Verticordia~enwiki, Alphachimpbot, AndreasWittenstein, Blaine Steinert, Yill577, Wasell, VoABot II, Tobogganoggin, Arrowcatcher, DAGwyn, Catgut, Web-Crawling Stickler, Ours18, Skylights76, Jke310, Stephenchou0722, R'n'B, J.delanoy, Captain panda, Trusilver, Eliz81, Dogstar11, CardinalDan, VolkovBot, Butwhatdoiknow, Fennmeister, Alphanon, Cheffoxx, Betanon, ARUNKUMAR P.R, Deansinclair, PaddyLeahy, Gaelen S., Travisbmoore, Gammanon, Zharradan.angelfire, Anakin101, De728631, ClueBot, NossB, EhJJ, Bhushan foryou, Versus22, Qwfp, Pandanator75, Arthur chos, Addbot, Cxz111, Bobtron5000, Bte99, MrOllie, Deepthought137, Favonian, AtheWeatherman, Whitematter, Scientryst, MuZemike, Yobot, TaBOT-zerem, AnomieBOT, 9258fahsflkh917fas, Kanat Abildinov, Materialscientist, JohnnyB256, Addihockey10, Shirik, Mathonius, Pereant antiburchius, Natural Cut, Shadowjams, Lovelylilian, A. di M., Paine Ellsworth, Ottokar~enwiki, Knowandgive, RandomStringOfCharacters, Grandunifier, Tennant uk, Miracle Pen, Afteread, RA0808, Tommy2010, Wikipelli, Hhhippo, Susfele, AvicAWB, S.Lenane, DanielBurnstein, Barendjacobus, Davidaedwards, Mkh025, Dudge1983, ClueBot NG, StevenPower, Theopolisme, MerllwBot, Mophedd, Orphadeus, Peter Donald Rodgers, Joe0x7F, Dilaton, Astralbound, LynnetteA11, GabeIglesia, Jamesmcmahon0, NfrHtp, KEVIN123456789, Froglich, BuilderE, Michelle1881, Balbinder1706, Stowcalj, KasparBot and Anonymous: 194

- **Symmetry breaking** *Source:* https://en.wikipedia.org/wiki/Symmetry_breaking?oldid=682384566 *Contributors:* Michael Hardy, Kku, Charles Matthews, Phys, Aleron235, Giftlite, C8to, Thincat, FT2, Xezbeth, Gauge, TheParanoidOne, BD2412, Rjwilmsi, Nihiltres, Debivort, Hairy Dude, Splash, Dan Gluck, Thermochap, Ben MacDui, Parthasarathy.kr, R'n'B, DadaNeem, NigelHarris, Moose-32, Brews ohare, Torage, Stephen Poppitt, Addbot, DOI bot, Dranorter, Download, Ptbotgourou, AnomieBOT, RibotBOT, 卷卷, AllCluesKey, Enredanrestos, Citation bot 1, DrilBot, Mary at CERN, RjwilmsiBot, EmausBot, WikitanvirBot, Nyxhadanielle, Shivsagardharam, Qx2020, HMman, Mark viking and Anonymous: 19

- **Fundamental interaction** *Source:* https://en.wikipedia.org/wiki/Fundamental_interaction?oldid=683385494 *Contributors:* AxelBoldt, Zundark, The Anome, Tarquin, AstroNomer~enwiki, William Avery, Roadrunner, Ellmist, Robert Foley, Heron, Isis~enwiki, Stevertigo, Patrick, Michael Hardy, Gdarin, CesarB, Looxix~enwiki, Cyp, William M. Connolley, Theresa knott, Mxn, Bemoeial, Reddi, Zoicon5, Finlay McWalter, Robbot, Lowellian, Brjaga, Roscoe x, Seth Ilys, Ancheta Wis, Giftlite, Christopher Parham, Herbee, Monedula, Xerxes314, Alison, Pearbonn, Beland, Melikamp, Karol Langner, AmarChandra, Mike Rosoft, Jørgen Friis Bak, JimJast, Discospinster, Guanabot, FT2, Harriv, Quietly, GoldenRing, Clement Cherlin, El C, Lycurgus, Joanjoc~enwiki, Alereon, Euyyn, Kanzure, Army1987, Rbj, Haham hanuka, Nsaa, Jumbuck, Foant, Dachannien, Kdau, ReyBrujo, Reaverdrop, BDD, Someoneinmyheadbutit'snotme, DV8 2XL, Kazvorpal, Woohookitty,

Linas, Mindmatrix, Sabejias, StradivariusTV, Mpatel, Miss Madeline, Isnow, Elvey, Chun–hian, Koavf, Strait, Jmcc150, RE, Gadha, FlaBot, DClement, ZoneSeek, Alfred Centauri, Lmatt, Rell Canis, Mstroeck, Chobot, Subtractive, Visor, GangofOne, Mysekurity, YurikBot, Ashleyisachild, Bambaiah, Lucinos–enwiki, Wavesmikey, Chaos, FFLaguna, Dbfirs, Trigger hippie77, Enormousdude, Shimei, RG2, Bweenie, Phr en, GrinBot–enwiki, SmackBot, Unyoyega, Andy M. Wang, Vvarkey, Jjalexand, Mithaca, Acipsen, DHN–bot–enwiki, Colonies Chris, Andy120290, Addshore, SundarBot, Jgwacker, LeoNomis, Sadi Carnot, TTE, SashatoBot, FrozenMan, Philosophus, A. Parrot, Fangfufu, GDallimore, Avanishsharma, CRGreathouse, Green caterpillar, McVities, MaxEnt, A. Exeunt, Scott.medling, LouisBB, Thijs!bot, Martin Hogbin, Mojo Hand, Headbomb, Dfrg.msc, Dodecahedron–enwiki, JAnDbot, The penfool, Fordskydog, MER-C, TheEditrix2, Fabrictramp, Leyo, Trusilver, Joshuaali, Idioma-bot, VolkovBot, TXiKiBoT, Anonymous Dissident, MackSalmon, Praveen pillay, BotKung, Gnomon13, Lamro, RMW42, EmxBot, Neparis, SieBot, WereSpielChequers, ToePeu.bot, Avargasm, RadicalOne, Dhatfield, SuperSpy00bob, Showers3, Beast of traal, Lightmouse, Nskillen, Sunrise, OKBot, Bpeps, C0nanPayne, StewartMH, Sfan00 IMG, ClueBot, MichaelVernonDavis, SuperHamster, Djr32, Sadiqsaleem09, PixelBot, Eeekster, Zamis45, Yonskii, 1ForTheMoney, Noctibus, Truthnlove, Addbot, Mabdul, LinkFA-Bot, F Notebook, Lightbot, Legobot, Clay Juicer, Luckas-bot, Yobot, Il MusLiM HyBRiD Il, Rifter0x0000, AnomieBOT, Glen Dillon, Girl Scout cookie, Cleroth, JackieBot, Piano non troppo, Flewis, AthenaO, Xqbot, Omnipaedista, RibotBOT, A. di M., Ironboy11, Goodbye Galaxy, Jmbenham, Unkownkid2400, Rameshagbot, Jschnur, RedBot, Σ, Frankjohnson123, IVAN3MAN, Right-wing genius, Lokentaren, Setsuna29, EngineerFromVega, RjwilmsiBot, Anuandraj, Beyond My Ken, Deadlyops, Carbo1200, Kbasford, ClueBot NG, Greedohun, MelbourneStar, Grannis3, Kasirbot, Kaos Magician, Einsteiner900, Widr, Shelbylv, CasualVisitor, Hz.tiang, JimmyMachineGunHand, Cengime, BattyBot, Prof. Squirrel, Dexbot, Mogism, Makecat-bot, Ryan.laff, Jamesx12345, Ttitts, CsDix, Kenanwang, GregRos, Prokaryotes, Robertpb97, Occurring, Basedrawnz, Learnerktm, Julietvbarbara, Barbarousbunch815, Brendapallister, Nicholaspurcellstudio, Pickleslover, Tetra quark, Claudio.nahmad.arcaraz, KasparBot, The oracle 2015, Mitzionne, Huritisho and Anonymous: 255

- **Gravity** *Source:* https://en.wikipedia.org/wiki/Gravity?oldid=683336844 *Contributors:* Bryan Derksen, Danny, Rmhermen, Caltrop, Heron, Montrealais, Stevertigo, Patrick, D, Michael Hardy, Ixfd64, TakuyaMurata, GTBacchus, Wintran, Snarfies, Ahoerstemeier, Mac, Dgaubin, William M. Connolley, Jeff Relf, TonyClarke, Smack, Ec5618, Tpbradbury, Phoebe, Fairandbalanced, BenRG, Pollinator, Lumos3, Jni, Northgrove, Donarreiskoffer, Robbot, Sander123, Jakohn, TimothyPilgrim, Academic Challenger, Caknuck, Bkell, Nerval, Aetheling, Ruakh, Dina, Alan Liefting, Cedars, Ancheta Wis, Giftlite, Mikez, Haeleth, Wolfkeeper, Inkling, Fropuff, Everyking, Frencheigh, Aechols, Bobblewik, Utcursch, Keith Edkins, Antandrus, Phe, Quarl, Elembis, Kiteinthewind, Jossi, Karol Langner, Lindberg G Williams Jr, Demiurge, Trevor MacInnis, Grstain, Mike Rosoft, &Delta, DanielCD, Shipmaster, JimJast, Discospinster, Zaheen, Supercoop, Vsmith, Smyth, Notinasnaid, Dbachmann, Bender235, Rubicon, Loren36, El C, Huntster, Kwamikagami, Aude, RoyBoy, Nrbelex, Gershwinrb, Bobo192, Smalljim, Elipongo, I9Q79oL78KiL0QTFHgyc, Larry V, MPerel, Danski14, Alansohn, JYolkowski, Anthony Appleyard, Davetcoleman, Atlant, Maya Levy, Paleorthid, Andrew Gray, Cjthellama, Riana, Goldom, Mac Davis, Wdfarmer, Hdeasy, Snowolf, Mononoke–enwiki, BRW, Yuckfoo, RainbowOfLight, Jesvane, Bookandcoffee, Kazvorpal, Falcorian, Stephen, Feezo, Richard Arthur Norton (1958-), OwenX, Linas, Camw, LOL, Benhocking, JFG, MONGO, Mpatel, Tabletop, GregorB, Eaolson, Isnow, Scm83x, SDC, Philbarker, TheAlphaWolf, Brownsteve, Radiant!, Dysepsion, Sin-man, Ashmoo, Chrispasquale, Chun–hian, Rjwilmsi, Tangotango, Nichiran, Fel64, Kazrak, Ligulem, Ems57fcva, Boccobrock, Bhadani, MarnetteD, Matt Deres, Sango123, Syced, Yamamoto Ichiro, Fish and karate, JanSuchy, Magmafox, Titoxd, RobertG, Old Moonraker, Nihiltres, Nivix, RexNL, Gurch, Fresheneesz, Alphachimp, Tardis, Srleffler, King of Hearts, DVdm, Guliolopez, Hall Monitor, Bomb319, Gwernol, EamonnPKeane, Raelx, The Rambling Man, Wavelength, Drdisque, Sceptre, Hairy Dude, Jimp, Hillman, Brandmeister (old), Ohwilleke, Red Slash, Musicpvm, Anonymous editor, Loom91, Bhny, Pigman, Epolk, Philip Hazelden, DanMS, Scott5834, CambridgeBayWeather, Wimt, Cemcem, NawlinWiki, Nowa, Wiki alf, Ytcracker, SigPig, SCZenz, Nick, Brandon, Katrina Graziano, Semperf, Aaron Schulz, Roy Brumback, Addps4cat, Jessemerriman, Rayc, Mgnbar, Dna-webmaster, Eurosong, Heeroyuy135, Enormousdude, 2over0, SFGiants, Chase me ladies, I'm the Cavalry, Lappado, Endomion, E Wing, Fmyhr, Smurrayinchester, Kevin, Geoffrey.landis, Anclation–enwiki, Emc2, Willtron, Kungfuadam, RG2, JDspeeder1, Bo Jacoby, Draicone, FyzixFighter, Mejor Los Indios, Sbyrnes321, DVD R W, Hide&Reason, That Guy, From That Show!, Shenhemu, Luk, TravisTX, Sardanaphalus, SmackBot, Ulterior19802005, Incnis Mrsi, Reedy, KnowledgeOfSelf, Hydrogen Iodide, NaiPiak, David Shear, Kilo-Lima, Jagged 85, Thunderboltz, Delldot, Hardyplants, Cessator, Syckls, BiT, Timotheus Canens, GraemeMcRae, HalfShadow, Typhoonchaser, Yamaguchi🔔🔔, Algont, Hmains, Ppntori, ERcheck, Andy M. Wang, Kmarinas86, The monkeyhate, Saros136, Bluebot, Cush, Keegan, Raymond arritt, Fplay, Silly rabbit, Lehkost, Complexica, Bbq332, Jeff5102, Sbharris, Hallenrm, CharonM72, Scwlong, Can't sleep, clown will eat me, Scott3, UNHchabo, MJCdetroit, Apostolos Margaritis, Lesnail, Cryocide, Rsm99833, Amazins490, Jmlk17, Cybercobra, Nakon, Steve Pucci, TedE, Red1–enwiki, Jiddisch–enwiki, Dreadstar, Dave-ros, Weregerbil, Cockneyite, Crd721, Bryanmcdonald, Jklin, DMacks, Wizardman, Where, LeoNomis, Risker, Sadi Carnot, Carlosp420, TTE, Yevgeny Kats, Will Beback, Jonpalmer, Nathanael Bar-Aur L., Beasterline, Geoffrey Wickham, Rklawton, Djeneba, Sophia, Dsantesteban, Kuru, Thefro552, Titus III, Richard L. Peterson, John, Scientizzle, Stephane Yelle, DocRocks1, Jaganath, Thegathering, Skoobieschnax, JorisvS, LestatdeLioncourt, Coredesat, Accurizer, Minna Sora no Shita, Mgiganteus1, A5y, Nonsuch, Ridersbydelta, Mr. Lefty, AtD, Jess Mars, Ben Moore, JHunterJ, MarkSutton, Slakr, Special-T, Momolee, LuYiSi, Mr Stephen, Samaster1991, Spiel496, Buttle, Novangelis, PSUMark2006, Inquisitus, DI2000, ShakingSpirit, Hgrobe, Ginkgo100, Vanished user, JMK, Craigboy, Lakers, Newone, MOBle, J Di, StephenBuxton, Matt Bernius, Igoldste, Taucetiman, Tofoo, Tawkerbot2, Lsskys, George100, Kurtan–enwiki, Lahiru k, CalebNoble, SkyWalker, JForget, CmdrObot, Tanthalas39, Gholson, Porterjoh, Ale jrb, Scohoust, Aherunar, Galo1969X, Picaroon, Shakespeare87, User92361, Zureks, Basawala, Ruslik0, GHe, Dgw, OMGsplosion, ShelfSkewed, WHATaintNOcountryIeverHEARDofDOtheySPEAKenglishINwhat, MarsRover, Hi There, Groosh, Myasuda, Anthony Bradbury, Gregbard, Logicus, Cydebot, Steel, Travelbird, Red Director, Jon Stockton, A Softer Answer, Adolphus79, Nicesai, Rracecarr, Codingmasters, Ch0rx, Tawkerbot4, Alexnye, Christian75, M a s, DumbBOT, DarkLink, Interwiki gl, FastLizard4, Optimist on the run, Jimip, Waxigloo, SteveMcCluskey, Omicronpersei8, Stoked, Gimmetrow, Sevenaces, Raoul NK, FrancoGG, Thijs!bot, Epbr123, DarlingFriend, Opabinia regalis, Pajz, Ishdarian, Jamesluster, Andyjsmith, 24fan24, Gamer007, ClosedEyesSeeing, Headbomb, John254, Bobblehead, Neil916, Pogogunner, Grayshi, EdJohnston, HistoryMaster 1, Zachary, The Hybrid, Nick Number, Lithpiperpilot, CarbonX, MichaelMaggs, Sam42, J.S.B.Anderson, Escarbot, Eleuther, Mentifisto, Hmrox, AntiVandalBot, Yonatan, Luna Santin, CodeWeasel, Themaxeditor, Prolog, Yay unto the Chicken, Dylan Lake, Gdo01, Myanw, Ioeth, JAnDbot, Leuko, Husond, Vorpal blade, ThomasO1989, Roman à clef, MER-C, Nevadacall, Andonic, Hut 8.5, 100110100, N shaji, Pablothegreat85, Magioladitis, Foobird107, Murgh, Bennybp, Bongwarrior, VoABot II, AuburnPilot, Xn4, Wikidudeman, Hendrixjoseph, Careless hx, Aerographer1981, Crazytonyi, 9holdss, ThomasThePolishMan, Bubba hotep, Mi6agent00g off, BatteryIncluded, Beetfarm Louie, Adrian J. Hunter, Alexei Kojenov, Kane1047, LorenzoB, Mollwollfumble, Scot.parker, Andykass, Talon Artaine, Chris G, DerHexer, MeEricYay, WLU, Seph Vellius, TheRanger, Patstuart, Seba5618, Oroso, NatureA16, FisherQueen, Hdt83, E.vondarkmoor, MartinBot, Sinfear, Shimwell, Flamingpanda, The Ubik, UnfriendlyFire, APT, Rettetast, Juansidious, Anaxial, Sm8900, David J Wilson, Mschel, R'n'B, GarrisonGreen, AlexiusHoratius, Pekaje, LittleOldMe old, PrestonH, J.delanoy, Filll, Trusilver, Tonmoy Chowdhury, Bloomingiris, 72Dino, MoogleEXE, Lhynard, Ginsengbomb, WarthogDemon, Willow123–enwiki, SubwayEater, Yeti Hunter, Hisagi, James Mead, M C Y 1008, Wandering Ghost, Redmotherfive, Vertigo900, Mr Rookles, Samtheboy, Gurchzilla, Supuhstar, Pyrospirit, AntiSpamBot, GhostPirate, Belovedfreak, Raichu Trainer, Ohms law, Mitchell is hollywood, SJP, Policron, Touch Of Light, Pwnasaurusrex38, MKoltnow, Mufka, FJPB, Blckavnger, Cmichael, Mohrflies, Stoned Proffesor, Kenneth M Burke, Cosmictinker, RB972, Tiggerjay, U.S.A.U.S.A.U.S.A., Treisijs, Mike V, Redrocket, Gtg204y, MtyQuinn, Darkfrog24, Jxzj, Annax3, Smartman10, Ronbo76, Micmic28, Yecril, Missphysics, GoldenGolem, Xiahou, Lorax835, Steel1943, Washboard6, Sheliak, Funandtrvl, Gravityc, Teesup, Black Kite, Chinneeb, Deor, VolkovBot, TreasuryTag, TJ Elliot Scott, Meaningful Username, Danwills, DSRH, Mtesm, Indubitably, Gseletko, Veddan, Bouex, Dominics Fire, Heerojyuy, Philip Trueman, Childhoodsend, TXiKiBoT, GVlleneuve27, Davecas0, Adamwang, BuickCenturyDriver, 99DBSIMLR, MeStevo, Lolerballer–enwiki, Andrius.v, Z.E.R.O., Anonymous Dissident, Jonnymagic, Ask123, Trente, MattCarterSurrealist, IPSOS, Sarthella, Seraphim, Fizzackerly, Wvogeler, The One Cause, Tallcreek, Markp93, Drappel, PDFbot, Freak104, Manticore55, Cremepuff222, Blackdragon 1002, Wikiisawesome, Vrsixfire, Liberal Classic, Witchzilla, Noel rebeira, Inductiveload, Gladiator2155, Sydweighz, Spiral5800, Larklight, RobertFritzius, Dirkbb, SQL, SallyBoseman, Frag1983, Synthebot, ChillDeity, Speria, Heroesrule17, Enviroboy, Hollop09, Sylent, Kaseman519, Ballsucker22, Sesshomaru, Brianga, Skins88, Chickyfuzz14, AlleborgoBot, GavinTing, Shaidar cuebiyar, Happyhacker101, NHRHS2010, Steven Weston, D. Recorder, Al.Glitch, Xarr, The Random Editor, Mr. dick 008, SieBot, Tosun, Crunchedfor6, High2lowo, K. Annoyomous, Thong123456789, Work permit, Scarian, Invmog, Gerakibot, Mazza uk04, Mickeyd24, LealandA, Caltas, BreakfastTom, Beethoven314, Pieman123456789, RJaguar3, Triwbe, Rangutan, Chs^3, Tubular bells83, The way, the truth, and the light, Garrett gagne, Keilana, Elliott Fontain, CaptainIraq, RadicalOne, Tiptoety, CombatCraig, Bhahn0125, Oda Mari, Sunayanaa, Cnormansen, Thin joe, M keshe, Jirachipokemon, Kcin213, Chives4life, Luskj, Oxymoron83, Faradayplank, Smilesfozwood, AngelOfSadness, Nuttycoconut, Edwardwittenstein, Zharradan.angelfire, John fromer, Lightmouse, AMCKen, Iain99, Techman224, Skateboards fly, Mr mimises, Blobbucket, BenoniBot–enwiki, Dillard421, Panicum, Fedosin, C'est moi, Hamiltondaniel, ShexRox, Dolphin51, M2Ys4U, Into The Fray, Canglesea, C0nanPayne, Albin&dani, Quinling, Muhends, Moony1000, Atif.t2, Tomasz Prochownik, ClueBot, Thejman123456789, Artichoker, PipepBot, The Thing That Should Not Be, JavaJesus, Mr.Pinklesworth, Meekywiki, Artist7337, Drmies, Firth m, Mild Bill Hiccup, CounterVandalismBot, Kitty9992, VandalCruncher, Harland1, Otolemur crassicaudatus, Jackey0105, Andwor9, Aik-Bkj, Another Matt, RogerEllman, Puchiko, Shocky95, Gakusha, DragonBot, Joshisamazing, Chris earl 89, Alexbot, Mccann tom, Icreaser, Robbie098, Sct5333, Shinkolobwe, Fsunka, Itel94, Ola Hansen, Ice Cold Beer, Jotterbot, PhySusie, Glacialvortex, Razorflame, Handcannonbeast, Applejacks47, Chaosdruid, Thingg, Venera 7, Wnt, Myagooshki666, Deproduction, MasterOfHisOwnDomain, Jaykay0424, Ioannes93, TimothyRias, Misterbeal, InternetMeme, Mhamhamha, XLinkBot, Royboturso, Gwandoya, Crustillicus, BodhisattvaBot, Rror, Ance.cdas, Baconlover13, Ost316, Srossman07, Noctibus, JinJian, Truthnlove, Ttimespan, Airplaneman, Infonation101, EEng, Freestyle-69, Kbdankbot, Xerbaycom, Addbot, Nyw195, Willking1979, NateDres23, Bobocheese, Betterusername, Olli Niemitalo, Cre84u, TutterMouse, Presentabsent, Eedlee, Veraptor, WFPM, Ashanda, MrOllie, Lost on Belmont, FerrousTigrus, Ld100, Delaszk, Glass Sword, Maddox1, Jasper Deng, Harvardstudent, Kicka, Tide rolls, EugeneKantarovich, Cesiumfrog, Krano, WikiDreamer Bot, Hartz, Narutolovehinata5, Legobot, Drpickem, Luckas-bot, Dov Henis, Senator Palpatine, Il MusLiM HyBRiD II, JustWong, Becky Sayles, THEN WHO WAS PHONE?, AnakngAraw, Solo Zone, Azcolvin429, Atqueamemus, MacTire02, AnomieBOT, Jdiyef, Ahmediq152, Jt16733, Gatoradeparade, Kanwaraj, Piano non troppo, Chaosmaker39, Gsd65, LlywelynII, Rhlowe, Unicornlad, VCleemput, Kanat Abildinov, Typesships, Are you ready for IPv6?, Dje 8, Citation bot, Oddball.bfi, Nyrox395, Maxis ftw, Persistent76, Chase4813, ArthurBot, Snorlax Monster, Gravityforce, Xqbot, TinucherianBot II, TechBot, Millahnna, Markell West, GrouchoBot, Celebration1981, Gsard, Amaury, Charvest, Doulos Christos, A. di M., Acannas, CES1596, Ninjainventor, LucienBOT, Paine Ellsworth, Tobby72, TedlyW, Lookang, 暗月星, Slawomir Bialy, ArkianNWM, Allen Jesus, Parvons, Cannolis, Orion 8, Citation bot 1, Pinethicket, Tom.Reding, RedBot, IceBlade710, SpaceFlight89, Rohitphy, Jauhienij, RockSolidCosmo, FoxBot, TobeBot, Yunshui, Gafferuk, Comet Tuttle, Le Docteur, Schwede66, MitchLay, Oms22, Earthandmoon, Tbhotch, RobertMfromLI, Brambleclawx, RjwilmsiBot, Androstachys, DASHBot, EmausBot, John of Reading, Mnkyman, Atwarwiththem, Qurq, Gfoley4, Kueller1, Dewritech, Baguettes, Syncategoremata, Jmencisom, Sheeana, The Blade of the Northern Lights, Solomonfromfinland, JSquish, Ryan.vilbig, ZéroBot, Crua9, GoldRenet, Dtfgd, Everard Proudfoot, Quondum, Wikfr, Confession0791, Ocaasi, Wtsbeynon, BrokenAnchorBot, Brandmeister, L Kensington, Zayzya, Ally1604, Checkmark56, Fluctuating metric, RockMagnetist, Teapeat, Laned130, Rememberway, ClueBot NG, Nebulosus, CocuBot, PoqVaUSA, Jj1236, Lilptrsn, Megalobingosaurus, MerllwBot, Helpful Pixie Bot, BG19bot, Negativecharge, Furkhaocean, GKFX, Cadiomals, Mr.viktor.stepanov, RGloucester, BattyBot, Tutelary, GravityForce, Padenton, Khazar2, JYBot, Davidlwinkler, Lightand-Dark2000, Mogism, Rudrene, Reatlas, CsDix, Everymorning, Yuan Jullian Morales, Gacman67, DavidLeighEllis, Prokaryotes, Jwratner1, My name is not dave, Mfb, Konveyor Belt, Mahusha, Monkbot, Filedelinkerbot, Gronk Oz, Oiyarbepsy, Stefania.deluca, Loraof, Ps20231131, Pishcal, Freedom2003, Tetra quark, James 123234, Inorout, Supdiop, KasparBot, The oracle 2015, Sweepy, Jeffryan123, ImperatorRomanorvm, Addycrisp, The golden colten, J.A.Witt (Tony), Hiimemilylol, Dan6233 and Anonymous: 1221

- **Strong interaction** *Source:* https://en.wikipedia.org/wiki/Strong_interaction?oldid=681830498 *Contributors:* AxelBoldt, Sodium, Bryan Derksen, RK, Andre Engels, Danny, Peterlin–enwiki, Heron, Xavic69, Tim Starling, EddEdmondson, Gdarin, Ixfd64, Wintran, Salsa Shark, AugPi, Timwi, Bemoeial, Wikiborg, Fuzheado, ElusiveByte, Populus, Phys, Omegatron, Bevo, PuzzletChung, Robbot, Mayooranathan, Henrygb, Giftlite, DocWatson42, Sj, Harp, Monedula, Xerxes314, Remy B, Jason Quinn, Utcursch, Zfr, AmarChandra, Sam Hocevar, Nicobn–enwiki, Jørgen Friis Bak, Discospinster, FT2, Vsmith, ArnoldReinhold, Trekie8472, Roybb95–enwiki, David Schaich, Gianluigi, Marknewlyn–enwiki, Drhex, CDN99, Army1987, Matt McIrvin, Danski14, VivaEmilyDavies, TenOfAllTrades, Vuo, DV8 2XL, Kazvorpal, Oleg Alexandrov, Superstring, Spettro9, BillC, Eras-mus, Tevatron–enwiki, Mendaliv, Zbxgscqf, Strait, Loudenvier, Donotresus, Yamamoto Ichiro, Siv0r, Lmatt, Srleffler, Chobot, DVdm, Wavelength, Borgx, Bambaiah, Phmer, Jimp, Supasheep, Limulus, Salsb, RazorICE, Expensivehat, Dhollm, Lobwedge, Superiority, Tetracube, Willtron, GrinBot–enwiki, Asterion, Finell, AndrewWTaylor, SmackBot, RDBury, Tom Lougheed, Trojo–enwiki, Jrockley, Dauto, Chris the speller, Jjalexand, Complexica, Sbharris, Colonies Chris, Richard001, TTE, Spiritia, Titus III, FrozenMan, Newone, Happy-melon, Conrad.Irwin, Rowellcf, Chrisahn, Cydebot, Danny Bierek, Mtpaley, Wannabe Runny, Irigi, Headbomb, Niduzzi, KP Botany, Tlabshier, Hanzoro5, Dougher, Steelpillow, JAnDbot, Supertheman, Roleplayer, Magioladitis, Kopovoi, Vssun, Hoverfish, Khalid Mahmood, Pan Dan, Robin S, Vortimer, Sujaybhu, Natsirtguy, Peter Chastain, Cpiral, Tygrrr, Treisijs, Alpvax, VolkovBot, JohnBlackburne, TXiKiBoT, Marskuzz, Muro de Aguas, Qxz, Sintaku, SieBot, Gerakibot, Escape Artist Swyer, Proton666, ObfuscatePenguin, ClueBot, GorillaWarfare, Jackey0105, DragonBot, Jefflayman, Nownownow, Cenarium, Razorflame, Zahnrad, Silvercromagnon, InternetMeme, Rreagan007, WikHead, Drogs630, Addbot, Guoguo12, Omega Squad, ThisIsMyWikipediaName, Seratna, CarsracBot, Purple Emu, CosmiCarl, AgadaUrbanit, Tide rolls, Lightbot, Luckas-bot, Timeroot, Donthedev, Rifter0x0000, Umnum, AnomieBOT, VanishedUser sdu9aya9fasdsopa, Orange Knight of Passion, Piano non troppo, Citation bot, Obersachsebot, Xqbot, DSisyphBot, Barelistido, Almabot, GrouchoBot, RibotBOT, SassoBot, Mnmngb, CES1596, Gummer85, Citation bot 1, Boulaur, RedBot, Jauhienij, Surf5270, ElPeste, Slon02, EmausBot, John of Reading, Mnkyman, JSquish, Cogiati, Bamyers99, Rexprimoris, Donner60, ChuispastonBot, ClueBot NG, Jj1236, Helpful Pixie Bot, Bolatbek, ElphiBot, J.wong.wiki, Glevum, Zedtwitz, Zedshort, Nishantkumar19, Kisokj, YFdyh-bot, Andyhowlett, Reatlas, CsDix, EvergreenFir, Aurelianjh, Jwratner1, Diggerh, Kshitizarora2993, Tetra quark, KasparBot and Anonymous: 171

- **Weak interaction** *Source:* https://en.wikipedia.org/wiki/Weak_interaction?oldid=679571094 *Contributors:* AxelBoldt, Chenyu, Sodium, Bryan Derksen, Tarquin, AstroNomer~enwiki, Andre Engels, XJaM, Heron, JohnOwens, Gdarin, Delirium, Andrewa, Andres, Emperorbma, Timwi, Fibonacci, Phys, Phil Boswell, Lowellian, Mayooranathan, Tobias Bergemann, Giftlite, Sj, Herbee, Xerxes314, Jcobb, Mckaysalisbury, Munkee, Toby Woodwark, Bbbl67, Icairns, AmarChandra, Lumidek, Jørgen Friis Bak, Discospinster, ArnoldReinhold, Roybb95~enwiki, Gianluigi, Joanjoc~enwiki, Shanes, AJP, AtomicDragon, Danski14, Alansohn, Arthena, Axl, SidneySM, Hwefhasvs, DV8 2XL, Nightstallion, Kazvorpal, Linas, StradivariusTV, Benbest, Bbatsell, Palica, Tevatron~enwiki, Graham87, BD2412, Ketiltrout, Rjwilmsi, Strait, Erkean, The wub, FlaBot, Naraht, Itinerant1, Srleffler, Chobot, Krishnavedala, YurikBot, Borgx, Bambaiah, Hairy Dude, Jimp, Sillybilly, Conscious, Epolk, JabberWok, Gaius Cornelius, Shaddack, SCZenz, Irishguy, Shimei, Willtron, RG2, Phr en, That Guy, From That Show!, Luk, SmackBot, David Kernow, Tom Lougheed, WookieInHeat, Dauto, Chris the speller, Philosopher, Moshe Constantine Hassan Al-Silverburg, Complexica, DHNbot~enwiki, Zircomscot, BIL, Wen D House, "alyosha", Maxwahrhaftig, Akriasas, Vina-iwbot~enwiki, Bdushaw, TTE, SashatoBot, Fontenello, Herr apa, Condem, Tony Fox, MottyGlix, JRSpriggs, Heartofgoldfish, Calmargulis, Green caterpillar, Joelholdsworth, Cydebot, Michael C Price, Mtpaley, Thijs!bot, ChKa, Kichwa Tembo, Headbomb, Hcobb, Icep, Escarbot, AntiVandalBot, Jimeree, Steelpillow, JAnDbot, Magioladitis, Swpb, بلال, Wormcast, DAGwyn, Giggy, Khalid Mahmood, Gah4, Tarotcards, 2help, Lighted Match, DorganBot, Halmstad, Idiomabot, VolkovBot, Jeuadros, Hilarious Bookbinder, TXiKiBoT, Rei-bot, CaptinJohn, Awl, Shenanegins, BotKung, Wingedsubmariner, Antixt, Xxxlilbritxxx, Ptrslv72, Monty845, AlleborgoBot, SieBot, Paolo.dL, Skyentist, Ptr123, ClueBot, Bondchic007, SuperHamster, Erudecorp, Rotational, Jackey0105, Alexbot, Cenarium, Zomno, Zahnrad, He6kd, TimothyRias, InternetMeme, Timo Metzemakers, Stephen Poppitt, Addbot, Some jerk on the Internet, Markdman, ChenzwBot, Ehrenkater, Tide rolls, Luckas-bot, Yobot, Les boys, Kilom691, THEN WHO WAS PHONE?, Rifter0x0000, Duping Man, Dickdock, Magog the Ogre, AnomieBOT, Materialscientist, Citation bot, Quebec99, Kreigiron, Xqbot, Drilnoth, BurntSynapse, GrouchoBot, Omnipaedista, RibotBOT, Workanode, Jaz1305, Mnmngb, Dave3457, FrescoBot, Charles.walker, LucienBOT, Ionutzmovie, Grandiose, Pinethicket, Boulaur, Rameshngbot, RedBot, 23790AD, Tea with toast, Jauhienij, FoxBot, Earthandmoon, RjwilmsiBot, Itamarhason, Newty23125, EmausBot, WikitanvirBot, GA bot, GoingBatty, Splibubay, StringTheory11, Braswiki, Git2010, Wayne Slam, Jsayre64, Maschen, ChuispastonBot, ClueBot NG, VinculumMan, Physics is all gnomes, Fjpyanez, Mouse20080706, Helpful Pixie Bot, Geo7777, Bibcode Bot, Junaid2754, Bolatbek, Phbarnacle, Neutral current, Glevum, Idenshi, Marioedesouza, Dexbot, Spray787, Reatlas, CsDix, Jamesmcmahon0, Ihatedirac2k13, Kharkiv07, Jwratner1, YimmyYohnson, Monkbot, BalderdashVonDrivel, ASCarretero, Malerisch, Lachlan Newland, Tetra quark, KasparBot and Anonymous: 155

- **Electromagnetism** *Source:* https://en.wikipedia.org/wiki/Electromagnetism?oldid=682255582 *Contributors:* AxelBoldt, Magnus Manske, Trelvis, Carey Evans, CYD, Mav, Bryan Derksen, Zundark, Szopen, The Anome, Malcolm Farmer, FreddyZ, Miguel~enwiki, William Avery, Maury Markowitz, Heron, Patrick, Tim Starling, Kku, Bcrowell, Delirium, Ahoerstemeier, William M. Connolley, Khorn, Babbo, Darkwind, Kevin Baas, Julesd, Glenn, Sray, Poor Yorick, Mm, Rawr, Mxn, Hike395, Emperorbma, Wikiborg, Reddi, Phys, Lumos3, Phil Boswell, Rob-bot, Hankwang, Pigsonthewing, ZimZalaBim, Arkuat, Hemanshu, Texture, Roscoe x, Sunray, Wikibot, Fuelbottle, Lupo, Diberri, Wile E.Heresiarch, Tobias Bergemann, Ancheta Wis, Decumanus, Giftlite, DocWatson42, Andries, Wolfkeeper, Lethe, Koyn~enwiki, Everyking, Snowdog, Dratman, Curps, Ssd, Tom-, Jason Quinn, Brockert, Sohanley, Karol Langner, APH, Maximaximax, Bodnotbod, Icairns, Lumidek, Iantresman, Tsemii, Slipstream (usurped), Adashiel, Mike Rosoft, EugeneZelenko, Rich Farmbrough, Bedel23, Pjacobi, Vsmith, Theiloth, MuDavid, Paul August, Bender235, El C, Shanes, Mkosmul, RoyBoy, Femto, Matt McIrvin, Bert Hickman, Physicistjedi, Sam Korn, SignorGiuseppe, Mareino, Ranveig, Jumbuck, Red Winged Duck, Alansohn, Gary, Pinar, ChristopherWillis, Arthena, Atlant, Andrewpmk, Lectonar, Snowolf, Melaen, BRW, Wtshymanski, Evil Monkey, Bob1817, Mikeo, Gene Nygaard, Ttownfeen, Oleg Alexandrov, Cimex, TheNightFly, MONGO, Nakos2208~enwiki, Macaddct1984, Eras-mus, Gimboid13, Rnt20, Graham87, Qwertyus, Ando228, Rjwilmsi, Collins.mc, Tan-gotango, Ligulem, SeanMack, Bhadani, Yamamoto Ichiro, Cjpuffin, FlaBot, Nihiltres, Nivix, RexNL, Ewlyahoocom, Gurch, Otets, Fresh-eneesz, Tea Drinker, Srleffler, Physchim62, Chobot, Roboto de Ajvol, YurikBot, Wavelength, Spacepotato, JabberWok, CambridgeBay-Weather, Salsb, Wimt, David R. Ingham, NawlinWiki, Bachrach44, Buster79, Tearlach, Anetode, Scottfisher, Figaro, Bota47, Nick123, FF2010, Light current, Orioane, Enormousdude, 21655, 2over0, JoanneB, Phil Holmes, Willtron, Sizarieldor, AGToth, Katieh5584, RG2, GrinBot~enwiki, Sbyrnes321, DVD R W, Luk, SmackBot, Manu0x0~enwiki, PEHowland, Prodego, KnowledgeOfSelf, McGeddon, Jagged85, Jrockley, Swerdnaneb, Rpmorrow, Skizzik, JAn Dudík, LinguistAtLarge, Master of Puppets, Complexica, CMacMillan, TheGerm, Frap, Ioscius, Avoidance, SundarBot, Stevenmitchell, Cybercobra, MichaelBillington, Blake-, Akriasas, Illnab1024, Daniel.Cardenas, LeoNomis, Sadi Carnot, Apoorvchebolu, Skinnyweed, TTE, WayKurat, Sarfa, DJIndica, Nmnogueira, Ozhiker, Wvbailey, Finejon, UberCryxic, Philoso-phus, Cronholm144, Ckatz, El Dahveed, Grapetonix, Alatius, Sinistrum, Dicklyon, Jon186, Waggers, Ryulong, Describer, KJS77, Newone, Adambiswanger1, Courcelles, Ziusudra, Tawkerbot2, Yanah, Xcentaur, Mosaffa, JForget, KyraVixen, Baiji, Vyznev Xnebara, Fjomeli, MarsRover, Musicalantonio, Marly88, Peripitus, Fifo, Ssilvers, UberScienceNerd, Thijs!bot, VoABot, Jb.schneider-electric, Headbomb, Marek69, GerryAshton, Leon7, D.H. MichaelMaggs, AntiVandalBot, Majorly, Seaphoto, Quintote, Gnixon, Jnyanydts, Tyco.skinner, Eleos, Steelpillow, JAnD-bot, Matthew Fennell, Mkch, Bongwarrior, VoABot II, Tails4, SHCarter, TxAlien, GBYork, WhatamIdoing, Vanished user ty12k189jq10, Adrian J. Hunter, 28421u2232nfenfcenc, Coldwarrier, User A1, Khalid Mahmood, Ruhihumphries, PrattTA1, InvertRect, MartinBot, Jar-gon777, LedgendGamer, J.delanoy, Sasajid, Abecedare, Bogey97, JohnPritchard, Maurice Carbonaro, Extransit, Tarotcards, NewEnglandYan-kee, Wesino, Shoesss, Cometstyles, ACBest, DorganBot, Treisijs, Bently34, Lights, 28bytes, Part Deux, Thedjatclubrock, Constant314, Philip Trueman, TXiKiBoT, The Original Wildbear, Guillaume2303, Anonymous Dissident, Kevin Steinhardt, Monkey Bounce, Captin-John, Imasleepviking, Mbarrieau, DoktorDec, Atomicswoosh, TongueSpeaker, Andy Dingley, Dirkbb, Lova Falk, @pple, Sylent, Doc James, PGWG, NHRHS2010, SieBot, Coffee, Tresiden, Graham Beards, Work permit, Hertz1888, Avargasm, Winchelsea, Dawn Bard, Caltas, Yin-tan, Zoragotcha, Keilana, Flyer22, Qst, Csblack, Klmeze~enwiki, Joseph Banks, Oxymoron83, Fbarw, Maelgwnbot, StaticGull, Dolphin51, TreeSmiler, Kanonkas, Anilina, ElectronicsEnthusiast, Llywelyn2000, WikipedianMarlith, Bschaeffer~enwiki, Atif.t2, Martarius, ClueBot, Stevekirst7, The Thing That Should Not Be, Jan1nad, Uncle Milty, Blanchardb, Twicemost, LizardJr8, Jackey0105, Electromagnetic, OttoTanaka, Excirial, Kocher2006, BlueLikeYou, Jusdafax, Abrech, Plastic Fish, MacedonianBoy, PhySusie, Friedlibend und tapfer, Thingg, Je-sus.murphy55567, Versus22, InternetMeme, XLinkBot, Emmette Hernandez Coleman, Jovianeye, Rror, Boratlike, HarlandQPitt, Rogimoto, Deineka, Addbot, Dustbin123, Allfor12008, CanadianLinuxUser, Download, EconoPhysicist, Redheylin, WikiDegausser, K Eliza Coyne, LinkFA-Bot, Kisbesbot, Agadz aUrbanit, VASANTH S.N., Tide rolls, Lightbot, Gail, Kurtis, Will.M.Thompson, Luckas-bot, Yobot, Ptbot-gourou, Lichen from Hell, ScienceMind, Tempodivalse, Orion11M87, AnomieBOT, Jim1138, Piano non troppo, AdjustShift, Penguinatortoo.

Materialscientist, Citation bot, Vuerqex, Xqbot, Konor org, Lolman33, Plumpurple, Melmann, DSisyphBot, GrouchoBot, Nayvik, Omnipaedista, RibotBOT, Jsleaby, Maplestory101, Slowart, A. di M., GliderMaven, Tobby72, Wikipe-tan, Sum33, Ian88800, Hippycaller, Steve Quinn, HamburgerRadio, Citation bot 1, Чахонін Уладзіслаў, Pinethicket, Elockid, Hard Sin, LittleWink, PvsKllKsVp, Tinton5, Yahia.barie, Jschnur, Ezhuttukari, Corinne68, FoxBot, Рыцарь поля, Retired user 0001, SchreyP, Hickorybark, ItsZippy, Lotje, Cjlim, LilyKitty, Antipastor, Reaper Eternal, DARTH SIDIOUS 2, Triden, RjwilmsiBot, Alph Bot, Agent Smith (The Matrix), WildBot, DASHBot, EmausBot, John of Reading, WikitanvirBot, Mnkyman, Never give in, ITshnik, IncognitoErgoSum, RA0808, Tommy2010, Amrator, Wikipelli, Thecheesykid, JSquish, ZéroBot, Harddk, Leptonoggin, Fæ, Maypigeon of Liberty, WFarver, H3llBot, Quondum, Git2010, Makecat, Sonygal, Coasterlover1994, L. Kensington, Rr2wiki, Maschen, Donner60, Rjowsey, RockMagnetist, Peter Karlsen, Sir sachin, Teapeat, Planetscared, Cgt, Xanchester, ClueBot NG, Jack Greenmaven, MelbourneStar, Hiperfelix, Hallaman3, Ant.acke, Enopet, Frietjes, Milikguay, Widr, Helpful Pixie Bot, Eemeginnis, Vagobot, Bolatbek, Manuelfeliz, MusikAnimal, Stephenwanjau, Rm1271, Thekillerpenguin, Cadiomals, Mariano Blasi, Whyamidampandalol, Franz99, Physicsch, Brad7777, Glucialfox, Neumannjo, Acul132, ChrisGualtieri, Dexbot, Marlowfrontier, Lugia2453, Frosty, Josophie, Hamerbro, SubratamindPal, Trillig, Abhaikumar10, CsDix, Ihatedirac2k13, Hell to earth88, CROY123, Zenibus, Jwratner1, YiFeiBot, JosephSpiral, SpecialPiggy, Kdmeaney, Yashshrotf97, Csutric, Mahusha, Venomous Cobra, Trackteur, Gronk Oz, Meski33, Hkeyser, Lolbob12345, Nanophysisct12345, Simonessnygg, Noobmagnet, Mediavalia, Slayeredwarrior, Tetra quark, Isambard Kingdom, SergioCruz2015, Aenfinger, KasparBot, Zeke Essiestudy, Kafishabbir, JJMC89, Cesarnajera56 and Anonymous: 702

- **Elementary particle** *Source:* https://en.wikipedia.org/wiki/Elementary_particle?oldid=683602242 *Contributors:* CYD, Mav, Bryan Derksen, XJaM, Heron, Stevertigo, Patrick, Fbjon, Looxix~enwiki, Александър, Julesd, Glenn, AugPi, Mxn, Timwi, Reddi, Tpbradbury, Furrykef, Bevo, Donarreiskoffer, Robbot, Craig Stuntz, Nurg, Papadopc, Wikibot, Jimduck, Anthony, Ancheta Wis, Giftlite, DavidCary, Mikez, Haselhurst, Monedula, Xerxes314, Alison, Guanaco, Greydream, Anythingyouwant, Bodnotbod, Kate, Brianjd, Mormegil, Urvabara, Rich Farmbrough, Guanabot, Qutezuce, Hidaspal, Dmr2, Goplat, RJHall, RoyBoy, Robotje, Neonumbers, דוד ל, Dirac1933, DV8 2XL, Azmaverick623, Blaxthos, Kay Dekker, Joriki, Simetrical, TomTheHand, Mpatel, Isnow, Ggonnell, Palica, Strait, Miserlou, Ligulem, Naraht, DannyWilde, Lmatt, Srleffler, Chobot, Cactus.man, Roboto de Ajvol, YurikBot, Hairy Dude, NTBot~enwiki, Ohwilleke, Albert Einsteins pipe, Stephenb, Chaos, Vibritannia, SCZenz, Edwardlalone, Larsobrien, Bota47, BraneJ, Dna-webmaster, Arthur Rubin, Oyvind, GrinBot~enwiki, SmackBot, Mrcoolbp, Bomac, GrGBL~enwiki, Chris the speller, MalafayaBot, George Rodney Maruri Game, Silly rabbit, Complexica, MovGP0, Fmalan, Scwlong, Amazins490, Cybercobra, EPM, Garry Denke, Drphilharmonic, Sadi Carnot, ArglebargleIV, Tktktk, NongBot~enwiki, WhiteHatLurker, Jonhall, Dekaels~enwiki, Jynus, Newone, Courcelles, Laplace's Demon, SchmittM, J Milburn, Fordmadoxfraud, Cydebot, Bvcrist, Kozuch, Thijs!bot, Lord Hawk, Headbomb, MichaelMaggs, Escarbot, Ssr, JAndbot, Eurobas, Acroterion, VoABot II, Appraiser, BatteryIncluded, R'n'B, Sgreddin, MikeBaharmast, Lk69, Acalamari, DraakUSA, TomasBat, Joshua Issac, Kenneth M Burke, Ken g6, Idioma-bot, VolkovBot, SarahLawrence Scott, Nxavar, JhsBot, Abdullais4u, Lejarrag, Antixt, PGWG, SieBot, Timb66, Sonicology, PlanetStar, Bamkin, Dhatfield, Byrialbot, Svick, Perfectapproach, Thorncrag, Big55e, ClueBot, Jmorris84, Maxtitan, Alexbot, Dekisugi, Paradoxalterist, Saintlucifer2008, Cockshut12345, Rreagan007, RP459, Truthnlove, Addbot, Yakiv Gluck, Draco 2k, Mac Dreamstate, Funky Fantom, CarsracBot, HerculeBot, Legobot, Blah28948, Luckas-bot, Zhitelew, KamikazeBot, Kulmalukko, Orion11M87, AnomieBOT, Girl Scout cookie, Templatehater, Icalanise, Citation bot, Onesius, Vuerqex, Bci2, ArthurBot, Rightly, Xqbot, Phazvmk, Kirin13, FrescoBot, Delphinus1997, Steve Quinn, Robo37, SuperJew, HRoestBot, Sthyne, Hellknowz, Yahia.barie, Skyerise, Tobi - Tobsen, FoxBot, Physics therapist, Think!97, Bj norge, RjwilmsiBot, Beyond My Ken, EmausBot, John of Reading, Mnkyman, GoingBatty, Mthorndill, ZéroBot, Bollyjeff, StringTheory11, Markinvancouver, Quantumor, Maschen, RolteVolte, Negovori, NTox, I hate whitespace, ClueBot NG, CocuBot, Widr, Micah.yannatos1, Helpful Pixie Bot, Guzman.c, Bibcode Bot, BG19bot, Spaceawesome, Rainbot, Leaverward, Let'sBuildTheFuture, Eduardofeld, Sha-256, Dr.RobertTweed, ZX95, Joeinwiki, Mark viking, Cephas Atheos, Yo butt, Snakeboy666, Psyruby42, Haminoon, Sardeth42, TaiSakuma, LadyCailin, Morph dtlr, Delbert7, Karam adel, Liance, Isambard Kingdom, KasparBot, Kurousagi and Anonymous: 187

- **Planck energy** *Source:* https://en.wikipedia.org/wiki/Planck_energy?oldid=682694906 *Contributors:* Bryan Derksen, Nnh, Henrygb, Jason Quinn, Superborsuk, Icairns, Lumidek, CoyneT, Hidaspal, Rbj, Jag123, Ardric47, Wendell, Keenan Pepper, Bucephalus, Robert K S, Grace Note, Joke137, Christopher Thomas, Thechamelon, HappyCamper, Chobot, SmackBot, Baad, JoeMarfice, A.R., Gruznov, Lambiam, Eassin, Zarex, Zginder, Z10x, Spartaz, Omeganian, Maurice Carbonaro, Thurth, TXiKiBoT, Spiral5800, Djr32, SchreiberBike, Palnot, Addbot, Lightbot, Luckas-bot, AnomieBOT, RibotBOT, Erik9bot, Ammodramus, User10 5, Anagogist, BG19bot, Timothy Gu, Yardimsever, Coreyemotela, Nertuop and Anonymous: 21

- **Supersymmetry** *Source:* https://en.wikipedia.org/wiki/Supersymmetry?oldid=681009502 *Contributors:* Bryan Derksen, Taw, Andre Engels, Roadrunner, Maury Markowitz, Ewen, Stevertigo, Edward, Michael Hardy, Arpingstone, Theresa knott, IMSoP, Jeandré du Toit, Samw, Smack, Charles Matthews, Maximus Rex, Phys, Raul654, BenRG, Rursus, Mor~enwiki, Ancheta Wis, Giftlite, Mporter, Ferkelparade, Monedula, Fropuff, Xerxes314, Anville, Gus Polly, Moyogo, Unconcerned, DO'Neil, Maarten van Vliet, Pharotic, LiDaobing, Sam Hocevar, Lumidek, Deglr6328, Arivero, Rich Farmbrough, Roybb95~enwiki, Bender235, El C, Nornagon~enwiki, Duk, Tweet Tweet, Russ3Z, LostLeviathan, Pearle, Gary, Francescog~enwiki, Wtmitchell, RJFJR, Reaverdrop, Blaxthos, Killing Vector, Jordan14, Ted BJ, MONGO, Mpatel, MFH, SeventyThree, Bodera, VermillionBird, Drbogdan, Rjwilmsi, Josiah Rowe, R.e.b., Bubba73, Maxim Razin, Drrngrvy, FlaBot, Cless Alvein, Nowhither, Itinerant1, Gparker, KFP, Lmatt, Chobot, Vyroglyph, YurikBot, Wavelength, RussBot, Ohwilleke, Bhny, Epolk, Sasuke Sarutobi, Maxim Leyenson, Chaos, Romanc19s, Bota47, Mgnbar, Closedmouth, Arthur Rubin, RG2, That Guy, From That Show!, A bit iffy, SmackBot, Mira, Kurochka, Wangjiaji, Gilliam, Bluebot, Cadmasteradam, Complexica, Bazonka, Colonies Chris, Can't sleep, clown will eat me, QFT, Ruff ilb, Wen D House, Solarapex, Radagast83, Jgwacker, TheMaster42, Martijn Hoekstra, Acjohnson55, Yevgeny Kats, Charleswestbrook, TriTertButoxy, Lambiam, Tktktk, Xiaphias, JarahE, Mdanziger, Dan Gluck, Newone, Marysunshine, Tawkerbot2, Cydebot, Hydraton31, Bazzargh, David edwards, Michael C Price, Crum375, Koeplinger, Headbomb, J.christianson, Escarbot, Salgueiro~enwiki, Kborland, Jpod2, Cgingold, Maliz, TimidGuy, C9, Kostisl, R'n'B, Zentropa77, Natsirtguy, Maurice Carbonaro, Kevin Hickerson, Shawn in Montreal, Idioma-bot, Sheliak, Cuzkatzimhut, Nxavar, Kawakameha, Cuboidal, Ptrslv72, PhysPhD, Kbrose, SieBot, Nn123645, ClueBot, Jcpilman, Chessmaster7m, Kitsunegami, Rhododendrites, Mastertek, Mishas42, Scrabby~enwiki, TimothyRias, WikHead, MystBot, Addbot, DOI bot, Zabd, Barak Sh, F Notebook, Lightbot, Windward1, Luckas-bot, Yobot, Ibayn, TaBOT-zerem, Amirobot, Nonnormalizable, AnomieBOT, Girl Scout cookie, Materialscientist, Citation bot, ArthurBot, Plumpurple, Tomwsulcer, Omnipaedista, Gsard, CES1596, FrescoBot, HaloStereo1, Paine Ellsworth, Xmikywayx, Citation bot 1, Gil987, Kikeku, Jonesey95, Eddie Nixon, MondalorBot, Aknochel, Tom1661, Gagoga ju, TobeBot, Puzl bustr, Andraas, EmausBot, Djloststylez, Ddimensões, Arbnos, Susy is it, ChuispastonBot, Isocliff, ClueBot NG, KagakuKyouju, IJVin, Frietjes, Helpful Pixie Bot, Bibcode Bot, BG19bot, Teika kazura, JayBeeEye, Ninmacer20, ChrisGualtieri, Dexbot, Logosun, AHusain314, NA48, Rfassbind, Katherine Pendleton, Lioinnisfree, Laplacemat, Liquidityinsta, TaiSakuma, Stamptrader, Kdmeaney, Qxxxxxq,

Almaionescu, Monkbot, Janhaithabu, Mammoth2011, Jwill530, Stacie Croquet, Cuttlas1 and Anonymous: 173

- **Superstring theory** *Source:* https://en.wikipedia.org/wiki/Superstring_theory?oldid=682066627 *Contributors:* Mav, Bryan Derksen, Stevertigo, Michael Hardy, Erik Zachte, Minesweeper, Looxix~enwiki, Ahoerstemeier, JWSchmidt, Cyan, Palfrey, Evercat, Schneelocke, Hashar, Charles Matthews, Tpbradbury, Motor, David Shay, Omegatron, Bevo, Bcorr, Robbot, Fredrik, Hadal, Vuara, Tobias Bergemann, Giftlite, Barbara Shack, Herbee, Fropuff, Anville, Maarten van Vliet, WalkinDownThirtyThree, Christopherlin, Steuard, Karol Langner, Lumidek, Prestonmarkstone, Rich Farmbrough, Igorivanov~enwiki, Autiger, Pavel Vozenilek, El C, Rgdboer, Shadow demon, Causa sui, Billymac00, BM, Gary, Pion, Tycho, Cal 1234, Redvers, Postrach, Supercool Dude, Mindmatrix, Mpatel, Joke137, Mandarax, Bill37212, Yamamoto Ichiro, Bubbleboys, Chobot, Ben Tibbetts, Wavelength, RussBot, Chris Capoccia, Chensiyuan, Cate, Chaos, NawlinWiki, Astral, Voidxor, TheMadBaron, Zerodamage, Allens, SmackBot, Android 93, Kurochka, McGeddon, Kintetsubuffalo, Cesoid, Silly rabbit, Stevage, Baronnet, Colonies Chris, Sewlong, Mesons, Kurrupt3d, Bjankuloski06en~enwiki, Makyen, MathStuf, Hu12, Iridescent, Kahalachan, Gatortpk, Mattbr, Neelix, Gregbard, Nauticashades, Cydebot, Davidanzaldua, ChKa, Headbomb, Escarbot, Jj137, Shlomi Hillel, Dougher, JAnDbot, 100110100, 28421u2232nfenfcenc, Aziz1005, Jean-Pierre Petit~enwiki, Hans Dunkelberg, Maurice Carbonaro, Bot-Schafter, SmilesALot, Student7, Cmichael, Sheliak, JayCo777, Calwiki, Andrius.v, Molinogi, Billinghurst, Lamro, Enviroboy, Seraphita~enwiki, Drschawrz, Henry Delforn (old), Lightmouse, Altzinn, Gratedparmesan, ClueBot, Arakunem, Vergil 577, Frdayeen, Vizzini101, Niceguyedc, Neverquick, Resoru, Mastertek, Kakofonous, Princess Janay, Alex123irish123, Madeinmexico567, Oldnoah, Madeinmexico566, Truthnlove, YeAaMsLtA, Addbot, Physicman123, CWatchman, Cuaxdon, Semdino, AnnaFrance, LinkFA-Bot, TaBOT-zerem, Evans1982, Gerixau, Eric-Wester, Magog the Ogre, AnomieBOT, DemocraticLuntz, ^musaz, Josh Guffin, Jim1138, Citation bot, Renaissancee, BLP-outrageous move logs, Omnipaedista, RibotBOT, Paine Ellsworth, Steve Quinn, Tom.Reding, Klavesin, Tkachyk, Dinamik-bot, Bj norge, ldh0854, Arbnos, Vramasub, L Kensington, Particle hep, Isocliff, ClueBot NG, Widr, Adminium, Delivernews, Bibcode Bot, Khanduras, Quarkgluonsoup, Flowerhat15, MythosMagic, OCCullens, Aldrich2122, Graphium, AHusain314, Jochen Burghardt, WorldWideJuan, Jakec, Liquidityinsta, E8xE8, Polytope24, FlaviusCorcoata, Cirksena, BakedLikaBiscuit, KasparBot, Rantonels and Anonymous: 173

- **M-theory** *Source:* https://en.wikipedia.org/wiki/M-theory?oldid=674732283 *Contributors:* AxelBoldt, CYD, Eloquence, BF, Bryan Derksen, Zundark, The Anome, Ap, Tim Chambers, Hari, Maury Markowitz, Stevertigo, Michael Hardy, Tim Starling, Gabbe, Tompagenet, Ixfd64, CesarB, Looxix~enwiki, JWSchmidt, Darkwind, Marco Krohn, Jeandré du Toit, Evercat, Schneelocke, Charles Matthews, Timwi, Reddi, Malcohol, Bevo, Jusjih, Slawojarek, Sander123, Fredrik, R3m0t, RedWolf, Blainster, DHN, Hadal, HaeB, Tobias Bergemann, David Gerard, Giftlite, DocWatson42, Jmnbpt, Barbara Shack, Fropuff, Moyogo, Sigfpe, Daen, Antandrus, Lumidek, ChrisCostello, Mike Rosoft, Spiffy sperry, Urvabara, Noisy, Discospinster, H0riz0n, Vsmith, Loren36, El C, Momotaro, Shanes, RoyBoy, Triona, Constantine, Smalljim, 19Q79oL78KiL0QTFHgyc, Giraffedata, Wolfrider~enwiki, Physicistjedi, MPerel, Gsklee, ShardPhoenix, Axl, Mac Davis, Kocio, Burn, Hu, Wtmitchell, SidP, DV8 2XL, Ringbang, Kazvorpal, Omnist, Sharkie, Joelpt, Angr, Firsfron, FeanorStar7, Pol098, WadeSimMiser, Mpatel, GregorB, Jugger90, Paxsimius, Mandarax, Chun-hian, Grammarbot, Rjwilmsi, Nightscream, Koavf, Zbxgscqf, Oblivious, Yug, Lionelbrits, Ruidlopes, The ARK, Latka, Mathbot, Diza, Phoenix2~enwiki, DVdm, Eric B, Bomb319, Loom91, Zafiroblue05, Bhny, Stephenb, KSchutte, Bovineone, Salsb, Erielhonan, Bobak, Asarelah, Dna-webmaster, Sandstein, Superdude99, Zzuuzz, Imaninjapirate, Arthur Rubin, Ilmari Karonen, Caco de vidro, DVD R W, Hide&Reason, Jmeden2000, Teo64x, Sardanaphalus, MartinGugino, RupertMillard, SmackBot, Kurochka, K-UNIT, Rwp, Rlbates99, Ajt, Ian Rose, Gilliam, Wlmg, DividedByNegativeZero, Mirokado, Bluebot, Cush, SMP, Ben.c.roberts, MalafayaBot, Nbarth, DHN-bot~enwiki, Colonies Chris, Joemah, N.MacInnes, Xiner, Nunocordeiro, Mbertsch, Addshore, EPM, Nakon, Kiplantt, Bigmantonyd, Martijn Hoekstra, Kabain52, Brdforallseasons, Sayden, Doug Bell, Jaganath, Shadowlynk, IronGargoyle, Jochietoch, Hu12, Jxh2154, Tawkerbot2, Valoem, Gebrah, Albertod4, Kurtan~enwiki, Harold f, Devourer09, Cyrusc, CRGreathouse, Olaf Davis, Lambertian, Friendlystar, Rowellcf, Bmk, Myasuda, DepartedUser2, Ekajati, Cydebot, Fluence, Mike Christie, Meno25, Gagueci, Kahananite, Michael C Price, Alexnye, IComputerSaysNo, Lord Satorious, Krowe, Mrockman, Thijs!bot, Epbr123, Daniel, Headbomb, NeilHalfway, James086, KrakatoaKatie, AntiVandalBot, Blue Tie, Alphachimpbot, J rowley, Shambolic Entity, SuperLuigi31, Buchhemi, Fetchcomms, 100110100, WolfmanSF, VoABot II, Madevin314, SHCarter, Rami R, Jqshenker, Just H, War wizard90, Rickard Vogelberg, Stephen Shenker, Theoretic, MartinBot, Kostisl, R'n'B, Euku, Numbo3, Maurice Carbonaro, Nly8nchz, Thucydides411, LordAnubisBOT, Janus Shadowsong, Peskydan, Isoko, Belovedfreak, Antony-22, Wesino, WJBscribe, Thomas795135, Blood Oath Bot, Idioma-bot, Sheliak, Gogobera, Jeff G., Reibot, Ask123, Pennstatephil, JhsBot, Mazarin07, Peace keeper II, Lamro, Antixt, Why Not A Duck, PhysPhD, Rknasc, Guystout, Drschawrz, YohanN7, SieBot, Robdunst, Paradoctor, Wing gundam, Holt27, Astroboyretro, Caidh, OKBot, Divinestuff, Wpac5, Ayleuss, Beofluff, Loren.wilton, ClueBot, Master Shake 9, The Thing That Should Not Be, Haemorrhage, Arakunem, Drmies, IMNTU, Yupjohnny, Huntthetroll, Patrik Andersson, Dank, Gardv, DumZiBoT, Jfosc, Maky, Truthnlove, Autocoast~enwiki, Albambot, Addbot, Uruk2008, Cuaxdon, CanadianLinuxUser, WikiUserPedia, Barak Sh, Tassedethe, Carapheonix, Togekiss101, Tide rolls, OlEnglish, Snaily, Legobot, Luckas-bot, Yobot, Fraggle81, Pcap, Foolo~enwiki, CinchBug, Tempodivalse, AnomieBOT, KDS4444, Götz, Charlesvi, Dalton h, Marcka, Alexzabbey, Jim1138, IRP, AdjustShift, Materialscientist, Citation bot, Quebec99, Ruike, TinucherianBot II, Ekwos, Techwiz2000, Omnipaedista, Peanuts4life, Pinethicket, Vicenarian, Tom.Reding, EDG161, Jusses2, Serols, ActivExpression, SkyMachine, Gerda Arendt, Tkachyk, 122589423KM, சந்திர சிவகுமார், Reaper Eternal, Apb91781, 786 zikhar, LcawteHuggle, Adam1217, EmausBot, GoingBatty, Pyschobbens, StringTheory11, Smiwi, Suslindisambiguator, SporkBot, PoisonGM, Besneatte, Maschen, SBaker43, Denholm Reynholm, RockMagnetist, ClueBot NG, Blueshift333, Rgwkenyon, Helpful Pixie Bot, Bibcode Bot, BG19bot, SharkinthePool, Msaunier, MusikAnimal, Copernicus01, Elginfball10, Qed3, Blaspie55, Zujua, Kooky2, Mediran, Chris5631, FEYKATD, Ecila3, Lugia2453, Frosty, AHusain314, Armanschwarz, Among Men, Faizan, Epicgenius, Diekilldie, EddieHugh, Dustin V. S., RaphaelQS, Beakr, DavidLeighEllis, Tedsanders, TFA Protector Bot, Vampre1122, Polytope24, Evandas, Oneidiotsavant, Pretickle, TheRealTheKoi, Shantsforeverandalways, QuantumMatt101, FACBot, Kh3368, Sizeofint, Jyhtgqwqsdfghjydwq, Mberkson12, KasparBot, Cmealo, Yadav.aakash.500 and Anonymous: 448

- **Hierarchy problem** *Source:* https://en.wikipedia.org/wiki/Hierarchy_problem?oldid=682553903 *Contributors:* The Anome, WhisperToMe, Phys, AnonMoos, Jni, Giftlite, Xerxes314, Thincat, Lumidek, Rich Farmbrough, FT2, Pt, Jag123, 19Q79oL78KiL0QTFHgyc, Mindmatrix, GregorB, VermillionBird, Coemgenus, Mattmartin, Strait, Salix alba, UkPaolo, Ugha, Bhny, Netrapt, Ephraim33, QFT, Jgwacker, NNemec, Ninjakannon, Shambolic Entity, Dr. Morbius, Drgnrave, X!, James Banogon, Megalekaitrane, D.scain.farenzena, Copyeditor42, Alexbot, Lalegria, Addbot, Mixen Dixon, TutterMouse, Debresser, Topquark22, Yobot, AnomieBOT, Yemibedu, Materialscientist, Citation bot, Neurolysis, ArthurBot, Pra1998, Omnipaedista, A. di M., Erik9bot, Banak, FrescoBot, Paine Ellsworth, Puzl bustr, Bj norge, Hauntedpz, RjwilmsiBot, EmausBot, AsceticRose, Arbnos, Suslindisambiguator, Quondum, Jbackroyd, Bibcode Bot, Ervin Goldfain, Drcooljoe, IluvatarBot, Ownedroad9, MSUGRA, Prokaryotes, Mfb and Anonymous: 45

- **String theory landscape** *Source:* https://en.wikipedia.org/wiki/String_theory_landscape?oldid=680316596 *Contributors:* Maury Markowitz,

Sho Uemura, Fropuff, Eequor, I9Q79oL78KiL0QTFHgyc, Kusma, DV8 2XL, Ringbang, GregorB, BlaiseFEgan, Joke137, YurikBot, Bhny, Archelon, IAMTHEEGGMAN, Gadget850, SmackBot, Kurochka, Colonies Chris, CmdrObot, Mbell, Headbomb, Peter Gulutzan, Landscape-enwiki, Tim Shuba, Glen, Dr. Morbius, IgorSF, Jrcla2, Gbawden, Martarius, Niceguyedc, Addbot, DOI bot, Cuaxdon, SeniorInt, Macquaire, Omnipaedista, Citation bot 2, Citation bot 1, Tom.Reding, Xaviertan, Ale And Quail, RA0808, ThePowerofX, Isocliff, Bibcode Bot, Polytope24, Tetra quark and Anonymous: 28

- **Loop quantum gravity** *Source:* https://en.wikipedia.org/wiki/Loop_quantum_gravity?oldid=683638404 *Contributors:* Bryan Derksen, The Anome, AstroNomer-enwiki, RK, Toby Bartels, Miguel-enwiki, Schewek, Ewen, Michael Hardy, TakuyaMurata, Islandboy99, GTBacchus, Mcarling, Looxix-enwiki, Ahoerstemeier, Cyp, Kimiko, Palfrey, Jordi Burguet Castell, Mxn, Charles Matthews, Sanxiyn, Maximus Rex, Phys, Omegatron, Finlay McWalter, Dmytro, Sdedeo, Astronautics-enwiki, Peak, Chris Roy, Mirv, Sverdrup, Kn1kda, Hadal, Jheise, Clementi, Connelly, Giftlite, Sj, Fastfission, Herbee, Anville, Dratman, Curps, JeffBobFrank, Jason Quinn, Gzornenplatz, C17GMaster, DÅ,ugosz, PhiloVivero, DefLog-enwiki, Gadfium, HorsePunchKid, Sam Hocevar, Lumidek, Tdent, Joyous!, M1ss1ontomars2k4, Eep², Poccil, Rich Farmbrough, Avriette, Pjacobi, Vsmith, MuDavid, Pavel Vozenilek, Bender235, ESkog, Clement Cherlin, Peter M Gerdes, Drhex, John Vandenberg, C S, Cmdrjameson, GTubio, Tweet Tweet, Slicky, Ral315, Lysdexia, Arthena, Xaphan9966, Wtmitchell, Greg Kuperberg, Count Iblis, Egg, Lee-Anne, Kazvorpal, Killing Vector, Linas, Merlinme, HFarmer, Sympleko, Hfarmer, Mpatel, GregorB, J M Rice, Ae7flux, Tjbk tjb, Alienus, Fleisher, Sjö, Rjwilmsi, Nightscream, Zbxgscqf, Bubba73, FlaBot, John Baez, Don Gosiewski, Smithbrenon, Chobot, Spasemunki, Bgwhite, Roboto de Ajvol, YurikBot, Wavelength, RobotE, Rt66lt, Hillman, DanMS, Chaos, Salsb, Welsh, Schmock, Crasshopper, Beanyk, Akashmitra, Bota47, JonathanD, Endomion, Modify, Petri Krohn, Ilmari Karonen, Caco de vidro, Benandorsqueaks, SmackBot, Bayardo, FlashSheridan, Unyoyega, Vald, JMiall, Chris the speller, IvanAndreevich, DHN-bot-enwiki, Colonies Chris, Chlewbot, Pepsidrinka, Chrylis, MegaHasher, TriTertButoxy, Lambiam, Vincenzo.romano, Loadmaster, Konklone, K, G-W, Kurtan-enwiki, Harold f, Will314159, Friendly Neighbour, Vyznev Xnebara, Ian Beynon, Myasuda, Gmusser, Rjm656s, Fournax, Headbomb, Nick Number, MichaelMaggs, Edokter, Byrgenwulf, Knotwork, Arch dude, Igodard, Yill577, WolfmanSF, Tonyfaull, Skylights76, Rickard Vogelberg, Gwern, AltiusBimm, Melamed katz, Vanished user 47736712, WJBscribe, Izno, KittyHawker, Sheliak, Maxzimet, AlnoktaBOT, Nxavar, Jackfork, Carlorovelli, Anotherak, SieBot, Keskival, AS, Robdunst, Hugh16, Senderista-enwiki, Bnsreenath, Caidh, Oxymoron83, Dcattell, Swiebodzice, Sk8hack, Danthewhale, Martarius, Sfan00 IMG, Shaded0, Djr32, CohesionBot, JavierReynaldo, Arjayay, SchreiberBike, Pqnelson, Mjaniec, DumZiBoT, Ianbay, Neuralwarp, XLinkBot, Fastily, Tenner47, Arthur chos, Avoided, Tenderbuttons, Benplusnumber, Balungifrancis, Addbot, DOI bot, 15Isoucy, Tarosie, Debresser, SamatBot, Yobot, Ibayn, 4th-otaku, AnomieBOT, VanishedUser sdu9aya9fasdsopa, Archon 2488, Francois33, Citation bot, Xqbot, Imushfiq, MIRROR, Pra1998, Dumontierc, Omnipaedista, Franco3450, Rr2000, FrescoBot, Paine Ellsworth, Nunc aut numquam, Martlet1215, Citation bot 1, Jonesey95, Tom.Reding, Schiefesfragezeichen, ROMVLVS, Casimir9999, RobinK, Meier99, Dinamik-bot, Bj norge, ElPeste, Afteread, EmausBot, Detogain, John of Reading, Racerx11, GoingBatty, XinaNicole, Ensabah6, Uploadvirus, ZéroBot, Arbnos, Zueignung, WaterCrane, Crown Prince, LaurentRDC, Isocliff, Vodkacannon, Raidr, Helpful Pixie Bot, Titodutta, Bibcode Bot, BG19bot, Spaligo, KateWishing, PhnomPencil, Sylvain.maurin, Keechina, Halfb1t, Brad7777, Fylbecatulous, Jimw338, MyTuppence, Mogism, LT-Woods, Andyhowlett, Jawa0, &reasNink, SomeFreakOnTheInternet, Tentinator, EvergreenFir, DimReg, Pedarkwa, Db9199 24, Anrnusna, Notspelly, Ntomlin1996, Monkbot, Isbromberg, Dspre, YeOldeGentleman, Tetra quark and Anonymous: 333

- **Gauge theory** *Source:* https://en.wikipedia.org/wiki/Gauge_theory?oldid=677681547 *Contributors:* The Anome, Michael Hardy, Tobias Bergemann, Ancheta Wis, TedPavlic, Xezbeth, MuDavid, Bender235, Pt, Phils, BD2412, Rjwilmsi, JocK, Modify, Teply, SmackBot, RDBury, Henning Makholm, Byelf2007, Michael C Price, Biblbroks, Headbomb, Nick Number, Fashionslide, VectorPosse, Magioladitis, Bakken, Email4mobile, JaGa, Policron, Squids and Chips, Cuzkatzimhut, VolkovBot, Red Act, Michael H 34, Setreset, Jwpitts, Tcamps42, Moonriddengirl, ClueBot, Mastertek, TimothyRias, XLinkBot, Addbot, Mortense, Eric Drexler, Bte99, Zorrobot, Luckas-bot, AnomieBOT, Christopher.Gordon3, Citation bot, Northryde, Xqbot, Pra1998, Gsard, A. di M., Erik9bot, FrescoBot, Fortdj33, Citation bot 1, Ganondolf, RedBot, RobinK, Mary at CERN, EmausBot, Brent Perreault, Slawekb, Cogiati, Maschen, Isocliff, ClueBot NG, Helpful Pixie Bot, Bibcode Bot, Dzustin, Brendan.Oz, ChrisGualtieri, SD5bot, Dexbot, Enyokoyama, Joeinwiki, Dath Thou Even Lift, Dhm4444, Dbw1976, KasparBot and Anonymous: 42

- **Dimensionless physical constant** *Source:* https://en.wikipedia.org/wiki/Dimensionless_physical_constant?oldid=682816547 *Contributors:* Bryan Derksen, The Anome, Roadrunner, Michael Hardy, Shellreef, Karada, Reddi, Maximus Rex, Pakaran, Fuelbottle, Diberri, Giftlite, Kennysh, Eequor, Wmahan, Gadfium, Karol Langner, Mschlindwein, Cacycle, Bender235, Sgeo, Army1987, Rbj, Russ3Z, I9Q79oL78KiL0QTFHgyc, Linas, Marudubshinki, Bytor, Ketiltrout, Nightscream, TheRingess, HappyCamper, Bubba73, Ian Dunster, John Baez, Wavelength, Open4D, Salsb, ArséniureDeGallium, 2over0, SmackBot, B.Wind, Dauto, TimBentley, Concerned cynic, Lottamiata, GDallimore, CapitalR, Chetvorno, Kehrli, AndrewHowse, DumbBOT, Mbell, EdJohnston, Azaghal of Belegost, Spartaz, Propaniac, Magioladitis, STBot, Leyo, Tdadamemd, Ohms law, STBotD, James Kidd, EverGreg, Kbrose, YohanN7, Paradoctor, HairyWombat, Farolif, Thegeneralguy, AntrygG.Revok, Palnot, Rror, Saeed.Veradi, Addbot, Guigny seck, Brentdeezee, Jack who built the house, Yobot, Legobot II, AnomieBOT, Materialscientist, Thehelpful bot, InternetFoundation, Thinking of England, Quondum, Cloudmichael, ClaudeDes, Happyboy2011, Rarelight, Mcarlie, Jimjohnson2222, TwoTwoHello, Mfb, Dgheimjoseph, NekoKatsun, ComicsAreJustAllRight, Inorout and Anonymous: 85

30.9.2   Images

- **File:A_Swarm_of_Ancient_Stars_-_GPN-2000-000930.jpg** *Source:* https://upload.wikimedia.org/wikipedia/commons/6/6a/A_Swarm_of_Ancient_Stars_-_GPN-2000-000930.jpg *License:* Public domain *Contributors:* Great Images in NASA *Description Original artist:* NASA, The Hubble Heritage Team, STScI, AURA

- **File:Acap.svg** *Source:* https://upload.wikimedia.org/wikipedia/commons/5/52/Acap.svg *License:* Public domain *Contributors:* Own work *Original artist:* F l a n k e r

- **File:AdS3.svg** *Source:* https://upload.wikimedia.org/wikipedia/commons/4/47/AdS3.svg *License:* CC BY-SA 3.0 *Contributors:* This file was derived from: AdS3 (new).png
Original artist:

- derivative work: Alex Dunkel (Maky)

30.9.3 Content license

www.ingramcontent.com/pod-product-compliance
Lightning Source LLC
Chambersburg PA
CBHW080757180526
45168CB00006B/2236